Neurotransmitter Methods

METHODS IN MOLECULAR BIOLOGY™

John M. Walker, Series Editor

73. **Neuropeptide Protocols,** edited by *G. Brent Irvine and Carvell H. Williams, 1997*
72. **Neurotransmitter Methods,** edited by *Richard C. Rayne, 1997*
71. **PRINS and *In Situ* PCR Protocols,** edited by *John R. Gosden, 1997*
70. **Sequence Data Analysis Guidebook,** edited by *Simon R. Swindell, 1997*
69. **cDNA Library Protocols,** edited by *Ian G. Cowell and Caroline A. Austin, 1997*
68. **Gene Isolation and Mapping Protocols,** edited by *Jacqueline Boultwood, 1997*
67. **PCR Cloning Protocols:** *From Molecular Cloning to Genetic Engineering,* edited by *Bruce A. White, 1996*
66. **Epitope Mapping Protocols,** edited by *Glenn E. Morris, 1996*
65. **PCR Sequencing Protocols,** edited by *Ralph Rapley, 1996*
64. **Protein Sequencing Protocols,** edited by *Bryan J. Smith, 1996*
63. **Recombinant Protein Protocols:** *Detection and Isolation,* edited by *Rocky S. Tuan, 1997*
62. **Recombinant Gene Expression Protocols,** edited by *Rocky S. Tuan, 1997*
61. **Protein and Peptide Analysis by Mass Spectrometry,** edited by *John R. Chapman, 1996*
60. **Protein NMR Protocols,** edited by *David G. Reid, 1997*
59. **Protein Purification Protocols,** edited by *Shawn Doonan, 1996*
58. **Basic DNA and RNA Protocols,** edited by *Adrian J. Harwood, 1996*
57. **In Vitro Mutagenesis Protocols,** edited by *Michael K. Trower, 1996*
56. **Crystallographic Methods and Protocols,** edited by *Christopher Jones, Barbara Mulloy, and Mark Sanderson, 1996*
55. **Plant Cell Electroporation and Electrofusion Protocols,** edited by *Jac A. Nickoloff, 1995*
54. **YAC Protocols,** edited by *David Markie, 1995*
53. **Yeast Protocols:** *Methods in Cell and Molecular Biology,* edited by *Ivor H. Evans, 1996*
52. **Capillary Electrophoresis:** *Principles, Instrumentation, and Applications,* edited by *Kevin D. Altria, 1996*
51. **Antibody Engineering Protocols,** edited by *Sudhir Paul, 1995*
50. **Species Diagnostics Protocols:** *PCR and Other Nucleic Acid Methods,* edited by *Justin P. Clapp, 1996*
49. **Plant Gene Transfer and Expression Protocols,** edited by *Heddwyn Jones, 1995*
48. **Animal Cell Electroporation and Electrofusion Protocols,** edited by *Jac A. Nickoloff, 1995*
47. **Electroporation Protocols for Microorganisms,** edited by *Jac A. Nickoloff, 1995*
46. **Diagnostic Bacteriology Protocols,** edited by *Jenny Howard and David M. Whitcombe, 1995*
45. **Monoclonal Antibody Protocols,** edited by *William C. Davis, 1995*
44. ***Agrobacterium* Protocols,** edited by *Kevan M. A. Gartland and Michael R. Davey, 1995*
43. **In Vitro Toxicity Testing Protocols,** edited by *Sheila O'Hare and Chris K. Atterwill, 1995*
42. **ELISA:** *Theory and Practice,* by *John R. Crowther, 1995*
41. **Signal Transduction Protocols,** edited by *David A. Kendall and Stephen J. Hill, 1995*
40. **Protein Stability and Folding:** *Theory and Practice,* edited by *Bret A. Shirley, 1995*
39. **Baculovirus Expression Protocols,** edited by *Christopher D. Richardson, 1995*
38. **Cryopreservation and Freeze-Drying Protocols,** edited by *John G. Day and Mark R. McLellan, 1995*
37. **In Vitro Transcription and Translation Protocols,** edited by *Martin J. Tymms, 1995*
36. **Peptide Analysis Protocols,** edited by *Ben M. Dunn and Michael W. Pennington, 1994*
35. **Peptide Synthesis Protocols,** edited by *Michael W. Pennington and Ben M. Dunn, 1994*
34. **Immunocytochemical Methods and Protocols,** edited by *Lorette C. Javois, 1994*
33. ***In Situ* Hybridization Protocols,** edited by *K. H. Andy Choo, 1994*
32. **Basic Protein and Peptide Protocols,** edited by *John M. Walker, 1994*
31. **Protocols for Gene Analysis,** edited by *Adrian J. Harwood, 1994*
30. **DNA–Protein Interactions,** edited by *G. Geoff Kneale, 1994*
29. **Chromosome Analysis Protocols,** edited by *John R. Gosden, 1994*
28. **Protocols for Nucleic Acid Analysis by Nonradioactive Probes,** edited by *Peter G. Isaac, 1994*
27. **Biomembrane Protocols:** *II. Architecture and Function,* edited by *John M. Graham and Joan A. Higgins, 1994*
26. **Protocols for Oligonucleotide Conjugates:** *Synthesis and Analytical Techniques,* edited by *Sudhir Agrawal, 1994*
25. **Computer Analysis of Sequence Data:** *Part II,* edited by *Annette M. Griffin and Hugh G. Griffin, 1994*
24. **Computer Analysis of Sequence Data:** *Part I,* edited by *Annette M. Griffin and Hugh G. Griffin, 1994*
23. **DNA Sequencing Protocols,** edited by *Hugh G. Griffin and Annette M. Griffin, 1993*
22. **Microscopy, Optical Spectroscopy, and Macroscopic Techniques,** edited by *Christopher Jones, Barbara Mulloy, and Adrian H. Thomas, 1993*
21. **Protocols in Molecular Parasitology,** edited by *John E. Hyde, 1993*
20. **Protocols for Oligonucleotides and Analogs:** *Synthesis and Properties,* edited by *Sudhir Agrawal, 1993*
19. **Biomembrane Protocols:** *I. Isolation and Analysis,* edited by *John M. Graham and Joan A. Higgins, 1993*

METHODS IN MOLECULAR BIOLOGY™

Neurotransmitter Methods

Edited by

Richard C. Rayne

Birkbeck College, London, UK

Humana Press **Totowa, New Jersey**

999 Riverview Drive, Suite 208
Totowa, New Jersey 07512

This publication is printed on acid-free paper. ∞
ANSI Z39.48-1984 (American Standards Institute)
Permanence of Paper for Printed Library Materials.

Cover illustration: Fig. 1C in Chapter 12, "In Situ Hybridization to Determine the Expression of Peptide Neurotransmitters," by Niovi Santama.

Cover design by Patricia F. Cleary.

For additional copies, pricing for bulk purchases, and/or information about other Humana titles, contact Humana at the above address or at any of the following numbers: Tel.: 201-256-1699; Fax: 201-256-8341; E-mail: humana@mindspring.com; www.humanapress.com

Printed in the United States of America. 10 9 8 7 6 5 4 3 2 1

Library of Congress Cataloging in Publication Data

Main entry under title:

Methods in molecular biology™.
Neurotransmitter methods/edited by Richard C. Rayne.
p. cm.—(Methods in molecular biology™; vol. 72)
Includes index.
ISBN 0-89603-394-5 (alk. paper)
1. Neurotransmitters—Laboratory manuals. I. Rayne, Richard C. II. Series: Methods in molecular biology™ (Totowa, NJ); 72.
[DNLM: 1. Neurotransmitters—physiology. 2. Histochemistry—methods. 3. In Situ Hybridization—methods. 4. Chromatography, Liquid—methods. 5. Spectrum Analysis, Mass—methods. W1 ME 9616J v. 72 1997/QV 126 N4939 1997]
QP364.7.N469 1997
573.8'64'0724—dc21
DNLM/DLC
for Library of Congress 96-49802
CIP

Preface

Neurotransmitter Methods is intended as a bench-side companion for researchers who seek to identify, localize, or measure neurotransmitters and/or to identify sites of neurotransmitter action. Each method is detailed in a user-friendly "recipe" format and the protocols are accompanied by extensive notes to highlight and explain crucial steps. Approaches utilizing an incredibly diverse array of modern techniques are presented: methods including HPLC, histochemistry, immunocytochemistry, *in situ* hybridization, mass spectrometry, microdialysis, and electrochemistry all make at least one appearance. In addition, protocols for associated methodologies, including the production of brain slices, dissociated neurons, synaptosomes/synaptoneurosomes, and neuronal plasma membranes are presented. Methods applicable to most of the recognized chemical types of neurotransmitter are to be found and, although you may find absent any mention of your favorite neurotransmitter, many of the protocols are sufficiently general to be adapted to alternative uses.

So, how does this book contribute usefully to the horde of methods volumes stampeding across our bookshelves? I hope that the strength of *Neurotransmitter Methods* lies in the variety of its content. The book provides in a single volume an array of techniques that could take a researcher from selection and preparation of a tissue source through to identification and measurement of neurotransmitter content and even onto characterization of neurotransmitter sites of action.

Furthermore, the diversity of approaches presented reflects the need for today's bench neuroscientist to be a bit of a "jack (or a "jill"!) of all trades." The authors of this volume have written not only explicit, step-by-step protocols, but also have provided detailed analyses of their methods. This explanatory approach is invaluable in troubleshooting for the experienced worker and, of course, an important education for the novice. My hope is that researchers expert in a given area will find *Neurotransmitter Methods* treatment of their speciality useful and fresh and that chapters on unfamiliar topics will inspire readers to find expert collaborators—or provide the courage to try something new in their own laboratories.

Finally, I would like to extend thanks to all the contributors for their hard efforts, to the publishers for their patience, and to the inventors of electronic mail (especially for making possible attached files) for making extinct the typical last minute dash to the post office!

Richard C. Rayne

Contents

Contributors

JONATHAN P. BACON • *Sussex Centre for Neuroscience, School of Biological Sciences, University of Sussex, Brighton, East Sussex, UK*

RICHARD A. BAINES • *Sussex Centre for Neuroscience, School of Biological Sciences, University of Sussex, Brighton, East Sussex, UK. Current Address: Department of Zoology, University of Cambridge, UK*

PETER J. BERGOLD • *Department of Pharmacology, Health Sciences Center, State University of New York, Brooklyn, NY*

ELLY BESSELSEN • *Institute of Neurobiology, University of Amsterdam, The Netherlands*

ALEXANDRA I. M. BREUKEL • *Institute of Neurobiology, University of Amsterdam, The Netherlands*

PATRIZIA CASACCIA-BONNEFIL • *Department of Anatomy and Cell Biology, Cornell University Medical College, New York, NY*

JAN DE VENTE • *Department of Psychiatry and Neuropsychology, University of Limburg, Maastricht, The Netherlands*

GUY EBINGER • *Department of Pharmaceutical Chemistry and Drug Analysis, Vrije Universiteit, Brussels, Belgium*

PETER EKSTROM • *Department of Zoology, Lund University, Lund, Sweden*

MAURICE R. ELPHICK • *School of Biological Sciences, Queen Mary and Westfield College, London, UK*

STEPHEN GENTLEMAN • *Department of Anatomy, Charing Cross and Westminster Medical School, London, UK*

WIJNAND P. M. GERAERTS • *Research Institute of Neurosciences, Vrije Universiteit, Amsterdam, The Netherlands*

WIM E. J. M. GHIJSEN • *Institute of Neurobiology, University of Amsterdam, The Netherlands*

TIMOTHY K. HAYES • *Department of Entomology, Biotechnology Instrumentation Facility: Peptide and Protein Technologies, Texas A&M University, College Station, TX. Current Address: Research and Development, Department of Bioanalytical Chemistry, Bayer Corporation, Clayton, NC*

HRVOJE HEĆIMOVIĆ • *Department of Pharmacology, University of Edinburgh, Scotland*

JACQUES J. H. HENS • *Rudolf Magnus Institute for Neurosciences, Department of Medical Pharmacology, Utrecht University, Utrecht, The Netherlands*

G. MARK HOLMAN • *Food Animal Protection Research Laboratory, USDA-ARS, College Station, TX*

IRA S. KASS • *Departments of Anesthesiology and Pharmacology, Health Sciences Center, State University of New York, Brooklyn, NY*

PHILLIP M. LARKMAN • *Department of Pharmacology, University of Edinburgh, Scotland*

ANDREW LEVY • *Department of Medicine, Bristol Royal Infirmary, University of Bristol, UK*

KA WAN LI • *Research Institute of Neurosciences, Vrije Universiteit, Amsterdam, The Netherlands*

YVETTE MICHOTTE • *Department of Pharmaceutical Chemistry and Drug Analysis, Vrije Universiteit, Brussels, Belgium*

JOHN M. MIDGLEY • *Department of Pharmaceutical Sciences, University of Strathclyde, Glasgow, Scotland*

JULIAN MILLAR • *Basic Medical Sciences, Queen Mary and Westfield College, London, UK*

DICK R. NÄSSEL • *Department of Zoology, Stockholm University, Stockholm, Sweden*

SUZANNA J. NEWMAN • *Department of Neuropathology, SmithKline Beecham Pharmaceuticals, Harlow, Essex, UK*

DAVID V. POW • *Department of Physiology and Pharmacology, Vision, Touch, and Hearing Research Centre, University of Queensland, Brisbane, Queensland, Australia*

NIOVI SANTAMA • *Gene Expression Program, EMBL, Heidelberg, Germany*

SOPHIE SARRE • *Department of Pharmaceutical Chemistry and Drug Analysis, Vrije Universiteit, Brussels, Belgium*

ILSE SMOLDERS • *Department of Pharmaceutical Chemistry and Drug Analysis, Vrije Universiteit, Brussels, Belgium*

HARRY W. M. STEINBUSCH • *Department of Psychiatry and Neuropsychology, University of Limburg, Maastricht, The Netherlands*

KATRIEN THORRE • *Department of Pharmaceutical Chemistry and Drug Analysis, Vrije Universiteit, Brussels, Belgium*

MIRIAM N. G. TITULAER • *Institute of Neurobiology, University of Amsterdam, The Netherlands*

TING WANG • *Department of Anesthesiology, Health Sciences Center, State University of New York, Brooklyn, NY*

DAVID G. WATSON • *Department of Pharmaceutical Sciences, University of Strathclyde, Glasgow, Scotland*

MARK ZUIDERWIJK • *Institute of Neurobiology, University of Amsterdam, The Netherlands*

1

Preparation of Brain Slices

Ting Wang and Ira S. Kass

1. Introduction

The brain-slice technique has been utilized in electrophysiological, morphological, biochemical, and pharmacological studies of almost all brain structures. A search of the literature between 1991 and 1995 on the Ovid Medline revealed 4387 entries that used the brain-slice technique; of these, 2038 are relevant to the study of receptors and neurotransmitters. This technique is widely used because it has many advantages over in vivo methods. It provides precise control over experimental conditions, such as temperature, pH, and drug concentration. It also allows the examination of metabolic parameters and electrophysiological properties without contamination from anesthetics, muscle relaxants, or intrinsic regulatory substances. The stability of electrophysiological recording is greatly improved as the heart beat and respiration of the experimental animal are eliminated. The cells being studied can be located, identified, and accessed easily. Use of the brain slice has greatly increased our knowledge of the mammalian central nervous system (CNS). This technique is continuously improving and will remain valuable for a long time.

Preparation of brain slices is an indispensable procedure for a variety of experiments. The goal of the slice preparation is to obtain a thin piece of brain tissue containing the cells of interest and to maintain the slice in a viable (although artificial) condition that is similar to its in vivo environment. In this chapter we describe procedures that are fundamental for brain slice preparation. Because the hippocampus is the most widely used tissue for brain slices, its isolation will be used here to illustrate the steps of the procedure. Two slicing methods, using either a vibratome or a tissue chopper, are described. Our discussion of these methods covers the steps from the decapitation of the animal to the storage of slices in artificial cerebrospinal fluid (aCSF). Some of the

From: *Methods in Molecular Biology, Vol. 72: Neurotransmitter Methods*
Edited by: R. C. Rayne Humana Press Inc., Totowa, NJ

notions and dogma mentioned throughout are derived from the authors' personal experience or anecdotal reports. There is no standard procedure for the preparation and the existing technique is not perfect. Beginners are encouraged to explore new approaches and test alternatives.

1.1. Preparation of Artificial Cerebrospinal Fluid

The aCSF mimics the extracellular environment of neurons and provides the necessary ingredients that allow neurons to survive in vitro for at least several hours. Because metabolic processes and the maintenance of the ionic gradient across the cell membrane constantly consume ATP, glucose and oxygen must be supplied to generate energy.

The common ingredients in aCSF are NaCl, KCl, KH_2PO_4, $NaHCO_3$, glucose, $MgSO_4$, and $CaCl_2$. Although the ideal composition of aCSF mirrors that of the cerebrospinal fluid in vivo, investigators often use an aCSF that differs from CSF. The concentration of the ionic components of the aCSF vary slightly from lab to lab to meet different experimental purposes. For example:

1. The response of the NMDA receptor to its agonist is reduced by raising the concentration of magnesium.
2. The excitability of neurons is increased by raising the potassium or lowering the calcium or magnesium concentrations.
3. Cells in slices survive longer and withstand anoxia better in solutions with higher glucose concentrations *(1)* or lower potassium concentrations *(2)*.

The alteration of the aCSF ingredients obviously affects the electrophysiological properties of the neurons in slices. Our experiments showed dramatic changes in the firing rate, the firing pattern, and the shape of action potentials of CA1 neurons when the ionic concentration of the aCSF was changed (unpublished observation). Caution must be exercised to keep the osmolarity and pH constant when one alters the composition of aCSF. It is important to bear in mind that multiple biological effects may be produced by changing a single component of the aCSF. For example, raising the concentration of potassium not only increases the excitability of the neuron, but also increases the activity of Na/K-ATPase *(3)*.

1.2. Anesthesia, Decapitation, and Removal of the Brain from the Skull

The animal should be deeply anesthetized before being sacrificed. Inhalation anesthetics, such as ether, halothane, isofluorane, or methoxyfluorane, are often used. However, some investigators prefer such anesthetics as ketamine or chloral hydrate, which need to be injected. For some experiments, the use of anesthetics may complicate the interpretation of the experimental results, especially in studies of anesthetics themselves. Decapitation with a small guillotine is one method of sacrificing animals without an anesthetic. In this case, guidelines for the use of animals in experiments must be followed carefully.

After the animal is sacrificed, the skull should be opened quickly and the brain removed into chilled aCSF (4–8°C). The low temperature slows down the metabolic rate of cells and reduces their energy consumption. These effects allow cells to survive the ischemic period during brain slice preparation. The brain also becomes more firm when the temperature is lowered, facilitating the dissection and slicing of the brain.

1.3. Dissection of the Brain Tissue

In order to facilitate slicing, it is best to prepare from the whole brain a tissue block containing the cells of interest. This is achieved by cutting away the excessive surrounding brain tissue with a sharp blade, but intentionally leaving part of the surrounding tissue intact to protect and support the region of interest when it is later sliced. The shape and size of the tissue block varies according to the brain region targeted for slicing. At least one flat surface should be created on the tissue block for mounting onto the slicing stage. Making the dissection on a chilled platform and repeatedly pouring chilled aCSF over the brain tissue will keep the temperature low and increase the viability of the slices.

1.4. Mounting onto the Slicing Stage

After the tissue block is made, it is mounted onto a stage for slicing. For the vibratome method, the tissue block must be glued to the slicing stage. Cyanoacrylate (Krazy glue, Super glue) or Histoacryl is commonly used for this purpose. Although a toxic effect of cyanoacrylate glue has been suspected *(4)*, no systematic study on its toxicity has been found in the literature.

To prevent the tissue block from leaning back because of the advancing vibratome blade, the tissue block may be supported by an agar gel block. Some authors use Plexiglas instead of agar, but the edge of the blade will be damaged if it touches the Plexiglas. When a tissue chopper is used, a piece of filter paper moistened with aCSF is placed between the tissue block and the slicing stage. This will minimize the variation in the slice thickness and orientation, because the filter paper reduces sliding of the tissue block during slicing.

1.5. Slicing

The vibratome and tissue chopper methods are two major techniques used to slice tissue blocks. The vibratome seems to produce better slices and is gaining popularity. The available commercial vibratomes include the Vibratome series from Technical Products International (St. Louis, MO), and the Vibroslice series from Campden Instruments (Loughborough, UK). One advantage of vibratome slicing is that the whole tissue block is submerged in cold aCSF; thus, the tissue temperature can be kept low and the slices gain immediate access to the oxygen-saturated aCSF during slicing.

The tissue chopper, on the other hand, has the advantage of low cost and faster slicing. The available commercial tissue choppers include the Tissue Slicer from Stoelting (Wood Dale, IL) and McIlwain Tissue Chopper (Mickle Laboratories Engineering, Gromshall, Surrey, UK). Slicing is improved if the subarachnoid membrane is removed.

Most commonly, slices of 300–500 μm are made. This thickness range provides good cell preservation without compromising the diffusion of oxygen into the core of the slice. Thinner slices are preferred for optical experimental techniques in order to visualize cells; however, the cells within about 50 μm from the cut surface may be damaged. Commercial razor blades are used for the slicing and are usually shipped covered with a thin film of oily material to protect against oxidation. Blades should be cleared of the film with 75% alcohol and rinsed thoroughly with distilled water before use.

1.6. Storage of the Slices

Brain slices can be kept viable at 20–37°C for several hours in an aCSF-filled storage chamber aerated with 95% O_2 and 5% CO_2. We routinely use slices for intracellular recording after up to 6 h of storage. Another lab has reported viability for 24 h *(5)*. The following factors may contribute to the viability of the slices:

1. The brain tissue should be handled delicately to reduce physical damage; it should not be squeezed or twisted.
2. The preparation time should be short; a prolonged ischemic period is associated with a fall in intracellular ATP and reduction in the recovery of population spike *(6–8)*. Although most investigators think that the brain slice preparation should be accomplished in <5 min, another group found that there was no significant change in the amplitude of the population spike even with 30 min postmortem delay at room temperature *(9)*.
3. Lowering the brain tissue temperature during the preparation is effective in reducing the sensitivity of neurons to ischemia.
4. In order to reduce the neurotoxicity of the glutamate that may be massively released during slice preparation, some authors use an aCSF containing 10 m*M* magnesium and 0.5 m*M* calcium for slicing and storage *(10,11)*.

Because the slices survive for a long period in the aCSF, and more than one slice can be obtained from most brain regions, multiple experiments can be performed with slices from one animal. This is especially useful in pharmacological studies, because the slices can be moved individually from the storage to the recording chamber and then discarded after each drug test. Each slice can, therefore, be examined without contamination by the previously tested drugs.

An incubation period of 1 h is necessary for cells to recover from the ischemia and physical trauma of preparation. The transfer of slices from the storage chamber to the experimental chamber should be made by using a Pasteur pipet with its narrow opening attached to the rubber bulb (*see* Section 3.1.5.). If the temperature or a component of aCSF in the experimental chamber is different from that in the storage chamber, 20 min should be allowed for the slice to adapt to the new environment in the experimental chamber.

The brain slices may be further treated to isolate individual neurons or may themselves be maintained in culture media for several weeks. For detailed descriptions of slice culture and neuronal isolation, please *see* Chapters 2 and 3.

1.7. Shortcomings of the Brain Slice Technique

No matter how healthy the slice may appear, and how closely the components of the aCSF mimic the in vivo cellular *milieu*, the slice is surrounded by an artificial environment. Many substances that are present in vivo and important in regulating neuronal function, such as trophic factors and amino acids, are not included in the aCSF. Although local neuronal circuits may be preserved, the synaptic connections of the slice to and from other brain regions are interrupted. The effect of the artificial extracellular environment and the interruption of synaptic input is illustrated by the change of the firing pattern of midbrain dopamine neurons recorded in slices. In contrast to the irregular firing pattern recorded in vivo, an extremely regular pacemaker-like firing pattern is a characteristic feature of the dopamine neurons recorded in vitro *(12,13)*. This pacemaker-like firing pattern has never been observed in vivo. The change of firing pattern may reflect a liberation of the cells from synaptic innervation, revealing their intrinsic activity. Alternatively, this may represent abnormal electrophysiological activity because of the altered cellular environment.

2. Materials

2.1. The Vibratome Method

2.1.1. Preparation of the aCSF

The artificial cerebrospinal fluid (aCSF) contains 124 m*M* NaCl, 2 m*M* KCl, 1.25 m*M* KH_2PO_4, 26 m*M* $NaHCO_3$, 10 m*M* D-glucose, 2 m*M* $MgSO_4$, and 2 m*M* $CaCl_2$. Dissolve the following in ultrapurified water and bring the final volume to 1000 mL: 7.25 g NaCl, 0.149 g KCl, 0.17 g KH_2PO_4, 2.18 g $NaHCO_3$, 1.8 g glucose, 0.24 g $MgSO_4$, and 0.294 g $CaCl_2$. Note that $CaCl_2$ should be added last and as a dissolved solution. The final pH of the aCSF is 7.4 after it is saturated with a 5% CO_2 and 95% O_2 gas mixture (*see* Notes 1, 2, and 3).

2.1.2. Anesthesia, Decapitation, and Removal of the Brain from the Skull

1. Halothane.
2. Small animal guillotine.
3. One pair of straight surgical scissors (sharp/blunt, 16 cm).
4. One bone cutter (16 cm).
5. Spatula (20 cm).
6. Beaker (20 mL).

2.1.3. Preparation of the Brain Tissue Block

1. One Petri dish (60 × 10 mm).
2. Filter paper (55 mm, diameter).
3. Razor blade (Gillette, superstainless).

2.1.4. Mounting onto the Slicing Stage

1. Cyanoacrylate glue (Krazy glue)
2. Agar gel block (*see* Note 4).
3. Two flat-end spatulas (20 cm).

2.1.5. Slicing

1. Vibratome (Vibratome Series 1000, Technical Products International).
2. Razor blade (Gillette, superstainless).
3. Glass Pasteur pipet with a rubber bulb.
4. Alcohol.

2.1.6. Storage of the Slices

1. Beaker (200 mL).
2. 95% O_2 and 5% CO_2 gas mixture.

2.2. The Tissue Chopper Method

2.2.1. Preparation of the aCSF

Identical to the vibratome method (*see* Section 2.1.1.).

2.2.2. Anesthesia, Decapitation, and Removal of the Brain from the Skull

Identical to the vibratome method (*see* Section 2.1.2.).

2.2.3. Preparation of the Brain Tissue Block

1. Petri dish (60 × 10 mm).
2. Filter paper (55 mm diameter).
3. Flat-end spatula (20 cm).

2.2.4. Mounting onto the Slicing Stage

1. Flat-end spatula (20 cm).
2. Filter paper (55 mm diameter).

2.2.5. Slicing

1. Tissue chopper apparatus (Stoelting Tissue Slicer).
2. Razor blade (Gillette, superstainless).
3. Sable #2 (fine) artist's brush.

2.2.6. Storage of the Slices

Identical to the vibratome method (*see* Section 2.1.6.).

3. Methods

3.1. Vibratome Method

3.1.1. Anesthesia, Decapitation, and Removal of the Brain from the Skull

1. Mount an agar gel block (*see* Note 4) on the vibratome stage with Krazy glue. A tissue block, when prepared, will be placed next to it (*see* Section 3.1.3.).
2. Anesthetize the rat with halothane in a glass desiccator (*see* Note 5).
3. When the rat falls asleep, decapitate it using a small animal guillotine (*see* Note 6).
4. Expose the skull and cut along the sagittal suture from the foramen magnum to the forehead with a pair of surgical scissors (*see* Section 2.1.2.); the sharp pointed blade of the scissors should be placed on the inner side of the skull.
5. Using a bone cutter, make one cut at the foramen magnum on each temporal side of the skull and expose the brain by carefully prying the skull open.
6. Hold the skull upside down and, using a spatula to sever the cranial nerves that hold the brain to the skull, allow the brain to fall into a 20 mL beaker containing chilled aCSF.

3.1.2. Preparation of a Brain Tissue Block that Contains the Hippocampus

1. Place the cooled brain with its dorsal side up and cerebellum toward you on a piece of filter paper (*see* Note 7) in a Petri dish. Place ice under the Petri dish to keep the brain cool.
2. Pour chilled aCSF into the Petri dish until the brain is half submerged.
3. The two hippocampal formations in the brain resemble an inverted V as viewed from above (*see* Fig. 1A). Each hippocampus is a U-shaped structure with the top opening of the letter U facing forward and medially (*see* Fig. 1B). If the hippocampus on the right side of the brain is used, the blade should be placed 45° across the longitudinal fissure at a point one-third from the posterior end (*see* Fig. 1A) and rotated 15° back from the vertical plane (*see* Fig. 1B).
4. The surface of the first cut is roughly perpendicular to the longitudinal axis of the dorsal portion of the right hippocampus (stippled area in Fig. 1B), and will be used for mounting onto the slicing stage.

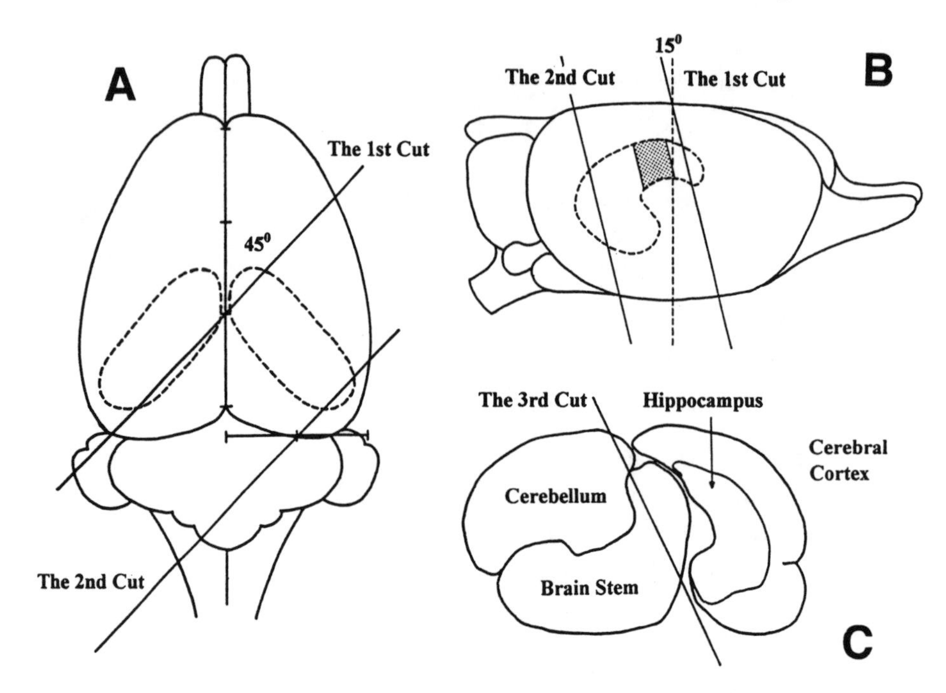

Fig. 1. Preparation of a tissue block containing a portion of hippocampus. **(A)** View of the rat brain from above. The hippocampus, represented by the dashed lines, is covered by the neocortex and is not visible from above. The two solid lines across the brain indicate the positions where the first and the second cuts are made. Note that the blade should be rotated 15° back from the vertical plane as shown in B. **(B)** Lateral view of the brain showing the right hippocampus. The straight dashed line is perpendicular to the surface of the neocortex. The solid lines across the brain are 15° offset from the dashed line and indicate where the first and second cuts are made. Five to six slices can be obtained from the stippled area that is perpendicular to the longitudinal axis of hippocampus. **(C)** View of the second cut surface. Make the third cut perpendicular to the surface of the second cut along the line shown in the figure.

5. The second cut is parallel to the first one, and crosses the midpoint of the posterior border of the right cerebral cortex (*see* Fig. 1A). The tissue block between the two cuts contains the dorsal portion of the hippocampus. The cutting angles can be adjusted to obtain other portions of the hippocampus.
6. The third cut should be made perpendicular to the surface of the second cut and along the line between the cerebellum and the cerebrum (*see* Fig. 1C). Make the third cut 1–2 mm away from the cerebrum to avoid damaging the hippocampus. The surface of the third cut will be placed against the agar block on the slicing stage (*see* Fig. 2). Some authors use different methods to make the hippocampal tissue block (*see* Note 8).

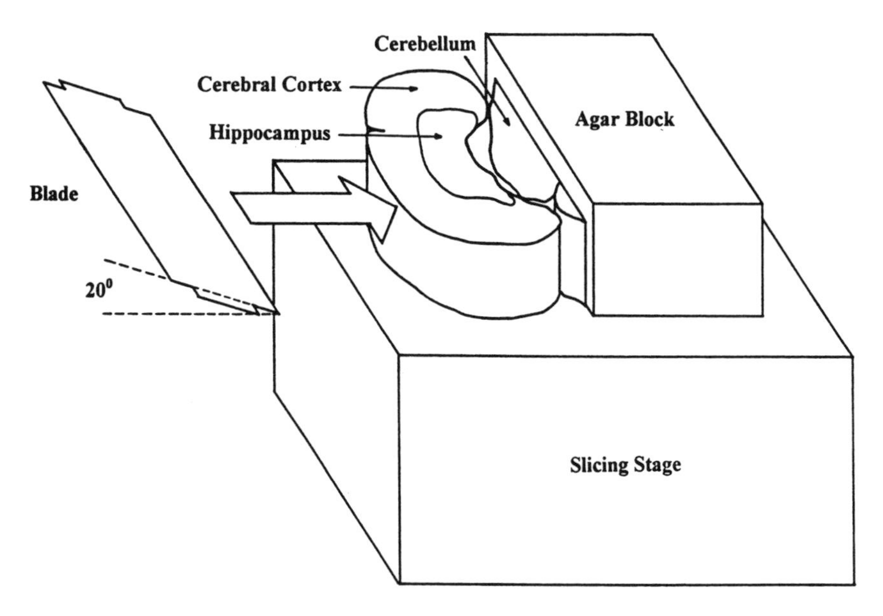

Fig. 2. Arrangement of the agar and tissue blocks on the slicing stage. Note that the surface of the third cut of the tissue block is placed against the agar block.

3.1.3. Mounting onto the Slicing Stage

1. Spread a thin layer of cyanoacrylate glue on the slicing stage (*see* Notes 4 and 9) to which an agar block has already been mounted (*see* Fig. 2).
2. Shovel the tissue block off the Petri dish with a spatula and place it on a piece of dry filter paper with the mounting surface down. Wait for about 2–3 s and allow the fluid on the mounting surface to be drained away by the filter paper. While waiting, dry the spatula by rubbing it on the filter paper.
3. Using the spatula, shovel the tissue block from the filter paper onto the slicing stage and place it next to the agar block (*see* Fig. 2). Withdraw the spatula while using another spatula to gently hold the tissue block in place.
4. Clamp the slicing stage onto the vibratome specimen vise with the agar block furthermost from the blade (*see* Fig. 2). The tissue block and the blade should be totally submerged in cold aCSF (4–8°C) that had been poured into the vibratome and aerated with 95% O_2 and 5% CO_2 before sacrificing the animal.

3.1.4. Slicing

1. Adjust the vibratome section thickness to 400 μm and the amplitude of vibration to maximum (Vibratome series 1000, Technical Products International).
2. Position the sectioning blade at a 20° angle and advance it forward, into the tissue block, at a speed that does not distort the tissue while slicing.
3. Discard the first few slices until a cutting plane that is roughly perpendicular to the longitudinal axis of the hippocampus is reached (shaded area in Fig. 1B). The

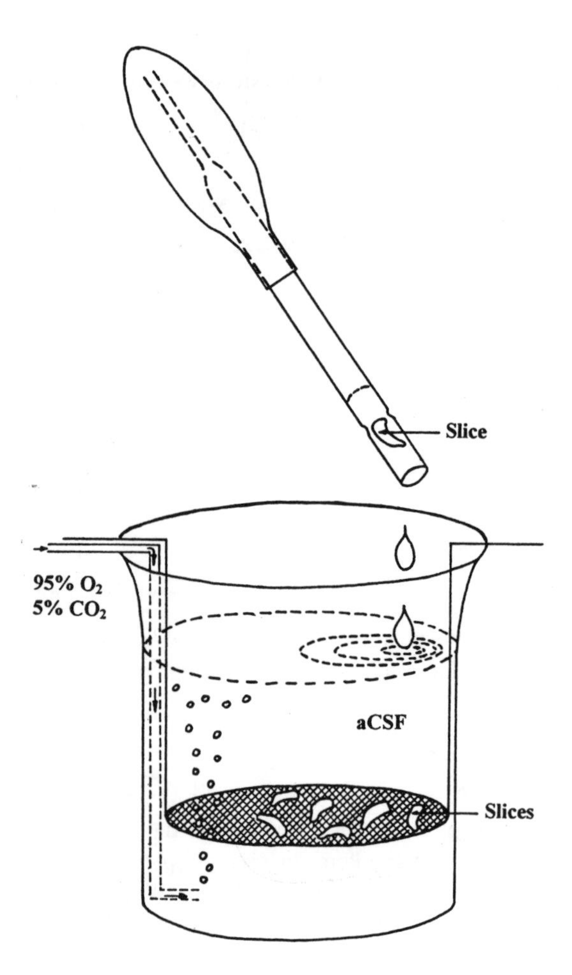

Fig. 3. Storage of brain slices. The slices are incubated in aCSF aerated with a 95% O_2 and 5% CO_2 gas mixture. The crosshatched area represents the nylon net on which the slices are placed. A Pasteur pipet with its narrow opening attached to the rubber bulb is used for transferring slices.

hippocampus is sliced along with part of its surrounding tissues. Separate the latter from the hippocampal slices using forceps and fine scissors after all the slices have been made.

3.1.5. Storage of the Slices

1. Immediately after slicing, move each slice to a storage chamber using a Pasteur pipet with the bulb placed over the narrow opening (*see* Fig. 3).
2. Incubate the slices in the storage chamber at room temperature (*see* Note 10) for 60 min before use. A 200-mL beaker containing aCSF bubbled with 95% O_2 and

5% CO_2 can be used as the storage chamber (*see* Note 11 and Fig. 3). Place the slices on a nylon net that is positioned halfway into the beaker to prevent the slices from adhering to the glass wall (*see* Note 11).

3.2. Tissue Chopper Method

3.2.1. Anesthesia, Decapitation, and Removal of the Brain from the Skull

Identical to the vibratome method (*see* Section 3.1.1.).

3.2.2. Isolation of the Hippocampus

1. Use an upside down Petri dish with a piece of moistened filter paper on it (*see* Note 7) as a dissecting platform.
2. Place the brain on the platform and cut it in half along the longitudinal fissure.
3. Expose the hippocampus by holding the brain stem in place and gently pushing the cortex away from the midbrain with a flat spatula.
4. While doing so, sever the connections that hold the cortex to the brain stem and the midbrain.
5. Separate the hippocampus from its dorsal and temporal connections to the cortex with a spatula.
6. Roll the hippocampus out gently and isolate it by cutting it free from the remaining cortex.

3.2.3. Slicing

1. Place a piece of filter paper (*see* Note 7) on the slicing stage and moisten it with chilled aCSF.
2. Lay the isolated hippocampus on the stage with its longitudinal axis perpendicular to the razor blade.
3. Make the slices by lowering the blade through the hippocampus. Eight to 12 slices of 400–500 μm thickness can be obtained.
4. Immediately after making each slice, transfer the slice to the storage chamber (*see* Fig. 3) using a fine brush. Take extreme care to avoid stretching slices during handling. The best strategy is to rotate the brush gently to accommodate the position of the slice.

3.2.4. Storage of the Slices

Identical to the vibratome method (*see* Section 3.1.5.).

4. Notes

1. All chemicals used for aCSF should be ACS grade reagents.
2. Contaminants in the water used to make the aCSF can lead to unhealthy slices and poorly controlled experiments. Deionized water with an electrical resistance >18 MΩ indicates high purity. Deionized water with a lower resistance should be distilled. Double-distilled water is also employed in many laboratories. Distilla-

tion and deionization do not remove some organic contaminants and many laboratories pretreat the water with a carbon filter.

3. The aCSF recipe presented here is the most commonly used for rat hippocampus slices (*see* refs. *9* and *14*). A large number of alternative recipes can be found in the research literature. A few optional recipes are listed below for the readers convenience.
   a. 126 m*M* NaCl, 3 m*M* KCl, 1.4 m*M* KH_2PO_4, 26 m*M* $NaHCO_3$, 4 m*M* glucose, 1.3 m*M* $MgSO_4$, 1.4 m*M* $CaCl_2$ *(10)*.
   b. 126 m*M* NaCl, 5 m*M* KCl, 1.26 m*M* KH_2PO_4, 26 m*M* $NaHCO_3$, 10 m*M* glucose, 1.3 m*M* $MgSO_4$, 2.4 m*M* $CaCl_2$ *(15)*.
   c. 125 m*M* NaCl, 2.5 m*M* KCl, 1.25 m*M* NaH_2PO_4, 13 m*M* HEPES, 31 m*M* $NaHCO_3$, 12.5 m*M* Glucose, 2 m*M* $MgSO_4$, 1.5 m*M* $CaCl_2$ *(16)*.

   The composition of the aCSF in a. most closely resembles the cerebrospinal fluid *in situ*.
4. To make an agar gel block:
   a. Dissolve 5 g of agar in 100 mL of distilled water by heating and stirring;
   b. Pour the melted agar into a Petri dish;
   c. Cut it into blocks (about 1 × 1 × 1 cm) after it hardens.

   The agar gel block is used to support and prevent the tissue block from leaning back when slicing. It is glued in place before sacrificing the animal to reduce the ischemic period during slice preparation.
5. Halothane should be used in a well-ventilated area, preferably under a hood.
6. If the rat must be sacrificed without anesthetics, handle it gently and do not make loud noises. This will make the decapitation easier for both the investigator and the animal. If anesthetics do not complicate the experiments, they are recommended before decapitation.
7. Placing a piece of filter paper on the platform will prevent the brain tissue from sliding during dissection or slicing.
8. Some authors isolate the hippocampus in a way as described in the tissue chopper method. The middle portion of the hippocampus is cut out and mounted onto the vibratome slicing stage with cyanoacrylate glue. In our opinion, the tissue block method described in this chapter is superior, since it reduces the direct handling of the hippocampus.
9. Using only a thin layer of cyanoacrylate glue is the key to a secure and clean mounting. A common mistake is to use an excessive amount of the glue, resulting in a messy mounting and poor slicing.
10. The incubation of the slices at 37°C seems to produce equally good results.
11. To make the storage chamber:
    a. Cut away the bottom of a 100 mL polypropylene beaker;
    b. Cover the bottom of the beaker by attaching a piece of nylon netting to it with Krazy glue;
    c. Place the modified polypropylene beaker into a 200 mL glass beaker filled with aCSF.

    Aerate the aCSF in the storage chamber using a gas dispersion tube placed below the nylon netting.

References

1. Schurr, A., West, C. A., Tseng, M. T., Reid, K. H., and Rigor, B. M. (1987) The role of glucose in maintaining synaptic activity in the rat hippocampal slice preparation, in *Brain Slices: Fundamentals, Applications and Implications* (Schurr, A., Teyler, T. J., and Tseng, M. T., eds.), Karger, Basel, pp. 39–44.
2. Reid, K. H., Schurr, A., and West, C. A. (1987) Effects of duration of hypoxia, temperature and aCSF potassium concentration on probability of recovery of CA1 synaptic function in the *in vitro* rat hippocampal slice, in *Brain Slices: Fundamentals, Applications and Implications* (Schurr, A., Teyler, T. J., and Tseng, M. T., eds.), Karger, Basel, pp. 143–146.
3. Kimelberg, H. K., Biddlecome, S., Narumi, S., and Bourke, R. S. (1978) ATPase and carbonic anhydrase activities of bulk-isolated neuron, glia and synaptosome fractions from rat brain. *Brain Res.* **141,** 305–323.
4. Deisz, R. A. (1992) The neocortical slice, in *Practical Electrophysiological Methods* (Ketterman, H. and Grantyn, R., eds.), Wiley-Liss, New York, pp. 45–50.
5. Madison, D. V. (1991) Whole-cell voltage-clamp techniques applied to the study of synaptic function in hippocampal slices, in *Cellular Neurobiology* (Chad, J. and Wheal, H., eds.), Oxford University Press, Oxford, pp. 132–149.
6. Kass, I. S. and Lipton, P. (1982) Mechanisms involved in irreversible anoxic damage to the *in vitro* rat hippocampal slice. *J. Physiol.* **332,** 459–472.
7. Kass, I. S. (1987) The hippocampal slice: an *in vitro* system for the studying irreversible anoxic brain damage, in *Brain Slices: Fundamentals, Applications and Implications* (Schurr, A., Teyler, T. J., and Tseng, M. T., eds.), Karger, Basel, pp. 105–117.
8. Fried, E., Amorim, P., Chambers, G., Cottrell, J. E., and Kass, I. S. (1995) The importance of sodium for anoxic transmission damage in rat hippocampal slices: mechanism of protection by lidocaine. *J. Physiol.* **489,** 557–566.
9. Leonard, B. W., Barnes, C. A., Rao, G., Heissenbuttel, T., and McNaughton, B. L. (1991) The influence of post-mortem delay on evoked hippocampal field potentials in the *in vitro* slice preparation. *Exp. Neurol.* **113,** 373–377.
10. Kass, I. S., Bendo, A. A., Abramowicz, A. E., and Cottrel, J. E. (1989) Methods for studying the effect of anesthetics on anoxic damage in the rat hippocampal slice. *J. Neurosci. Meth.* **28,** 77–82.
11. Feig, S. and Lipton, P. (1990) N-methyl-D-aspartate receptor activation and Ca^{2+} account for poor pyramidal cell structure in hippocampal slices. *J. Neurochem.* **55,** 473–483.
12. Grace, A. A. and Onn, S.-P. (1989) Morphology and electrophysiological properties of immunocytochemically identified rat dopamine neurons recorded *in vitro.* *J. Neurosci.* **9,** 3463–3481.
13. Wang, T., O'Connor, W. T., Ungerstedt, U., and French, E. D. (1994) N-Methyl-D-aspartic acid biphasically regulates the biochemical and electrophysiological response of A10 dopamine neurons in the ventral tegmental area: *in vivo* microdialysis and *in vitro* electrophysiological studies. *Brain Res.* **666,** 255–262.

14. Berg-Johnsen, J., Grondahl, T. O., Langmoen, I. A., Haugstad, T. S., and Hegstad, E. (1995) Changes in amino acid release and membrane potential during cerebral hypoxia and glucose deprivation. *Neurol. Res.* **17,** 201–208.
15. Hori, N., Doi, N., Miyahara, S., Shinoda, Y., and Carpenter, D. O. (1991) Appearance of NMDA receptors triggered by anoxia independent of voltage *in vivo* and *in vitro*. *Exp. Neurol.* **112,** 304–311.
16. Perouansky, M. and Yaari, Y. (1993) Kinetic properties of NMDA receptor-mediated synaptic currents in rat hippocampal pyramidal cells versus interneurons. *J. Physiol.* **465,** 223–244.

2

Preparation of Organotypic Hippocampal Slice Cultures Using the Membrane Filter Method

Peter J. Bergold and Patrizia Casaccia-Bonnefil

1. Introduction

Thin slices of brain are commonly used to study regulation of neurotransmission in the central nervous system (CNS). In most of these studies, experiments are performed using brain slices soon after their preparation. Thin brain slices have limited viability: Unless special culture conditions are employed, slices die after 1 d in vitro (DIV) (*see* Chapter 1). Longer viability is often desired to permit study of more extended processes, such as long-term modulation of synaptic plasticity, apoptosis, process outgrowth, and development. For this reason, neuroscientists have developed a variety of methods to place brain slices into long-term culture. All of these methods successfully preserve much of the tissue architecture and synaptic connections. Two early methods—using roller tubes *(1)* or Maximov chambers *(2)*—are no longer commonly used since they are technically cumbersome and do not permit optimum access to the slice. Most studies now employ variations of the membrane filter method *(3)*. Brain slices are placed on a porous membrane filter that provides a suitable substratum and allows a thin film of media to cover the surface of the slice by capillary action (Fig. 1). The membrane inserts are placed into standard tissue culture dishes for long-term culture. The slice remains well-oxygenated without drying out and adequate nutrition is provided through the membrane. Slices from a variety of brain regions have been successfully cultured. This chapter will concentrate on the culturing of hippocampal slices. Culturing of slices from other brain regions requires little modification of this method.

Cultured hippocampal slices prepared using the membrane insert method have been used to study process outgrowth *(4–6)*, synaptogenesis *(4,5)*, epileptogenesis *(7)*, delayed neuronal loss *(7–12)*, calcium homeostasis *(12)*,

From: *Methods in Molecular Biology, Vol. 72: Neurotransmitter Methods*
Edited by: R. C. Rayne Humana Press Inc., Totowa, NJ

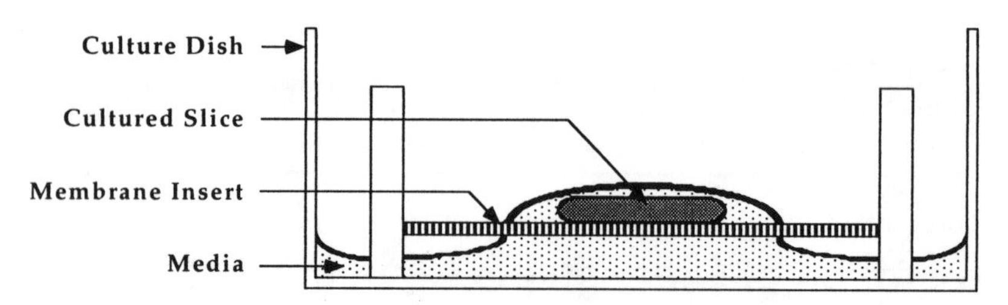

Fig. 1. Schematic drawing of the culture of brain slices on membrane inserts. The culture medium forms a thin layer on the top and sides of the slice by capillary action.

gene transfer by viral vectors *(13,14)*, neurotrophin action *(15)*, and modulation of gene expression *(15,16)*. The diverse nature of these studies testifies to the general usefulness of the cultured slice method in modern neuroscience. This chapter will describe a slightly modified version of the original membrane insert method that has been successfully used in the author's laboratory.

2. Materials

2.1. Animals

Perform all experiments using Sprague-Dawley postnatal d 9–11 rat pups (*see* Notes 1 and 2).

2.2. Instruments

All surgical instruments should be of the highest quality.

1. Large scissors (Fisher, cat. no. 13-804-6).
2. Large forceps (Fisher, cat. no. 08-887).
3. Medium scissors (Fine Science Tools, cat. no. 14002-12).
4. Microdissection scissors (Roboz, cat. no. RS-5676).
5. Microsuturing needle holder (Roboz, cat. no. RS-6410).
6. Large spatula (Biological Research Instruments, cat. no. 16-1630).
7. Microspatulae (Fine Science Tools, cat. no. 10091-12).
8. Barraquer (Roboz, cat. no. RS-6456).
9. 200-μm Polyester film (Pearl Paint, New York, NY).

2.3. Tissue Culture Reagents

Unless indicated, all tissue culture solutions are obtained from Gibco/Life Technologies. If you are purchasing from other vendors, use tissue culture grade or the highest analytical grade reagents.

1. Gey's balanced salt solution, supplemented with glucose (GBSS-G): Gey's balanced salt solution containing 25 m*M* sodium *N*-[2-hydroxyethyl]piperazine-

N'-[2-ethanesulfonic acid] (HEPES), pH 7.2 (Sigma, St. Louis, MO), and D-glucose (6.5 mg/mL; Sigma).
2. Millipore-CM tissue culture inserts (Millipore, Bedford, MA) (*see* Note 3).
3. Six-well culture dishes (Falcon 3046) or 35-mm Petri dishes (Falcon 1008).
4. Slice culture medium (SCM): 50% Eagle's basal medium, 25% heat-inactivated horse serum, 25% Earle's balanced salt solution, supplemented with glutamine (1 m*M*; Sigma), glucose (6.5 mg/mL; Sigma), and sodium HEPES, (25 m*M*, pH 7.2; Sigma; *see* Notes 4 and 5).
5. Antimitotic medium: SCM containing 100 n*M* 5'-flurodeoxyuridine, 100 n*M* cytosine arabinoside, 100 n*M* uridine (Sigma).

3. Method

Employ rigorous sterile technique throughout the following procedures because the use of antibiotics is discouraged (*see* Note 6). Use GBSS-G to bathe tissues throughout the dissection procedure.

3.1. Dissection

1. Anesthetize a postnatal d 9–11 rat by injection of ketamine (100 μL, 10 mg/mL, Sigma) and decapitate with large scissors.
2. Make an incision along the midline of the head and peel the skin away to expose the skull. Using the small scissors, remove the neck muscles and cut along the midline of the skull from the foramen magnum to the interhemispheric sulcus; follow this with two lateral cuts.
3. Peel away the skull using the microsuturing needle holder. Invert the head and remove the brain by placing a spatula between the brain and the skull and cutting the cranial nerves and olfactory bulb.
4. Gently drop the brain into a Petri dish containing 1 mL of GBSS-G that has been precooled to 4°C.
5. With the scalpel, make two sagittal cuts at the interhemispheric sulcus. Using the large spatula, gently pull away the cortex and the hippocampus from the thalamus and basal ganglia and sever the septo-hippocampal connections.
6. Using the scalpel, make an incision at the level of the hippocampal fissure. With the small spatula, scoop the hippocampus from the underlying cortex. If the dissection is sufficiently rapid (<10 min), isolate the second hippocampus and slice along with the first hippocampus.

3.2. Preparation of Slices

1. Using a sterile, wide-bore, plastic transfer pipet, transfer the hippocampi to a polyester film immobilized on the stage of a McIllwain tissue chopper (Brinkman, Westbury, NJ; *see* Note 7).
2. Align the hippocampus so the temporal end is perpendicular to the chopper blade (*see* Note 8).
3. Chop rapidly: Approx 1 chop/s. Typically, prepare transverse sections of 400–500 μm.

4. Using a sterile, wide-bore, plastic transfer pipet, transfer sliced hippocampi to a 35-mm dish containing 5 mL of ice cold GBSS-G and separate the slices by shaking the dish or by use of microspatulae. **Use great care** since the slices are easily damaged. Incubate the separated slices for 1 h at 4°C.
5. Place the sliced hippocampus under a dissection microscope. Select slices having even margins and clear, uniform, well-defined pyramidal cell layers and plate on membrane inserts.
6. Pipet 1 mL of SCM into each well of a six-well culture dish. Place a Milli-Cell CM filter insert into each well (*see* Note 9).
7. Using a sterile, wide-bore transfer pipet, transfer 1–3 slices to the wet surface of the filter insert and remove excess fluid using a micropipet. Plate the slices close to the center of the insert, but keep slices separated by more than 2 mm (*see* Note 10).
8. Maintain cultures at 35°C in a humidified, 95% air, 5% CO_2 atmosphere (*see* Note 11). Completely replace the SCM every 3–5 d (*see* Note 12).
9. After 4–5 d in vitro (DIV), remove actively mitotic cells by treatment with antimitotic medium for 24 h (*see* Notes 13 and 14).
10. Examine the cultures after 10–12 DIV to check for the preservation of pyramidal and granule cell layers. Discard cultures that display selective neuronal loss of a hippocampal region (*see* Note 15). Cultures may be used up to 28 DIV (*see* Note 16).
11. Make electrophysiological recordings from each slice culture preparation to ensure that spontaneous activity does not arise prior to 28 DIV. Discard the entire dissection if spontaneous activity >0.5 mV in amplitude is observed (*see* Fig. 2 and Note 17).

4. Notes

1. Virtually all hippocampal cultures have been prepared from rats. No differences among different rat strains have been reported. Because of the smaller size of the mouse brain as compared to the rat, it has been very difficult to prepare cultures from postnatal mice.
2. Although there has been one report of slice cultures prepared from rats as old as postnatal d 22 *(3)*, most investigators have reported using cultures prepared from rats prepared between postnatal d 8 and 12. Slice culture plating efficiency is highest using postnatal d 10 to postnatal d 11 rats. Slices older than postnatal d 10 can be cultured, yet the number of useful cultures drops sharply for unknown reasons. Preparation of cultures from animals younger than postnatal d 8 is more difficult since the brain is smaller and softer. The hippocampus from younger animals is less developed and will likely develop more unwanted synapses in vitro. This will be most pronounced in the dentate gyrus since it is has the slowest development of all hippocampal regions.
3. Despite similar products from other vendors, most investigators have had the greatest success using Milli-Cell CM inserts.
4. Slice cultures are grown in high concentrations of horse serum. Different horse sera have large effects on the quality of slice cultures. Multiple lots of horse sera should to be tested to find sera that work well.

0 DIV

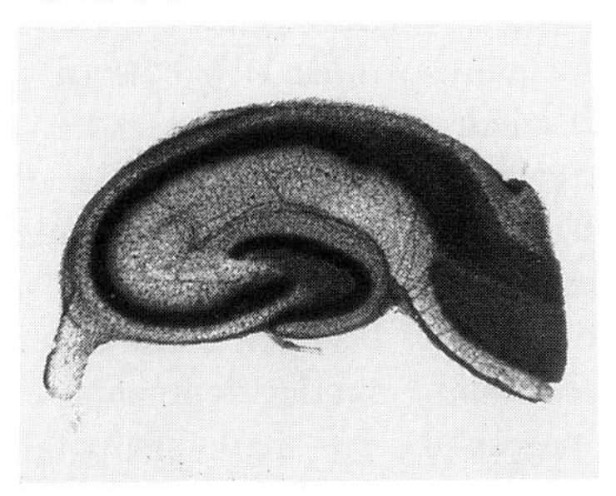

21 DIV

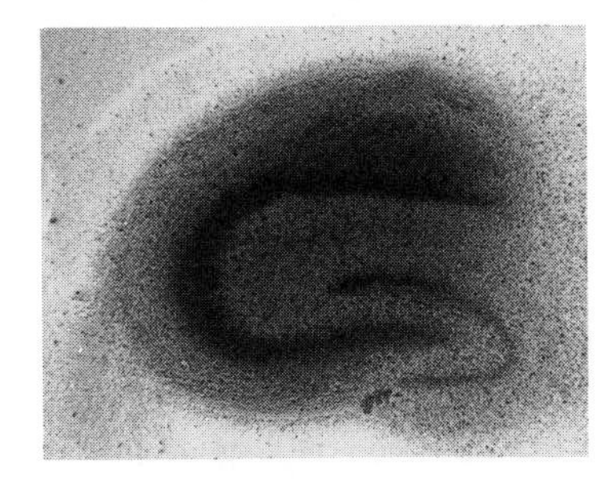

Fig. 2. Appearance of hippocampal slices from postnatal d 10 rat pups immediately after slicing and after 21 DIV. Transverse slices were immediately prepared (0 DIV) or cultured (21 DIV). Slices were fixed and stained with cresyl violet. The neuronal layers of the dentate gyrus, CA1, CA3, the subiculum, and the entorhinal cortex are evident immediately after slicing. After 21 DIV, CA3, CA1 and the dentate gyrus are well preserved whereas the entorhinal cortex and subiculum have degenerated. Neuronal layers in the 21DIV culture are wider because of the thinning of the slice in vitro. Superior preservation of the slice is indicated by a similar density of staining between the CA3 and CA1 pyramidal cell layers or between the upper and lower limbs of the dentate gyrus.

5. All slice culture solutions must be well buffered since small shifts in pH can be very toxic. pH thresholds have never been reported, but both alkalization and acidification are selectively neurotoxic. Exposure to pH 6.6 for as little as 15 min induces profound neuronal loss *(11)*.
6. Wipe all tissue culture incubators, hoods, tissue choppers, and microscope stages with 70% ethanol on a regular basis. It is advisable to have a dedicated tissue chopper that remains in a sterile tissue culture hood. Most contamination problems occur during the preparation of the slice cultures. Contaminants are generally of two types: yeast and bacteria. Control bacteria by adding chloramphenicol

(to 25 μg/mL; Sigma) to slice culture media. Control yeast by discarding all suspicious media and cleaning all hoods and incubators with 0.5% sodium dodecyl sulfate followed by a second wash with 70% ethanol.

7. The hippocampus is sliced on a 200 μm polyester film of 2 × 3 cm (Pearl Paint) that has been extensively washed with 70% ethanol and air dried in a sterile hood. Any plastic film can be used to support the hippocampus. The film must be sterile and must resist being sliced by the tissue chopper blade.
8. There are consistent differences between cultures of the septal and temporal hippocampus that likely result from developmental gradients in the immature hippocampus *(17)*. At postnatal d 10–11, the temporal end of the hippocampus is more mature than the septal end. Therefore, consistent differences can be observed between slice cultures originating from the temporal end or septal end. In particular, the mossy fiber projection differs between septal and temporal cultures *(17)*. In septal cultures, there is mossy fiber sprouting; mossy fiber terminals are found in the inner molecular layer of the dentate gyrus and the CA3 pyramidal cell layer. In temporal cultures, the mossy fiber projection is normal. This likely results in differences in slice culture excitability since epileptiform activity is more readily induced in septal cultures than temporal cultures *(17)*.
9. The filter should become completely wetted within seconds. Slow or incomplete wetting indicates a poor lot of membrane inserts.
10. Up to ten slice cultures can be plated on a single filter insert. Slices may innervate each other when plated closer than 2 mm, particularly if they have been damaged.
11. Temperatures above 37°C are selectively neurotoxic to slice culture neurons. This may be a result of spontaneous epileptiform activity. Temperatures between 31 and 36°C are tolerated quite well.
12. The insert may be placed in a 100 mm tissue culture dish (Falcon 1029) containing 8 mL of slice culture medium to avoid frequent feedings.
13. This antimitotic treatment is not obligatory, but it limits glial outgrowth at the periphery of the culture.
14. After plating, the neuronal layers will widen modestly. This likely results from the removal of cells that were damaged during culture preparation and antimitotic treatment. Thinning of the slice is most evident in regions enriched in glia. Neuronal layers remain quite thick, thus permitting recording of extracellular field potentials using standard slice technology.
15. A common problem is that a portion of the cultured hippocampus consistently degenerates. Localized damage is often diagnostic. Damage resulting from inferior dissection technique can be consistent and highly localized. Loss of CA1 may result from hypoxia during slice preparation. CA3 is susceptible to excitotoxic damage resulting from spontaneous epileptiform activity. Cutting of the fimbria too close to the CA3 pyramidal layer results in damage of CA3 pyramidal cells closest to the fimbria. Excessive blade force during chopping results in the absence of the lower blade of the dentate gyrus or displacement of granule cells out of stratum granulosum.

It is important to consider the optimal time to perform an experiment. Slice cultures change continuously in vitro. The development of the hippocampal slice in vitro has not been extensively studied. Extracellular recordings have shown that synaptic responses are substantially larger at 21 than at 7 DIV *(3)*. The mossy fiber pathway connecting dentate granule cells with CA3 pyramidal cells and hilar interneurons forms entirely in vitro between 14 and 21 DIV *(4,5,7)*. After 28 DIV, a steadily increasing proportion of cultures will be spontaneously epileptic (*see* Note 14). As a result, we perform most experiments between 21–28 DIV. This permits mossy fiber development, yet avoids spontaneous epileptiform activity.

16. Even if properly prepared, most investigators observe spontaneous epileptiform activity in slice cultures, beginning at 5 wk in vitro *(17–19)*. If slices are properly prepared, no spontaneous neurodegeneration should be observed until 5–6 wk in vitro. Spontaneous neuronal loss begins soon after spontaneous epileptiform activity, suggesting that the two are related. This is consistent with a recent report that spontaneous neuronal loss in slice cultures can be suppressed with glutamate receptor antagonists *(20)*.
17. Epileptiform activity will appear at any time in poorly prepared slice cultures. Between 1 and 21 DIV, slices with normal excitability and epileptic slices appear identical. Spontaneous neuronal loss will begin at approx 21 DIV.

References

1. Gahwiler, B. H. (1981) Organotypic monolayer cultures of nervous tissue. *J. Neurosci. Meth.* **4,** 329–342.
2. Lindner, G. and Grosse, G. (1982) Morphometric studies of the rat hippocampus after static and dynamic cultivation. *Z. Mikros.-Anatom. For.* **96,** 485–496.
3. Stoppini, L., Buchs, P.-A., and Muller, D. (1991) A simple method for organotypic culture of nervous tissue. *J. Neurosci. Meth.* **37,** 173–182.
4. Dailey, M. E., Buchanan, J., Bergles, D. E., and Smith, S. J. (1994) Mossy fiber growth and synaptogenesis in rat hippocampal slices *in vitro*. *J. Neurosci.* **14,** 1060–1078.
5. Robain, O., Barbin, G., Billette de Villemeur, T., Jardin, L., Jahchan, T., and Ben-Ari, Y. (1994) Development of mossy fiber synapses in hippocampal slice culture. *Brain Res. Dev. Brain Res.* **80,** 244–250.
6. Diekmann, S., Nitsch, R., and Ohm, T. G. (1994) The organotypic entorhinal-hippocampal complex slice culture of adolescent rats. A model to study transcellular changes in a circuit particularly vulnerable in neurodegenerative disorders. *J. Neural Trans.* **44(Suppl.),** 61–71.
7. Casaccia-Bonnefil, P., Benedikz, E., Rai, R., and Bergold, P. J. (1993) Excitatory and inhibitory pathways modulate kainate toxicity in hippocampal slice cultures. *Neurosci. Lett.* **154,** 5–8.
8. Tasker, R. C., Coyle, J. T., and Vornov, J. J. (1992) The regional vulnerability to hypoglycemia-induced neurotoxicity in organotypic hippocampal culture: protection by early tetrodotoxin or delayed MK-801. *J. Neurosci.* **12,** 4298–4308.

9. Vornov, J. J., Tasker, R. C., and Coyle, J. T. (1994) Delayed protection by MK-801 and tetrodotoxin in a rat organotypic hippocampal culture model of ischemia. *Stroke* **25,** 457–464.
10. Vornov, J. J., Tasker, R. C., and Park, J. (1995) Neurotoxicity of acute glutamate transport blockade depends on coactivation of both NMDA and AMPA/Kainate receptors in organotypic hippocampal cultures. *Exp. Neurol.* **133,** 7–17.
11. Shen, H., Chan, J., Kass, I. S., and Bergold, P. J. (1995) Transient acidosis induces delayed neurotoxicity in cultured hippocampal slices. *Neuroscience Lett.* **185,** 115–118.
12. Petrozzino, J. J., Pozzo Miller, L. D., and Connor, J. A. (1995) Micromolar Ca^{2+} transients in dendritic spines of hippocampal pyramidal neurons in brain slice. *Neuron* **14,** 1223–1231.
13. Bergold, P. J., Casaccia-Bonnefil, P., Liu, Z.-X., and Federoff, H. J. (1993) Transsynaptic neuronal loss induced in hippocampal slice cultures by a herpes simplex virus vector expressing the GluR6 subunit of the kainate receptor. *Proc. Natl. Acad. Sci. USA* **90,** 6165–6169.
14. Casaccia-Bonnefil, P., Benedikz, E., Shen, H., Stelzer, A., Edelstein, D., Geschwind, M., Bownlee, M., Federoff, H. J., and Bergold, P. J. (1995) Localized gene transfer into organotypic hippocampal slice cultures and acute hippocampal slices. *J. Neurosci. Meth.* **50,** 341–351.
15. Collazo, D., Takahashi, H., and McKay, R. D. (1992) Cellular targets and trophic functions of neurotrophin-3 in the developing rat hippocampus. *Neuron* **9,** 643–656.
16. Rivera, S., Gold, S. J., and Gall, C. M. (1994) Interleukin-1 beta increases basic fibroblast growth factor mRNA expression in adult rat brain and organotypic hippocampal cultures. *Brain Res. Mol. Brain Res.* **27,** 12–26.
17. Casaccia-Bonnefil, P., Stelzer, A., Federoff, H. J., and Bergold, P. J. (1995) A role for mossy fiber activation in the loss of CA3 and hilar neurons induced by transduction of the GluR6 kainate receptor subunit. *Neurosci. Lett.* **191,** 67–70.
18. McBain, C. J., Boden, P., and Hill, R. G. (1988) The kainate/quisqualate receptor CNQX, blocks the fast component of spontaneous epileptiform activity in organotypic culture of rat hippocampus. *Neurosci. Lett.* **93,** 341–345.
19. McBain, C. J., Boden, P., and Hill, R. G. (1989) Rat hippocampal slices "*in vitro*" display spontaneous epileptiform activity following long-term organotypic culture. *J. Neurosci. Meth.* **27,** 35–49.
20. Pozzo Miller, L. D., Mahanty, N. K., Connor, J. A., and Landis, D. M. (1994) Spontaneous pyramidal cell death in organotypic slice cultures from rat hippocampus is prevented by glutamate receptor antagonists. *Neurosci.* **63,** 471–487.

3

The Preparation and Use of Brain Slices and Dissociated Neurons for Patch-Clamp Studies of Neurotransmitter Action

Philip M. Larkman and Hrvoje Hećimović

1. Introduction

The development of in vitro brain slice and isolated neuron techniques has greatly facilitated detailed studies of the electrophysiology of a wide range of neuronal types in the adult and neonatal vertebrate central nervous system (CNS). Particularly advantageous are the greater mechanical stability that these preparations provide over in vivo models and the control allowed over the composition of the extracellular environment. In addition, the development of the patch-clamp technique has opened up the possibility of direct access to the intracellular environment via internal patch pipet solutions. In combination, these approaches have enabled detailed investigations of neuronal membrane properties, the cellular actions of neurotransmitters, and synaptic mechanisms.

Commonly a range of complimentary preparations and techniques may be needed to uncover the complexity of neurotransmitter actions. For example, we have used conventional brain slices (300–500 µm thick) to investigate the central actions of 5-hydroxytryptamine (5-HT) using sharp intracellular microelectrodes *(1)*. Nevertheless, slices of conventional thickness present several problems when attempting to study ion channels and neurotransmitter mechanisms in relative isolation. Several barriers, including uptake mechanisms and degradative enzymes, can slow or prevent the diffusion of drugs and hence hamper accurate quantitative pharmacological analysis. Pre- and postsynaptic components can be difficult to isolate when applying exogenous ligand (e.g., by iontophoresis or perfusion application). It is also clear that not all conventional brain slice preparations readily allow the formation of high resistance "giga-ohm" seals essential for the success of the patch-clamp tech-

From: *Methods in Molecular Biology, Vol. 72: Neurotransmitter Methods*
Edited by: R. C. Rayne Humana Press Inc., Totowa, NJ

nique (although "blind-patching" is highly successful in some brain regions; *see* ref. *2*).

In patch-clamp recordings the formation of "giga-ohm" seals requires that the cell membrane, to which a glass recording electrode will be sealed, be free of surrounding connective tissue. This requirement may be fulfilled using the organotypic slice culture method that provides a thin layer of neurons accessible for patch-clamping (*see* Chapter 2). Alternatively, it may be necessary to isolate cells from the tissue and maintain them in a dissociated cell culture. The relative ease with which many neuronal types can be maintained in cell culture makes this method attractive and useful. Nevertheless, these preparations do have disadvantages. Conditions during cell growth can affect the expression of receptors or ion channels and may also alter the "normal" release of neurotransmitters.

To circumvent these difficulties, acute in vitro preparations not involving the use of culture conditions but that provide clean cell membranes, may be preferred. One such preparation is a variation of the conventional brain slice technique. Brain slices are cut thinly enough (100–140 μm) to allow neurons to be visualized using a conventional upright microscope, water-immersion objective, and Nomarski optics *(3)*. Connective tissue and debris is cleaned from the cell surface by gentle positive and negative pressure through a wide-tipped pipet filled with external solution, thereby providing a clean cell surface for giga-ohm seal formation.

This chapter describes an alternative approach involving the preparation and use of acutely dissociated rat brain neurons maintained in a physiological state as close to normal as possible. The procedure is a fast and reliable method for the isolation and dissociation of cells with "clean" and intact membranes that facilitate the formation of "giga-ohm" seals and allow stable and long patch-clamp recordings. The method, derived from Kay and Wong *(4)* and used in our laboratory *(5)*, allows dissociation of neurons from defined regions of the young and adult rat brain, preserving their proximal dendritic structure as well as their electrical characteristics. This methodology has been used to isolate neurons of the dorsal raphé nucleus and the CA1 region of the hippocampus and is generally applicable for the dissociation of other cell types.

Although mechanical dissociation, in the absence of any enzymes, is possible *(6)*, mammalian cells usually require enzyme treatment to isolate them from the surrounding tissue. Most of the gentlest procedures use trypsin or collagenase as the major or only enzyme. Protease, hyaluronidase, and elastase can also be used alone or in combination with other enzymes. DNAase is sometimes added to the dissociation medium to hydrolyze DNA released from damaged cells that is sticky and may cause cells to clump. In general, the best procedure should be gentle and reasonably rapid leaving the cell membrane intact. The dissociation procedure can be divided into three sections: the preparation of brain slices, incubation of brain tissue in enzyme solution, and dissociation of the cells by trituration. The preparation of brain slices is similar to that described in detail in Chapter 1.

2. Materials

2.1. Brain Slice Preparation

1. Artificial cerebrospinal fluid (aCSF), for dorsal raphé and hippocampal slices: 126 m*M* NaCl, 5 m*M* KCl, 2 m*M* $MgSO_4$, 2 m*M* $CaCl_2$, 26 m*M* $NaHCO_3$, 1.25 m*M* NaH_2PO_4, 10 m*M* D-glucose, pH 7.4, after equilibration with 95% oxygen, 5% carbon dioxide. Osmolarity adjusted to 320 mosM with sucrose (*see* Notes 1–3).
2. 95% Oxygen/5% carbon dioxide (BOC).
3. Large scissors or small animal guillotine.
4. Small scissors (pointed tip).
5. Microdissection scissors (e.g., Aesculap 80-mm straight; cat. no. OC498).
6. Small stainless steel spatula.
7. Small brush.
8. Forceps (e.g., Dumont #5; Fine Science Tools Inc.).
9. Cyanoacylate glue (e.g., Cyanolit super glue, Eurobond).
10. Razor blades (e.g., Wilkinson Sword).
11. Tissue slicer (e.g., Vibroslice 752, Cambden Inst. or DSK Microslicer DTK-1000, Dosaka EM Co.).
12. Agar: 2% (w/v) in aCSF.
13. Filter paper (e.g., Whatman, Qualitative 1).
14. 100-mL Pyrex beaker.
15. Petri dish, Pyrex, (diameter to match filter paper).
16. Wide-tipped, fire-polished, Pasteur pipet.

2.2. Slice Incubation

1. aCSF: *see* Section 2.1., item 1.
2. PIPES solution: 120 m*M* NaCl, 5 m*M* KCl, 1 m*M* $MgCl_2$, 1 m*M* $CaCl_2$, 25 m*M* D-glucose, 20 m*M* PIPES (piperazine-*N,N*'-bis-[2-ethanesulfonic acid]), pH 7.0, adjusted with NaOH. Osmolarity adjusted to 320 mosM with sucrose (*see* Notes 3 and 4).
3. Trypsin (Type XI, Sigma T-1005).
4. 95% Oxygen/5% carbon dioxide (BOC).
5. 100% Oxygen (BOC).
6. Petri dish, Pyrex.
7. Razor blades.
8. Wide-tipped, fire-polished, Pasteur pipet.
9. 50-mL Polypropylene conical tube (e.g., Falcon Opticul 2070).
10. Incubation chamber (*see* Fig. 1).
11. Water heater/pump.

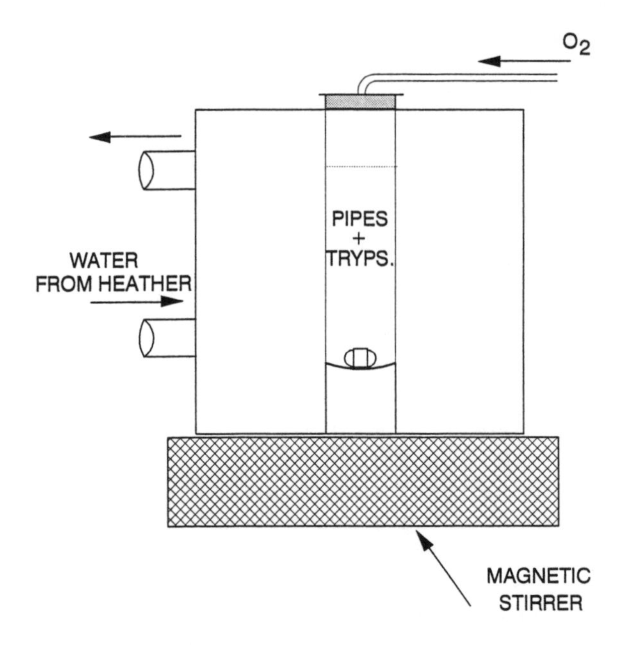

Fig. 1. Schematic for the design of the tissue incubation chamber. The outer chamber is made from Perspex and sits on a magnetic stirrer. The top has a circular hole cut to the diameter of the polypropylene tube containing the PIPES/trypsin solution whereas the base has a small podium machined to accommodate the base of the tube. Oxygen is continuously supplied by a tube attached to a loose-fitting cap. Water maintained at the required incubation temperature is pumped in the direction shown.

2.3. Cell Dissociation

1. Pasteur pipets with heat-polished tips of decreasing diameter (1, 0.75, and 0.5 mm).
2. Glass vial (~10 mL).
3. Polystyrene Petri dish (e.g., Corning 35-mm tissue culture dish).
4. Dulbecco's Modified Eagle's Medium (DMEM, Gibco), **or**
5. Extracellular recording solution: 138 m*M* NaCl, 1 m*M* KCl, 1 m*M* $MgCl_2$, 5 m*M* $CaCl_2$, 10 m*M* sucrose, 10 m*M* D-glucose, 20 m*M* HEPES, pH 7.4, adjusted with NaOH. Osmolarity adjusted to 320 mosM with sucrose.
6. Inverted phase-contrast microscope (e.g., Nikon Diaphot).

3. Methods

3.1. Brain Slice Preparation

1. Transfer 25 mL of PIPES solution, presaturated with 100% oxygen, into a polypropylene tube and add 5000 U/mL (0.07%, w/v) of trypsin. Place the tube in

the incubation chamber (*see* Fig. 1) and prewarm to 33°C under a continuous supply of 100% oxygen (*see* Notes 4 and 5).

2. Decapitate the animal either with large scissors or a small animal guillotine depending on animal size (*see* Note 6).
3. Using small pointed scissors, cut the bone lengthways on either side of the cranium and then gently lift away the top of the skull taking care to cut any connective tissue to prevent tearing the brain tissue. Wash the brain frequently with ice-cold aCSF during this dissection.
4. Quickly and gently remove the brain from the cranium by running a small spatula underneath it, taking care to sever the optic and cranial nerves (*see* Note 7).
5. Place the brain in a 100-mL beaker containing aCSF cooled on ice to ~4°C and continuously bubbled with 95% oxygen/5% carbon dioxide. Allow to cool for approx 30–45 s.
6. Place a disk of filter paper on an upturned Petri dish sitting on a bed of ice. After moistening the paper with aCSF, place the brain on the filter paper and gently dissect away the dura, blood vessels, and other connective tissue using fine forceps and microscissors (*see* Note 8).
7. Using a razor blade, carefully prepare a small block of tissue containing the desired brain region (the exact size will be determined by the region of interest; *see* Note 9).
8. Using a small amount of cyanoacrylate glue, stick the tissue block to the stage of a tissue slicer supporting the side furthest away from the slicer blade with a small block of agar (*see* Notes 10 and 11).
9. Place the slicer stage into the slicer chamber filled with ice-cold (~4°C) aCSF and cut slices at approx 400 μm thickness (*see* Note 12).
10. Transfer the slices using a wide-tipped Pasteur pipet to a Pyrex Petri dish filled with aCSF. Continuously bubble the aCSF with 95% oxygen/5% carbon dioxide (*see* Note 13).
11. Using two sharp razor blades cut tissue pieces of approx 2 × 2 mm from the region of the slice containing the cells of interest (*see* Notes 9 and 13).

3.2. Slice Incubation

1. Remove the polypropylene tube filled with the PIPES/trypsin solution from the incubation chamber and, using a wide-tipped Pasteur pipet, transfer the 2 × 2 mm pieces of tissue into the tube.
2. Place the tube back in the incubation chamber and stir for 90 min using a magnetic stirrer and bar with just sufficient vigor to keep the pieces of tissue suspended. The temperature should be maintained at 33°C and the atmosphere above the PIPES/trypsin solution continuously supplied with 100% oxygen.
3. Terminate the incubation by removal of the PIPES/trypsin solution using a Pasteur pipet and wash the tissue three times (5, 5, and 10 mL) with PIPES solution.
4. Keep the pieces of tissue stirring in the final wash of PIPES solution for at least another 60 min at room temperature (18–22°C) prior to commencing the cell dissociation procedure.

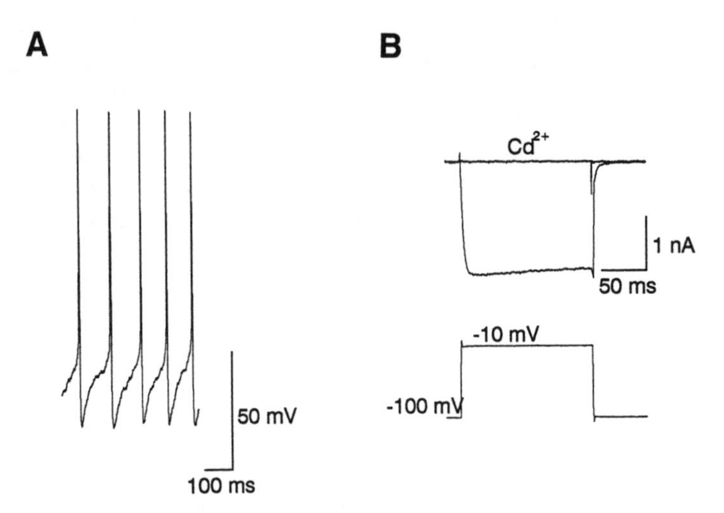

Fig. 2. Electrophysiological recordings from an acutely dissociated dorsal raphé neuron. Action potentials **(A)** and calcium currents **(B)** were recorded. In (B) calcium currents were evoked with a step potential from V_H –100 to –10 mV, and the currents were blocked in the presence of cadmium.

3.3. Cell Dissociation

1. Remove two pieces of tissue at a time into a small glass vial containing ~1.5 mL of DMEM or extracellular recording solution (*see* Note 14). The remaining pieces of tissue can be left stirring in the final wash PIPES solution for up to 8 h and still be suitable for cell dissociation.
2. Gently triturate the tissue with three fire-polished pipets of decreasing tip diameter (1 to 0.5 mm; *see* Note 15).
3. Use a Pasteur pipet to transfer the dissociated cells to a 35-mm polystyrene Petri dish placed in the holder on the stage of the inverted phase contrast microscope (*see* Note 5).
4. Allow 15–20 min for the cells to settle to the bottom of the dish before attempting to start patch-clamp recording (*see* Fig. 2 and Notes 16 and 17).

4. Notes

1. aCSF and external recording solutions are best made fresh on the day of dissociation using distilled, deionized water, although they can be stored overnight at ~4°C. All solutions should be saturated with 95% oxygen/5% carbon dioxide by continuously bubbling for about 30 min before and throughout the slicing, dissociation, and recording procedures. This is an attempt to reduce hypoxic insults to the tissue and dissociated neurons.

2. Although this aCSF is suitable for obtaining good slices of most higher brain regions, for brainstem slices we have found the sucrose replacement methodology of Aghajanian and Rasmussen *(7)* to greatly enhance the number of viable slices containing motoneurons obtained over a series of experiments. Thus for motoneuron-containing slices our aCSF composition is: 248 m*M* sucrose, 5 m*M* KCl, 2 m*M* $MgSO_4$, 2 m*M* $CaCl_2$, 26 m*M* $NaHCO_3$, 1.25 m*M* HEPES, 10 m*M* D-glucose, pH 7.4, after equilibration with a 95% oxygen, 5% carbon dioxide gas mixture. Note that some Na^+ remains (~27 m*M*) providing a substrate for the proton transporter so the intracellular pH may not be dramatically altered. Before electrophysiological recordings can be made, the sucrose-based aCSF should be switched to a recording aCSF containing 124 m*M* NaCl instead of the 248 m*M* sucrose. This exchange should be slow and graded. We routinely allow a further 1 h for complete exchange and equilibration of the recording aCSF.
3. The osmolarity of all external solutions (aCSF, *see* Section 2.1.1.; PIPES solution, *see* Section 2.2.2.; extracellular recording solution, *see* Section 2.3.5.) should be adjusted as necessary with 0.5*M* sucrose to 320 mosM. This prevents cell swelling during whole-cell recording when using internal (patch-pipet) solutions adjusted to 310 mosM.
4. PIPES solution can be stored frozen at –20°C for up to 3 mo. Prior to use it should be thawed, saturated by bubbling with 100% oxygen for about 30 minutes and prewarmed in the incubation chamber to 33°C. Trypsin is added just before commencing the slicing procedure allowing 10–15 min for equilibration before the tissue is ready for incubation.
5. Polypropylene tubes and polystyrene Petri dishes from the suggested supplier are essential because some types contain substances that may be toxic to the dissociated neurons. The polystyrene Petri dish used here also provides a suitable surface for cell adhesion without the need for coating with adhesive substrates.
6. The animal may be decapitated with or without prior anesthesia or it may be sacrificed by neck dislocation. Neck dislocation is not recommended when preparing brainstem slices.
7. For brainstem slices, only the brainstem with attached cerebellum need be removed. This may be achieved by cutting at the level of the inferior colliculus with a razor blade.
8. Tissue compression and tearing during slicing may occur if removal of connective tissue and blood vessels is incomplete (such tissue will not be cleanly cut by the blade). Compression and tearing will compromise slice viability.
9. Razor blades must be precleaned in an ethanol/acetone mixture (50:50) followed by rinsing in distilled water. This removes any lubricating fluids that may have toxic effects on the neurons.
10. The block of tissue can be easily transferred and glued to the cutting stage using a small rectangular piece of filter paper attached to the top of the tissue by capillary attraction. The filter paper is gently removed after submerging the tissue in the aCSF-filled cutting chamber. Agar should be made prior to slicing and stored

in a refrigerator at ~4°C. A small block is stuck to the slicer stage using cyanoacrylate glue before sacrificing the animal.

11. Remove as much fluid as possible by absorbing on a piece of tissue paper. This is essential for firm attachment; otherwise, the tissue is likely to come unstuck during slicing.
12. Blade advance should be slow and smooth, but blade oscillation should be as high as possible. Deleterious effects as a result of tissue compression can be minimized by cutting from the side of the tissue block closest to the location of the cells of interest. Slicing is best observed under a dissecting microscope.
13. Slice dissection can be carried out either in a Pyrex Petri dish or on a bed of agar. In both cases dissection is performed in aCSF cooled to ~4°C on ice. A binocular dissecting microscope is helpful for accurate dissection.
14. The dissociated cells should not be exposed to medium containing zero calcium (Ca^{2+}) although the whole tissue may be exposed to such solutions. Low Ca^{2+}/high Mg^{2+} solutions improve the viability of some slices, an effect probably related to deleterious effects of calcium entry through NMDA receptor channels under hypoxic conditions. Once cells are dissociated, the presence of calcium is probably essential for membrane stability. Thus, the external solution used for establishing the "giga-ohm" seal contains 5 m*M* calcium instead of 1 m*M* in the aCSF.
15. The dissociated cells and tissue are subjected to only gentle agitation or trituration. Trituration pipets are prepared by heat-polishing to the desired diameter. Larger diameter trituration pipets (1–0.75 mm diameter) produce low yields (~10 cells/tissue block) of viable neurons with some primary dendritic structure. The number of neurons can be increased by trituration with smaller diameter pipets although the amount of dendritic amputation is increased, resulting in dendrite-free soma.
16. Two forms of criteria are used to identify viable neurons: morphological and electrophysiological. The cell membrane should be smooth, clearly visible, and "phase-bright" under phase contrast microscopy, the soma showing a bluish-green tinge with the dendrites more blue. Cells meeting these criteria generally then fulfill the electrophysiological criteria. Unhealthy cells have somas spotted with dark, granular patches or a uniform dark appearance. For isolated dorsal raphé neurons, the electrophysiological criteria include membrane potentials more negative than –55 mV and the ability to fire regular action potentials of brief duration that overshoot 0 mV (*see* Fig. 2A). Under voltage-clamp, evoked calcium currents should have peak amplitudes around 2 nA and be completely blocked by cadmium (*see* Fig. 2B).
17. A final word of warning concerning the identification of isolated cells. The neuronal nature of a dissociated cell can be confirmed by its electrical properties. Identifying the specific neuronal type, however, may prove more difficult and require the use of other techniques. For example, immunohistochemical methods were used to establish that the dorsal raphé neurons isolated in this laboratory were those that contain 5-HT *(8)*.

References

1. Larkman, P. M. and Kelly, J. S. (1995) The use of brain slices and dissociated neurons to explore the multiplicity of actions of 5-HT in the central nervous system. *J. Neurosci. Meth.* **59,** 31–39.
2. Blanton, M. G., Lo Turco, J. J., and Kriegstein, A. R. (1989) Whole-cell recording from neurons in slices of reptilian and mammalian cerebral cortex. *J. Neurosci. Meth.* **30,** 203–210.
3. Edwards, F. A., Konnerth, A., Sakmann, B., and Takahashi, T. (1989) A thin slice preparation for patch-clamp recordings from neurons of the mammalian central nervous system. *Pflügers Arch.* **414,** 600–612.
4. Kay, A. R. and Wong, R. K. S. (1986) Isolation of neurons suitable for patch-clamping from adult mammalian central nervous systems. *J. Neurosci. Meth.* **16,** 227–238.
5. Penington, N. J. and Kelly, J. S. (1990) Serotonin receptor activation reduces calcium current in an acutely dissociated adult central neuron. *Neuron* **4,** 751–758.
6. Hećimović, H., Rusin, K. I., Ryu, P. D., and Randic, M. (1990) Modulation of excitatory amino-acid induced currents by tachykinins and opioid peptides in rat spinal cord neurons. *Soc. Neurosci. Abstr.* **16,** 853.
7. Aghajanian, G. K. and Rasmussen, K. (1989) Intracellular studies in the facial nucleus illustrating a simple new method for obtaining viable motoneurons in adult rat brain slices. *Synapse* **3,** 331–338.
8. Lawrence, J., Penington, N. J., and Kelly J. S. (1989) Identification of acutely dissociated neurons from the slices of adult rat brainstem as projecting neurons of the Nucleus Raphe Dorsalis (NRD). *Neurosci. Lett.* **36(Suppl.),** 99.

4

Synaptosomes

A Model System to Study Release of Multiple Classes of Neurotransmitters

Alexandra I. M. Breukel, Elly Besselsen, and Wim E. J. M. Ghijsen

1. Introduction

Synaptosomes were first isolated by Whittaker *(1)* in 1958 and identified by electron microscopy as detached synapses 2 yr later *(2)*. Synaptosomes are sealed particles that contain small, clear vesicles and sometimes larger dense-core vesicles, indicating their presynaptic origin. Occasionally, fractions of electron-dense, postsynaptic membranes remain attached to the synaptosomes, facing the active zone. Evidently, synaptosomes contain all the components necessary to store, release, and retain neurotransmitters, the biochemical messengers that effect synaptic transmission. In addition, synaptosomes contain viable mitochondria, enabling production of ATP and active energy metabolism. Ca^{2+}-buffering capacity is retained, as evidenced by the ability of synaptosome preparations to maintain resting internal Ca^{2+}-concentrations of 100–200 n*M* in the presence of 2 m*M* extracellular Ca^{2+} *(3)*. In addition, synaptosomes maintain a normal membrane potential (which is regulated by a Na^+/K^+-ATPase) and express functional uptake carriers and ion-channels in their plasma-membranes. On application of diverse depolarizing stimuli (e.g., potassium, veratridine, and 4-aminopyridine), Ca^{2+} enters synaptosomes via high voltage-sensitive Ca^{2+} channels and triggers exocytosis of docked vesicles. Through this process, multiple neurotransmitters (including amino acids, peptides, and catecholamines) are released into the extracellular medium *(4)*.

The metabolic state or viability of synaptosomes is largely dependent on the methodology employed to isolate them. Most of the methods are based on the

From: *Methods in Molecular Biology, Vol. 72: Neurotransmitter Methods*
Edited by: R. C. Rayne Humana Press Inc., Totowa, NJ

"classical" hyperosmotic sucrose density gradient techniques of Gray and Whittaker *(2)*, but both the lengthy preparation time and the hyperosmolarity of the sucrose solutions may adversely affect the metabolic integrity of synaptosomes. Some workers have attempted to circumvent the disadvantages of this preparation by using iso-osmotic density gradients of Ficoll *(5)* or Percoll *(3,6–8)*. Synaptosomes purified on Ficoll gradients are contaminated with other cell structures and Ficoll itself has the further disadvantage of adhering to membranes *(5,6,9)*. Percoll density gradient isolation of synaptosomes, on the other hand, offers a number of advantages (as enumerated in ref. *6*):

1. The isolating medium is iso-osmotic, so stabilizing the synaptosomes;
2. The procedure is relatively short;
3. Percoll itself is inert and nontoxic to synaptosomes; and
4. Synaptosome preparations of very high purity are possible.

Furthermore, Dunkley et al. *(7,8)* loaded directly the supernatant, obtained after low-speed centrifugation of a brain-homogenate, on top of the gradient. In contrast, other workers *(6)* added an additional centrifugation step to pellet this supernatant, introducing another possibly damaging mechanical action on the tissue, and yielding less protein than the former. Therefore, the procedure of Dunkley is more appropriate for purifying synaptosomes from small brain (sub-)regions than the latter.

As the aforementioned discussion indicates, the synaptosome preparation provides an excellent model in which to study regulation of neurotransmitter release. Differential regulation of neurotransmitter release may play important roles in synaptic plasticity (underlying such processes as learning and memory) as well as in pathophysiological processes (such as epilepsy and ischemia). In the hippocampus—a brain region widely studied with respect to these processes—multiple transmitters have been identified, of which glutamate, GABA, and certain neuropeptides are the most prominent. In this chapter, we describe the purification of synaptosomes from rat hippocampus, using Percoll density gradients and differential centrifugation. In addition, we describe two methods to characterize isolated synaptosomes: electron microscopy to determine purity and structural integrity of synaptosomes, and a Ca^{2+}-dependent release assay to measure regulated vesicular secretion of diverse transmitters (amino acids, catecholamines, and neuropeptides) from synaptosomes.

2. Materials

2.1. Stock Solutions

Make all stock solutions using Milli-Q (or equivalent quality) water. Solutions 1–7 are stable for 3–5 mo when stored at 4°C.

1. Sucrose stock solution: 2.5*M* sucrose.
2. EDTA stock solution: 0.1*M* EDTA.
3. Percoll, 100% (Pharmacia, Uppsala, Sweden).
4. NaCl stock solution: 1.32*M* NaCl.
5. KCl stock solution: 1.5*M* KCl.
6. $CaCl_2$ stock solution: 0.1*M* $CaCl_2$.
7. 5X concentrated artificial cerebrospinal fluid (5X aCSF): 50 m*M* HEPES, 10 m*M* $MgSO_4$, and 6 m*M* NaH_2PO_4, pH 7.4. Adjust the pH with 3*M* NaOH or 1*M* Tris-hydroxy-methyl-amino-methane (*see* Note 1).
8. Bio-Rad protein assay reagent (store at 4°C; *see* ref. *10*).
9. EGTA solution: 200 m*M* EGTA, pH 7.4, freshly prepared.

2.2. Percoll Gradient Preparation

The quantities given are sufficient to produce three Percoll gradients (*see* Note 2).

1. Sucrose iso-osmotic Percoll (SIP): 9 vol of stock Percoll (100%) + 1 vol of sucrose stock solution. Before adding to the sucrose, filter the Percoll through a Millipore (AP 15) filter to remove crystals that may interfere with cell membranes.
2. Gradient buffer: 0.32*M* sucrose, 5 m*M* HEPES, 0.1 m*M* EDTA in Milli-Q water adjusted to pH 7.2 with 3*M* NaOH or 1*M* Tris, also in Milli-Q water. Store at 4°C and use in <3–4 d. About 25 mL will be required to make three gradients (*see* Note 1).
3. 23% Percoll/sucrose solution: Add 4.455 mL of gradient buffer to 1.545 mL of SIP and mix (6.0 mL, total). Store at 4°C and use within 24 h.
4. 20% Percoll/sucrose solution: Add 3.495 mL of gradient buffer to 1.005 mL of SIP and mix (4.5 mL, total). Store at 4°C and use within 24 h.
5. 10% Percoll/sucrose solution: Add 15.99 mL of gradient buffer to 2.01 mL of SIP and mix (18 mL, total). Store at 4°C and use within 24 h.

2.3. Synaptosome Preparation

1. Homogenization buffer: 0.32*M* sucrose, 5 m*M* HEPES, 0.1 m*M* EDTA, adjusted to pH 7.5 with 3*M* NaOH or 1*M* Tris. Store at 4°C and use within 3–4 d (*see* Note 1).
2. Low-Ca^{2+} aCSF: 0.02 m*M* $CaCl_2$, 132 m*M* NaCl, 3 m*M* KCl, 2 m*M* $MgSO_4$, 1.2 m*M* NaH_2PO_4, 10 m*M* HEPES, 10 m*M* glucose, pH 7.4. Prepare by diluting 20 mL of 5X aCSF stock, 10 mL of NaCl stock, 0.2 mL of KCl stock, 0.02 mL of $CaCl_2$ stock, and 0.18 g of glucose in Milli-Q water to a total volume of 100 mL. Stored at 4°C the solution is stable for 2–3 d.
3. Potter Teflon-glass homogenizer: A motor-driven homogenizer with 5- and 1-mL mortar/pestle combinations.
4. Superspeed centrifuge (e.g., Sorvall RC-5B) and 8 × 50 mL rotor (e.g. Sorvall SS-34) with adapters for 16 × 100 mm tubes.

5. Ultracentrifuge capable of 450,000*g* (e.g., Beckman L5-65).
6. Ultracentrifuge rotors: A swing-out rotor with six places for 13.2-mL buckets (e.g., Beckman SW 41) and a fixed-angle rotor with eight places for 39-mL tubes (e.g., Beckman Ti 60).
7. Polycarbonate centrifuge tubes: 11 mL capacity, 16 × 100 mm (for the Superspeed rotor).
8. Open-top ultra-clear tubes: 13.2 mL capacity, 14 × 89 mm (for the swing-out rotor).
9. Polycarbonate centrifuge bottles: 26.3 mL capacity, 25 × 89 mm with caps (for the ultracentrifuge fixed-angle rotor).

2.4. Electron Microscopy (EM)

1. Electron microscope (e.g., Philips 201C).
2. Ultra-microtome (e.g., Reichert OMU 2).
3. EM fixative (*see* Note 3): 4% (w/v) paraformaldehyde, 5% (v/v) glutaraldehyde, 0.1*M* sodium cacodylate, 3.4 m*M* $CaCl_2$, pH 7.2. Depolymerize the paraformaldehyde by dissolving 4 g in 50 mL of Milli-Q water, warmed to 60°C on a hotplate or in a water bath. Add 1*N* NaOH dropwise until the solution is dissolved. Allow to cool and add 50 mg of $CaCl_2 \cdot 2H_2O$, 20 mL of 25% (v/v) glutaraldehyde, 20 mL of 0.5*M* sodium cacodylate, and adjust to pH 7.2–7.3 with HCl. Add Milli-Q water to a final volume of 100 mL. Store under liquid nitrogen.
4. EM wash buffer: 0.1*M* sodium cacodylate, 3.4 m*M* $CaCl_2$, pH 7.2.
5. Osmium tetroxide fixative: 2% (w/v) OsO_4 in 0.12*M* sodium phosphate, pH 7.2 (**Take care:** poisonous!).
6. Ethanol series: 50, 70, 90, 96, 100% EtOH, v/v, in Milli-Q water.
7. Ethanol/propylene oxide (EtOH/PPO): absolute EtOH/PPO, 1:1. (**Caution:** PPO is toxic).
8. Epon mixture (e.g., from Electron Microscopy Sciences): Prepare by mixing 53 vol of EMbed 812, 23 vol of DDSA (dodecenyl succinic anhydride), 24 vol of MNA (Nadic methyl anhydride), and 1–1.5 vol of DMP 30 ([2,4,6,{tri}-dimethylaminomethyl] phenol) according to ref. *11*.
9. PPO/Epon solutions: 1:1 (v/v) and 1:3 (v/v) PPO:Epon mixture.
10. Lead citrate solution (*see* Note 4): Boil approx 250 mL of Milli-Q water for 5 min. Cool to 20°C in a closed container. Immediately dissolve 0.5 g of NaOH in 125 mL of this water. Dissolve 0.2 g of lead citrate $Pb_3(C_6H_5O_7)_2 \cdot 3H_2O$ in 100 mL of NaOH-solution.
11. Uranyl acetate solution: Dissolve uranyl acetate to 3.5% (w/v) in Milli-Q water and adjust to pH 3.5 acetic acid. Filter the solution through a 0.2-μm Millipore filter.

2.5. Transmitter Release Assay

1. Microcentrifuge capable of 15,000*g*.
2. Shaking water bath at 37°C.
3. High-Ca^{2+} aCSF: 2 m*M* $CaCl_2$, 132 m*M* NaCl, 3 m*M* KCl, 2 m*M* $MgSO_4$, 1.2 m*M* NaH_2PO_4, 10 m*M* HEPES, 10 m*M* glucose, pH 7.4. Prepare by diluting 20 mL of 5X aCSF stock, 10 mL of NaCl stock, 0.2 mL of KCl stock, 2 mL of $CaCl_2$ stock,

and 0.18 g of glucose in Milli-Q water to a total volume of 100 mL. Stored at 4°C, the solution is stable for 2–3 d.
4. Ca^{2+}-free aCSF: the same as item 3 (high-Ca^{2+} aCSF) *except* add water instead of $CaCl_2$ stock. Add EGTA from a freshly prepared stock solution (200 m*M*, pH 7.4) to a final concentration of 0.05 m*M*. Stored at 4°C, the solution is stable for 2–3 d.
5. Oil mixture: Analytical grade Silicone-oil (DC550, Dow Corning, Mavom B.V., Alphen a/d Ryn, the Netherlands) and dinonylphthalate (Merck, Amsterdam, the Netherlands) at 45:55% (v/v).

3. Methods

3.1. Preparation of Percoll Gradients (On Day of Use)

1. Add 2 mL of 23% Percoll/sucrose solution to each of three or four open-top, ultra-clear centrifuge tubes (*see* Note 2).
2. Using a Pasteur pipet, carefully layer 6 mL of 10% Percoll/sucrose solution onto the 23% Percoll/sucrose solution in each tube. A thin discontinuity will become apparent between the 23 and 10% layer. Store the gradients at 4°C until use.

3.2. Preparation of Synaptosomes

Perform all steps at 0–4°C.

1. Anesthetize two adult male Wistar rats (200–225 g) with ether and decapitate them (*see* Note 5).
2. Remove the brains and transfer them to ice-cold homogenization buffer.
3. Dissect the hippocampi on a cooled Petri dish and transfer them immediately to a 5-mL Teflon-glass homogenizer containing ice-cold homogenization buffer (2.5 mL of buffer/rat; *see* Note 6).
4. Homogenize gently (avoiding air-bubbles) with 10 strokes up and down at 700 rpm. Keep the Teflon-glass homogenizer on ice during homogenization.
5. Using a Pasteur pipet, transfer the homogenate (5 mL) to a polycarbonate centrifuge tube. Add 2.5 mL of homogenization buffer to the tube (total volume now 7.5 mL) and centrifuge for 10 min at 1,500*g* in a Sorvall SS-34 rotor (*see* Note 7).
6. Collect the supernatant—designated S_1—and discard the pellet (*see* Fig. 1).
7. Mix the supernatant (7.5 mL) gently with 4.5 mL of 20% Percoll/sucrose solution.
8. Using a Pasteur pipet, carefully layer 4 mL of this mixture onto each of three 10/23% Percoll gradients (*see* Notes 2 and 5).
9. Centrifuge the gradients for 30 min at 23,000*g* in a swing-out rotor at 4°C (rotor brake off).
10. Remove upper layers of the gradients and leave a small band of elucident solution on top of the C-band (*see* Fig. 1 and Note 8).
11. Collect synaptosomes from the 10/23% Percoll interfaces (the C-band; *see* Fig. 1), consolidating them in a polycarbonate ultracentrifuge tube.
12. Dilute the synaptosomes with low-Ca^{2+} aCSF until the tube is completely filled (approx 25 mL).

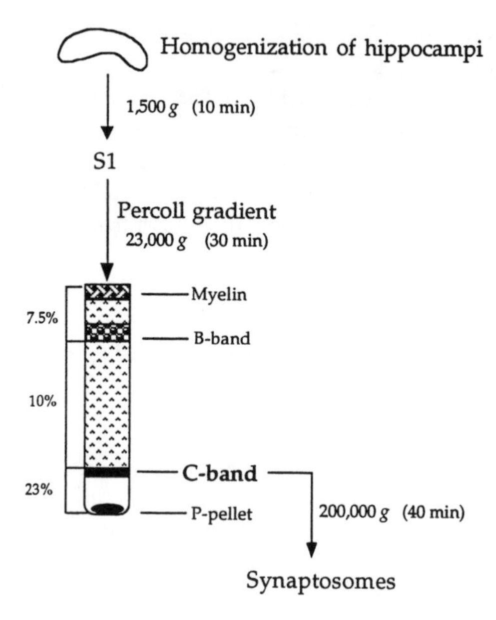

Fig. 1. Synaptosome isolation procedure. *See* Sections 3.1., 3.2., and Notes 2, 5, and 7–9 for further discussion.

13. Centrifuge this preparation at 200,000*g* for 40 min (fixed angle rotor; rotor brake on) to remove the Percoll from the synaptosome suspension. The synaptosomes sediment as a fluffy layer on a tight, glassy Percoll pellet (*see* Note 9).
14. Use a Pasteur pipet to remove and discard the supernatant. Avoid disturbing either the synaptosomal pellet or the pelleted Percoll.
15. Without producing air bubbles, use a pipet to carefully resuspend the synaptosomal pellet in high-Ca^{2+} aCSF to a volume of 0.1 mL/rat (*see* Note 10).
16. Without producing air bubbles, carefully homogenize the suspension by hand using a 1-mL Teflon-glass homogenizer (five strokes).
17. Allow the synaptosomes to rest on ice for at least 30 min prior to starting the experiments. The synaptosomes will remain viable in this state for 3–4 h (*see* Notes 10 and 11 and refs. *12* and *13*).
18. Remove an aliquot to measure the total protein content according to Bradford et al. (*see* Note 12 and ref. *10*).

3.3. Characterization of Fractions by Electron Microscopy (see Note 13)

1. Follow the protocol for synaptosome preparation, setting aside each fraction depicted in Fig. 1: S_1, B- and C-bands, P-pellet, and synaptosomal suspension (*see* Note 8 and Fig. 1).
2. Add 1 vol of EM fixative to each collected fraction and fix for 1 h at 4°C.

3. Transfer the fixed fractions to polycarbonate tubes and centrifuge for 10 min at 4,500*g* in a fixed angle rotor at 4°C.
4. Aspirate the supernatants and rinse the pellets once by adding 1 mL of EM wash buffer to each tube.
5. Transfer carefully the intact pellets to clean tubes containing 1 mL of EM wash buffer and hold them on ice for 15 min. If the pellets crumble, resuspend them with EM wash buffer and centrifuge again (intact pieces of approx 1 mm^3 are required; *see* Note 14).
6. Aspirate the wash buffer and add 1 mL of fresh EM wash buffer to the loose but intact pellets. Incubate the pellets for 15 min on ice, then repeat this wash step (*see* Note 15).
7. After the second wash, store the pellets in fresh EM wash buffer (1 mL) overnight at 4°C.
8. Aspirate the wash buffer and fix approx 1 mm^3 pieces of each pellet in 1 mL of osmium tetroxide fixative for 60 min at 20°C. This procedure will further harden the pellets.
9. Wash the pellets for 10–15 min in Milli-Q water, then dehydrate by consecutively incubating the pellets for 10–15 min in each solution of an ethanol series (1 mL of each solution for each pellet; *see* Section 2.4., item 6).
10. Incubate the pellets in 1 mL of EtOH/PPO for 10–15 min. Aspirate the EtOH/PPO, then incubate the pellets in 1 mL of PPO for 3 × 10–15 min.
11. Incubate the pellets for at least 1 h (but not longer than 2 h) in 1 mL of PPO:Epon (1:1, v/v) at 20°C.
12. Repeat step 11 using PPO:Epon (1:3, v/v).
13. Repeat step 11 using Epon mixture.
14. Embed the pellets in fresh Epon mixture by polymerizing for 24 h at 35°C followed by 24 h at 45°C, then 24 h at 60°C.
15. Cut ultrathin sections (800 Å) on an (Reichert OMU 2) ultramicrotome. Collect the sections on copper grids.
16. Stain the sections with uranyl acetate for 30 min and lead citrate for 10–60 s.
17. Examine the grids in an electron microscope (Philips 201C; *see* Fig. 2, and Notes 8 and 16).

3.4. Release Batch Assay

1. Dispense approx 10 mL of high-Ca^{2+} aCSF into each of two disposable tubes. Place one tube at 37°C and the other at 4°C. Repeat using Ca^{2+}-free aCSF.
2. Divide the synaptosome preparation equally between two microfuge tubes, on ice: (a) synaptosomes for measuring total release (high-Ca^{2+} aCSF), and (b) synaptosomes for measuring Ca^{2+}-independent release (Ca^{2+}-free aCSF; *see* Note 10).
3. Centrifuge the synaptosome batches (a) and (b) at 15,000*g* for 1 min in a microcentrifuge at 4°C.
4. Resuspend synaptosomes with either (a) the 4°C high-Ca^{2+} aCSF, or (b) the 4°C Ca^{2+}-free aCSF, adding sufficient buffer to each so that the final protein concentration is approx 3 mg/mL (*see* Note 17).

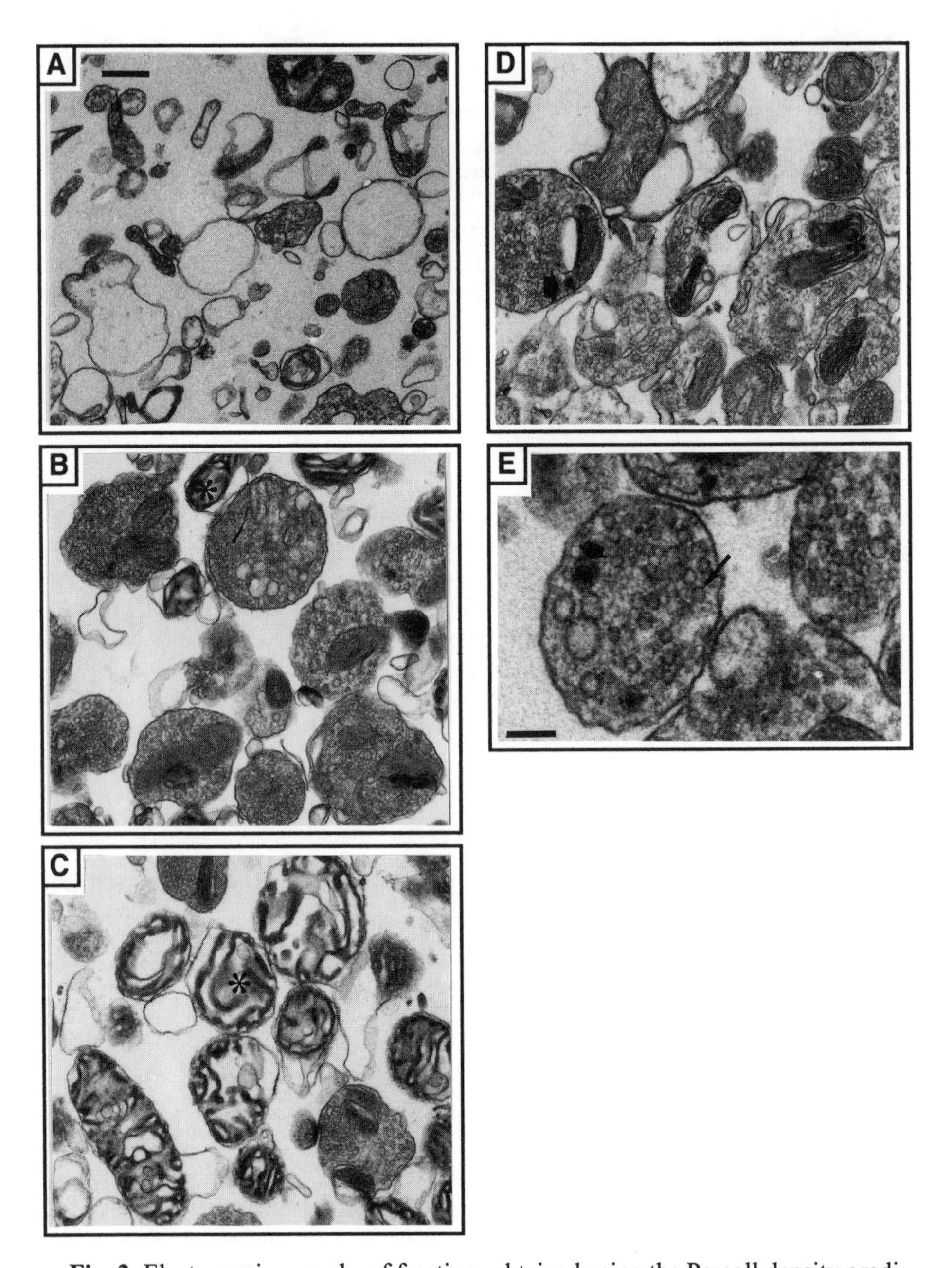

Fig. 2. Electron micrographs of fractions obtained using the Percoll density gradient procedure. Shown are the B-band **(A)**, the synaptosome-containing C-band **(B)**, the mitochondrial pellet, P **(C)**, and the final high-speed centrifuged synaptosomal fraction **(D)**. (A–D) are magnified 28,500×; the scale bar in each panel represents 0.35 μm. **(E)** Final synaptosome fraction in detail (85,500×; scale bar, 0.12 μm). The broad arrows in (D) and (E) indicate large, dense-core vesicles; thin arrows point to small, clear-core vesicles. Mitochondria are indicated by asterisks in (B–D).

5. Pipet 20 µL aliquots of each synaptosome batch into respective series of Eppendorf tubes (10–15 tubes for each batch, depending on the volume required to produce 3 mg/mL in step 4). Keep the tubes on ice until they are required for the assays.
6. Begin an assay ($t = 0$) by adding 100 µL of either (a) 37°C high-Ca^{2+} aCSF or (b) 37°C Ca^{2+}-free aCSF to corresponding synaptosome aliquots (batch a or b, respectively, containing 20 µL synaptosomes). Mix gently and preincubate in a 37°C water bath for 2–3 min (*see* Note 18).
7. At the end of the preincubation ($t = 2–3$ min), add 3 µL of a 40–50X depolarizing agent stock solution to each tube. Mix gently and place in the 37°C water bath for an additional 3–5 min (*see* Notes 18 and 19).
8. At the end of the incubations, quickly pipet the contents of the assay tubes (125–150 µL) into respective microfuge tubes containing 50 µL of the oil mixture. Centrifuge the tubes at 15,000*g* for 1 min to terminate the assays (*see* Note 20).
9. Remove 100 µL aliquots from the supernatant (avoid the oil!) for transmitter analysis. (Aliquots may be subdivided permitting measurements of multiple transmitters from a single sample). Treat the samples in the manner appropriate for the type of transmitter ultimately to be measured: For amino acids, add TCA to 10% (v/v); for neuropeptides or catecholamines, add 3 vol of methanol.
10. Store the samples at –20°C until further analysis (*see* Note 21). Measurements of diverse transmitters released from hippocampal synaptosomes in a typical release batch assay are shown in Fig. 3.

4. Notes

1. We use 3*M* NaOH to adjust the pH of buffers when preparing synaptosomes to study release of amino acid transmitters. In such cases, Tris should not be used as a pH buffer because it interferes with the amino acid analysis after *o*-phthaldialdehyde (OPA)-derivatization.
2. For a typical synaptosome preparation, we use hippocampi from two rat brains. To avoid overloading the Percoll gradient at the 10% layer, the hippocampal homogenate from two rats should be distributed over three Percoll gradients (*see* Section 3.2., steps 7 and 8). We have given instructions in Section 2.2. for setting up three Percoll gradients and these instructions may be scaled up if additional gradients are required. For example, depending on the centrifuge rotor that is used, four gradients may be required (three to receive homogenate samples and one "blank" to be used as a balancer in the centrifuge) although in some rotors, three tubes may be balanced in a triangle arrangement. If a blank gradient is required, gently mix 2.5 mL of HB with 1.5 mL of 20% Percoll/sucrose solution and layer this (4 mL) onto a 10/23% Percoll gradient.
3. Do not heat the paraformaldehyde above 60°C, otherwise, the paraformaldehyde will decompose.
4. When preparing the lead citrate solution, CO_2 will precipitate with Pb^{2+}. Therefore, work quickly and perform each step, as much as possible, under CO_2-free conditions. Boiling the water from which the solution is to be made will drive off dissolved CO_2.

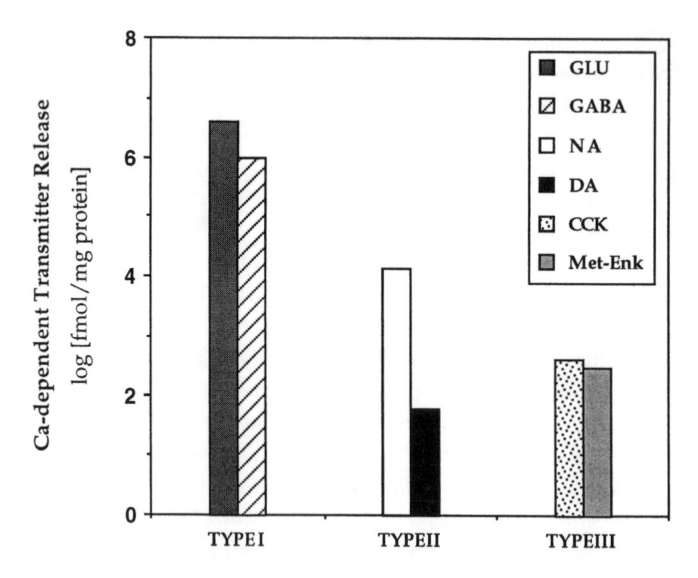

Fig. 3. Ca^{2+}-dependent (vesicular) release of multiple neurotransmitter types from rat hippocampal synaptosomes. To evoke release, the extracellular K^+ concentration was raised to 30 m*M* by adding KCl. The basis of the neurotransmitter type categories (Type I, and so forth) is described in Note 19. The measurements are expressed in log[fmol/mg protein]. *See* Section 3.4. and Note 21 for further discussion. GLU = glutamate, NA = noradrenaline, DA = dopamine, CCK = cholecystokinin, met-Enk = met-enkephalin.

5. We usually use brains from two rats (200–225 g) for preparing hippocampus synaptosomes, but this number may be varied depending on the amount of protein required for the experiments. Synaptosomes may be prepared by our method from various species, including the rat, guinea pig, and mouse *(1)*. Synaptosomes also can be isolated from other brain areas using this procedure, but with slight modifications. The forebrain of one rat, for example, is homogenized in 10 mL of homogenization buffer and further prepared to be distributed on six Percoll density gradients (*see also* Note 2).
6. Briefly, the hippocampus is dissected from the brain as follows: Cut the brain along the midline into two halves to obtain the two hemispheres. While performing the dissection on one hemisphere, return the other to the ice-cold homogenization buffer. The white, moon-shaped hippocampus is easily visible. Gently push the cerebral cortex away from the hippocampus by placing a fine brush near the surrounding ventricles. Using a scalpel, make a small cut into the upper, rostral part of the hippocampus and "roll" the entire hippocampus out of the brain. Make sure that the hippocampus is free of blood vessels. Place the hippocampus into the Teflon-glass homogenizer and repeat the procedure for the second brain hemisphere.

7. The 1,500*g*, 10 min spin will sediment large fragments in the homogenate (e.g., blood cells, cell nuclei, and some of the myelin).
8. Three distinct bands of material will separate on the Percoll gradients. At the top, a band comprised primarily of myelin will form. A second band, comprised of undefined cell fragments and empty sacs (the B-band), will form at the 7.5/10% Percoll boundary. The C-band, appearing at the 10/23% Percoll boundary, contains the synaptosomes. Mitochondria and large synaptosomes comprise the pellet (designated P; *see* Figs. 1 and 2).
9. The high-speed centrifugation (200,000*g*) in the final step is used to separate the synaptosomes and the Percoll, thereby purifying the nerve terminal preparation. Percoll, which is present in the gradient-purified synaptosomal fraction, may interfere with membrane proteins, such as reuptake carriers or presynaptic receptors. These proteins are often the subject of study and, therefore, this step is required to avoid possible artifacts. Interestingly, EM studies show that this step effects a further elimination of extrasynaptosomal mitochondria (not pelleted at the lower speed) from the Percoll C-band (compare Fig. 2B and 2D).
10. Because Ca^{2+} appears to be necessary to maintain stability of synaptosomes, the preparation is held in high-Ca^{2+} aCSF (physiological extracellular concentration) at the end of the isolation. Storage in low-Ca^{2+} aCSF (0.02 m*M* Ca^{2+} or 0.05 m*M* EGTA) for more than 1 h increases LDH activity, indicating increased membrane leakage (*see* Note 11). Because the synaptosomes will have been stored in Ca^{2+}-containing buffer, when performing the release batch assay (e.g., Section 3.4., step 2) it is necessary to divide the synaptosomes, pellet them, and resuspend in and either high- or low-Ca^{2+} aCSF, as required.
11. Both EM and biochemical markers are used to monitor synaptosomal viability. For example, we monitor extracellular activity of the cytosolic enzyme lactate dehydrogenase (LDH) to indicate membrane leakage. LDH, which converts exogenously applied pyruvate and NADH to lactate and NAD^+, may be estimated using saponin as a 100% LDH-activity reference level, according to ref. *12*. Our synaptosomal preparation showed <5% leakiness. To obtain further information regarding the metabolic state of the synaptosomes (*see also* the intra-synaptosomal mitochondria in Fig. 2D), ATP and ADP levels may be estimated using the luciferase ATP assay kit (LKB/ITL, Heemstede, The Netherlands) according to ref. *13*. Luciferase luminescence is used to monitor ATP. By this procedure, synaptosomal ADP levels were measured indirectly, by adding the substrate phosphoenol-pyruvate and the enzyme pyruvate kinase to the synaptosomal extract. In our synaptosomal preparation, ATP and ADP levels are about 6.5 and 1.4 nmol/mg protein, respectively. The ATP/ADP ratio that reflects oxidative phosphorylation is about 5 (*see* ref. *13*).
12. We use the assay according to Bradford *(10)* with BSA as a standard. An aliquot of 0.005 mL from a preparation of two rats (end volume 0.2 mL) is usually sufficient to yield total protein in the range of 6–8 mg/mL.
13. A routine EM check of synaptosomes is performed in our lab about two times a year. The synaptosome preparation method is very robust if carried out carefully

and consistently, so more frequent (time consuming!) EM checks are unnecessary unless significant changes in control values for transmitter release are noted. Four hippocampi (two rat brains) are needed for the EM procedure. The S_1 (±0.35 mL), the Percoll gradient fractions "B-band" (±15 mL), "C-band" (±3 mL), "P-pellet" (±1.5 mL), and the final synaptosomal suspension (±0.2 mL) are all prepared for EM (*see* Figs. 1 and 2).

14. The size of each fixed pellet should not significantly exceed 1 mm^3, otherwise the osmium tetroxide fixative (*see* Section 3.3., step 8) will not be able to penetrate effectively to the center of the pellet.
15. In this and subsequent steps, it is not necessary to transfer the pellets from tube to tube, but rather to aspirate carefully the existing solution, leaving the pellets intact. Add the next solution, as directed in the protocol, to the same tube.
16. Classical neurotransmitters, like glutamate and GABA, are stored in small clear-core vesicles, whereas neuropeptides are assumed to be stored in large dense-core vesicles. Catecholamines may be stored in either type of vesicle *(14,15)*. The detailed electron micrograph depicted in Fig. 2E shows a synaptosome containing both small, clear-core and large, dense-core vesicles, as indicated by the arrows. Random viewing by EM of hippocampal synaptosome preparations showed that about 10% of synaptosomes typically contained—in addition to a large population of small clear-core vesicles—two to three large, dense-core vesicles. This may indicate that hippocampal terminals contain a spectrum of coexisting neurotransmitter types. This observation is in agreement with observed colocalization of amino acids (GABA) and certain neuropeptides (somatostatin, CCK) *(16)*.
17. The volume used in resuspending the synaptosome pellets will depend on the results of the protein assay performed in Section 3.2., step 18. Typically, 200–300 μL of the respective aCSF solutions will be required to produce suspensions of approx 3 mg/mL.
18. To ensure accurate timing of experimental treatments, perform the procedure in stages with a maximum of four tubes at one time. With four tubes, the operations in steps 6 and 7 of Section 3.4. require about 30 s each to perform.
19. Depolarization of synaptosomes to evoke exocytotic transmitter release may be accomplished by one of the following methods:
    a. Clamping all K^+ channels by increasing extracellular potassium concentration (10–30 m*M*; this method was employed to generate the data shown in Fig. 3);
    b. Inducing repetitive action potentials via the K^+_A channel blocker 4-aminopyridine (1–10 m*M*); or
    c. Preventing Na^+ channel inactivation using veratridine (as reviewed by Nicholls in ref. *17*).

    Prepare stock solutions of depolarizing agents so that adding of 3 μL of the stock solution to a release batch assay tube (containing 125–150 μL) achieves the desired final concentration of depolarizing agent (i.e., 40–50X stocks).

    The hippocampal transmitters belong to the three different types (type I, II, and III) identified by McGeer et al. *(15)*, and are released in quantities proportional to the respective quantities stored within the neurons. Type I transmitters include the amino acids that are present at about 1–10 nmol/mg protein. Type II refers to the catecholamines (0.1–10 pmol/mg protein) and type III to the neuropeptides (100–1000 fmol/mg protein).

20. The oil method permits rapid separation of synaptosomes from the aqueous phase, thereby terminating the reaction. The synaptosomal fraction, owing to its density, goes through the oil layer on centrifugation and is pelleted. The supernatant, containing the released neurotransmitters, remains on top of the oil layer, in the aqueous phase.
21. Ca^{2+}-dependent (vesicular) release of diverse transmitter types (i.e., amino acids, catecholamines, and neuropeptides) from hippocampus synaptosomes is calculated by subtracting the extracellular neurotransmitter levels in the absence of Ca^{2+} (Ca^{2+}-free aCSF) from the total levels in the presence of 2 m*M* Ca^{2+} (high-Ca^{2+} aCSF). Ca^{2+}-dependent, exocytotic release of different classes of neurotransmitters is shown in Fig. 3.

    One should realize that by application of the release batch-assay net release values are measured. To avoid interference by recycling of released amino acid or catecholamine transmitters by their reuptake carriers, or breakdown of neuropeptide transmitters by active peptidases, the appropriate reuptake or peptidase inhibitors, respectively, should be included in the assay. For blocking aspartate and glutamate uptake, we use 25 μ*M* L-*trans*-pyrrolidine-2,4-dicarboxylate (L-*trans*-PDC) and for blocking GABA uptake, we include 10 μ*M* *N*-(4,4-diphenyl-3-butenyl)-3-piperidine carboxylic acid 89976-A (SK&F 89766-A). Pargyline (100 μ*M*) inhibits the endogenous monoamine oxidase (MAO) activity, which metabolizes catecholamines. Bacitracin (0.1%) is used as an inhibitor of peptidases, in the case of monitoring neuropeptide release.

    Measurement and analysis of the various transmitters may be performed according to previously published procedures. The Type I (amino acid) transmitters, L-glutamate and GABA, may be analyzed by reversed-phase HPLC after precolumn derivatization with OPA in mercaptoethanol, using fluorimetric detection *(18)*. (Glutamate and GABA may alternatively be determined using a microbore HPLC method as described in Chapter 15.) The Type II transmitters (catecholamines), noradrenaline and dopamine may be quantified by HPLC using fluorimetric detection (*4,19*; *see also* Chapter 14, which includes microbore HPLC methods for analysis of monoamine transmitters). The neuropeptides, cholecystokinin, and met-enkephalin (Type III) may be analyzed by radioimmunoassays *(20)* or by HPLC (*see* Chapter 16).

Acknowledgments

We would like to thank G. Scholten for her excellent assistance in the EM-studies. We are thankful to Dr. Verhage for developing the release assays for the different transmitters. We appreciate the cooperation with V. Wiegant and A. Frankhuizen for the neuropeptide RIAs, and Dr. Boomsma for the catecholamine analysis. T. N. O. Zeist is appreciated for packing the HPLC columns for amino acid analysis.

References

1. Whittaker, V. P. (1993) Thirty years of synaptosome research. *J. Neurocytol.* **22,** 735–742.
2. Gray, E. G. and Whittaker, V. P. (1960) The isolation of synaptic vesicles from the central nervous system. *J. Physiol.* **153,** 35–37.
3. Verhage, M., Besselsen, E., Lopes da Silva, F. H. and Ghijsen, W. E. J. M. (1988) Evaluation of the Ca^{2+} concentration in purified nerve terminals: relationship between Ca^{2+} homeostasis and synaptosomal preparation. *J. Neurochem.* **51,** 1667–1674.
4. Verhage, M., McMahon, H. T., Ghijsen, W. E. J. M., Boomsma, F., Scholten, G., Wiegant, V. M., and Nicholls, D. G. (1991) Differential release of amino acids, neuropeptides, and catecholamines from isolated nerve terminals. *Neuron* **6,** 517–524.
5. Booth, R. F. G. and Clark, J. B. (1978) A rapid method for the preparation of relatively pure metabolically competent synaptosomes from rat brain. *Biochem. J.* **176,** 365–370.
6. Nagy, A. and Delgado-Escueta, A. V. (1984) Rapid preparation of synaptosomes from mammalian brain using nontoxic isoosmotic gradient material (Percoll). *J. Neurochem.* **43,** 1114–1123.
7. Dunkley, P. R., Heath, J. W., Harrison, S. M., Jarvie, P. E., Glenfield, P. J., and Rostas, J. A. P. (1988) A rapid Percoll gradient procedure for isolation of synaptosomes directly from an S1 fraction: homogeneity and morphology of subcellular fractions. *Brain Res.* **441,** 59–71.
8. Harrison, S. M., Jarvie, P. E., and Dunkley, P. R. (1988) A rapid Percoll gradient procedure for isolation of synaptosomes directly from an S1 fraction: viability of subcellular fractions. *Brain Res.* **441,** 72–80.
9. Dagani, F., Zanada, F., Marzatico, F., and Benzi, G. (1985) Free mitochondria and synaptosomes from single rat forebrain. A comparison between two known subfractionating techniques. *J. Neurochem.* **45,** 653–656.
10. Bradford, M. M. (1978) A rapid and sensitive method for the quantification of microgram quantities of protein utilizing the principle of protein dye binding. *Anal. Biochem.* **72,** 248–252.
11. Luft, J. H. (1961) Improvement in epoxy resin embedding procedures. *J. Biophys. Biochem. Cytol.* **9,** 409.
12. Koh, J. Y. and Choi, D. W. (1987) Quantitative determination of glutamate mediated cortical neuronal injury in cell culture by lactate dehydrogenase efflux assay. *J. Neurosci. Meth.* **20,** 83–90.
13. Kauppinen, R. A. and Nicholls, D. G. (1986) Synaptosomal bioenergetics, the role of glycolysis, pyruvate oxidation and responses to hypoglycaemia. *Eur. J. Biochem.* **158,** 159–165.
14. Verhage, M., Ghijsen, W. E. J. M., and Lopes da Silva, F. H. (1994) Presynaptic plasticity: the regulation of Ca^{2+}-dependent transmitter release. *Prog. Neurobiol.* **42,** 539–574.
15. McGeer, P. L., Eccles, J. G., and McGeer, E. G. (1987) *Molecular Neurobiology of the Mammalian Brain.* Plenum, New York.

16. Somogyi, P., Hodgson, A. J., Smith, A. D., Nunzi, M. G., Gorio, A., and Wu, J.-Y. (1984) Different populations of GABAergic neurons in the visual cortex and hippocampus of cat contain somatostatin- or cholecystokinin-immunoreactive material. *J. Neurosci.* **4,** 2590–2602.
17. Nicholls, D. G. (1993) The glutamatergic nerve terminal. *Eur. J. Biochem.* **212,** 613–631.
18. Verhage, M., Besselsen, E., Lopes da Silva, F. H., and Ghijsen, W. E. J. M. (1989) Ca^{2+}-dependent regulation of presynaptic stimulus-secretion coupling. *J. Neurochem.* **53,** 1188–1194.
19. Van der Hoorn, F. A. J., Boomsma F., Man in 't Veld, A. J., and Schalekamp, M. A. D. H. (1989) Determination of catecholamines in human plasma by high-performance liquid chromatography: comparison between a new method with fluorescent detection and an established method with electrochemical detection. *J. Chromatogr.* **487,** 17–28.
20. Verhage, M., Ghijsen, W. E. J. M., Nicholls, D. G., and Wiegant, V. M. (1991) Characterization of the release of cholecystokinin from isolated nerve terminals. *J. Neurochem.* **56,** 1394–1400.

5

Synaptoneurosomes

A Preparation for Studying Subhippocampal $GABA_A$ Receptor Activity

Miriam N. G. Titulaer and Wim E. J. M. Ghijsen

1. Introduction

Although synaptosomes are a widely accepted preparation for studying regulation of transmitter release at the isolated presynaptic level (*see also* Chapter 4), they are not suited for the investigation of synaptic events mediated by postsynaptic mechanisms. In the past, isolated membrane fractions enriched in postsynaptic densities have been used for this purpose *(1)*. However, this fraction of postsynaptic densities is not appropriate to study receptor-mediated signal transduction because of the absence of intact sealed structures in this preparation. For this reason, a subcellular preparation enriched in resealed presynaptic structures (synaptosomes) with attached sealed postsynaptic entities (neurosomes) has been developed. These composite structures are called synaptoneurosomes. Biochemical characterization of synaptoneurosome preparations has revealed the presence of a number of receptor-mediated properties and a maintained electrochemical gradient *(2,3)*. Since its first description, this preparation has been used for the investigation of numerous phenomena, including neurotransmitter release *(4)*, inositol phospholipid turnover *(5,6)*, cAMP accumulation *(2)*, as well as in functional studies of various neurotransmitter receptors *(7)*.

In most studies, synaptoneurosomes have been prepared from whole rat forebrains in order to obtain sufficient material for investigation. However, the usefulness of such a preparation is compromised by the functional heterogeneity within this whole brain area. To avoid such heterogeneity, we have chosen to prepare synaptoneurosomes from more functionally and anatomically

From: *Methods in Molecular Biology, Vol. 72: Neurotransmitter Methods*
Edited by: R. C. Rayne Humana Press Inc., Totowa, NJ

homogenous brain regions. We have, because of our interest in the hippocampus, focused our efforts on discrete subregions within this region of the rat brain.

The hippocampal region of the brain has been the focus of many studies on synaptic plasticity under physiological and pathophysiological conditions. Being the most important transmitters in the hippocampal region, the amino acids glutamate and γ-amino butyric acid (GABA) play a decisive role in the synaptic transmission in this area, acting via their ionotropic and metabotropic receptors. Among the receptor types, the $GABA_A$ receptor has been particularly well characterized in electrophysiological studies. $GABA_A$ receptors are typically comprised of three different subunit classes (α, β, and γ), and it is well established that diverse $GABA_A$ receptor subunits are expressed in the central nervous system (CNS) *(8)*. Interestingly, it has been demonstrated by histochemistry that $GABA_A$ receptors in different brain areas can be assembled from different combinations of subunits *(8,9)*. In addition, there are some indications of region-specific variability in responses after activation of $GABA_A$ receptors by modulators of this receptor complex, such as the benzodiazepines *(9)* or GABA itself *(10)*. An example of a pathophysiological condition where nonhomogeneous GABAergic responses in the CNS take place is the hippocampal model of kindling epileptogenesis *(11)*, which has been studied in detail in our laboratory over the last decade. Several studies have revealed that in GABAergic signaling systems in two major hippocampal subareas—the cornus ammonis (CA_{1-3}) area and dentate gyrus (DG)—responses to kindling stimulations are different, and in some cases in the opposite direction *(12–14)*. This indicates that, at least for this model, investigations using whole brain homogenates, or even homogenates of the hippocampus proper, could lead to highly inaccurate conclusions.

Here we will describe a modification of the synaptoneurosome preparation method that permits the isolation of synaptoneurosomes from restricted brain areas, as demonstrated by the isolation of synaptoneurosomes from the CA_{1-3} area and DG, respectively. Subsequent filtration and low-speed centrifugation of subhippocampal homogenates yield viable synaptoneurosomal preparations within 45 min after tissue homogenization. In addition, we will describe an assay for muscimol-stimulated $^{36}Cl^-$-uptake as a measure of $GABA_A$-receptor activity in these subhippocampal synaptoneurosomal preparations.

2. Materials

2.1. Isolation of Synaptoneurosomes

2.1.1. Preparatory Steps

1. Bio-Rad (Hercules, CA) protein assay reagent *(15)*.
2. Homogenization buffer (HB): 137 m*M* NaCl, 2.7 m*M* KCl, 0.44 m*M* KH_2PO_4, 4.2 m*M* $NaHCO_3$, 10 m*M* HEPES, adjusted to pH 7.4 with 1*M* Tris. This buffer is

made fresh on the day of the experiment, and D-glucose is added just before use to a concentration of 10 m*M*.

3. Picrotoxin (PTX) stock: 5 m*M* PTX in H_2O, aliquoted in 1 mL Eppendorf tubes and stored at –20°C.
4. Stop buffer: HB (item 2) containing 100 μ*M* PTX (1:50 dilution of stock PTX).
5. Two 10-mL measuring cylinders.
6. Two Dounce glass-glass homogenizers (5 mL, Braun Salm en Kipp, The Netherlands) each with loose-fitting pestle and tight-fitting pestle.
7. Two 10-mL beakers and two 12-mL centrifuge tubes.
8. Nylon mesh, 60 μm pore size (Plankton, Diemen, The Netherlands).
9. Two 25-mL plastic syringes with the entire needle-attachment end of the syringe cut off.
10. Rubber bands.
11. Glass Petri dishes: 19 and 14 cm diameter.
12. Tissue paper.
13. Filter paper (Whatman [Maidstone, UK], 9 cm diameter).
14. Tissue slicer with 800 μm spacers.
15. Razor blades (Gillette).

2.1.2. Isolation Step

1. Male Wistar rats (250–300 g).
2. Ether pot.
3. Small animal guillotine.
4. Luer Ronguer.
5. Two Millipore (Bedford, MA) filter holders of 2.5-cm diameter (type Swinnex-25) connected to 10-mL syringes.
6. Millipore filters, type SCWP 02500, of 8 μm pore size.
7. Polycarbonate centrifuge tubes, 12-mL capacity.
8. High-speed refrigerated centrifuge and fixed-angle rotor (e.g., Sorvall SS-34).

2.2. Characterization of Synaptoneurosomes by Electron Microscopy

All required materials are listed in Section 2.4. of Chapter 4.

2.3. Measurement of Muscimol-Stimulated $^{36}Cl^-$ Uptake

1. Muscimol stock solution: 10 m*M* muscimol in H_2O, aliquoted in 1-mL Eppendorf tubes and stored at –20°C.
2. $^{36}Cl^-$, 84 Ci/mmol (NEN Dupont de Nemours, The Netherlands).
3. Chloride uptake medium: add 0.085 μCi of $^{36}Cl^-$ and 0.28 μL of 10 m*M* muscimol/40 μL of HB (Section 2.1.1., item 2).
4. Nonspecific uptake medium: Add 0.085 μCi of $^{36}Cl^-$, 0.28 μL of 10 m*M* muscimol, and 1.6 μL of 5 m*M* PTX/40 μL of HB.
5. Heated water bath.
6. Vacuum pump.

7. Stop buffer: *see* Section 2.1.1., item 4.
8. Bottle fitted with volumetric dispenser.
9. Whatman GF/C filters.
10. Stainless-steel Millipore filter-holders for 2.5-cm diameter filters.
11. Vials and fluid for liquid scintillation counting.

3. Methods

3.1. Isolation of Synaptoneurosomes from CA_{1-3} and DG (see Note 1)

3.1.1. Preparatory Steps

1. Filter the Bio-Rad protein assay reagent through Schleicher and Schuell (Dassel, Germany) folded filters (150-mm diameter) for the protein assay.
2. Prepare 25 mL of the homogenization buffer (HB; *see* Section 2.1.1., item 2).
3. Prepare 500 mL of the stop buffer (*see* Section 2.1.1., item 4).
4. Label respective 10 mL measuring cylinders "CA_{1-3}" and "DG" and place them on ice. Pour 10 mL of HB into each.
5. Label respective 5 mL glass-glass Dounce homogenizers "CA_{1-3}" and "DG" and place them on ice.
6. Transfer 3 mL of the HB from the "CA_{1-3}" measuring cylinder to the corresponding homogenizer. Repeat for the "DG" measuring cylinder and homogenizer, except transfer 1.5 mL of HB.
7. Place two 10-mL beakers and two 12-mL polycarbonate centrifuge tubes (one of each for the CA_{1-3} and one for the DG, labeled appropriately) in the same ice bucket.
8. Take two 25-mL syringes with ends cut off (*see* Section 2.1.1., item 9) and using rubber bands, fasten tightly over each cut end a double layer of nylon mesh.
9. Fill a large Petri dish with ice and place tissue paper on top of the ice. Place a small Petri dish upside-down on the paper and place a Whatman filter paper on top of the dish.
10. Moisten the filter paper with ice-cold HB just before isolation of the hippocampi (*see* Section 3.1.2., step 3).
11. Clean the razor blades with ethanol and rinse them with water. Prepare the tissue slicer for cutting transverse slices of 800 μm thickness.

3.1.2. Isolation of Synaptoneurosomes (see Note 1 and refer to Fig. 1)

1. Anesthetize the rat with ether.
2. Decapitate the rat using a guillotine and open the skull carefully using a Luer Ronguer to avoid damaging the hippocampus. Remove the forebrain.
3. Place the forebrain on the Petri dish/filter paper (*see* Section 3.1.1., step 10).
4. Pour ice-cold HB over the forebrain to remove residual blood.
5. Using a scalpel, cut the forebrain through the midline into two hemispheres.
6. Dissect carefully both hippocampi by cutting away the surrounding brain tissue with a sharp blade.

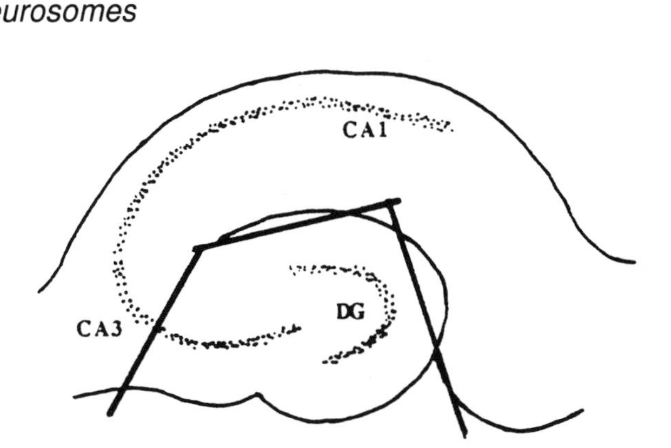

Fig. 1. Schematic view of a transverse section of the hippocampus. Subslices from the CA_{1-3} area and the DG are isolated by three cuts: two just above the fissure separating the DG and CA_3, and one through the CA_3 area, as indicated by the thick lines (reprinted from ref. *14*, with permission).

7. Make transverse slices (800 µm) of each isolated hippocampus using the tissue slicer.
8. Isolate hippocampal subareas by three cuts in each slice: two just above the fissure separating the DG and CA_1, and one through the CA_3 area, as indicated in Fig. 1.
9. Collect the CA_{1-3} and DG slices in the respective, labeled Dounce homogenizers.
10. Homogenize the slices by hand using seven strokes with a loose-fitting pestle followed by four strokes with a tight-fitting pestle (*see* Note 2).
11. Add 3 mL of HB to the CA_{1-3} homogenate and 1.5 mL of HB to the DG homogenate.
12. Pour the homogenates into respective mesh-covered 25-mL syringes and filter by gentle application of positive pressure. Collect the filtrates in the ice-cold 10-mL beakers designated "CA_{1-3}" and "DG".
13. Pass the filtrates through 8 µm Millipore syringe filters by gentle application of positive pressure. Collect the filtrates in the respective, ice-cold 12–mL centrifuge tubes.
14. Centrifuge the filtrates at 1000*g* for 20 min at 4–6°C.
15. Gently resuspend the resulting pellets in 1 mL of ice-cold HB using a 1-mL pipeter.
16. Centrifuge the suspensions at 1000*g* for 10 min at 4–6°C.
17. Resuspend the pellets in 200 µL (CA_{1-3}) and 100 µL (DG) of HB. Determine the protein content using the Bio-Rad total protein assay (*see* Notes 3 and 4 and ref. *15*).
18. Use the hippocampal synaptoneurosome preparations for receptor function studies within 90 min.

3.2. Characterization of Synaptoneurosomes by Electron Microscopy

1. Prepare synaptoneurosomes as outlined in Section 3.1.2. and proceed according to the method described in Section 3.3. of Chapter 4.

2. Figure 2 shows an electron micrograph of a synaptoneurosome preparation isolated from the hippocampus $CA_{1–3}$ region. *See* the caption and Note 5 for further description.

3.3. Measurement of Muscimol-Stimulated $^{36}Cl^-$ Uptake (16,17)

3.3.1. Preparatory Steps (see Note 6)

1. Prepare the chloride uptake medium and nonspecific uptake medium.
2. Prewarm the uptake media by incubation in a 36°C water bath for 10 min.
3. Turn on the vacuum pump connected to the filter holder.
4. Place a GF/C Whatman filter in the filter holder.
5. Fill a 500 mL auto dispenser with cold stop buffer and place it on ice. Set the dispenser to deliver 3 mL of stop buffer.
6. Set up a rack for 5-mL scintillation vials and make available a ready supply of vials and caps.

3.3.2. $^{36}Cl^-$-Uptake Assay (see Note 7)

1. Pipet 40 μL of synaptoneurosomes (containing 32–48 μg total protein; *see* Note 4) into a 1.5-mL Eppendorf tube. Begin a 2-min preincubation in a 36°C water bath.
2. Pipet and hold 1 mL of ice-cold stop buffer in a 1-mL pipeter; set this aside.
3. At 1 min 30 s, pre-wash the GF/C filter with 3 mL of stop buffer under vacuum.
4. At 1 min 45 s, pipet and hold 40 μL of chloride uptake medium.
5. At exactly 2 min elapsed time, initiate $^{36}Cl^-$ uptake by adding to the synaptoneurosomes the 40 μL of chloride uptake medium pipeted in step 4.
6. Vortex the mixture gently and briefly!
7. Terminate the reaction after exactly 5 s by adding 1 mL of ice-cold stop buffer (from the 1-mL pipeter set aside in step 2). Vortex the suspension briefly (*see* Note 8).
8. Pipet 1 mL of the stopped suspension onto the center of the Whatman GF/C filter and immediately filter under vacuum. Set aside the remainder (approx 80 μL) of the stopped suspension (do not discard!; *see* step 13).
9. Use the auto dispenser to wash the filter four times with 3 mL of stop buffer.
10. Using forceps, remove the filter and place it into a 5-mL scintillation vial.
11. Begin a second 2 min preincubation, using another 40 μL aliquot of synaptoneurosomes (as in step 1). Rinse the filter holder with water, install a new filter, and proceed from step 2 to step 10 of this protocol, except this time add 40 μL of nonspecific uptake medium to the synaptoneurosomes (*see* steps 4 and 5; and Note 7).
12. Determine background $^{36}Cl^-$ binding to the filter, in triplicate, following the protocol as above (i.e., steps 1–10) except use 40 μL of homogenization buffer instead of the 40 μL of synaptoneurosomes (*see* Note 7).
13. Pipet 10 μL (in duplicate) of the remaining stopped solutions from each reaction (saved from step 8) into a series of scintillation vials (*see* Note 9).
14. When the assays are completed, add 4 mL of scintillation fluid to all of the vials.

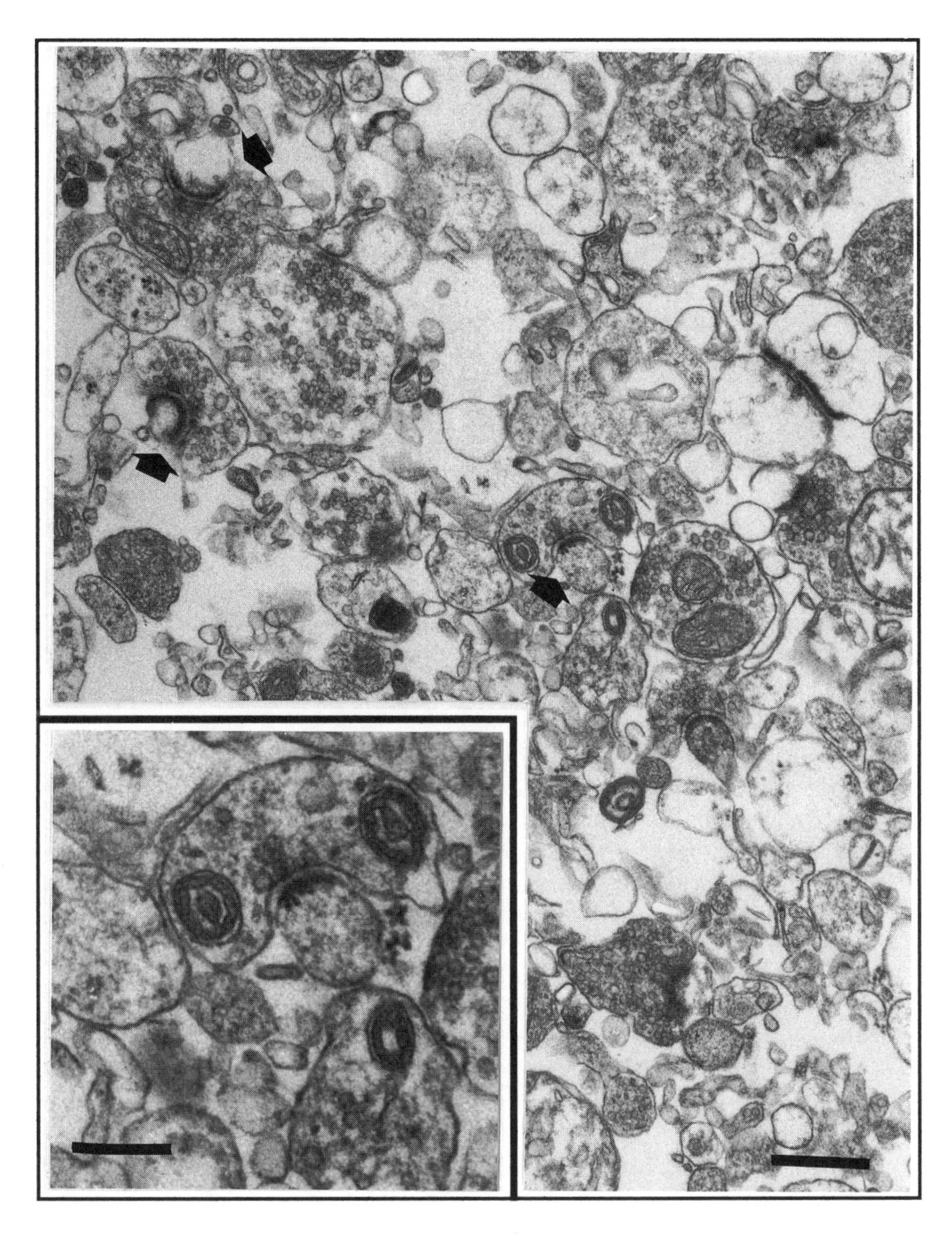

Fig. 2. Electron micrograph of a synaptoneurosome preparation isolated from the hippocampus CA_{1-3} region. The preparation contains sealed presynaptic structures containing vesicles and mitochondria. Some of these structures are attached to intact postsynaptic structures (synaptoneurosomes; *see* arrows). Magnification: ×32,500; scale bar = 0.21 μm (Inset: ×55,000; scale bar = 0.36 μm).

15. Determine the quantity of radioactivity trapped on the filters by liquid scintillation counting (*see* Note 9).
16. Figure 3 shows muscimol-stimulated $^{36}Cl^-$ uptake into synaptoneurosomes, isolated from the CA_{1-3} and DG areas. *See* the figure caption and Note 10 for further discussion.

4. Notes

1. It is essential that all steps of the isolation procedure are carried out very quickly and accurately, and that all solutions and glassware are chilled on ice (4°C). Furthermore, the synaptoneurosome preparation must be used for study within 90 min. After this period the muscimol-mediated $^{36}Cl^-$-uptake decreases rapidly indicating loss of viability of the preparation.
2. Avoid production of air bubbles during homogenization to limit oxidation of the synaptoneurosomes. After two strokes the suspension should be homogeneous, with no visible pieces of the slices present.
3. Protein content is determined by the Bio-Rad assay according to Bradford *(15)*, using the micro assay protocol given in the manufacturer's instructions.
4. The final protein concentration should be in the range of 0.8–1.2 mg/mL, which is the minimum required for the $^{36}Cl^-$ uptake experiments. Depending on the protein concentration measured initially, adjust the final volume to 200–300 μL CA_{1-3} or 100–200 μL DG with HB. Keep the preparations in Eppendorf tubes on ice.
5. In addition to the vesicle-containing presynaptic structures with sealed postsynaptic sacs attached (synaptoneurosomes), isolated presynaptic structures (synaptosomes), free mitochondria, and undefined cellular fragments are observed in the electron micrographs. About 20% of the presynaptic structures contain these attached postsynaptic structures *(17)*.
6. Keep the synaptoneurosome preparations on ice until use. The preparatory steps for the $^{36}Cl^-$ uptake assay described in Section 3.3.1. should be completed well before beginning the assay protocol in Section 3.3.2. to ensure performance of the experiment within 90 min (*see* Note 1).
7. As a measure of $GABA_A$-receptor mediated $^{36}Cl^-$ uptake, the specific receptor agonist muscimol is used and added in a concentration (35 μ*M*) that maximally stimulates $^{36}Cl^-$ uptake in hippocampal synaptoneurosomes *(14)*. Only this concentration of muscimol could be tested in the described experiments, because of the limited yields of synaptoneurosomes from the CA_{1-3} area and DG area isolated from one rat.

   To estimate specific $GABA_A$-receptor-mediated $^{36}Cl^-$ flux, nonspecific $^{36}Cl^-$ uptake determined in the presence of the $GABA_A$-receptor blocker PTX should be subtracted from the $^{36}Cl^-$ uptake assayed in the absence of the blocker. It is essential that the specific and nonspecific $^{36}Cl^-$ uptake be determined in tandem (i.e., alternating) to circumvent time-dependent differences because of longer series. Measurements should be carried out at least in duplicate. Only those $^{36}Cl^-$ uptake values that amount to 100% or more above the background binding of $^{36}Cl^-$ to the filters are to be taken into account for the uptake measurements.

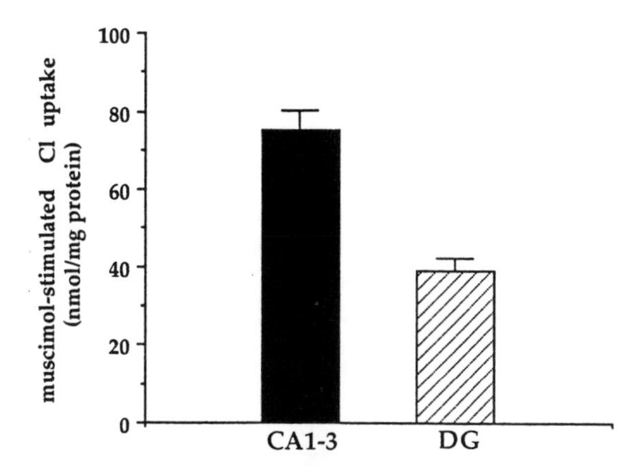

Fig. 3. Muscimol-stimulated $^{36}Cl^-$ uptake into synaptoneurosomes isolated from the CA_{1-3} area and the dentate gyrus (DG). The concentration of muscimol used in the experiment was 35 *μM*. The assay procedure is described in Section 3.3.2. and these results are further discussed in Note 10.

8. In control studies, the relation between the muscimol concentration and $^{36}Cl^-$ uptake into hippocampal synaptoneurosomes was determined after 5 s of incubation. In a few experiments, we also measured muscimol-stimulated $^{36}Cl^-$ uptake in hippocampal synaptoneurosomes after 10 s of incubation. We found that by doubling the incubation time, the muscimol-stimulated $^{36}Cl^-$ uptake tended to decrease whereas the variability increased markedly (data not shown). This may be explained by increases in nonspecific $^{36}Cl^-$ uptake with increasing incubation-time *(18)*, resulting in a decrease of GABA-mediated uptake of more than 95% after 3 min of incubation *(17)*. We stress, therefore, that muscimol-stimulated $^{36}Cl^-$ uptake **must** be determined using incubation times of 5 s or less.

   Studying $GABA_A$ receptor-mediated $^{36}Cl^-$ flux using an incubation time of 5 s, a rapid decrease in the measurement of ion fluxes in the synaptoneurosome preparation would be expected as a result of desensitization of the $GABA_A$ receptor and of equilibration of $^{36}Cl^-$ across the membrane. Electrophysiological measurements of $GABA_A$ receptor desensitization on hippocampal neurons at room temperature *(19)* showed a decay ($t_{1/2}$) of the response decrement to iontophoretically applied GABA and muscimol of 1.1 and 3.3 s, respectively. Schwartz et al. *(16)* investigated desensitization of the $GABA_A$ receptor in cerebrocortical synaptoneurosomes and observed a $t_{1/2}$ of approx 6 s for the muscimol (50 *μM*) response. The fact that muscimol-stimulated $^{36}Cl^-$ uptake can still be measured after several seconds of incubation may be explained by the presence of a furosemide and DIDS-(4,4'-di-isothiocyano-2,2'-disulfonic acid stilbene) sensitive Cl^- pump in the cerebrocortical synaptoneurosome preparation *(3,20)*, that may be involved in maintaining a Cl^- gradient across the membrane after prolonged $GABA_A$ receptor activation.

9. Count duplicate 10 μL aliquots of the stopped suspensions (specific and nonspecific uptake and background) and calculate the total amount of chloride in each incubation, assuming that 10 μL represents 10/1080, or 0.93% of the total dpm. The amount of $^{36}Cl^-$ accumulated in the synaptoneurosomes is expressed in nmol Cl^- by comparison with the total amount of Cl^- in the assay, and normalized to the amount of synaptoneurosomal protein (in mg).
10. The 35 μ*M* muscimol-stimulated $^{36}Cl^-$ uptake is about two times larger in CA_{1-3} synaptoneurosomes than in synaptoneurosomes from the DG. This may be because of a different proportion of synaptoneurosomes in the preparations of the two hippocampal subareas per milligram protein. No differences in morphology between the synaptoneurosome preparations from the two subregions could be observed *(14)*.

    Muscimol-stimulated $^{36}Cl^-$ uptake takes place in synaptoneurosomes, but it is likely that undefined, resealed dendritic fragments may also be involved in $^{36}Cl^-$ uptake (*see* Note 5). Because the synaptoneurosome preparation contained both intact pre- and postsynaptic structures, we cannot exclude the possibility that muscimol-stimulated $^{36}Cl^-$ uptake partly occurs presynaptically, although the evidence for presynaptic $GABA_A$ receptors is rather scarce *(21)*.

References

1. Siekevitz, P. (1981) Isolation of postsynaptic densities from cerebral cortex. *Res. Meth. Neurochem.* **5,** 75–89.
2. Hollingsworth, E. B., McNeal, E. T., Burton, J. L., Williams, R. J., Daly, J. W., and Creveling, C. R. (1985) Biochemical characterization of a filtered synaptoneurosome preparation from guinea pig cerebral cortex: cyclic adenosine 3',5'-monophosphate-generating systems, receptors, and enzymes. *J. Neurosci.* **5,** 2240–2253.
3. Schwartz, R. D., Jackson, J. A., Weigert, D., Skolnick, P., and Paul, S. M. (1985) GABA- and barbiturate-stimulated chloride efflux from rat brain synaptoneurosomes. *J. Neurosci.* **5,** 2963–2970.
4. Ebstein, R. P., Seamon, K., Creveling, C. R., and Daly, J. W. (1982) Release of norepinephrine from brain vesicular preparations: effects of an adenylate cyclase activator, forskolin, and a phosphodiesterase inhibitor. *Cell. Mol. Neurobiol.* **2,** 179–192.
5. Chandler, J. L. and Crews, F. T. (1990) Calcium-versus G protein-mediated phosphoinositide hydrolysis in rat cerebral cortical synaptoneurosomes. *J. Neurochem.* **55,** 1022–1030.
6. Gusovsky, F. and Daly, J. W. (1988) Formation of inositol phosphates in synaptoneurosomes of guinea pig brain: stimulatory effects of receptor agonists, sodium channel agents and sodium ionophores. *Neuropharmacology* **27,** 95–105.
7. Harris, R. A. and Allan, A. M. (1985) Functional coupling of γ-amino-butyric acid receptor in chloride channels in brain membranes. *Science* **228,** 1108,1109.
8. Schofield, P. R. (1989) The $GABA_A$ receptor: molecular biology reveals a complex picture. *TIPS* **10,** 476–478.

9. Persohn, E., Malherbe, P., and Richards, J. G. (1992) Comparative molecular neuroanatomy of cloned $GABA_A$ receptor subunits in the rat CNS. *J. Comp. Neurol.* **326,** 193–216.
10. Schönrock, B. and Bormann, J. (1993) Functional heterogeneity of hippocampal $GABA_A$ receptors. *Eur. J. Neurosci.* **5,** 1042–1049.
11. Lopes da Silva, F. H., Kamphuis, W., and Wadman, W. J. (1992) Epileptogenesis as a plastic phenomenon of the brain, a short review. *Acta Neurol. Scan.* **86,** 34–40.
12. Kamphuis, W. and Lopes da Silva, F. H. (1990) The kindling model of epilepsy: the role of GABAergic inhibition. *Neurosci. Res. Comm.* **6,** 1–10.
13. Titulaer, M. N. G., Kamphuis, W., Pool, C. W., van Heerikhuize, J. J., and Lopes da Silva, F. H. (1994) Kindling induces time-dependent and regional specific changes in the [^{3}H]muscimol binding in the rat hippocampus: a quantitative autoradiographic study. *Neuroscience* **59,** 817–826.
14. Titulaer, M. N. G., Ghijsen, W. E. J. M., Kamphuis, W., De Rijk, T. C., and Lopes da Silva, F. H. (1995) Opposite changes in $GABA_A$ receptor function in the CA_{1-3} area and fascia dentata of kindled rat hippocampus. *J. Neurochem.* **64,** 2615–2621.
15. Bradford, M. M. (1976) A rapid and sensitive method for the quantification of microgram quantities of protein utilizing the principle of protein dye binding. *Anal. Biochem.* **72,** 248–252.
16. Schwartz, R. D., Suzdak, P. D., and Paul, S. M. (1986) γ-Aminobutyric (GABA)- and barbiturate-mediated $^{36}Cl^-$ uptake in rat brain synaptoneurosomes: evidence for rapid desensitization of the GABA receptor-coupled chloride ion channel. *Mol. Pharm.* **30,** 419–426.
17. Verheul, H. B., de Leeuw, F.-E., Scholten, G., Tulleken, C. A. F., Lopes da Silva, F. H., and Ghijsen, W. E. J. M. (1993) $GABA_A$ receptor function in the early period after transient forebrain ischaemia in the rat. *Eur. J. Neurosci.* **5,** 955–960.
18. Cupello, A. and Rapallino, M. V. (1992) Components of basal and GABA activated $^{36}Cl^-$ influx in rat cerebral cortex microsacs. *Int. J. Neurosci.* **62,** 35–43.
19. Thallmann, R. H. and Hershkowitz, N. (1985) Some factors that influence the decrement in the response to GABA during its continuous iontophoretic application to hippocampal neurons. *Brain Res.* **342,** 219–233.
20. Engblom, A. C. and Åkerman, K. E. O. (1991) Effect of ethanol on γ-amino-butyric acid and glycine receptor-coupled Cl^- fluxes in rat brain synaptoneurosomes. *J. Neurochem.,* **57,** 384–390.
21. Starke, K., Göthert, M., and Kilbinger, H. (1989) Modulation of neurotransmitter release by presynaptic autoreceptors. *Physiol. Rev.* **69,** 864–989.

6

Preparation of Synaptosomal Plasma Membranes by Subcellular Fractionation

Jacques J. H. Hens

1. Introduction

Brain tissue is extremely heterogeneous—being comprised of neurons (which are often myelinated) and a number of glial cell types—and, therefore, requires some purification in order to study neuronal and glial functioning in isolation at the molecular level. The introduction of the ultracentrifuge by Svedberg in 1925 offered neurochemists for the first time the ability to fractionate homogenized brain tissue into its subcellular components. Further development of the preparative ultracentrifuge in the early 1940s and the pioneering work by Claude, De Duve, and Palade (who shared a Nobel Prize in 1974 for their work on tissue fractionation; *see* refs. *1–3*), led to the establishment of a variety of techniques for subcellular fractionation. Today, using variations on these original methods, neurochemists are able to isolate from brain homogenates, relatively pure fractions of subcellular components that retain their metabolic activity and structural integrity.

Currently, several methods are available for the bulk isolation of purified plasma membrane fractions. Each of these plasma membrane isolation procedures involves three basic, successive steps: tissue homogenization, subcellular fractionation, and assessment of the yield and/or purity of the isolated membrane fraction. During homogenization of brain tissue, nerve terminals are pinched off from their axonal connections and broken away from surrounding glial cell elements. Because nerve terminals are relatively resistant to mechanical stress, they can reseal to form synaptic bodies that are known as synaptosomes. Whittaker *(4)* and deRobertis *(5)* were the first to describe a method to purify these pinched-off nerve terminals. In their procedure, a crude

From: *Methods in Molecular Biology, Vol. 72: Neurotransmitter Methods*
Edited by: R. C. Rayne Humana Press Inc., Totowa, NJ

synaptosomal/mitochondrial suspension in isotonic sucrose is loaded onto discontinuous sucrose gradients consisting of 0.8*M* sucrose layered on top of 1.2*M* sucrose. These workers observed that the synaptosomes (which were harvested from the 0.8/1.2*M* sucrose interface after centrifugation) consisted primarily of plasma membranes with inclusions of vesicles, mitochondria, and assorted soluble components (for further discussion and for protocols to prepare synaptosomes, *see* Chapter 4).

It appeared, furthermore, that synaptosomal plasma membranes (SPMs) could easily be obtained after lysis and further subcellular fractionation of such synaptosomal preparations. Essentially, this procedure permits isolation of synaptosomal plasma membranes with minimal contamination by glial cell elements. Throughout the years, this method has been modified and improved in many labs to permit detailed studies of neuronal signal transduction using isolated synaptosomal plasma membranes *in vitro*. Recently, synaptosomal plasma membranes have been used to study protein phosphorylation and dephosphorylation events *(6–9)*, polyphosphoinositide metabolism *(10)*, and protein–protein interactions using chemical crosslinkers *(9,11)* and immunoprecipitation techniques *(6,12)*. Synaptosomal plasma membranes have also been used as source material for isolation of membrane proteins *(13)*.

In this chapter, the preparation of synaptosomal plasma membranes using centrifugation techniques will be described in detail. The method here is based on that described previously by Kristjansson et al. *(14)* with only slight modifications. Section 3.1. outlines protocols for dissection and homogenization of the brain, indicating the parameters most important to obtain synaptosomal plasma membrane preparations of reproducibly high quality. Section 3.2. describes the subcellular fractionation procedure itself, and Section 3.3. outlines a protocol for assessment of yield of the synaptosomal plasma membranes employing a protein determination assay described by Bradford *(15)*.

2. Materials

2.1. Brain Dissection and Homogenization

1. Male Wistar rats, weighing 100–120 g.
2. Small animal guillotine.
3. Petri dishes filled with ice for dissection.
4. Whatman 3MM filter circles.
5. Dissection instruments: One pair of scissors, two pairs of forceps, a single-edge razor blade, and optionally, a claw and a scalpel.
6. Isotonic sucrose solution: 0.32*M* sucrose in Milli-Q or double-distilled water. This solution is stable for 1 wk when stored at 4°C.
7. Motor-driven, 25-mL Potter-Elvehjem (Teflon-glass) homogenizer with a pestle clearance of 0.125 mm.

2.2. Synaptosomal Plasma Membrane Preparation

1. Refrigerated high speed centrifuge and fixed angle rotor: e.g., Beckman SS34 rotor, 8 × 40 mL.
2. Ultracentrifuge and swing-out ultracentrifuge rotor: e.g., Beckman SW-28 rotor, 6 × 16 mL.
3. 0.4 and 1.0*M* sucrose solutions: Dissolve sucrose in Milli-Q or double-distilled water. These solutions are stable for 1 wk when stored at 4°C.
4. Potter-Elvehjem (Teflon-glass) homogenizers, 10 and 1 mL capacity, each with a pestle clearance of 0.125 mm.
5. Magnetic stirring plate.
6. SPM buffer: 10 m*M* Tris-HCl, 10 m*M* $MgCl_2$, 0.1 m*M* $CaCl_2$, pH 7.4. Prepare fresh daily from 1*M* stock solutions of each component. Stock solutions are stable for 1 mo when stored at 4°C.
7. Glycerol/glycerin (Merck).

2.3. Assessment of Yield of Synaptosomal Plasma Membranes

1. Spectrophotometer allowing absorption measurements at 595 nm and either glass or quartz cuvets.
2. Bovine serum albumin (BSA) stock solution of exactly 0.5 mg/mL in SPM buffer (*see* Section 2.2., item 6). This solution is stable for several months when stored at –20°C.
3. Disposable Eppendorf tubes (1.5 mL).
4. Bradford protein assay stock solution: dissolve 150 mg of Coomassie Brilliant blue-G250 in 50 mL of ethanol (96%, HPLC grade). Mix extensively using a magnetic stirrer. Filter the stock solution using Whatman No. 1 filter paper and store the filtrate in the dark at 4°C. This solution is stable for several months (*see* Note 1).
5. Bradford protein assay working reagent: Add 3 mL of Bradford stock solution to 97 mL of 7% phosphoric acid in water. Prepare the working reagent at least 1 d in advance of use. Bradford working reagent is stable for 1 mo when stored in the dark and at room temperature (*see* Note 1).

3. Methods

3.1. Brain Dissection and Homogenization

1. Decapitate a rat and place the head on a chilled surface until dissection (*see* Notes 2 and 3).
2. On the dorsal surface of the head, make a rostral incision (toward the nose) using a pair of scissors or scalpel. Peel aside the skin on top of the skull.
3. To open the skull, make a median incision using scissors from the occiput toward the nose. Carefully pry off the top of the skull using a pair of scissors or claw. (The brain will remain untouched at the bottom of the skull; i.e., reminiscent of an avocado nut).

4. Using forceps, remove carefully the pia mater surrounding the brain. (In most cases the pia mater will have already been partially removed during the opening of the skull at step 3).
5. Remove the brain from the skull by making a spoon-like movement with a pair of forceps, starting at the rostral end. Ideally, use forceps that are slightly curved at the tip.
6. Transfer the brain immediately (usually within 30 s to 1 min after decapitation) to a beaker containing ice-cold isotonic sucrose solution. Continue to chill the brain in the sucrose solution for at least 30 s before performing any further dissection (*see* Note 4).
7. Transfer the dissected brain(s) to a prechilled glass Potter-Elvehjem homogenizer containing ice-cold isotonic sucrose solution. Use a 1:10 w/v ratio of tissue to sucrose solution (*see* Notes 5 and 6).
8. Keeping the homogenizer on ice, homogenize the tissue using 7 up-and-down strokes at 700 rpm (*see* Note 7).

3.2. Preparation of Synaptosomal Plasma Membranes

Perform all centrifugations at 4°C and keep all fractions on ice between steps. A flow diagram to illustrate this procedure is shown in Fig. 1.

1. Prepare discontinuous sucrose gradients by transferring 8 mL of ice cold 1.0*M* sucrose into 16 mL ultracentrifuge tubes and layering 4 mL of 0.4*M* sucrose on top. Set the gradients aside until step 8 (*see* Note 8).
2. Pour the brain homogenate into prechilled centrifuge tubes on ice and centrifuge at 1000*g* for 10 min using a fixed-angle rotor. Carefully collect the supernatant (S1) and discard the pellet (P1) that contains myelin, nuclei, cell debris, and blood cells (*see* Note 9).
3. Centrifuge the S1 fraction at 10,000*g* for 20 min using a fixed-angle rotor. Discard the supernatant (S2) and retain the P2 fraction, which is the crude mitochondrial/synaptosomal pellet (*see* Note 10).
4. Add 5 mL of ice-cold Milli-Q or double-distilled water to the P2 fraction, resuspend it, and pipet it into a prechilled Potter-Elvehjem homogenizer. Rinse the centrifuge tube with 1 mL of ice-cold water, giving a final volume of approx 7 mL in the homogenizer. Manually homogenize the P2 pellet using 5 up-and-down strokes.
5. Transfer the suspension to a small beaker and stir with a magnetic stirrer for 5–10 min on ice or in a 4°C climate chamber (*see* Note 11).
6. To prepare a total synaptosomal plasma membrane fraction (T-SPM), proceed to step 8 (*see* Note 12).
7. To prepare a light synaptosomal plasma membrane fraction (L-SPM), centrifuge the P2 lysate at 10,000*g* for 20 min using a fixed-angle rotor. Collect the supernatant; approx 7 mL will be recoverable (*see* Note 12).
8. Layer either the total P2 lysate (from step 5) or the supernatant of the P2 lysate (from step 7) onto two sucrose gradients (3–4 mL/gradient).

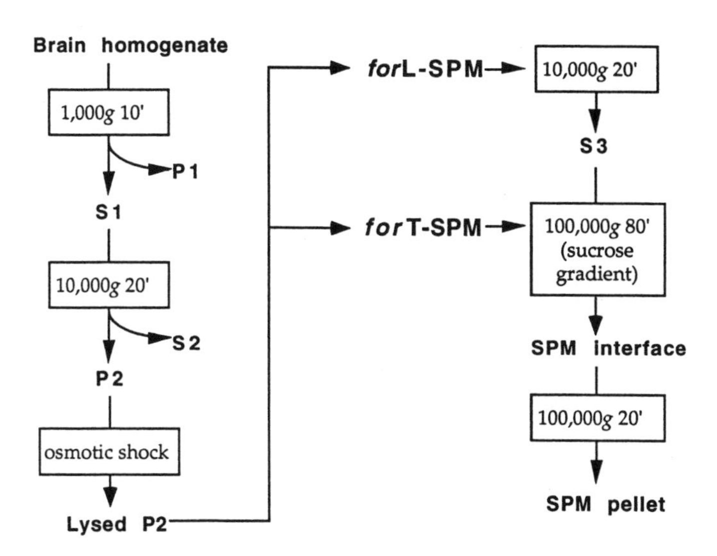

Fig. 1. Preparation of synaptosomal plasma membranes. *See* Section 3.2. for complete details.

9. Centrifuge the gradients for 80 min at 100,000g_{max} in an ultracentrifuge using a swing-out rotor.
10. Collect the 0.4/1.0*M* sucrose interface from each gradient (approx 2 mL from each) and transfer each to a 16-mL ultracentrifuge tube. Gently wash by dilution to 16 mL using ice-cold SPM buffer (*see* Note 13).
11. Centrifuge the diluted sucrose interface for 20 min at 100,000g_{max} using a swing-out rotor.
12. Discard the supernatant and resuspend the resulting SPM pellet in a volume of ice-cold SPM buffer sufficient to make an approx 5 mg protein/mL solution (*see* Note 14) and transfer the suspension to a 1 mL Potter-Elvehjem homogenizer.
13. Homogenize the suspension manually, on ice, using 5 up-and-down strokes.
14. Add glycerol to 10% (v/v) and store the synaptosomal plasma membrane preparation at –80°C (*see* Notes 15 and 16).
15. Measure the protein content of the preparation using the protocol outlined in Section 3.3.

3.3. Assessment of the Yield of Synaptosomal Plasma Membranes

1. Dilute the BSA stock solution with SPM buffer to make duplicate BSA samples of 0, 2, 4, 6, 8, and 10 µg of protein/tube, each in a total volume of 20 µL.
2. Make dilutions of the synaptosomal plasma membrane preparation (20 µL total vol per tube, in triplicate) such that the protein concentrations will fall within the range of the BSA standard series (*see* Section 3.2., step 12 and Note 14).

3. Add 0.5 mL of Bradford reagent to each tube, mix, and wait 3 min for the blue color to develop.
4. Measure absorbance of each sample at 595 nm (*see* Note 17).
5. The protein yield of the synaptosomal plasma membrane preparation is interpolated from a plot of absorbance at 595 nm against protein concentration for the BSA standards.

4. Notes

1. A number of kits to determine protein content are commercially available, some of which make use of a Bradford-type method (e.g., Pierce, Sigma, Bio-Rad). In general, the Bradford method is most convenient, because it is reasonably sensitive and simple and rapid to perform.
2. Decapitation without using anesthetics is preferable, because anesthetics may affect the biological properties of the synaptosomal plasma membrane preparation. The brain should be dissected on ice immediately after decapitation. Brains from species other than rat can also be used to prepare synaptosomal plasma membranes using this protocol.
3. Inverted Petri dishes containing ice make ideal chilled surfaces for manipulating the brain. Sprinkling kitchen salt on top of the ice just before dissection will ensure that the cooling capacity of the ice is maximal.
4. For further dissection, prepare a chilled dissection surface by placing a circular Whatman filter paper, soaked with isotonic sucrose solution, on top of an inverted, ice-filled Petri dish (*see* Note 3). Dissect the brain, as required, on this ice-cold surface (e.g., if preparing synaptosomal plasma membranes from forebrain only, set the brain on the filter paper and remove the brainstem and cerebellum).
5. The wet weight of one rat forebrain is approx 1 g.
6. In some instances, it might be necessary to supplement the 0.32*M* sucrose solution, pH 7.4, with 1 m*M* EDTA, 0.25 m*M* dithiothreitol (DTT), and/or a cocktail of protease inhibitors. Membrane-bound phospholipases and proteases can be activated during cell disruption. EDTA chelates metal ions (calcium and magnesium) that activate certain phospholipases. DTT is a reducing agent that prevents oxidation of functionally important sulfhydryl groups. Cell disruption may also cause release of proteases from lysosomes. Protease attack on membrane proteins can be prevented by addition of protease inhibitors, such as phenylmethylsulfonyl fluoride (PMSF; an inhibitor of serine proteases) and/or E-64 (an inhibitor of cysteine proteases), to the sucrose solution.
7. In general, each homogenization procedure yields a broad spectrum of plasma membranes differing in size and shape. The size and shape of the predominantly formed plasma membrane fragments is strongly affected by the homogenization technique. Accordingly, the method of homogenization determines the behavior of the membranes during subsequent fractionation procedures. Thus, homogenization is one of the most important steps in the SPM isolation, and should be controlled very carefully in order to get reproducible results. Pestle clearance and rotation speed, homogenization volume, and number of strokes are the most critical parameters.

To reduce protein denaturation during homogenization, always place the homogenizer within an ice-filled container and chill the Teflon-coated pestle before homogenization by storing it in a refrigerator. The time delay between homogenization and the first centrifugation step should be kept as short as possible in order to prevent losses in enzyme activity of the final SPM preparation. Minimize foam production (a result of protein denaturation) during homogenization by leaving the pestle always inside the homogenization solution (i.e., avoid taking the pestle out completely after each up stroke; otherwise air may be sucked into the homogenate).

8. Sucrose gradients can be made using a peristaltic pump or glass pipet. Gradients should be made preferably fresh or within 2 h before use.
9. Myelin is not firmly attached to the pellet P1 and appears as a white tongue on decantation of the S1. To minimize the myelin contamination of the mitochondrial P2 fraction, younger animals (2–3 wk of age) may be used or the white matter may be dissected from the gray matter as much as possible before homogenization *(16)*.
10. At these low speeds, the yield of synaptosomes—and thus of synaptosomal plasma membranes—will be decreased by a loss of smaller-sized particles into the so-called microsomal fraction (S2). On the other hand, to use these speeds minimizes microsomal contamination of the crude mitochondrial/synaptosomal P2 fraction. Forces as high as 17,000*g* for 1 h have been used to maximize the yield of synaptosomes in the P2 fraction, at the expense of increasing the microsomal contamination.
11. The addition of water to a crude P2 fraction changes the osmotic pressure of the medium. The synaptosomes will easily lyse as a result of this osmotic shock, whereas the mitochondria and vesicles are largely unaffected. The lysis efficiency of the synaptosomes determines the yield of the synaptosomal plasma membrane preparation.
12. Step 7 in the procedure of synaptosomal plasma membranes is optional and only required if L-SPM is desired. To produce T-SPM, this centrifugation step is omitted. Besides a difference in protein yield, T-SPM and L-SPM show a difference in the sizes of the membrane fragments: Membrane sheets of L-SPM are smaller than those of T-SPM. In general, L-SPM is more homogenous and slightly more accessible for exogenous treatments (in particular after heat-inactivation of SPM) than T-SPM.
13. The synaptosomal plasma membranes layer at the 0.4/1.0*M* sucrose interface. The mitochondria are found in the pellet and the synaptic vesicles are found on top of the 0.4*M* sucrose fraction. Use a clean, disposable Pasteur pipet to harvest the SPMs, after first removing the top fractions (removing about one-half of the 0.4*M* sucrose to prevent contamination of the SPM fraction).
14. An L-SPM preparation yields about 1 mg of protein/rat forebrain whereas a T-SPM preparation, also from one rat forebrain, will yield about 4 mg of protein, although the precise amounts may vary. Thus, bring up L-SPM pellets in 150 μL and T-SPM pellets in 600 μL of SPM buffer to achieve ≥5 mg/mL. Dilutions of

these solutions (approx 1:10 to 1:40) can be tested in a Bradford protein assay to determine more precisely the protein concentrations.

15. Glycerol serves as a cryoprotectant. When stored at –80°C, SPMs retain their enzymatic activity for at least 1 yr.
16. To remove glycerol or to replace the buffer medium of the synaptosomal plasma membrane preparation, wash the preparation by centrifugation at 15,000*g* for at least 5 min at 4°C in SPM buffer.
17. To save time, start the absorbance measurements with the 0 μg protein standard and proceed in sequence to the 10 μg standard. This way the cuvet does not need rinsing between measurements.

Acknowledgment

Jacques J. H. Hens is supported by grant 903-53-091 of the Netherlands Organization for Scientific Research (MW-NWO).

References

1. De Duve, C. (1975) Exploring cells with a centrifuge. *Science* **189,** 186–194.
2. Palade, G. (1975) Intracellular aspects of the process of protein synthesis. *Science* **189,** 347–358.
3. Claude, A. (1975) The coming of age of the cell. *Science* **189,** 433–435.
4. Gray, E. G. and Whittaker, V. P. (1962) The isolation of nerve endings from brain: an electron microscopic study of the cell fragments of homogenisation and centrifugation. *J. Anat.* **96,** 79.
5. deRobertis, E., delraldi, A. P., Rodriguez de Lores Arnaiz, G., and Salganicoff, L. (1962) Cholinergic and non-cholinergic nerve endings in rat brain. I. Isolation and subcellular distribution of acetylcholine and acetyl cholinesterase. *J. Neurochem.* **9,** 23.
6. De Graan, P. N. E., Dekker, L. V., Oestreicher, A. B., Van der Voorn, L., and Gispen, W. H. (1989) Determination of changes in the phosphorylation state of neuron-specific protein kinase C substrate B-50 (GAP-43). *J. Neurochem.* **52,** 17–23.
7. Han, Y., Wang, W., Schlender, K. K., Ganjeizadeh, M., and Dokas, L. A. (1992) Protein phosphatases 1 and 2A dephosphorylate B-50 in presynaptic plasma membranes from rat brain. *J. Neurochem.* **59,** 364–374.
8. Hens, J. J. H., De Wit, M., Dekker, L. V., Boomsma, F., Oestreicher, A. B., Margolis, F., Gispen, W. H., and De Graan, P. N. E. (1993) Studies on the role of B-50 (GAP-43) in the mechanism of Ca^{2+}-induced noradrenaline release: lack of involvement of protein kinase C after the Ca^{2+} trigger. *J. Neurochem.* **60,** 1264–1273.
9. Hens, J. J. H., De Wit, M., Boomsma, F., Mercken, M., Oestreicher, A. B., Gispen, W. H., and De Graan, P. N. E. (1995) N-terminal-specific anti-B-50 (GAP-43) antibodies inhibit Ca^{2+}-induced noradrenaline release, B-50 phosphorylation and dephosphorylation, and calmodulin binding. *J. Neurochem.* **64,** 1127–1136.
10. De Graan, P. N. E., Dekker, L. V., De Wit, M., Schrama, L. H., and Gispen, W. H. (1988) Modulation of B-50 phosphorylation and polyphosphoinositide metabolism in synaptic plasma membranes by protein kinase C, phorbol diesters and ACTH. *J. Receptor Res.* **8,** 345–361.

11. De Graan, P. N. E., Oestreicher, A. B., De Wit, M., Kroef, M., Schrama, L. H., and Gispen, W. H. (1990) Evidence for the binding of calmodulin to endogenous B-50 (GAP-43) in native synaptosomal plasma membranes. *J. Neurochem.* **55,** 2139–2141.
12. Söllner, T., Whiteheart, S. W., Brunner, M., Erdjument-Bromage, H., Geromanos, S., Tempst, P., and Rothman, J. E. (1993) SNAP receptors implicated in vesicle targeting and fusion. *Nature* **362,** 318–324.
13. De Graan, P. N. E., Moritz, A., De Wit, M., and Gispen, W. H. (1993) Purification of B-50 by 2–mercaptoethanol extraction from rat brain synaptosomal plasma membranes. *Neurochem. Res.* **18,** 875–881.
14. Kristjansson, G. I., Zwiers, H., Oestreicher, A. B., and Gispen, W. H. (1982) Evidence that the synaptic phosphoprotein B-50 is localized exclusively in nerve terminals. *J. Neurochem.* **39,** 371–378.
15. Bradford, M. M. (1976) A rapid and sensitive method for the quantification of microgram quantities of protein utilizing the principle of dye-binding. *Anal. Biochem.* **76,** 248–254.
16. Autilio, L. A., Appel, S. H., Pettis, P., and Gambetti, P. (1968) Biochemical studies of synapses *in vitro*. I. Protein synthesis. *Biochemistry* **7,** 2615.

7

Detection of Neuropeptides by Immunocytochemistry

Dick R. Nässel and Peter Ekström

1. Introduction

Neuropeptides constitute the largest and most diverse class of signaling substances known in metazoans. Over the last 20 yr it has become apparent that neuropeptides have important roles as neurohormones, neuromodulators, cytokines, morphogenetic factors, and possibly in some cases, as true neurotransmitters. Each neuropeptide may even be multifunctional and exist in several isoforms in a given animal species. In the search for functions of neuropeptides, it has been critical to be able to localize sites of synthesis and release. Immunocytochemistry (ICC) has been instrumental in the accurate mapping of the cellular and subcellular distribution of neuropeptides in tissue. Other immunological assays, such as radioimmunoassay (RIA) and immunoenzymatic assay (ELISA) provide powerful complements for quantification of neuropeptides. Several important discoveries related to neuropeptides have relied on ICC, for example: Different neuropeptides have very specific distributions in small populations of neurons *(1–3)*, neuropeptides are commonly colocalized with low-mol-wt neurotransmitters or other neuropeptides *(4)*, the chemical diversity of neurons is far greater than previously suspected *(2,3)*, and neuropeptide synthesis and release can be episodic *(5)*.

In spite of its power, ICC is a technique with certain limitations in its specificity *(6,7)*, and in most cases, the technique needs to be supplemented with rigorous specificity testing and independent nonimmunological assay techniques. Neuropeptides, like other peptides and proteins, because of their complex chemical structure, are especially problematic to detect specifically. The antigens used for immunization to obtain neuropeptide antibodies are commonly synthetic oligopeptides coupled to larger carrier proteins. It is known

From: *Methods in Molecular Biology, Vol. 72: Neurotransmitter Methods*
Edited by: R. C. Rayne Humana Press Inc., Totowa, NJ

that immunoglobulins can recognize antigenic sites or epitopes that may consist of the surface structure of as few as 1–6 amino acids of a peptide or protein or, when present, 1–6 carbohydrate moieties *(8)*. Thus, the different antibodies in a polyclonal antiserum may recognize several epitopes in the synthetic oligopeptide used as hapten at immunization (disregarding antibodies recognizing the carrier protein). This is not a problem if similar epitopes do not exist in other peptides or proteins. Often, however, one finds that epitopes of neuropeptides are shared with other related or unrelated molecules, leading to a problem in the specific detection of defined antigens in complex tissue. The difference between analysis of antibody crossreactivity with, on one hand, defined peptide sequences in RIA and ELISA or synthetic peptide matrices in ICC, and on the other, tissue sections and extracts, is that in the model tests we have defined peptides and in tissue we do not know the structures of most of the native peptides or proteins. Thus, crossreactivity with undefined tissue antigens is the major problem. A second problem, already indicated, is that neuropeptides often exist in closely related isoforms in a given species and antisera raised against specific synthetic holopeptides tend to recognize all of the isoforms. The complications just mentioned can be circumvented to some extent by clever antigen design and by different specificity controls.

It is beyond the scope of this chapter to provide protocols for antigen and antiserum production, but some general guidelines are given here since these are among the most critical steps in peptide immunocytochemistry. In cases where several isoforms of a neuropeptide exist, there is often a conserved carboxy-terminal sequence. When generating antisera to the entire peptide, epitopes of the carboxy-terminus are commonly recognized by the antibodies, because most conjugation methods link the carrier protein to the N-terminal portion of the peptide (exposing more of the C-terminus). To distinguish similar isoforms it is thus advisable to generate antibodies to synthetic sequences of the variable N-terminus. N-terminus-specific antibodies applied together with ordinary C-terminus-specific antisera provide a double probe for a specific antigen (neuropeptide) *(9–12)*. For longer peptides, it is possible to generate antisera to additional portions of the same molecule to obtain multiple probes *(13)*. Region-specific antisera can also be produced by synthesizing specific sequences to a branched lysine matrix, so-called multiple antigenic peptides (MAPs), a strategy that also eliminates problems caused by the carrier protein *(14)*.

When the structure of the neuropeptide precursor protein is known, there is the option to raise antisera to nonneuropeptide portions (spacer sequences) of the precursor or additional neuropeptide products coded on the precursor *(15,16)*. If such antisera react with the same neurons as antisera raised to different portions of the processed neuropeptide, there is a likelihood that the neurons synthesize and store the neuropeptide in question. In cases where the precursor structure is known, it has commonly been deduced from molecular cloning. This information provides an opportunity to perform nonimmunological screening for localization of the precursor: *in situ* hybridization histochemistry using oligonucleotide probes to localize sites of transcription.

Monoclonal antibodies can be raised to defined antigens and provide very specific probes *(17)*; practical protocols are provided by Cuello and Côté *(18)*. It is important to keep in mind that monoclonal antibodies—in spite of their monospecificity—do not circumvent the problem of crossreactivity with unidentified tissue antigens. The same rigorous testing needs to be performed for monoclonals. To increase the specificity of polyclonal antisera, one can perform affinity purification and/or separate IgGs from the whole serum. For such purifications, commercial columns are available. Details on antiserum/antibody production have been given in many accounts (e.g., *13,18–23)*.

Once good antisera or antibodies to well-defined antigens are available, specificity tests have to be performed on the tissue. Such tests commonly consist of mixing the antiserum with a surplus of appropriate neuropeptide antigen (or unconjugated neuropeptide), as well as mixing it with related neuropeptides (and their conjugates) for crossreactivity tests. A complete abolition of immunolabeling after preincubation of antiserum with antigen is required to establish that the epitope is present in the tissue, but this alone does not prove the presence of the specific antigen (neuropeptide). If immunolabeling is not abolished, even after preincubation of antiserum with large amounts of antigen, there are reasons to suspect that the antiserum contains antibodies recognizing epitopes unrelated to the antigen utilized at immunization. For more detailed discussions on antiserum specificity tests, the reader is referred elsewhere *(6,7,24)*.

The objective of the present chapter is to provide a few standard protocols for immunocytochemical detection of neuropeptides, including double and triple labeling techniques that utilize different primary antisera/antibodies. It is very important to remember that ICC protocols have been modified in almost every laboratory that employs the technique, and that almost every antiserum or tissue requires specific modifications. Thus, it is wise to consult the original papers describing initial characterization of the antiserum and its application to specific tissues and animal species. For commercially available antisera, standard protocols are commonly supplied. The protocols presented here should provide a starting point for peptide ICC, and we suggest points at which modifications can be made.

The ICC technique can be divided into several important steps that will be dealt with in some detail: Fixation, treatments preceding immunocytochemistry, and employment of antibody detection systems. Because we have not provided a comprehensive coverage of the ICC technique, the reader is referred to some of the more recent treatises on the subject for more details *(22,25–30)*.

1.1. Fixation Methods

Proper tissue fixation before further processing serves both to arrest the antigen in its proper location and to preserve the cytology. Fixation of neuropeptides in tissue for immunocytochemistry may be achieved by chemical fixation, physical fixation by freeze-substitution, or combinations of the two. Chemical fixation can be performed by administration of the fixative via the vascular system, i.e., vascular perfusion, or by simply immersing the tissue of interest in fixative solution. It appears as if choice of fixative for neuropeptides is less critical than for other neuroactive substances *(24)*, and for most light microscopy purposes, 4% paraformaldehyde in sodium phosphate buffer is a good fixative. Many investigators routinely use the picric acid-containing fixatives devised by Stefanini et al. *(31)* and Zamboni and deMartino *(32)*. It is, however, well recognized that the optimal fixative has to be found empirically for each antigen and each tissue. The ideal fixative should yield an optimal preservation of antigenicity, a good preservation of tissue integrity, and allow antibody access to the antigen. All these features cannot be obtained maximally with any single fixation protocol, and testing will be required to reach some compromise. In our experience, buffered fixatives containing more than 0.1% glutaraldehyde are less appropriate for most neuropeptides. The nonbuffered mixture of glutaraldehyde, picric acid, and acetic acid (GPA), however, can be utilized.

Fixation by vascular perfusion ensures rapid access and even distribution of fixative in the tissue. It is usually performed via the heart, i.e., as a transcardial perfusion, under deep anesthesia. Details of this procedure depend on the species (e.g., cardiac anatomy, body temperature, optimal anesthetic), and it is outside the scope of this chapter to treat this issue in detail. For most invertebrate preparations, vascular perfusion is not necessary, or is even impossible to perform.

Sometimes fixation by immersion of tissue directly in the fixative is advisable. This is particularly true for small pieces of tissue, poorly vascularized tissue, and tissue that is surrounded by large volumes of body fluid (the fixative administered by vascular perfusion tends to be diluted in the body fluids). Also, when deep anesthesia is known or suspected to affect antigen levels, it is advised to kill the animal under light anesthesia, rapidly expose the tissue, and fix it by immersion. To preserve structural integrity, it is preferable to fix the tissue *in situ* rather than to remove the tissue prior to fixation.

The time in the fixative varies depending on the size of the tissue, the type of fixative used, and the antigen under study. Generally, smaller tissues need to be immersed for 2–4 h in aldehyde-based fixatives, about 12 h in carbodiimide fixatives, and about 2 h in parabenzoquinone. Thorough buffer washes are essential before tissue can be taken through ICC. After strong aldehyde fixation, a treatment with $NaBH_4$ is also recommended.

In some cases, it is necessary to store the fixed tissue for prolonged periods before immunohistochemical processing. Storage is possible either in refrigerated or frozen form. Deep-freeze storage is usually the method of choice for long-term storage. To preserve antigenicity and histological quality, the tissue is frozen in a cryoprotectant solution.

1.2. Pretreatments

Pretreatments of the tissue are generally required before application of antisera/antibodies; the nature of the pretreatment(s) depends on how the tissue has been prepared. Tissue can be processed for immunohistochemistry in toto, i.e., as wholemounts or after histological sectioning. Cryostat sections, vibratome sections, or sections of tissue embedded in paraffin, resin, or other embedding media may be used. Paraffin and resins must be removed from the sections before immunohistochemical processing. To enhance antibody penetration and antigen exposure, the tissue may be treated with detergents and/or various proteolytic enzymes. (This is most often required for wholemounts and thick vibratome sections.) When immunoenzymatic detection methods are to be used, endogenous enzyme activity has to be quenched. Free reactive aldehyde groups in the tissue should be blocked and tight crosslinks (e.g., because of glutaraldehyde fixation) have to be reduced (e.g., using $NaBH_4$ treatment). For wholemounts it may be imperative to remove connective tissue and the neural sheath to enable the immunoglobulins and other reagents to penetrate into the nervous tissue. When mechanical removal is problematic, enzymatic disruption methods may be useful.

1.3. Detection Systems

Direct and indirect immunocytochemical detection strategies have been developed. The indirect methods are most commonly used. For indirect immunocytochemical detection of antigens, there are numerous commercially available detection systems based on detection of the first antibody with a second layer of IgGs. The secondary IgGs are either directly tagged with a fluorophore or enzyme marker, or conjugated with biotin for flexible detection with avidin-fluorophore or avidin-enzyme conjugates. A third type of detection system is based on introducing a second layer of unlabeled IgG that in turn is detected with a peroxidase antiperoxidase complex.

There are situations where a direct method is useful, notably when primary antisera to different antigens have been raised in the same animal species and are to be used for double labeling ICC. Simultaneous immunocytochemical detection of two antigens is a powerful tool for demonstrating colocalization of neuroactive compounds as well as for analysis of neural connectivity and spatial relationships of neuronal systems utilizing different neurotransmitters or neuropeptides. Different fluorophore combinations are useful and different combinations of primary antisera raised in different animal species can be utilized.

There are numerous commercially available enzyme-tagged secondary antisera, which in most cases are quite useful. It is, however, more common to employ three-step methods like the peroxidase antiperoxidase (PAP) technique and the ABC method. The latter utilizes biotin-streptavidin bridging (IgGs of secondary antiserum are biotinylated). Peroxidase detection systems have some advantages over immunofluorescence: The sensitivity of detection is superior, the resulting preparations are permanent and can be viewed repeatedly at convenience even after long storage (no fading of label), and with some modifications peroxidase detection can be used for electron microscopy.

2. Materials

2.1. Buffers and Fixatives (see Note 1)

1. Sodium phosphate buffer (Sörensen's phosphate buffer): 0.1*M* sodium phosphate, pH 7.4. Prepare stock solutions A (0.2*M* NaH_2PO_4) and B (0.2*M* Na_2HPO_4). Add about 19 mL of A to 81 mL of B to make pH 7.4; dilute with an equal volume of distilled water to obtain 0.1*M*, final concentration.
2. Cacodylate buffer: 0.16*M* sodium cacodylate, pH 7.2, in water. Dissolve 3.42 g of sodium cacodylate ($NaAsO_2(CH_3)_2$, trihydrate, mol wt 214.0) in 100 mL of distilled water. Adjust the pH with 1*M* HCl.
3. Phosphate-buffered saline (PBS): 0.01*M* sodium phosphate, 0.88% (w/v) NaCl, 0.02% (w/v) KCl, pH 7.4. Mix 100 mL of 0.1*M* sodium phosphate buffer (*see* item 1) with 8.8 g of NaCl and 0.2 g of KCl. Add distilled water to make 1000 mL.
4. Tris buffer: 0.05*M* Tris-HCl, pH 7.6. Dissolve 6.05 g of Tris (Sigma, St. Louis, MO) in distilled water, adjust the pH with HCl, and bring the volume to 1000 mL.
5. Paraformaldehyde fixative: 4% (w/v) paraformaldehyde in phosphate buffer (*see* item 1). Dissolve 80 g of paraformaldehyde in 1000 mL of distilled water while heating on a stir plate to about 60°C under a well-ventilated fume hood. Add 1*M* NaOH dropwise until the solution is clear (to completely dissolve the aldehyde). Cool on ice, then mix this 8% (w/v) paraformaldehyde solution with an equal volume of 0.2*M* sodium phosphate buffer (*see* item 1) to obtain the final fixative. Adjust the pH to 7.2. This fixative can be aliquoted and stored frozen in vials. After thawing, mix the solution well and check the pH.
6. PBS-heparin: PBS containing 1000–2000 U/mL of heparin (Sigma).
7. Isopentane.
8. Acetone.
9. Dry ice.

10. PBS-azide: PBS containing 0.01% (w/v) sodium azide.
11. DMSO.

2.2. Treatment Prior to Immunohistochemical Processing

These treatments are optional. All of the enzymes are available from Sigma.

2.2.1. Pretreatments to Unmask Antigenic Sites

1. Pepsin solution: 0.4% (w/v) pepsin in 0.01*N* HCl, pH 2.0.
2. Pronase (Protease VI) solution: 0.05–0.1% pronase (w/v) in PBS (*see* Section 2.1., item 3).
3. Tris-saline: 0.05*M* Tris-HCl, pH 7.6, containing 0.9% (w/v) NaCl and 0.1% (w/v) $CaCl_2$.
4. Tris-sucrose: Tris-saline containing 25% (w/v) sucrose.
5. Trypsin solution: 0.1% trypsin (w/v) in Tris-saline.

2.2.2. Pretreatments to Enhance Antibody Penetration

1. PBS: *See* Section 2.1., item 3.
2. Buffered ethanol treatment: 10, 25, and 40% (v/v) ethanol in PBS.
3. Harsh unbuffered ethanol-xylene treatment: 30, 50, 70, 96, and 100% (v/v) ethanol in water and xylene.
4. Detergent treatment: 1% (v/v) Triton X-100 in PBS (PBS-TX-1.0; *see* Note 2) or DMSO.
5. Freeze-thaw treatment: 25% (w/v) sucrose in PBS (cryoprotection solution).

2.3. Immunofluorescence Detection Methods

1. PBS: *See* Section 2.1., item 3.
2. PBS-TX-1.0: 1% (v/v) Triton X-100 in PBS (*see* Note 2).
3. PBS-TX: 0.25% (v/v) Triton X-100 in PBS (*see* Note 2).
4. PBS-TX-NS: PBS-TX containing 10% (v/v) normal serum (*see* item 10). Add the normal serum to the PBS-TX just before use (*see* Note 2).
5. PBS-BSA: PBS containing 0.25% (w/v) bovine serum albumin (BSA) (*see* Note 3).
6. Tris buffer: *See* Section 2.1., item 4. Adjust the pH to 7.6.
7. Tris-TX: Tris buffer containing 1% Triton X-100 (*see* Note 4).
8. Tris-TX-Na: Tris-TX containing 2–4% (w/v) NaCl (*see* Note 4).
9. Tris-Na: Tris buffer containing 2–4% (w/v) NaCl (*see* Note 4).
10. Normal serum: may be rabbit, rat, goat, or donkey to match the choice of primary and secondary antisera. Available from DAKO, Carpinteria, CA.
11. Mounting medium: Glycerol:PBS, 2:1 (v/v) (*see* Note 5).
12. Ethanol, 80% (v/v) in water.
13. 2% Paraformaldehyde fixative: Dilute paraformaldehyde fixative (Section 2.1., item 5) with an equal volume of phosphate buffer (Section 2.1., item 1).
14. Secondary antisera for single-labeling methods: Fluorophore-labeled secondary antisera; whole Ig, F(ab), or F(ab')2 fragments (*see* Notes 6–8).

15. Secondary antisera cocktail for double-labeling methods: Affinity-purified FITC-conjugated donkey antimouse IgG F(ab')2 fragments and affinity-purified LRSC-conjugated donkey antirabbit IgG F(ab')2 fragments (*see* Notes 6–8).
16. Secondary antisera cocktail for triple-labeling methods: Affinity-purified AMCA-conjugated donkey antimouse IgG F(ab')2 fragments, affinity-purified FITC-conjugated donkey antirabbit IgG F(ab')2 fragments, and affinity-purified LRSC-conjugated donkey antirat IgG F(ab')2 fragments (*see* Notes 6–8).
17. For double labeling with two primary antisera raised in one species: Biotinylated primary antiserum to antigen A, regular primary antiserum to antigen B; streptavidin-Texas Red (Amersham, Arlington Heights, IL) or other fluorescently tagged streptavidin (e.g., streptavidin Cascade Blue; Molecular Probes); swine antirabbit-FITC (DAKO) or swine antirabbit TRITC (DAKO) (*see* Note 6).

2.4. Immunoperoxidase Detection Methods: PAP and ABC

1. PBS, PBS-TX-1.0, PBS-TX, PBS-TX-NS, PBS-TX-BSA: *See* Section 2.3., items 1–5 (*see* Note 2).
2. Tris buffer, Tris-TX-Na, Tris-Na: *See* Section 2.3., items 6, 8, and 9 (*see* Note 4).
3. Peroxidase quenching solution: 80–100% methanol containing 0.5% (v/v) H_2O_2 (*see* Notes 9 and 10).
4. Secondary antiserum: Unlabeled swine antirabbit (DAKO, Inc.) or unlabeled goat antirabbit F(ab')2 fragments (Jackson Immunoresearch Labs).
5. Peroxidase antiperoxidase (PAP) complex: e.g., rabbit PAP complex raised in the same species as primary antiserum (DAKO).
6. Peroxidase substrate: 0.03% (w/v) 3,3'-diaminobenzidine tetrahydrochloride (DAB) in Tris buffer. Aliquots of this solution may be stored frozen and are stable for several months. Add H_2O_2 to a final concentration of 0.01% (v/v) just before use (*see* Note 11).
7. H_2O_2 or Perhydrite tablets (Merck, Darmstadt, Germany) (*see* Note 9).
8. Mounting media: Glycerol, methyl salicylate, Permount or Canada balsam (*see* Notes 5 and 56).
9. Osmium tetroxide solution: 0.1–0.5% osmium tetroxide in cacodylate buffer (*see* Section 2.1., item 2).
10. Biotinylated secondary antiserum (e.g., Goat antirabbit or universal goat antimouse/rabbit; Vector Laboratories, Burlingame, CA).
11. ABC kit Vectastain (Vector Laboratories).

2.5. Alternative Fixatives and Other Solutions (see Note 1)

1. Buffered paraformaldehyde-picric acid (Zamboni's fixative; *see* refs. *31,32*): 2% (w/v) paraformaldehyde and 15% (v/v) picric acid. Add 20 g of paraformaldehyde to 150 mL of filtered (0.45 μm), saturated aqueous picric acid. Heat the solution to about 60°C under a fume hood and add 1*M* NaOH dropwise to dissolve the paraformaldehyde. Cool, filter (0.45 μm), and add 850 mL of 0.1*M* phosphate buffer, bringing the total volume to 1000 mL. Adjust the pH to 7.3

using concentrated HCl. This fixative should be freshly made or stored frozen in aliquots (*see* Note 12).
2. Bouin's fixative: Mix 15 mL of aqueous, saturated picric acid, 5 mL of formalin, and 1 mL of glacial acetic acid. This fixative can be stored at room temperature for months (*see* Note 12).
3. GPA (*see* ref. *33*): Mix 5 mL of glutaraldehyde (25%, v/v), 15 mL of aqueous, saturated picric acid, and 0.1 mL of glacial acetic acid. Use freshly made fixative (*see* Note 13).
4. EDCDI fixative: 4% (w/v) 1-ethyl-3-(3-dimethylaminopropyl) carbodiimide (EDCDI) in phosphate buffer (Section 2.1., item 1). Dissolve 4 g of EDCDI (from Sigma) in 100 mL of phosphate buffer. Use freshly made fixative (*see* Note 14).
5. EDCDI/paraformaldehyde fixative: Dissolve 4 g of EDCDI in 100 mL of 4% paraformaldehyde fixative (*see* Section 2.1., item 5) Use freshly made fixative (*see* Note 15).
6. Parabenzoquinone (PBQ) fixative (*see* ref. *34*): 0.6% (w/v) PBQ in cacodylate buffer (*see* Section 2.1., item 2). Dissolve 0.6 g of *p*-benzoquinone in 100 mL of well-oxygenated cacodylate buffer, prewarmed to 37°C. Use freshly made fixative (*see* Note 16).

3. Methods

*3.1. Fixation (*see *Note 1)*

3.1.1. Fixation by Vascular Perfusion

1. Induce deep anesthesia and perform surgical exposure of the heart and aorta.
2. Perfuse with PBS-heparin (*see* Note 17).
3. Perfuse with paraformaldehyde fixative. As formaldehyde is a toxic and hazardous chemical, perform perfusion in a well-ventilated area or preferably, under a fume hood (*see* Note 18).
4. Dissect to expose the tissue to be studied, or excise the tissue.
5. Immerse the tissue overnight (16 h) at 4–8°C in freshly prepared paraformaldehyde fixative (*see* Notes 19 and 20).
6. Rinse the tissue thoroughly in several changes of phosphate buffer during 24–48 h (*see* Note 21).
7. After thorough rinsing, process the tissue immediately for immunocytochemistry. If the tissue must be stored after fixation and before ICC, refer to Section 3.1.4. and *see* Note 22.

3.1.2. Immersion Fixation

1. Immerse dissected or partly exposed tissue in large volume of fixative (4% paraformaldehyde) for 3–16 h, depending on tissue and subsequent treatment (*see* Notes 19 and 20 concerning fixation parameters and Note 23 concerning optional removal of the neural sheath prior to fixation).
2. Follow steps 6 and 7 of 3.1.1.

3.1.3. Rapid-Freeze Fixation and Freeze Substitution

Freeze substitution yields optimal preservation of antigenicity, but is useful only for small pieces of tissue (*see* Note 24).

1. Excise the tissue(s) of interest, blot carefully to remove excess moisture, and mount the tissue(s) on blotting paper.
2. Freeze the tissue in isopentane cooled by a mixture of dry ice and acetone (*see* Note 25).
3. Transfer frozen tissue to acetone cooled by a mixture of dry ice and acetone (*see* Note 26).
4. Perform freeze substitution in acetone cooled by a mixture of dry ice and acetone over 3 d.
5. Place the vial containing acetone and tissue in the refrigerator to raise the temperature. The tissue can now be directly stored (*see* Note 22), infiltrated for resin embedding, or rehydrated. After rehydration, it can be processed for immunohistochemistry as a wholemount, or cryoprotected and sectioned in a cryostat, prior to ICC (*see* Sections 3.3. and 3.4.).

3.1.4. Storage of Fixed Tissue (see Note 22)

1. For refrigerator storage: Store tissues at 4–8°C in fixative or, preferably, in PBS-azide to prevent bacterial and fungal growth (*see* Note 27).
2. For frozen storage, follow steps 3–7.
3. Transfer the fixed and well-rinsed tissue to 0.01*M* PBS.
4. Add pure DMSO dropwise to the PBS, while stirring, over 4–8 h, until the final concentration of DMSO is 25% (v/v) (*see* Note 28).
5. Transfer tissue plus cryoprotectant solution (that is, the PBS/DMSO or other solution as in Note 28) to Eppendorf vials.
6. Freeze the sealed Eppendorf vials in liquid nitrogen.
7. Store frozen tissue at –70°C.

3.2. Treatment Prior to Immunohistochemical Processing (see Note 29)

3.2.1. Pretreatment to Unmask Antigenic Sites (see Notes 29 and 30)

Choose from the three alternatives listed below.

1. Pepsin pretreatment of sectioned tissue (*see* Note 30):
   a. Prewarm slides with tissue sections to 37°C in PBS.
   b. Incubate the slides in pepsin solution at 37°C for 10–30 min.
   c. Rinse in PBS at room temperature, 3 × 5 min.
2. Pronase pretreatment of sectioned tissue (*see* Note 30):
   a. Incubate sections in pronase solution at room temperature for 10 min.
   b. Rinse in ice-cold PBS for 5 min.
   c. Rinse in PBS at room temperature for 2 × 5 min.

3. Trypsin pretreatment of sectioned tissue (*see* Note 30):
   a. Prewarm slides with sections to 37°C in Tris-saline.
   b. Incubate in trypsin solution at 37°C for 5 min to 2 h.
   c. Rinse in Tris-sucrose at room temperature for 3 × 5 min.

3.2.2. Pretreatment to Enhance Antibody Penetration (see Notes 29 and 31)

Choose from the alternatives listed below.

1. Buffered ethanol treatment: Take tissue through the following solutions, 10 min in each, at room temperature: 10, 25, 40, 25, 10% (v/v) ethanol in PBS. Follow by rinsing in PBS 3 × 10 min.
2. Harsh unbuffered ethanol-xylene treatment (*see* Note 32):
   a. Take the tissue through 10 min in each of the following solutions, at room temperature: 30, 50, 70, 96, 100% ethanol.
   b. Rinse the tissue in xylene, 20 min.
   c. Rinse the tissue in 100% ethanol, 2 × 10 min.
   d. Rinse the tissue in for 10 min in each of the following solutions: 96, 70, 50, 30% ethanol.
   e. Rinse the tissue well with PBS.
3. Detergent pretreatment (*see* Note 33): Incubate wholemounts overnight in PBS-TX-1.0.
4. Freeze-thaw pretreatment (*see* Note 34):
   a. Incubate tissue in cryoprotection solution for 16 h.
   b. Quick-freeze tissue, in cryoprotection solution, on dry ice, or in liquid nitrogen.
   c. Thaw tissue by bringing it to room temperature and rinse well in PBS.

3.3. Immunofluorescence Protocols (see Note 35)

3.3.1. Wholemount Immunofluorescence (see Notes 23 and 36)

1. Wash fixed tissue several times using the same buffer as used to dissolve the fixative.
2. Incubate overnight at 4°C in PBS-TX-1.0 to permeabilize the tissue (*see* Note 2).
3. Dehydrate the tissue in a graded ethanol series (50–70–96–100% ethanol) and xylene (10 min each) followed by rehydration back to PBS-TX-1.0 (The dehydration step is optional because it may reduce antigenicity; *see* Note 37.)
4. Wash the tissue several times (10 min each) in Tris-TX-Na (*see* Note 4).
5. Incubate the tissue for 3–5 d (depending on size of tissue) at 4–8°C in primary antiserum (or monoclonal antibody) during continuous, gentle agitation. Dilute the antiserum in PBS-TX-NS using normal serum from the same species as the secondary antiserum. The optimal primary antiserum/antibody dilution has to be empirically tested in a dilution series (*see* Note 38).
6. Wash the tissue several times in PBS-TX (over a total time of at least 1 h at room temperature) followed by an overnight wash in Tris-Na at 4°C.
7. Incubate the tissue for 24 h at 4°C in the dark in fluorophore-labeled secondary antiserum diluted in PBS-BSA-0.25 (dilution as recommended by supplier; often in the range of 1:30-1:100) (*see* Notes 7, 8, and 38).

8. Wash the tissue in PBS, 4 × 10 min, at room temperature.
9. Mount the tissue under a coverslip in a mounting medium containing an antifading agent (*see* Notes 5 and 39).

3.3.2. Rapid Wholemount Immunofluorescence Method (see Note 40)

1. Replace the fixative directly with 80% ethanol and wash the tissue 6 × 10 min in 80% ethanol.
2. Transfer tissue directly from 80% ethanol to primary antiserum diluted in PBS-TX-NS. Incubate overnight (16 h) at room temperature (*see* Note 38).
3. Wash the tissue in PBS, 4 × 10 min, at room temperature.
4. Incubate the washed tissue for 2–4 h at room temperature in the dark in fluorophore-labeled secondary antiserum diluted in PBS-BSA-0.5 as recommended by the supplier (*see* Notes 7, 8, and 38).
5. Wash the tissue in PBS, 4 × 10 min, at room temperature.
6. Mount the tissue under a coverslip in a mounting medium containing an antifading agent (*see* Notes 5 and 39).

3.3.3. Immunofluorescence Method or Slide-Mounted Tissue Sections (see Note 41)

1. Wash slide-mounted, fixed sections in Tris-TX-Na for 10 min (*see* Note 4). For this and subsequent steps, slides can be washed using 50–80 mL Coplin jars.
2. Place the slides in a humid chamber. Incubate the slides for 24–48 h at 4°C in primary antiserum (or monoclonal antibody). Dilute the antiserum in PBS-TX-NS (use normal serum from same species as secondary antiserum) (*see* Notes 38, 42).
3. Wash the slides for at least 1 h in several changes of PBS-TX.
4. Incubate the slides for 1–2 h at room temperature in the dark in fluorophore-labeled secondary antiserum diluted in PBS-BSA-0.25 using the dilution as recommended by the supplier (*see* Notes 7, 8, and 38).
5. Wash the slides in PBS, 3 × 10 min, at room temperature.
6. Mount the slides under a coverslip in a mounting medium containing an antifading agent (*see* Notes 5 and 39).

3.3.4. Immunofluorescence Double Labeling (see Note 43)

1. Rinse the fixed, slide-mounted sections for 15 min at room temperature in Coplin jars containing PBS-TX-0.25 (*see* Note 2).
2. Place the slides in a humid chamber (*see* Note 42). Expose the sections for 30 min at room temperature to PBS-TX-NS (using donkey serum). Drain off the solution, but do not rinse in buffer before the next step.
3. Incubate the sections overnight (about 16 h, at 4–8°C) in primary antibody cocktail: primary monoclonal antibody raised in mouse against antigen A and primary antiserum raised in rabbit against antigen B, diluted in PBS-TX-B (*see* Notes 38, 44, and 45).
4. Rinse the slides in PBS-TX, 1 × 10 min, at room temperature.

5. Rinse the slides in PBS, 2 × 10 min, at room temperature.
6. Incubate the slides for 45 min at room temperature in the dark in secondary antibody cocktail (diluted in PBS-BSA-0.5; *see* Notes 6, 38, and 45).
7. Rinse the slides in PBS, 3 × 10 min, in the dark at room temperature.
8. Mount the sections under a coverslip in a mounting medium containing an antifading agent (*see* Notes 5 and 39).

3.3.5. Immunofluorescence Triple Labeling

1. Rinse the fixed, slide-mounted sections for 15 min at room temperature in PBS-TX (*see* Note 2 and 41).
2. Incubate the sections for 30 min at room temperature in PBS-TX-NS (using donkey serum). Drain off the solution, but do not rinse the sections in buffer before the next step.
3. Incubate the sections overnight (about 16 h, at 4–8°C) in primary antibody cocktail: primary monoclonal antibody raised in mouse against antigen A, primary antiserum raised in rabbit against antigen B, and primary antiserum raised in rat against antigen C, diluted in PBS-TX-BSA (*see* Notes 38 and 46).
4. Rinse the slides in PBS-TX, 1 × 10 min, at room temperature.
5. Rinse the slides in PBS, 2 × 10 min, at room temperature.
6. Incubate the tissue for 45 min at room temperature in the dark in secondary antibody cocktail diluted in PBS-BSA-0.5 (*see* Notes 8, 38, 46, and 47).
7. Rinse the slides in PBS, 3 × 10 min, in the dark at room temperature.
8. Mount the sections under a cover slip in a mounting medium containing an antifading agent (*see* Notes 5 and 39).

3.3.6. Immunofluorescence Double Labeling with Two Primary Antisera Raised in the Same Species (see Notes 41 and 48)

1. Place the slides with sections in a humid chamber (*see* Note 42). Incubate the slide-mounted sections for 48 h at 4–8°C in unbiotinylated primary rabbit antiserum against antigen A, diluted in PBS-TX-BSA (*see* Notes 2 and 38).
2. Wash the slides in PBS, 4 × 10 min.
3. Incubate the sections for 1 h at room temperature in FITC-conjugated secondary antiserum (e.g., swine antirabbit-FITC), diluted in PBS-BSA-0.5 according to supplier (commonly about 1:50).
4. Wash the slides in PBS, 4 × 10 min.
5. Fix the sections in 2% paraformaldehyde for 10 min at room temperature (*see* Note 49).
6. Wash the sections in phosphate buffer at room temperature, 2 × 10 min.
7. Wash the slides in Tris buffer at room temperature, 1 × 10 min.
8. Incubate the sections in normal rabbit serum (diluted in PBS 1:30) for 1 h at room temperature to saturate putative remaining antirabbit binding sites.
9. Place the slides into a humid chamber. Incubate the sections for 48 h at 4–8°C in biotinylated primary rabbit antiserum against antigen B diluted in PBS-TX-BSA-0.5 (*see* Notes 38 and 50 concerning biotinylation).

10. Wash the slides in PBS, 4 × 10 min, at room temperature.
11. Incubate the slides for 1 h with streptavidin-Texas red (or other appropriate avidin-fluorophore conjugate), diluted 1:100 in PBS-TX-BSA (or other dilution as recommended by the supplier) (*see* Note 45).
12. Wash the slides thoroughly in PBS.
13. Mount the slide-mounted sections under a coverslip in a mounting medium containing an antifading agent (*see* Notes 5 and 39).

3.4. PAP and ABC Protocols (see Note 51)

3.4.1. Wholemount PAP Technique for Light Microscopy

1. Wash fixed tissue several times using the same buffer as used to dissolve the fixative (*see* Note 52).
2. Wash the tissue overnight in PBS-TX-1.0 at 4–8°C (*see* Note 2).
3. Dehydrate the tissue by passing it through a graded ethanol series (50–100% ethanol; 10 min in each). Then, incubate the tissue in xylene for 10 min, followed by rehydration back to PBS-TX (*see* Note 53).
4. Block endogenous peroxidase activity by immersing the tissue in peroxidase-quenching solution for about 20–30 min (*see* Note 10).
5. Wash the tissue four times (10 min each) in Tris-TX-1.0-Na (*see* Note 4).
6. Incubate the tissue for 3–5 d at 4–8°C in primary antiserum (depending on the size of the tissue and antiserum properties) during gentle shaking. Dilute the antiserum in PBS-TX-NS using normal serum from the same species as secondary antiserum (*see* Notes 2 and 38).
7. Wash the tissue four times in PBS-TX (over at least 1 h) followed by a wash overnight in Tris-Na.
8. Incubate the tissue for 24 h at 4–8°C in unlabeled secondary antibody. Dilute antiserum in PBS-TX-BSA-0.5 (*see* Notes 3 and 38).
9. Wash the tissue in PBS-TX, 4 × 10 min, at room temperature.
10. Incubate the tissue for 24 h at 4–8°C in PAP complex, raised in same species as primary antiserum/antibody (diluted as in step 8).
11. Wash the tissue several times, over 3–4 h, in PBS-TX.
12. Wash the tissue in Tris buffer, 1 × 10 min.
13. Incubate the tissue in the dark for 1 h at 4–8°C in peroxidase substrate, without H_2O_2. **Warning:** DAB in the substrate solution is hazardous (*see* Notes 11 and 54).
14. Perform the peroxidase reaction in fresh peroxidase substrate containing H_2O_2 for 20–30 min, in the dark, at 4–8°C. Progress of the peroxidase reaction can be inspected microscopically. **Warning:** DAB is hazardous (*see* Notes 9, 11, and 55).
15. Wash the tissue thoroughly in Tris buffer.
16. Dehydrate the tissue in graded ethanol series and mount (*see* Note 56).

3.4.2. Preembedding Osmium Intensified PAP Method (see Note 57)

1. Follow steps 1, 2, and 5–15 from Section 3.4.1., above.
2. Wash the whole tissue (or thick vibratome sections) in cacodylate buffer, 1 × 10 min.
3. Fix the tissue for 2 h at 4–8°C in the dark in osmium tetroxide solution.

4. Wash the tissue in cacodylate buffer, 4 × 10 min, at room temperature.
5. Dehydrate and embed in Durcopan ACM or similar resin and cut on sliding microtome at thickness of 25 μm (*see* Note 58 concerning further options).

3.4.3. PAP Technique for Tissue Sections

1. Wash the slide-mounted tissue sections for 10 min in PBS-TX (*see* Notes 2, 41, and 52).
2. Block endogenous peroxidase activity by immersing the sections in peroxidase-quenching solution for about 5 min (*see* Note 10).
3. Wash the sections in Tris-Na, 2 × 10 min, at room temperature (*see* Note 4).
4. Place the slides with sections in a humid chamber. Incubate the sections in primary antiserum/antibody for 16–48 h (depending on the antiserum) at 4–8°C. Dilute the antiserum in PBS-TX-NS using normal serum from the same species as the secondary antiserum (*see* Note 38).
5. Wash the sections in PBS-TX, 3 × 10 min, at room temperature.
6. Incubate the sections for 1 h at room temperature in unlabeled secondary antibody (e.g., goat antirabbit). Dilute the antiserum according to supplier's instructions (typically 1:50–1:100) in PBS-TX-BSA-0.5.
7. Wash the sections in PBS-TX, 3 × 10 min, at room temperature.
8. Incubate the sections for 1 h at room temperature in PAP complex (raised in the same species as the primary antiserum and diluted as in step 6).
9. Wash the sections in PBS-TX, 2 × 10 min, at room temperature.
10. Wash the sections in Tris buffer, 1 × 10 min, at room temperature.
11. Perform the peroxidase reaction in fresh peroxidase substrate with H_2O_2 for 10–15 min at room temperature, in darkness. Progress of the peroxidase reaction can be inspected microscopically. **Warning:** DAB is hazardous (*see* Notes 9, 54, 55, 59, and 60).
12. Wash the sections thoroughly in Tris buffer.
13. Dehydrate in graded ethanol series and mount in Permount or similar mounting medium (*see* Note 56).

3.4.4. ABC Technique for Tissue Sections (see Note 41)

A general outline is given here for using the commercially available Vectastain kit from Vector Laboratories.

1. Steps 1–5 are the same as in Section 3.4.3. The primary antiserum may be diluted to a greater extent than for the PAP method because of the higher sensitivity of the ABC method (*see* Note 61).
2. Incubate the slide-mounted sections in biotinylated secondary antiserum for 30–60 min at room temperature (using a dilution buffer as in Section 3.4.3).
3. Wash the slides in PBS-TX, 2 × 10 min, at room temperature.
4. Incubate the sections in Vectastain ABC reagent (Avidin and Biotinylated horseradish peroxidase macromolecular Complex) for 30–60 min at room temperature.
5. Wash the slides in PBS-TX, 2 × 10 min, at room temperature.

6. Perform steps 10–13 from Section 3.4.3., above. The duration of the peroxidase reaction may have to be shortened due to higher sensitivity of the method.

4. Notes

1. Section 3.1. outlines several protocols for fixation using as an example the fixative paraformaldehyde, and the recipes for the relevant solutions are given. Numerous other fixatives are available—and others indeed may be preferable—depending on the tissue and the characteristics of the antigen. A list of several alternative fixatives and recipes appears in Section 2.5. and appropriate notes relating to their use are provided although use of these alternative fixatives is not explicitly detailed in the protocols.
2. Triton X-100 and NaCl may be added to the buffer to reduce hydrophobic and ionic interactions to minimize unspecific binding of IgGs. One may also choose to raise the pH of these buffers (to between 7.6 and 8.6), which may further reduce unspecific binding. The nomenclature PBS-TX-1.0 refers to PBS containing 1.0% TX-100. PBS-TX contains 0.25% TX-100; the "0.25" is not appended to the name in the interest of clarity. Addition of "-Na" to the buffer name indicates the presence of additional NaCl, which may be added so that a 2–4% final w/v concentration is achieved. Take care to dissolve completely the TX-100 when making the buffer.
3. The BSA content may in some cases be varied to 0.5% (w/v), in which case the buffer is referred to as PBS-BSA-0.5.
4. Tris-TX refers to Tris buffer, as defined in Section 2.3., containing 1% v/v TX-100. Additional comments, found in Note 2, regarding alteration of pH and addition of NaCl are applicable here. Take care to dissolve completely the TX-100 when making the buffer.
5. To reduce fading of fluorophores during microscopy, various antioxidant compounds may be added to the mounting medium. There are several commercially available antifading mounting media, e.g., Vectashield (Vector), Slowfade, and ProLong (Molecular Probes, Eugene, OR). An easily made antifade mounting medium consists of 0.1% (v/v) *p*-phenylenediamine in 90% (v/v) glycerol/10% (v/v) PBS, pH 8.5 *(35)*. It should be noted that different antifading agents differentially affect the various fluorophores in common use, both with respect to fading and fluorescence intensity *(36)*.
6. A large number of fluorophore-tagged secondary antisera raised against rabbit IgGs are available. Swine antirabbit or goat antirabbit coupled to fluorescein isothiocyanate (FITC; green) are available from DAKO Corporation (Copenhagen, Denmark), Jackson Immunoresearch Laboratories Inc. (West Grove, PA), and Southern Biotechnology Associates, Inc. (Birmingham, AL). Swine and goat antirabbit coupled to tetramethylrhodamine isothiocyanate (TRITC; red) are available from the same companies. Goat antirabbit coupled to cyanine 3.18 (Cy-3; red) is available from Molecular Probes.

   Antirabbit F(ab')2 fragments are useful alternatives as secondary antisera. FITC-conjugated swine antirabbit Ig F(ab')2 fragments are available from DAKO

Corporation. Lissamine Rhodamine Sulfonyl Chloride (LRSC; red) conjugated donkey antirabbit Ig F(ab')2 fragments are available from Jackson Immunoresearch Laboratories.

For fluorophore-tagged antimouse secondary antisera (for detection of mouse monoclonals): TRITC-conjugated goat antimouse is available from Sigma, Jackson Immunoresearch Labs, and Southern Biotechnology Associates. FITC-conjugated swine antimouse is available from DAKO; FITC-conjugated goat antimouse may be purchased from Sigma, Jackson Immunoresearch, and from Southern Biotechnology. As alternatives, FITC-conjugated donkey antimouse Ig F(ab')2 fragments and aminomethyl coumarin acetic acid (AMCA; blue) conjugated donkey antimouse Ig F(ab')2 fragments from Jackson Immunoresearch can be used.

For antirat antisera: FITC conjugated goat antirat is available from Sigma; swine antirat FITC may be purchased from DAKO. Affinity-purified LRSC-conjugated donkey antirat Ig (F(ab')2 fragments are available from Jackson Immunoresearch.

We routinely use affinity-purified secondary fluorophore-conjugated antibodies to minimize the risk of crossreactions. A list of additional suppliers is given in Cuello *(26)*.

7. For secondary antisera, one chooses IgGs directed against IgGs of species in which primary antiserum or antibody was raised. Large numbers of different secondary antisera tagged with different fluorophores are commercially available. For conventional single labeling ICC, it is recommended to use secondary antisera with a strongly fluorescent fluorophore, such as fluorescein isothiocyanate (FITC; green), tetramethylrhodamine isothiocyanate (TRITC; red), or cyanine 3.18 (Cy-3; red) (*see also* Note 45).
8. Use of F(ab) and F(ab')2 IgG fragments coupled to fluorophores usually increases the fluorescence signal through:
   a. Better tissue penetration capability of the fragments than of whole IgGs,
   b. Less unspecific tissue binding, and
   c. Less steric hindrance, resulting in a larger number of fluorophores per unit tissue volume.

   We routinely use affinity-purified secondary fluorophore-conjugated antibodies to minimize the risk of crossreactions.
9. Instead of using concentrated (30%) liquid H_2O_2 to make dilute solutions, one can make 0.01% H_2O_2 from Perhydrate tablets (Merck, Darmstadt, Germany) dissolved in water. This eliminates the unpleasant handling of concentrated H_2O_2.
10. Endogenous peroxidase activity may also be quenched by immersing sections in 0.1% phenylhydrazine in PBS for 5 min at room temperature. This is reportedly the most gentle treatment *(37)*, and is thus recommended for labile antigens and cryostat sections.

    Alternatively, an alkaline phosphatase-conjugated antibody may be used. To quench endogenous phosphatase activity, incubate the sample in 20% (v/v) acetic acid in water at room temperature for 5 min. Afterward, rinse the material thoroughly using water.

11. It is important to keep the pH of the DAB solution around 7.6. At lower or higher pH values endogenous oxidases or other heme-containing molecules with pseudo-peroxidase activity in the tissue may oxidize the DAB. The DAB solution should be freshly made and kept in darkness to avoid photo-oxidation (drastic changes in pH, chemical contaminants, and exposure to light leads to discoloration, i.e., oxidation, of DAB solution). The storage of the undiluted DAB is also critical. It is recommended to make small aliquots of DAB (enough for each experiment) at the first opening of the freshly purchased reagent and to store the aliquots in the freezer. Repeated freezing-thawing and opening of container with DAB leads to deterioration of the product. The DAB can be inactivated after use by mixing with sodium hypochlorite (bleach).
12. Surplus picric acid should be removed from specimens fixed in, e.g., Zamboni's and Bouin's fixatives. This may be done by rinsing in several changes of buffer until the buffer no longer turns yellow. If the tissue is to be embedded in paraffin or resin, it is convenient to rinse out the picric acid during dehydration, in 80% ethanol with 10% ammonium acetate.
13. Tissue fixed with glutaraldehyde needs to be treated with $NaBH_4$ for successful detection of neuropeptide immunoreactivity. $NaBH_4$ acts by reduction of residual aldehyde groups (thus reducing nonspecific binding of antibodies to the sections), and by reduction of double bonds in Schiff bases created by glutaraldehyde-protein crosslinking into single bonds (thus partially restoring the tertiary structure of the antigen and the matrix, leading to improved antigen recognition and availability). Treatment with 1% $NaBH_4$ sometimes may be deleterious to the quality of cryostat sections. In this case, try using 0.1% $NaBH_4$ (or even lower concentration) and prolong the treatment.

    Prepare the $NaBH_4$ solution in PBS and immerse the tissue for 30 min at room temperature. Follow this by rinsing the tissue (4 × 15 min) in PBS before proceeding with further processing.
14. The carbodiimide-(EDCDI)-based fixatives should be freshly made. The preservation of tissue fixed in buffered EDCDI alone is poor. Thus, one can either perform a post fixation in 4% buffered paraformaldehyde (Section 2.1., item 5), or fix the tissue in a mixture of paraformaldehyde and EDCDI. Fixation should be overnight (16 h) at 4–8°C.
15. For fixation of neuropeptides, it is recommended to make the buffered EDCDI 4% with respect to paraformaldehyde.
16. PBQ is a bifunctional fixative that reacts with primary amines and amino acids of peptide *(34)*. The PBQ should be used freshly prepared, oxygenated, and at slightly alkaline pH. Fixations should be in the dark at ambient temperature for about 90 min.
17. Vascular perfusion also serves to rinse out erythrocytes, which have endogenous peroxidase activity and, thus, may seriously disturb peroxidase-based detection systems (*see* Section 3.4.). The temperature of the perfusate solution should be at the normal body temperature of the animal or ice-cold. The volume of the perfusate should be larger than that of the total blood volume. Both hydrostatic and

osmotic pressure should ideally be close to the normal blood pressure in the tissue to be fixed, provided that this is known. Inclusion of heparin prevents clotting of the blood, which would impede distribution of fixative. The concentration of heparin may need to be varied depending on the species being perfused.

In our experience, this PBS rinse step may be omitted in fixation of small brains of cold-blooded vertebrates. This approach has the advantage that the duration of deep anesthesia is shorter, i.e., fixation may be commenced more rapidly. However, when using rapidly reacting fixatives as Bouin's fixative or its variants, we have found it important to flush the vascular system with a PBS rinse, even in small animals.

18. All fixative solutions listed in Sections 2.1. and 2.5. may be used for vascular perfusion. To predict the optimal fixative is difficult because it depends on the nature of the antigen to be detected. For some neuropeptides, fixation parameters are not particularly critical, whereas for others, it is necessary systematically to test for optimal conditions.

    In some cases, sequential perfusion with different fixatives has been used to aid preservation of peptide immunoreactivity. Typical examples include sequential perfusion with 4% paraformaldehyde in slightly acidic buffer, pH 6.5, followed by 4% paraformaldehyde in basic buffer, pH 9.0–11.0, or sequential perfusion with carbodiimide followed by paraformaldehyde alone or a carbodiimide/paraformaldehyde mixture *(38)*.

    In order to enhance penetration of the fixative, one may add a detergent (e.g., 0.25% v/v DMSO, 0.25% v/v Triton X-100, or 0.25% v/v Saponin) to the fixation solution. Care should be taken to use a detergent that does not severely affect the structure and retention of the antigen under study.
19. The optimal duration of the postfixation has to be empirically determined for each antigen and each particular tissue. Storage in fixative for too long a period generally causes loss of antigenicity, whereas overly short fixation times leads to inadequate retention of antigen and suboptimal tissue preservation. If long fixation times are required, it is advisable to replace with fresh fixative each day.
20. Similar to the approach using sequential vascular perfusion with different fixatives, it may be useful to postfix the tissue by immersion in fixative other than that used for perfusion. This is certainly adequate when working with small pieces of nervous tissue, provided that the second fixative gains immediate and even access to the tissue. Sequential fixation (4% paraformaldehyde, pH 7.4, followed by 4% paraformaldehyde, pH 9.0) improved histological quality and preservation of enkephalin immunoreactivity in the goldfish retina *(39)*.
21. A timesaving approach, which certainly is useful when there is reason to believe that antigenicity may be lost during the buffer rinses, is to rinse only briefly, section the tissue, and rinse the sections thoroughly. This drastically reduces the time in buffer. If the tissue is to be cryosectioned, one may even add cryoprotectant to the fixative and section the tissue taken directly from the fixative.
22. Storage of tissue is not recommended, but if it is necessary, follow directions in Notes 27 and 28. One may also choose to process the fixed tissue for cryostat-,

paraffin-, or resin sectioning and store the sections at 4–8°C. Resin- and paraffin-embedded tissues can also be stored without sectioning.

23. For wholemounts, it is advisable to remove the neural sheath of nervous tissue before fixation. This can be done either mechanically or chemically. Chemical disruption may be accomplished as follows: Prepare a collagenase/dispase/hyaluronidase solution by dissolving 1 mg of collagenase/dispase and 1 mg of hyaluronidase in 1 mL of 0.05*M* Tris, pH 7.4, containing 0.1% (w/v) $CaCl_2$. Incubate the fixed tissue for 15–30 min at 36°C in the enzyme solution. Wash the tissue thoroughly in Tris-HCl. It is very important to test the duration of the chemical digestion for each tissue to avoid excess damage to underlying neural structures. If this treatment is too harsh, the hyaluronidase can be excluded from the mixture.
24. It has been reported that for a variety of antigens, rapid freezing and fixation by freeze substitution in polar solvents with or without chemical fixatives yields improved retention of antigenicity as well as tissue preservation *(40,41)*. Because the methodology demands special instrumentation for the rapid freezing, and is only applicable on small pieces of tissue, it will probably never be a routine methodology. However, when applicable, the methodology offers exciting prospects. Consider the not uncommon situation, when an antigen can be readily demonstrated with RIA, ELISA, immunoblotting, or other detection systems that recognize the unfixed antigen, but is undetectable in fixed tissue sections, regardless of the chemical fixative used. If it is possible to retain the antigen in the tissue by freeze substitution in a polar solvent, it may be predicted that the antigen in the tissue will be recognized by the antibodies used for RIA, and thus be detectable by immunocytochemistry. For a detailed account of freeze-substitution, *see* Steinbrecht and Zierold *(42)*.
25. Tissue may also be frozen in propane cooled by liquid nitrogen (–180°C).
26. Methanol (100%) may be used as polar solvent during freeze-substitution instead of acetone. The tissue may also be prefixed in aldehyde fixatives, rapidly frozen, and then subjected to freeze-substitution. This approach has proven excellent for precise tissue localization of neuropeptides *(43)*.

    Chemical fixatives may be added to the polar solvent, e.g., 3% v/v glutaraldehyde. Because water-free glutaraldehyde is not available, use high purity 50% glutaraldehyde in sealed vials. The water has then to be absorbed on silica gel, which is added to the mixture *(42)*.
27. For refrigerator storage, PBS may be replaced by other buffers (e.g., sodium phosphate buffer, cacodylate buffer, PIPES buffer), or by buffered Ringer solutions. During long-term storage in refrigerator, the storage buffer should be replaced with fresh buffer at least every 2 wk. Long-term storage (>2 wk) reduces antigenicity, but actual loss has to be determined empirically for each tissue/antigen combination.

    Addition of paraformaldehyde solution (final concentration, 0.4% w/v) or sucrose (final concentration, 5% w/v) to the storage buffer may enhance preservation of histological quality.

28. Sucrose (25%) or glycerol (30–50%) may also be used as cryoprotectants for deep-freeze storage.
29. Tissue can be processed for immunohistochemistry in toto, i.e., as wholemounts, or after histological sectioning. Cryostat sections, vibratome sections, or sections of tissue embedded in paraffin, resin, or other embedding media may be used for immunohistochemistry. Paraffin and resins must be removed from the sections before immunohistochemical processing. Space does not permit detailed descriptions of various embedding and sectioning techniques, and the reader is referred to standard textbooks in histology and electron microscopy. In Note 41, some directions are, however, given on the preparation and treatment of tissue sections.

    The whole tissues or sections often have to be treated in various ways before ICC in order to optimize the immunohistochemical protocol. To enhance antibody penetration and antigen exposure, the tissue may be treated with detergents and/or various proteolytic enzymes. When immunoenzymatic detection methods are to be used, endogenous enzyme activity has to be quenched. Free reactive aldehyde groups in the tissue should be blocked and tight crosslinks (because of, e.g., glutaraldehyde fixation) have to be reduced (*see* Note 13). For wholemounts it may be imperative to remove connective tissue and the neural sheath to enable the immunoglobulins and other reagents to penetrate into the nervous tissue. When mechanical removal is problematic, the enzymatic disruption outlined in Note 23 may be useful.
30. Enzymatic digestion has to be used with caution to reveal neuropeptide antigens. Optimal digestion is usually only achieved within a limited range of enzyme concentrations/digestion durations, and these parameters have to be empirically determined. In some cases, proteolytic digestion may even decrease neuropeptide immunoreactivity *(44)*. Digestion of tissue from cold-blooded vertebrates and invertebrates may be more successfully performed at temperatures matching their normal body temperatures. Proteolytic digestion should preferably be performed during gentle agitation in order to enhance circulation of the enzyme solution.
31. Pretreatments to allow better access of antibodies to antigenic sites are critical when using whole tissues and thick vibratome sections. Some tissues are less penetrable than others as can be revealed empirically.
32. This approach is harsher than the treatment in Section 3.2.2. (option 1), and may be tried when that one does not give satisfactory results.
33. The protocol for long-term storage in 25% DMSO at –70°C (Section 3.1.4.) also works as a mild detergent treatment. When stronger detergent effects are needed, we have successfully submitted wholemounts to overnight incubation in 1% Triton X-100 in 0.01*M* PBS. Note, however, that many detergents (including Triton X-100) reportedly dissolve neuropeptides from fixed tissue.
34. By submitting tissue to repeated freeze-thaw cycles, cell membranes are disrupted and antibody penetration is markedly enhanced. Especially when combined with detergent treatment or inclusion of detergents in the immunohistochemical protocol, freeze-thawing may result in loss of antigenicity and should be used with care.

35. Different detection systems are available for localizing sites of antibody binding, including detection of multiple antigens. These protocols start with fixed tissue that is to be processed either in toto as wholemounts, as free-floating vibratome sections, or as cryostat/paraffin sections (or resin embedded sections that have been etched to remove resin). Here we give some protocols for indirect methods: immunofluorescence and enzyme (e.g., peroxidase-antiperoxidase and ABC) techniques. The reader is referred to refs. *45–47* for additional protocols and variations on the methodology.
36. Another set of useful protocols for wholemount preparations are given by Costa and Furness *(46)*.
37. Step 3 can be supplemented by freezing and thawing the tissue, e.g., on dry ice as outlined in Section 3.2.2., option 4. The xylene can be replaced by n-heptane or acetone.
38. The dilution of the antiserum should be thoroughly tested. Overly high antibody titer may prevent optimal formation of bridges between the primary and secondary antibodies *(48)*. When the antibody titer is too low, there is, of course, weak immunolabeling. For the same reasons, one should also optimize the titer of secondary antisera (and PAP complex).
39. For mounting of whole tissues, one can choose from a variety of techniques. For small or flat tissues, ordinary microscope slides can be used for support and the coverslip mounted with mounting medium directly or after placing "spacers" (e.g., partly straightened out paper clips or strands of Vaseline) around the tissue to prevent its compression under the weight of the coverslip. Using thin sheets of aluminum or brass, holders can be made to the size and thickness of microscope slides. Holes large enough to hold several tissues can be bored through the aluminum slides and a coverslip glued with epoxy to one side of the slide. The cavity now created can be filled with mounting medium, the tissue placed in the cavity, and a second coverslip applied to cover the cavity (which can also be glued if so desired). Thus, the tissue is oriented between two coverslips in an aluminum holder that can be conveniently placed on the microscope stage and viewed from both sides. The same techniques can be used for embedding dehydrated tissue with mounting media of resin type (Permount, Canada Balsam, Durcopan ACM).
40. With this rapid method *(47)*, it is possible to take tissue from the animal to final microscopic analysis in <24 h. The method is especially well suited for smaller tissues, such as flat-mounted intestine and retina, as well as portions of the peripheral nervous system.
41. This method is useful for cryostat sections and paraffin and resin sections, but embedding media must be removed prior to immunocytochemical processing. To remove paraffin from sections, immerse the slides in two changes of xylene (10 and 5 min), then in two changes of 100% ethanol (2 × 5 min), and take the slides through a descending graded ethanol series (96, 70, and 50%) to phosphate-buffered saline. To remove resin (e.g., Epon, Araldite) from sections before immunocytochemical processing, dissolve 40 g of KOH slowly in a mixture of 200 mL of methanol and 100 mL of propylene oxide and filter the solution *(49)*. Immerse

the microscope slides for 6 min (for 5 μm-thick sections) in the mixture. The mixture must be cooled on ice during preparation and during resin removal.

42. Simple humid chambers can be made from large Petri dishes with thin glass rods glued to the bottom for support of microscope slides. Strips of moistened filter paper can be placed between the glass rods. After application of reagents (antisera) on the slides, the Petri dishes are simply covered. Alternatively, flat Perspex boxes with tight-fitting lids can be assembled (or obtained commercially) for the same purpose.
43. Simultaneous immunocytochemical detection of two antigens is a powerful tool for demonstration of colocalization of neuroactive compounds as well as for analysis of neural connectivity and spatial relationships of neuronal systems utilizing different neurotransmitters or neuropeptides. Different fluorophore combinations are useful, and different combinations of primary antisera raised in different animal species can be utilized. Presented is a basic protocol with all incubations performed at room temperature in humid chambers, utilizing primary antibodies/antisera raised in mouse and rabbit.
44. In our hands, this "cocktail" protocol yields the most consistent results. However, when beginning attempts to detect a new combination of antigens, it is important to try a sequential protocol in which first the one antigen is detected as in protocol 3.3.4., the sections rinsed, and then the other antigen is detected again following protocol 3.3.4., but using a different fluorophore. It is, of course, also possible to apply only the primary antibodies as a cocktail, followed by sequential application of fluorophore-conjugated secondary antibodies, or vice versa (*see also* Note 38).
45. Other combinations of fluorophores are suitable for double labeling. It is imperative to choose combinations where the excitation and emission spectra, respectively, of the two fluorophores show minimal overlap, and thus allows selective visualization of one fluorophore at the time. Useful blue-red fluorescence combinations are AMCA (aminomethylcoumarin acetic acid; 350 nm ex, 445 em) or Cascade blue (374, 398 nm ex, 420-nm em) with TRITC (tetramethylrhodamine isothiocyanate; 550 nm ex, 573 em), Cy-3 (cyanine 3.18; 554 nm ex, 566 nm em), LRSC (Lissamine rhodamine sulfonyl chloride; 570 nm ex, 590 nm em), or Texas red (590 nm ex, 610 nm em). Useful green-red fluorescence combinations are FITC (fluorescein isothiocyanate; 490 nm ex, 520 nm em), DTAF (dichlorotriazinyl fluorescein; 490 nm ex, 520 nm em), or BODIPY FL (503 ex, 512 em) with LRSC or Texas red. Double labeling using FITC together with TRITC or Cy-3 should be avoided, since their excitation and emission spectra overlap too much.
46. In our hands the "cocktail" protocol yields the most consistent results. There may be reasons, however, to try sequential visualization; for example, if there is reason to believe that one antigen is present in small amounts and that it may be masked by strong labeling of the other antigens.

    The choice of fluorophore combinations for triple labeling is more restricted than for double labeling. AMCA, FITC (or DTAF or BODIPY FL), and LRSC (or

Texas red) is a good combination, with adequate separation of excitation spectra as well as emission spectra (*see also* Note 47).

47. The choice of fluorophores for confocal laser scanning microscopy (CLSM) should match the emission lines of the laser light source. Moreover, the fluorophores should be stable, i.e., not bleach rapidly, and have a high quantum yield. The air-cooled Ar-ion laser has an emission line at 488 nm that is well-suited for fluorescein derivatives. The water-cooled Ar-ion laser has, in addition to its 488 nm line, lines at 334, 351, and 364 nm that may be used to excite, e.g., AMCA or Cascade blue. However, CLSM with UV light faces special optical problems and is not to be recommended when there is an alternative. The Ar/Kr-ion laser with 488, 568, and 647 nm emission lines is particularly well suited for single, double, or triple fluorophore detection. In our hands, the optimal combination for triple labeling is FITC (DTAF or BODIPY FL), LRSC, and Cy-5 (cyanine 5.18; 649 nm ex, 666 nm em). Ar-ion lasers are not suited for analysis of multiple labeling.
48. Commonly primary antisera are raised in rabbits and for most neuropeptides there are no antisera raised in other animal species. This precludes double and triple labeling with the protocols listed in Sections 3.3.4. and 3.3.5. Three main strategies are available for double labeling with antisera raised in the same species:
    a. Immunolabeling of alternating adjacent sections on different slides with the different antisera;
    b. Direct tagging of the primary antisera with different fluorophores (or enzymes); and
    c. Biotinylation of one of the primary antisera (and detection with streptavidin-fluorophore) in combination with conventional detection of the other primary antiserum.

    The first method relies on cutting cryostat-, paraffin-, or resin-embedded sections that are mounted alternately on two (or more) different microscope slides. Preferably, the sections should have a thickness of about 5 μm or thinner to increase incidence of portions of cell bodies in the adjacent sections. The approach with alternating sections can utilize any standard ICC protocol for each of the sets of sections. One of the advantages with alternating sections is that each slide can be exposed to totally different ICC protocols optimized for the peptides to be detected.

    The second method is also very convenient and rapid once the labeled primary antibodies have been produced. A cocktail of the two (or more) fluorophore-tagged primary antisera is applied to sections or whole tissues, and after washes and mounting they are ready for microscopical analysis.

    The third approach, which employs a biotinylated primary antiserum *(50)* in combination with unlabeled primary antiserum (or antisera), is quite versatile *(51)*, and is described in detail in Section 3.3.6.
49. The application of 2% paraformaldehyde is to further link the swine antirabbit-FITC (or other fluorophore-tagged IgG) to its binding site.
50. The following protocol can be used for biotinylation of primary antibodies (from ref. *22*):

a. Prepare a solution of 10 mg of biotinamidocaproate N-hydroxy succinimide ester (Sigma) in 1 mL of dimethyl sulfoxide.
b. Dissolve 1–3 mg of antibody/mL of sodium borate buffer (0.1*M*, pH 8.8).
c. Add the biotin ester to antibody solution at a ratio of about 25–250 μg of ester/mg of antibody. Mix well and incubate at room temperature for 4 h.
d. Add 20 μL of 1*M* NH_4Cl/250 μg of biotin ester. Incubate for 10 min at room temperature.
e. Dialyze the antibody solution against PBS to remove uncoupled biotin. Extensive dialysis is required.
f. The biotinylated antibody is stored and used as regular antibody.

Although the biotinylation is harmless to the binding properties of the IgGs, the biotinylated antibodies may have to be applied in a higher titer than unbiotinylated ones (a loss of titer occurs during the biotinylation process).

51. Instead of using fluorophore-based detection systems, one can apply the very sensitive detection systems relying on the enzyme marker horseradish peroxidase *(52)*. Also, other enzyme markers, alkaline phosphatase, and glucose oxidase, and particle-based markers, such as ferritin and colloidal gold, are alternatives and provide complementary detection systems useful for detection of multiple antigens. Here we shall deal only with those techniques that employ peroxidase markers.

    There are commercially available enzyme-tagged secondary antisera, which in most cases are quite useful. It is, however, more common to employ three-step methods like the PAP technique *(53)* and the ABC method *(54)*. The latter utilizes biotin-streptavidin bridging (IgGs of secondary antiserum are biotinylated). Peroxidase detection systems have some advantages over immunofluorescence: The sensitivity of detection is superior, the resulting preparations are permanent and can be viewed repeatedly at convenience even after long storage (no fading of label), and with some modifications, enzyme detection can be used for electron microscopy. We provide three PAP protocols and one ABC protocol. One of the PAP protocols can, with some modifications, be utilized for electron microscopy.

    These methods may also be used to preserve specimens that have been used for immunofluorescence. Remove the coverslip, rinse well in PBS-TX, and subject the sections to steps 7–13 in Section 3.4.3. Obviously, the PAP complex has to be raised in the same species as the primary antibodies. The fluorophore-conjugated secondary antibodies have to be whole IgGs or F(ab')2 fragments; F(ab) fragments cannot bind the PAP complex.
52. Tissues should be well dissected with neural sheath removed surgically or enzymatically (*see* Note 23).
53. This step can be omitted if preservation of antigenicity is very critical. Instead the freeze-thaw treatment of Section 3.2.2. (option 4) can be tried out.
54. The preincubation step in DAB solution is important since the tissue penetration rate of DAB is very slow *(55)*, and it is desirable to infiltrate the tissue completely with DAB before exposure to H_2O_2.

55. The DAB reaction product may be intensified, thus allowing a more sensitive immunocytochemical detection. There are several approaches, most of them based on heavy metal enhancement of the reaction. Here, we mention only modifications of two widely used techniques.
    a. Osmium intensification: After the DAB reaction, rinse the sections in PBS and postfix in 0.1% osmium tetroxide in cacodylate buffer, 1–5 min at room temperature.
    b. Nickel intensification: Here, Ni^{2+} ions are included in the DAB reaction medium: 0.025% DAB, 0.25% $(NH_4)_2Ni(SO_4)_2$, and 0.015% hydrogen peroxide, in 0.05*M* Tris buffer, pH 7.6. The reaction product is blue-black.

    Note that inclusion of Ni^{2+} ions dramatically increases the sensitivity. The titer of primary antibodies, PAP complex, DAB, and hydrogen peroxide should be lower than when using a standard DAB incubation: Approximately half the titer of each reagent is a good starting point. Optimal intensification without increase in background labeling usually requires careful adjustment of these parameters. Co^{2+} ions (from, e.g., $CoCl_2$) yields similar signal enhancement.
56. Mounting of immuno-PAP wholemounts can be done as in Note 39 using glycerin, methylsalicylate, Permount, or resin, such as Durcopan ACM (depending on whether they are to be permanently mounted). Metal holders with coverslips on each side (outlined in Note 39) are practical since preparations can be viewed from both sides. Glycerin or methylsalicylate are useful nonpermanent mounting media if one wants to reorient the preparations or take them to subsequent treatments.
57. This very sensitive method produces preparations with high contrast and good resolution of very thin neural processes. The tissue preservation is also very good with a high contrast as a result of osmium staining of membranes. The method takes advantage of the osmophilic nature of the DAB reaction product *(55)* and the fixative properties of osmium tetroxide. With some modifications, the technique can be employed for electron microscopy *(56,57)*. If the tissue is intended for electron microscopy (EM) analysis the tissue should be fixed in 4% paraformaldehyde with 0.1% glutaraldehyde and 3–5% sucrose in 0.1*M* sodium phosphate buffer. Furthermore, for EM, one should skip steps 2–5 in Section 3.4.1., and also minimize the exposure to Triton X-100 (lower its concentration or remove it completely). For EM it is recommended to start with fixed vibratome sections rather than whole tissues. The protocol is based on Nässel and O'Shea *(58)*.
58. Sections can be stretched in xylene and slide-mounted under coverslips with Permount or similar mounting medium. If intended for EM use, the sections can be mounted in resin, such as Durcopan ACM, between sheets of overhead film (acetate); after light microscopic inspection they can be remounted in resin and cut ultrathin for EM.
59. A rapid and sensitive alternative to the PAP/ABC protocols presented in Section 3.4. is to use peroxidase-conjugated F(ab) or F(ab')2 fragments of IgGs recognizing the primary antibodies. In our hands, this approach has been especially valuable for wholemounts and thick vibratome sections in which antibody penetration is limited.

60. It is possible simultaneously to visualize two antigens with immunoenzymatic protocols. The two basic approaches are visualizing one antigen with a peroxidase-conjugate and the other with an alkaline phosphatase-conjugate, with differently colored reaction products of the two enzyme reactions *(59)*; and visualizing both antigens with peroxidase conjugates, one with a conventional DAB reaction and the other with a Ni-enhanced DAB reaction (*60*; *see also* ref. *30* for further combinations).
61. Although it is generally recognized that ABC protocols allow more sensitive detection of antigens than PAP protocols, we have encountered specific situations for which this does not hold true. In cryostat sections of paraformaldehyde-fixed CNS tissue, utilizing whole IgG primary antibodies followed by any avidin-biotin-based detection protocol, unspecific tissue labeling is consistently very high. After careful comparison of various avidin-biotin protocols and systematic adjustment of blocking procedures and the protocols, we were able to abolish the unspecific labeling, but only at the price of sensitivity: A conventional PAP protocol was then at least as sensitive and yielded more consistent results. The problem may be because of unspecific binding by the Fc fragment, which is strongly amplified by the sensitive ABC technique. However, in the same tissue and using the same primary antibodies, excellent results were obtained with ABC protocols on paraffin sections.

References

1. Guillemin, R. (1978) Peptides in the brain: the new endocrinology of the neurone. *Science* **202,** 390–402.
2. Hökfelt, T., Johansson, O., Ljungdahl, Å., Lundberg, J. M., and Schultzberg, M. (1980) Peptidergic neurones. *Nature* **284,** 515–521.
3. Krieger, D. T. (1983) Brain peptides: what, where, why? *Science* **222,** 975–985.
4. Lundberg, J. M. and Hökfelt, T. (1983) Coexistence of peptides and classical neurotransmitters. *Trends Neurosci.* **6,** 325–333.
5. Hökfelt, T. (1991) Neuropeptides in perspective: the last ten years. *Neuron* **7,** 867–879.
6. Van Leeuwen, F. (1987) Immunocytochemical techniques in peptide localization. Possibilities and pitfalls, in *Neuromethods 6. Peptides* (Boulton, A. A., Baker, G. B., and Pittman, Q. G., eds.), Humana, Clifton, NJ, pp. 73–111.
7. Larsson, L. I. (1993) Antibody specificity in immunocytochemistry, in *Immunohistochemistry II.* IBRO Handbook Series: Methods in Neurosciences, vol. 14 (Cuello, A. C., ed.), Wiley, Chichester, UK, pp. 79–106.
8. Burrin, D. H. (1986) Immunochemical techniques, in *A Biologist's Guide to Principles and Techniques of Practical Biochemistry* (Wilson, K. and Goulding, K. H., eds.), Edward Arnold, London, UK, pp. 116–152.
9. Schooneveld, H., Romberg-Privee, H. M., and Veenstra, J. A. (1986) Immunocytochemical differentiation between adipokinetic hormone (AKH)-like peptides in neurons and glandular cells in the corpus cardiacum of *Locusta migratoria* and

Periplaneta americana with C-terminal and N-terminal specific antisera to AKH. *Cell Tissue Res.* **243,** 9–14.
10. Diederen, J. H. B., Maas, H. A., Pel, H. J., Schooneveld, H., Jansen, W. F., and Vullings, H. G. B. (1987) Co-localization of the adipokinetic hormones I and II in the same glandular cells and in the same granules of corpus cardiacum of *Locusta migratoria* and *Schistocerca gregaria*. *Cell Tissue Res.* **249,** 379–389.
11. McCormick, J. and Nichols, R. (1993) Spatial and temporal expression identify dromyosuppressin as a brain-gut peptide in *Drosophila melanogaster*. *J. Comp. Neurol.* **338,** 279–288.
12. Tibbetts, M. F. and Nichols, R. (1993). Immunocytochemistry of sequence-related neuropeptides in *Drosophila*. *Neuropeptides* **24,** 321–325.
13. Boersma, W. J. A., Haaijman, J. J., and Claasen, E. (1993) Use of synthetic peptide determinants for the production of antibodies, in *Immunohistochemistry II*. IBRO Handbook Series: Methods in Neurosciences, vol. 14 (Cuello, A. C., ed.), Wiley, Chichester, UK, pp. 1–77.
14. Posnett, D. N. and Tam, J. P. (1989) Multiple antigenic peptide method for producing antipeptide site specific antibodies, in *Methods in Enzymology* (Langone, J. J., ed.), Academic, New York, pp. 739–746.
15. Hekimi, S. and O'Shea, M. (1989) Antisera against AKHs and AKH precursors for experimental studies of an insect neurosecretory system. *Insect Biochem.* **19,** 79–83.
16. Schneider, L. E., Sun, E. T., Garland, D. J., and Taghert, P. H. (1993) An immunocytochemical study of the FMRFamide neuropeptide gene products in *Drosophila*. *J. Comp. Neurol.* **337,** 446–460.
17. Köhler, G. and Milstein, C. (1976) Derivation of specific antibody producing tissue culture and tumor lines by cell fusion. *Eur. J. Immunol.* **6,** 511–522.
18. Cuello, A. C. and Côté, A. (1993) Preparation and application of conventional and non-conventional monoclonal antibodies, in *Immunohistochemistry II*. IBRO Handbook Series: Methods in Neurosciences, vol. 14 (Cuello, A. C., ed.), Wiley, Chichester, UK, pp. 107–145.
19. Eckert, M. and Ude, J. (1983) Immunocytochemical techniques for demonstration of peptidergic neurons, in *Functional Neuroanatomy* (Strausfeld, N. J., ed.), Springer, Berlin, pp. 267–301.
20. Benoit, R., Ling, N., Brazeau, P., Lavielle, S., and Guillemin, R. (1987) Peptides. Strategies for antibody production and radioimmunoassays, in *Neuromethods 6. Peptides* (Boulton, A. A., Baker, G. B., and Pittman, Q. G., eds.), Humana, Clifton, NJ, pp. 43–72.
21. Catty, D. and Raykundalia, C. (1988) Production and quality control of polyclonal antibodies, in *Antibodies: A Practical Approach*, vol. 2 (Catty, D., ed.), IRL, Oxford, UK, pp. 19–79.
22. Harlow, E. and Lane, D. (1988) *Antibodies: A Laboratory Manual*. Cold Spring Harbor Laboratory, Cold Spring Harbor, NY.
23. Claassen, E., Zegers, N. D., Laman, J. D., and Boersma, W. J. A. (1993) Use of synthetic peptides for the production of site (amino acid) specific polyclonal and monoclonal antibodies, in *Generation of Antibodies by Cell and Gene Immortal-*

ization: Immunology, vol. 7 (Terhorst, C., Malavasi, F., and Albertini, A., eds.), Karger, Basel, pp. 150–161.
24. Nässel, D. R. (1996) Advances in the immunocytochemical localization of neuroactive substances in the insect nervous system. *J. Neurosci. Methods*, in press.
25. Cuello, A. C., ed. (1983) *Immunohistochemistry*. IBRO Handbook Series: Methods in Neurosciences, vol. 3, Wiley, Chichester, UK.
26. Cuello, A. C., ed. (1993) *Immunohistochemistry II*. IBRO Handbook Series: Methods in Neurosciences, vol. 14, Wiley, Chichester, UK.
27. Elde, R. (1983) Immunocytochemistry, in *Brain Peptides*. (Krieger, D. T., Brownstein, M. J., and Martin, J. B., eds.), Wiley, New York, pp. 485–494.
28. Polak, J. M. and Van Norden, S., eds. (1983) *Immunocytochemistry. Practical Applications in Pathology and Biology*. Wright, Bristol, UK.
29. Beesley, J. E., ed. (1993) *Immunocytochemistry. A Practical Approach*. IRL, Oxford, UK.
30. Patel, N. H. (1994) Imaging neuronal subsets and other cell types in whole-mount *Drosophila* embryos and larvae using antibody probes, in Drosophila melanogaster: *Practical Uses in Cell and Molecular Biology*. Methods in Cell Biology Series, vol. 44 (Goldstein, L. S. B. and Fyrberg, E. A., eds.), Academic, San Diego, CA, pp. 445–487.
31. Stefanini, M., de Martino, C., and Zamboni, L. (1967) Fixation of ejaculated spermatozoa for electron-microscopy. *Nature* **216,** 173,174.
32. Zamboni, L. and de Martino, C. (1967) Buffered picric-acid formaldehyde: a new rapid fixative for electron-microscopy. *J. Cell Biol.* **35,** 148A.
33. Boer, H. H., Schot, L. P. C., Roubos, E. W., Maat, A., Lodder, J. C., Reichelt, D., and Swaab, D. F. (1979) ACTH-like immunoreactivity in two electrotonically coupled giant neurons in the pond snail *Lymnaea stagnalis*. *Cell Tissue Res.* **202,** 231–240.
34. Bu'Lock, A. J., Vaillant, C., Dockray, G. J., and Bu'Lock, J. D. (1982) A rational approach to the fixation of peptidergic nerve cell bodies in the gut using parabenzoquinone. *Histochem.* **74,** 49–55.
35. Platt, J. L. and Michael, A. F. (1983) Retardation of fading and enhancement of intensity of immunofluorescence by *p*-phenylenediamine. *J. Histochem. Cytochem.* **31,** 840–842.
36. Florijn, R. J., Slats, J., Tanke, H. J., and Raap, A. K. (1995) Analysis of antifading reagents for fluorescence microscopy. *Cytometry* **19,** 177–182.
37. Ormerod, M. G. and Imrie, S. F. (1992) Enzyme-antienzyme method for immunohisto-chemistry, in *Methods in Molecular Biology, vol. 10: Immunochemical Protocols* (Manson, M. M., ed.), Humana, Totowa, NJ, pp. 117–124.
38. Moffett, J. R., Namboodiri, M. A. A., and Neale, J. H. (1993). Enhanced carbodiimide fixation for immunohistochemistry—application to the comparative distributions of N-acetylaspartylglutamate and N-acetylaspartate immunoreactivities in rat brain. *J. Histochem. Cytochem.* **41,** 559–570.
39. Eldred, W. D., Zucker, C., Karten, H. J., and Yazulla, S. (1983) Comparison of fixation and penetration enhancement techniques for use in ultrastructural immunocytochemistry. *J. Histochem. Cytochem.* **31,** 285–292.

40. Murray, G. J. (1992) Enzyme histochemistry and immunohistochemistry with freeze-dried or freeze-substituted resin embedded tissue. *Histochem. J.* **24,** 399–408.
41. Yamashita, S. and Yasuda, K. (1992) Freeze-substitution fixation for immunohistochemistry at the light microscopic level: effects of solvent and chemical fixatives. *Acta Histochem. Cytochem.* **25,** 641–650.
42. Steinbrecht, R. A. and Zierold, K. (1987) *Cryotechniques in Biological Electron Microscopy*. Springer-Verlag, Berlin, Heidelberg.
43. Zandbergen, M. A., Peute, J., Verkley, A. J., and Goos, H. J. Th. (1992) Application of cryosubstitution in neurohormone- and neurotransmitter-immunocytochemistry. *Histochemistry* **97,** 133–139.
44. Finley, J. C. V. and Petrusz, P. (1982) The use of proteolytic enzymes for improved localization of tissue antigens with immunocytochemistry, in *Techniques in Immunocytochemistry* (Bullock, G. R. and Petrusz, P., eds.), Academic, London, UK, pp. 239–249.
45. Costa, M. and Furness, J. B. (1983) Immunohistochemistry on whole mount preparation, in *Immunohistochemistry*. IBRO Handbook Series: Methods in Neurosciences, vol. 3 (Cuello, A. C., ed.), Wiley, Chichester, UK, pp. 373–397.
46. Côté , A., Ribeiro-Da-Silva, A., and Cuello, A. C. (1993) Current protocols for light microscopy immunocytochemistry, in *Immunohistochemistry II*. IBRO Handbook Series: Methods in Neurosciences, vol. 14 (Cuello, A. C., ed.), Wiley, Chichester, UK, pp. 147–168.
47. Klemm, N., Hustert, R., Cantera, R., and Nässel, D. R. (1986) Neurons reactive to antibodies against serotonin in the stomatogastric nervous system and in the alimentary canal of locust and crickets (Orthoptera, insecta). *Neurosci.* **17,** 247–261.
48. Bigbee, J. W., Kosek, J. C., and Eng, L. F. (1977) Effects of primary antiserum dilution on staining of antigen rich tissues with the peroxidase anti-peroxidase technique. *J. Histochem. Cytochem.* **25,** 443–447.
49. Maxwell, M. H. (1978) Two rapid and simple methods used for the removal of resins from 1.0 μm-thick epoxy sections. *J. Microscop.* **112,** 253–255.
50. Bayer, E. A., Skutelsky, E., and Wilchek, M. (1980) The avidin-biotin complex in affinity cytochemistry. *Methods Enzymol.* **62,** 308–315.
51. Würden, S. and Homberg, U. (1993) A simple method for immunofluorescent double staining with primary antisera from the same species. *J. Histochem. Cytochem.* **41,** 627–630.
52. Nakane, P. K. (1971) Application of peroxidase-labeled antibodies to intracellular localization of hormones. *Acta Endocrinol. (Kbh.) Suppl.* **153,** 190–204.
53. Sternberger, L. A. (1979) *Immunocytochemistry*. 2nd ed. Wiley, New York.
54. Hsu, S. M., Raine, L. and Fanger, H. (1981) The use of avidin-biotin-peroxidase complex (ABC) in immunoperoxidase techniques: a comparison between ABC and unlabeled antibody (PAP) procedures. *J. Histochem. Cytochem.* **29,** 577–580.
55. Nässel, D. R. (1983) Horseradish peroxidase and other heme proteins as neuronal markers, in *Functional Neuroanatomy* (Strausfeld, N. J., ed.), Springer, Berlin, pp. 44–91.

56. Nässel, D. R. and Elekes, K. (1984) Ultrastructural demonstration of serotonin immunoreactivity in the nervous system of an insect (*Calliphora erythrocephala*). *Neurosci. Lett.* **48,** 203–210.
57. Somogyi, P., Hodgson, A. J., and Smith, A. D. (1979) An approach to tracing neuron networks in the cerebral cortex and basal ganglia. Combination of Golgi-staining, retrograde transport of horseradish peroxidase and anterograde degeneration of synaptic boutons in the same material. *Neuroscience* **4,** 1805–1852.
58. Nässel, D. R. and O'Shea, M. (1987) Proctolin-like immunoreactive neurons in the blowfly central nervous system. *J. Comp. Neurol.* **265,** 437–454.
59. Green, J. A. and Manson, M. M. (1992) Double label immunohistochemistry on tissue sections using alkaline phosphatase and peroxidase conjugates, in *Methods in Molecular Biology, vol. 10: Immunochemical Protocols* (Manson, M., ed.), Humana, Totowa, NJ, pp. 125–129.
60. Boorsma, D. M. and Steinbusch, H. W. M. (1988) Immunoenzyme double staining, in *Molecular Neuroanatomy* (Van Leeuwen, F. W., Buijs, R. M., Pool, C. W., and Pach, O., eds.), Elsevier Science, Amsterdam, pp. 289–299.

8

Immunocytochemical Detection of Amino Acid Neurotransmitters in Paraformaldehyde-Fixed Tissues

David V. Pow

1. Introduction

It is generally thought that up to 90% of all synapses in the mammalian brain use amino acids as neurotransmitters *(1)*. Neurotransmitter amino acids may either be those that are also used directly as protein constituents (such as glutamate) or modified amino acids, such as γ-amino butyric acid (GABA). In addition, there are other amino acid-derived transmitters, such as the tryptophan derivative 5-hydroxytryptamine (serotonin), the tyrosine-derived transmitters, such as dopamine, and the histidine derivative, histamine. Although these latter groups are not normally considered to be amino acid neurotransmitters *per se* (because of the removal of the acidic carboxyl group), the methodologies outlined in this chapter are still directly applicable to their immunolocalization.

The general strategies in performing immunocytochemistry of amino acids are essentially the same as those for any other molecule. Immunolabeling may be performed on sections of embedded tissues, on frozen or Vibratome sections, and on wholemounts, with each approach having its own advantages and disadvantages.

Amino acids are small and highly mobile molecules, thus their efficient fixation in tissues is an essential prerequisite for their subsequent immunocytochemical detection. The main fixatives that can be used for immobilizing amino acids in tissues are glutaraldehyde, carbodiimide, and formaldehyde (formaldehyde is referred to here as paraformaldehyde, to distinguish it from formalin). Glutaraldehyde has traditionally been the fixative of choice for retaining amino acids in tissues prior to their immunocytochemical detection. Unfortunately, glutaraldehyde is incompatible with immunofluorescence-

From: *Methods in Molecular Biology, Vol. 72: Neurotransmitter Methods*
Edited by: R. C. Rayne Humana Press Inc., Totowa, NJ

labeling techniques, because of induced tissue autofluorescence. In addition, the heavy crosslinking of tissues produced by this dialdehyde results in poor penetration of immunoreagents into the fixed tissues. Moreover, in studies where fixation-sensitive antigens (such as neurotransmitter receptors) or histochemical reactions (such as the NADPH-diaphorase reaction) are also being investigated, glutaraldehyde is usually contraindicated. Carbodiimide has been used as a mild fixative for the immunocytochemical detection of glutamate *(2)*, but has not been more widely applied. Paraformaldehyde has received little attention as a fixative for retaining amino acids in immunocytochemical studies despite its efficacy as a fixative for retaining molecules, such as radiolabeled proline (in autoradiographic studies), serotonin (a tryptophan derivative), noradrenaline, and dopamine (tyrosine derivatives) *(3)*.

Antibodies to amino acids are generally raised by chemically coupling the amino acid to a carrier protein. Antibodies raised against such conjugates are specific for the amino acid plus the linking agent, and do not normally recognize the free amino acid *(4)*; thus the chemical used to couple the conjugate for immunization is an essential part of the immunoreactive epitope. Accordingly, the first essential feature in determining the success of immunocytochemistry for amino acids is to match the fixation paradigm used to prepare tissues with that originally used to generate the antibodies. Our recent studies *(5)* have demonstrated that all the classical amino acid transmitters can be efficiently fixed and retained in tissues when paraformaldehyde is used as a fixative.

Immunolabeling for amino acids in paraformaldehyde-fixed tissues is usually relatively straightforward, although requiring compliance with a few basic rules, particularly to use freshly prepared fixative and to keep the tissues cool. Its main uses include colocalization studies in which fixation-sensitive antigens are being examined and in which penetration of antibodies might otherwise be difficult. More recently, in our laboratory, the technique has been widely used in conjunction with intracellular injections of the tracer Neurobiotin to permit the neurochemical identity of morphologically defined classes of neurons to be elucidated *(6)*. This chapter outlines the specific procedures required to optimize the immunocytochemical detection of amino acids in paraformaldehyde-fixed tissues. Figures 1–3 illustrate immunolabeling for amino acids in a variety of paraformaldehyde-fixed tissues. An overall schemata is depicted in Fig. 4.

2. Materials

2.1. Raising Antisera

1. Phosphate buffer A: 0.1*M* sodium phosphate, pH 7.2.
2. Phosphate buffer B: 0.2*M* sodium phosphate, pH 7.2.
3. Sodium hydroxide solution: 5*M* sodium hydroxide in distilled water.

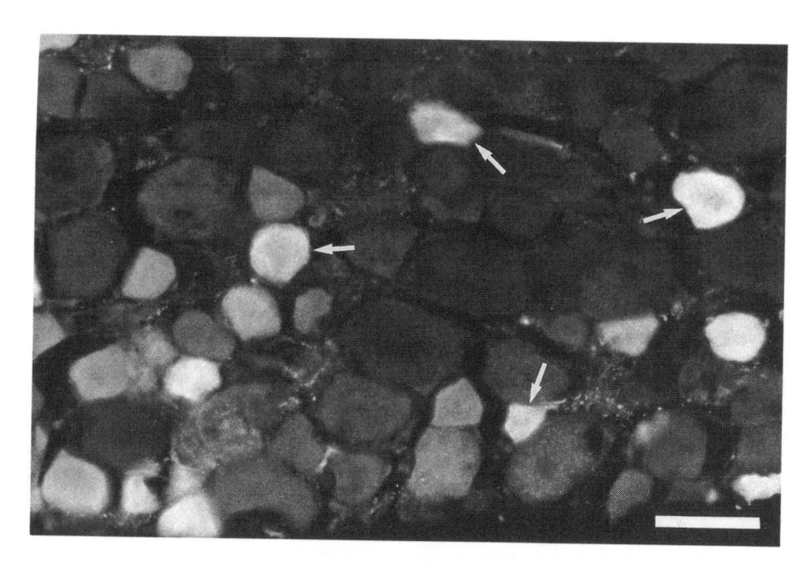

Fig. 1. Aspartate-immunoreactive somata (arrows) in a section of a rat L6 dorsal root ganglion, labeled using a fluorescently tagged secondary antibody. The small, brightly labeled cells are predicted to supply the viscera. Scale bar, 50 μm. Micrograph kindly provided by Janet Keast, University of Queensland.

4. Paraformaldehyde fixative (*see* Note 1): 4% (w/v) paraformaldehyde (extra pure, Merck, Darmstadt, Germany) in 0.1*M* phosphate buffer, pH 7.2–7.3. Add 16 g of paraformaldehyde to 100 mL of distilled water and warm to 60°C, with stirring. Slowly add drops of sodium hydroxide solution until the paraformaldehyde dissolves (depolymerizes) completely. Add distilled water, bringing the volume to 200 mL. To this add 200 mL of phosphate buffer B. The fixative must be prepared in and only used in a fume hood because it is toxic by inhalation and skin contact.
5. Freund's adjuvant: complete and incomplete (Sigma, St. Louis, MO).
6. Syringe and hypodermic (19-gage) needles.
7. Gold chloride solution: 1% (w/v) gold chloride (Sigma, G4022) in double-distilled water.
8. Sodium citrate solution: 1% (w/v) sodium citrate in double-distilled water.
9. Colloidal gold solution: Add 1 mL of gold chloride solution to 80 mL of double-distilled water, and warm to 60°C. Quickly add 1 mL of sodium citrate solution while stirring. Boil the solution until the volume has been reduced to approx 10 mL; allow to cool. This will yield approx 50-nm diameter gold particles. The pH of the colloidal gold solution should be about 6.5. Adjust the pH with small amounts of 1*M* NaOH or 1*M* HCl as required.
10. Thyroglobulin solution: 6 mg/mL thyroglobulin in 0.1*M* sodium phosphate, pH 7.2. Dissolve 30 mg of porcine thyroglobulin (Sigma, T1126) in 5 mL of phosphate buffer A.

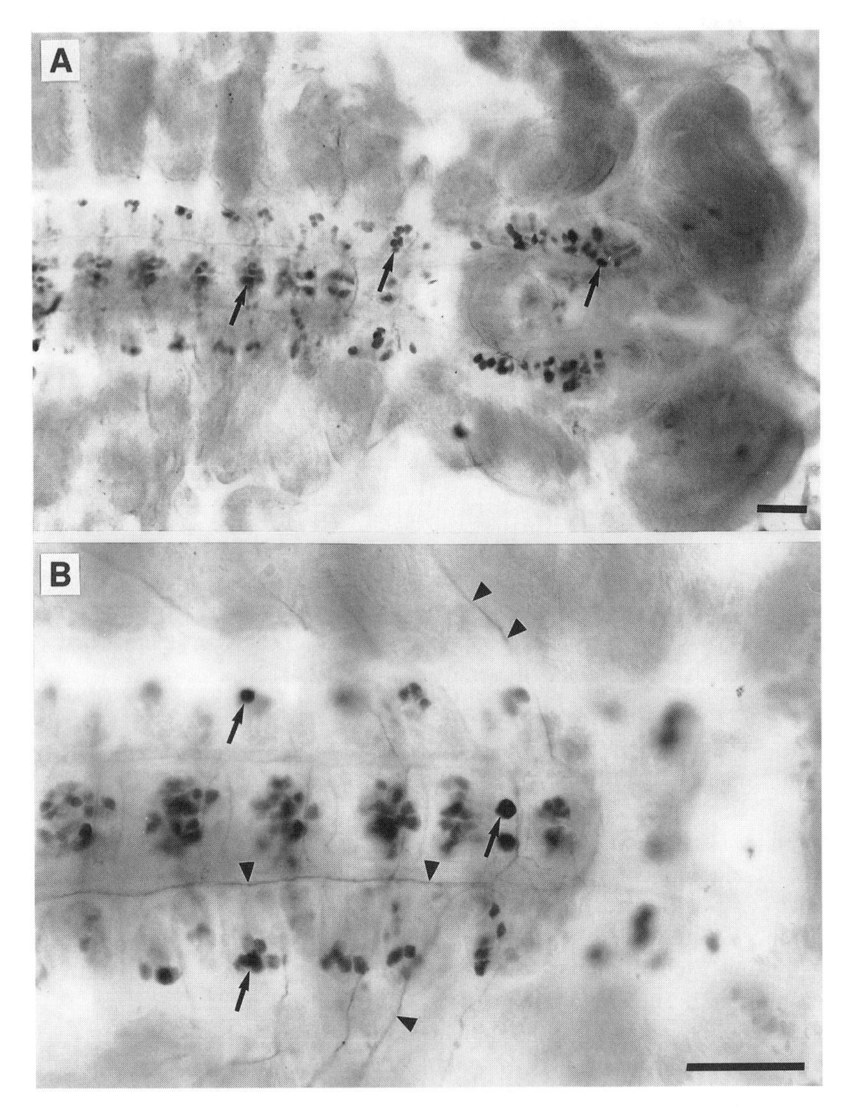

Fig. 2. Low **(A)** and high **(B)** magnification views of a wholemount of a *Chorex destructor* ("Yabbie") embryo (65% development), immunolabeled for GABA, using DAB as a chromogen. Arrows indicate labeled neuronal somata; arrowheads indicate labeled axons. Many labeled elements are out of the plane of focus of the microscope. Scale bars, 100 µm. Micrographs kindly provided by Renate Sandeman, University of New South Wales.

11. Colloidal gold/muramyl dipeptide solution: To each milliliter of colloidal gold solution add 20 µg of muramyl dipeptide (adjuvant peptide; Sigma, A9519). Stored at 4°C the solution is stable for several weeks.

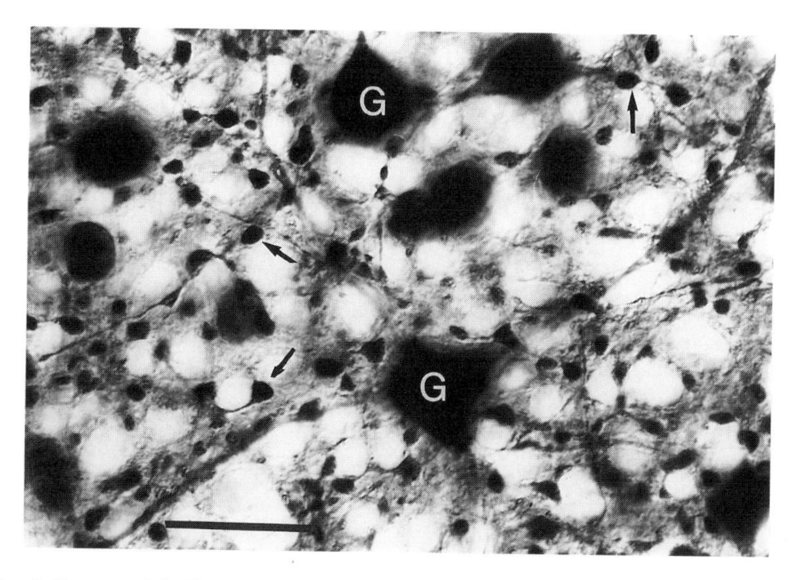

Fig. 3. Immunolabeling of a retinal wholemount for glutamate. Labeling is associated with somata of glutamatergic ganglion cells (G) and the glutamatergic bipolar cell terminals (arrows). Scale bar, 25 μm.

12. Amino acids: Purchased from Sigma, as required. If possible use the hydrochloride salts in preference to the free bases (the former are generally easier to dissolve).

2.2. Tissue Fixation

1. Physiological saline: May be Ames media or other suitable media. Dissolve the Ames media powder (Sigma, A1420) in double-distilled water to a final volume of 1 L. Add 1.9 g of $NaHCO_3$ and bubble the solution with 5% carbon dioxide, 95% oxygen.
2. Paraformaldehyde fixative (*see* Note 1): *See* Section 2.1., item 4.
3. Sodium pentobarbital (Nembutal; Boehringer Ingelheim).
4. 2-way Luer-lock plastic tap (Sigma, S7521).
5. Plastic tubing: Tygon, internal diameter 1/8 in. (Sigma, T2539).
6. Cannula: 12-gage hypodermic needle (internal bore, 2 mm; length 50 mm).

2.3. Immunocytochemistry

1. Phosphate-buffered saline (PBS): 0.9% (w/v) sodium chloride in 0.1*M* phosphate buffer, pH 7.2. Dissolve 9 g of NaCl in 1 L of phosphate buffer A (*see* Section 2.1., item 1). Store at 4°C for up to 1 wk.
2. Sucrose solution: 20% (w/v) sucrose in 0.1*M* sodium phosphate buffer, pH 7.2. Dissolve 20 g of sucrose in 100 mL of 0.1*M* sodium phosphate buffer (*see* Section 2.1., item 1). Store at 4°C for up to 1 wk.

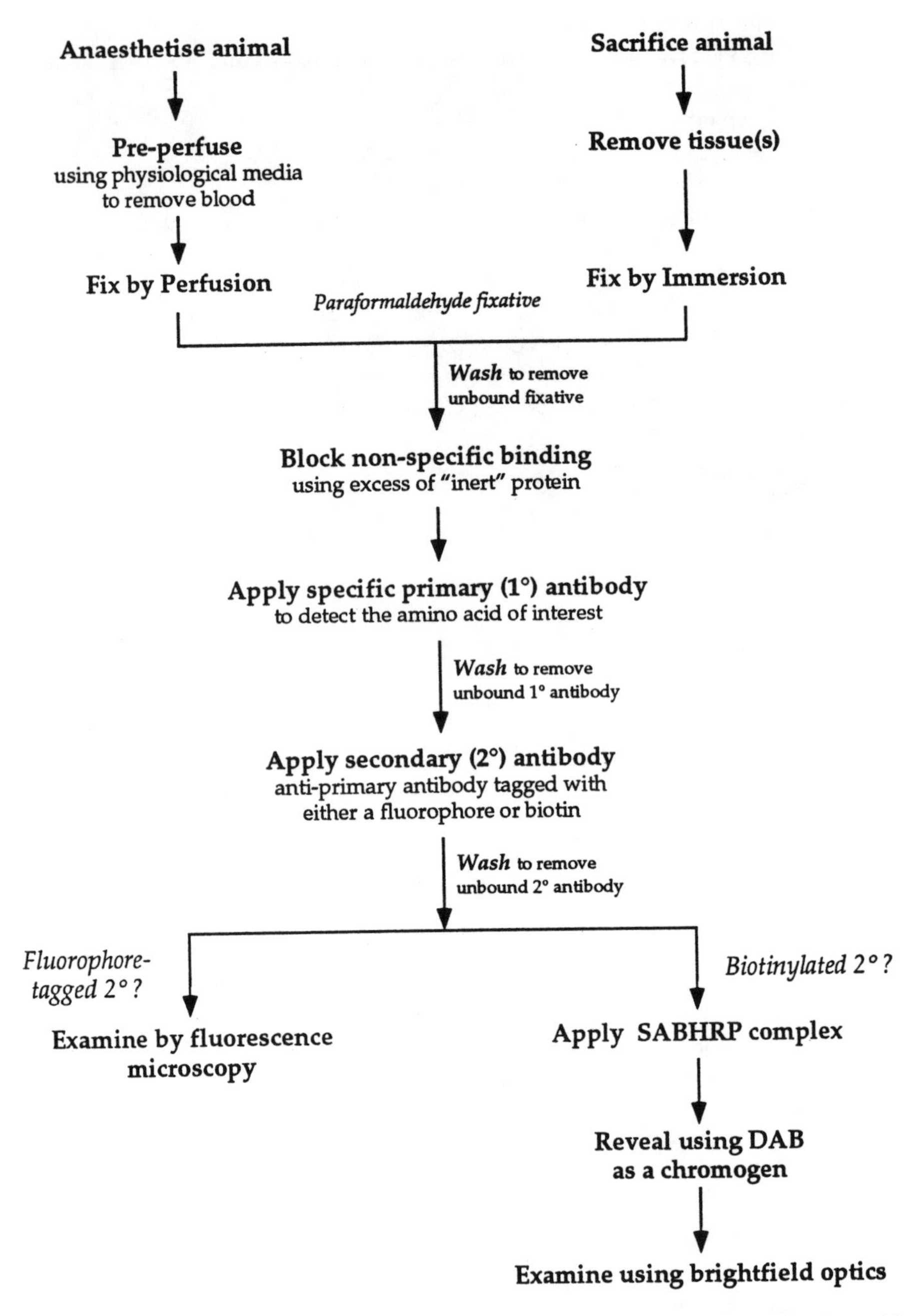

Fig. 4. A summary of steps involved in the immunocytochemical detection of amino acids in paraformaldehyde-fixed tissues. Details are given in the protocols and Notes. SABHRP = streptavidin-biotin-horseradish peroxidase complex; DAB = 3,3' diaminobenzidine.

3. Blocking solution (PBS-BSA): 0.5% (w/v) bovine serum albumin (BSA; Sigma, A3425) in PBS. Dissolve 0.5 g of BSA in 100 mL of PBS. Store at 4°C for up to 2 d.
4. OCT compound (Miles, Elkhart, IN).
5. Dry ice.
6. Isopentane (BDH, Poole, UK).
7. Triton X-100 (Sigma, T9284).
8. Primary antibodies used in this laboratory are generated in house (*see* Section 3.1. and refs. *5* and *7*). Antibodies that recognize GABA in paraformaldehyde-fixed tissues can also be purchased (Eugene Tech. International, Allendale, NJ).
9. Fluorescently tagged secondary antibodies: Jackson Immunoresearch Laboratories (West Grove, Pennsylvania).
10. Biotinylated secondary antibodies and streptavidin-biotin-horseradish peroxidase complex (SABHRP; Amersham, UK).
11. PBS-glycerol mountant: 50% (v/v) glycerol in PBS.
12. Clear nail varnish.
13. Peroxidase substrate solution (*see* Note 2): Immediately prior to use, dissolve 5 mg of DAB (3,3' diaminobenzidine (DAB; Sigma, D5637) in 100 µL of distilled water. Bring the solution up to 10 mL with PBS, then add 5 µL of hydrogen peroxide.
14. PBS-BSA-TX/0.2: 0.2% (v/v) Triton X-100 dissolved in PBS-BSA (*see* item 3, in this section).
15. PBS-BSA-TX/1.0: 1.0% (v/v) Triton X-100 dissolved in PBS-BSA.

3. Methods

3.1. Raising Antisera (see Note 3)

3.1.1. Preparation of Immunogenic Amino Acid Conjugates

1. Dissolve 10 mg of the amino acid of interest in 4.5 mL of phosphate buffer A.
2. Add 5 mL of thyroglobulin solution, mix, and then add 0.5 mL of freshly prepared paraformaldehyde fixative.
3. Mix the solution well and allow it to incubate for 18 h at room temperature.
4. Divide the thyroglobulin-conjugated amino acid into 0.5 mL aliquots and store them at –20°C until ready for use.

3.1.2. Conventional Immunization

1. Mix an aliquot of the amino acid conjugate (0.5 mL) with an equal volume of Freund's adjuvant. Use complete adjuvant for the first injection and incomplete adjuvant for subsequent injections.
2. Emulsify the mixture: Draw up both components into a syringe, cap the syringe tightly, and tape it to a Vortex mixer. Vigorously vortex the mixture for 10–15 min.
3. Using a 19-gage needle, inject animals subcutaneously at two to four sites, taking care not to enter any blood vessels. Dispense roughly equal volumes of the emulsion at each site.

4. Repeat the injections at 3–4 wk intervals and take blood samples 7 d after the second and subsequent injections. For rabbits, collect blood from the rear marginal ear vein.

3.1.3. Colloidal Gold Immunization (see Note 4)

1. Prior to beginning the procedure, prepare colloidal gold/muramyl dipeptide solution as described in Section 2.1., item 11.
2. Divide a 0.5 mL aliquot of the amino acid conjugate (immunogen) into two (*see* Note 5).
3. Emulsify 0.25 mL of the immunogen with Freund's adjuvant as above (Section 3.1.2., step 2).
4. Absorb the remaining 0.25 mL of immunogen onto colloidal gold particles by rapidly mixing the immunogen with an equal volume of colloidal gold/muramyl dipeptide solution using a Vortex mixer.
5. Inject this mixture into the rear marginal ear vein of rabbits (*see* Note 5).
6. Bleed rabbits 1 wk after immunization, collecting blood from the rear marginal ear vein (*see* Note 6).
7. Centrifuge the blood immediately after collection and immediately store the serum in aliquots at –80°C.
8. Use the serum for immunolabeling studies without any further purification steps.

3.2. Tissue Fixation (see Note 7)

3.2.1. Perfusion Fixation of Mammalian Tissues

1. Prepare 2 × 1 L bottles, containing, respectively, 500 mL of a physiological saline solution (such as Ames media) and 500 mL of paraformaldehyde fixative.
2. Place both bottles 1.2 m above the animal to be perfused (this height corresponds to average blood pressure for most animals) and connect the bottles by plastic tubing to a two-way tap, leading, in turn, to a cannula (*see* Notes 8 and 9).
3. Anesthetize the animal with a large (120 mg/kg) dose of sodium pentobarbital (Nembutal) given by ip injection (*see* Note 10).
4. As soon as deep anesthesia has been induced, open the thorax and insert the cannula (connected to the perfusion apparatus via the two-way tap) into the heart via the left ventricle. Push the cannula forward until its tip is visible in the ascending aorta.
5. Clamp the cannula in place with a pair of hemostats, sever the right atrium, and begin preperfusion (*see* Notes 11 and 12).
6. Preperfuse using warm (37°C) Ames media or other suitable physiological saline until all the blood has been flushed out of the animal.
7. After preperfusion turn the two-way tap to permit perfusion with fixative. Administer a total volume of fixative equivalent to about 2 mL of fixative/g of body weight (e.g., about 500 mL for one rat).
8. After fixation by perfusion, the tissues of interest should be rapidly removed, cut into small pieces (e.g., slices of 1–2 mm in thickness) and placed into fresh fixative at 4°C, overnight (*see* Notes 13 and 14). Thereafter, tissues may remain in fixative at 4°C for up to 1 wk, until commencement of the immunolabeling steps.

3.2.2. Immersion Fixation of Tissues (see Note 15)

1. Remove the tissues from the animal as rapidly as possible and place the tissues into a Petri dish that is half-full of the fixative.
2. Without delay, cut the tissue into pieces ≤1 mm in thickness (*see* Note 16).
3. Transfer the small pieces of tissue into fresh fixative using at least 30 vol of fixative/vol of tissue (*see* Note 17).
4. Fix the tissues at room temperature for 1–2 h. Continuously agitate the tissues to maximize the diffusion of fixative into the tissue.
5. Place the tissues in fresh fixative and continue to fix overnight at 4°C (*see* Note 14). Thereafter, tissues may remain in fixative at 4°C for up to 1 wk, until commencement of the immunolabeling steps.

3.3. Immunolabeling Cryostat Sections (see Notes 18 and 19)

1. Wash the fixed tissue with PBS for 30 min at 4°C to ensure that the fixative has been thoroughly rinsed away (*see* Notes 20 and 21).
2. Cryoprotect the fixed tissues by immersing them in sucrose solution at 4°C overnight, or until the tissues sink to the bottom of the sucrose solution (*see* Note 22).
3. Place the fixed, cryoprotected tissues into a mold (fashioned from aluminum foil), containing OCT compound.
4. Freeze the mold (containing the OCT compound and tissues) by placing it into a beaker containing a 1:1 mixture of dry ice and isopentane (*see* Note 23).
5. Cut sections using a cryostat or freezing microtome. Sections should be 5–20 μm thick.
6. Mount the sections onto slides that have been coated with gelatin or that have been silanated (*see* Note 24).
7. Perform steps 8–11 (and subsequent steps described in Section 3.7.) in a humid chamber (*see* Note 25).
8. Incubate the sections with blocking solution for 15 min (*see* Note 26).
9. Wash off the blocking solution with PBS and blot the slides using filter paper to remove excess PBS.
10. Apply a sufficient volume of primary antibody (diluted in PBS-BSA) to cover the sections and incubate at 4°C for between 30 min and 2 h (*see* Notes 27 and 28).
11. Wash the sections for 15 min in PBS to remove unbound primary antibody.
12. Treat the sections with secondary antisera and reveal immunolabeling using an appropriate detection system (*see* Section 3.7).

3.4. Immunolabeling Vibratome Sections (see Notes 18 and 29)

1. Wash the fixed tissue with PBS for 30 min at 4°C to ensure that the fixative has been thoroughly rinsed away (*see* Notes 20 and 21).
2. Cut sections of fixed tissue at a thickness of 25–100 μm.
3. Incubate the sections in glass scintillation vials (or similar containers) containing the primary antibody (diluted in PBS-BSA-TX-1.0) for 12–48 h (*see* Notes 29 and 30).

4. Wash the sections with three changes of with PBS over 1 h at 4°C to remove unbound primary antibody.
5. Treat the sections with secondary antisera and reveal immunolabeling using an appropriate detection system (*see* Section 3.7).

3.5. Immunolabeling Wholemounts (see Note 31)

3.5.1. Wholemounts of Retina

1. Wash the fixed tissue with PBS for 30 min at 4°C to ensure that the fixative has been thoroughly rinsed away (*see* Notes 20, 21 and 31).
2. Incubate the wholemount in primary antibody, diluted with PBS-BSA-TX-1.0 for approx 2 wk (*see* Note 30).
3. Wash the sections with three changes of with PBS over 1 h at 4°C to remove unbound primary antibody.
4. Treat the wholemounts with secondary antisera and reveal immunolabeling using an appropriate detection system (*see* Section 3.7).

3.5.2. Wholemounts of Invertebrate Ganglia (see Note 31)

1. Wash the fixed tissue with PBS for 30 min at 4°C to ensure that the fixative has been thoroughly rinsed away (*see* Notes 20 and 21).
2. Remove the collagenous neural sheath by dissection and/or by incubation in a solution of collagenase (*see* Note 32).
3. Incubate the wholemount in primary antibody, diluted with PBS-BSA-TX-1.0, for approx 2 wk (*see* Notes 30 and 33).
4. Wash the sections with three changes of with PBS over 1 h at 4°C to remove unbound primary antibody.
5. Treat the wholemounts with secondary antisera and reveal immunolabeling using an appropriate detection system (*see* Section 3.7.).

3.6. Immunolabeling Cultured Cells (see Note 34)

1. Wash the fixed cells with PBS for 30 min at 4°C to ensure that the fixative has been thoroughly rinsed away (*see* Notes 20 and 21).
2. Permeabilize the fixed cells (*see* Note 34).
3. Incubate the fixed cells in primary antibody, diluted with PBS-BSA containing 0.2% Triton X-100, for 30 min to 24 h (*see* Note 35).
4. Wash the cells with three changes of with PBS over 1 h at 4°C to remove unbound primary antibody.
5. Treat the cells with secondary antisera and reveal immunolabeling using an appropriate detection system (*see* Section 3.7.).

3.7. Detection Systems

The procedures outlined in this section may be applied without modification when immunolabeling cryostat sections. Notes are cited to explain modifications of the protocols that may be applicable when immunolabeling vibratome sections, wholemounts of retina, wholemounts of invertebrate ganglia, or cultured cells.

3.7.1. Detection with Fluorescently Tagged Secondary Antibodies

1. Select a suitable fluorescently tagged secondary antibody (*see* Notes 36 and 37).
2. Dilute the secondary antibody in PBS-BSA to the supplier's recommended working dilution (typically approx 1:200).
3. Apply a sufficient volume of diluted secondary antibody to cover the sections and incubate in a humid chamber for 30 min–2 h (*see* Notes 25 and 38).
4. Wash the sections with PBS for 15 min to remove unbound secondary antibody (*see* Note 39).
5. Blot excess wash buffer off the slides and cover slip the slides using a PBS-glycerol mountant (*see* Note 40). Seal the edges of the coverslip using clear nail varnish to prevent leakage of the mountant.
6. Store the sections in the dark at 4°C until ready for viewing by fluorescence microscopy (*see* Note 41).

3.7.2. Detection Using a Biotin-Streptavidin-Fluorophore System

1. Select a suitable biotinylated secondary antibody (*see* Notes 36 and 42).
2. Dilute the secondary antibody in PBS-BSA to the supplier's recommended working dilution (typically approx 1:200–1:300).
3. Apply a sufficient volume of diluted secondary antibody to cover the sections and incubate in a humid chamber for 30 min (*see* Notes 25 and 38).
4. Wash the sections with PBS for 15 min to remove unbound secondary antibody (*see* Note 43).
5. Apply a solution containing the streptavidin-fluorophore to the sections for 30 min (*see* Note 44).
6. Wash the sections with PBS for 15 min to remove unbound streptavidin-fluorophore (*see* Note 43).
7. Cover slip the slides and treat thereafter as if they have been labeled with a fluorescently tagged antibody (*see* Section 3.7.1., step 5; and Notes 40 and 41).

3.7.3. Detection Using the SABHRP System

1. Select a suitable biotinylated secondary antibody (*see* Notes 36 and 42).
2. Dilute the secondary antibody in PBS-BSA to the supplier's recommended working dilution (typically approx 1:200–1:300).
3. Apply a sufficient volume of diluted secondary antibody to cover the sections and incubate in a humid chamber for 30 min (*see* Notes 25 and 38).
4. Wash the sections with PBS for 15 min to remove unbound secondary antibody (*see* Note 43).
5. Dilute the SABHRP complex in PBS-BSA to 1:200 to 1:300 unless clearly contra-indicated by the supplier's instructions (*see* Note 45).
6. Apply a sufficient volume of diluted SABHRP complex to cover the sections and incubate in a humid chamber for 30 min (*see* Note 38).
7. Wash the sections with PBS for 15 min to remove unbound SABHRP complex (*see* Note 43).
8. Apply peroxidase substrate solution to the slides for 10 min or until the sections have turned brown (*see* Notes 46 and 47).

9. Wash the sections with PBS for 15 min to rinse away excess peroxidase substrate solution (*see* Note 43).
10. Cover slip the slide using PBS-glycerol mountant and seal the edges of the coverslip using clear nail varnish to prevent leakage of the mountant (*see* Note 40).
11. View the sections using a microscope equipped for brightfield observation (*see* Note 41).

3.8. Immunocytochemical Controls

Despite their essential nature, adequate controls are present in only a small proportion of immunocytochemical studies that are published each year. To ensure the validity of immunocytochemical results, the following controls should be performed.

1. Use immunoblotting to confirm the identity of the amino acid that the antiserum detects (*see* Note 48).
2. Use the preimmune serum from each animal that has been used to raise specific antibodies in place of the corresponding primary antiserum. Perform such control experiments using a range of preimmune serum dilutions.
3. Absorb the immune sera with paraformaldehyde-conjugates of the appropriate amino acid (*see* Note 49).
4. Omit the primary or secondary antibody.

4. Notes

1. Extra pure grade paraformaldehyde from Merck is preferred, because some other brands do not give adequate fixation, in turn giving rise to poor immunolabeling. It is vital to check the pH of the fixative. A frequent error is to add too much sodium hydroxide during the depolymerization of the paraformaldehyde. If the pH is above 7.4, then fixation is likely to be poor. The fixative should be used within 1 h of preparation. Longer delays between depolymerization and use as a fixative (4–12 h) will result in poor fixation of most neurotransmitter amino acids as a result of repolymerization of the formaldehyde.
2. DAB will not dissolve directly in PBS. In our experience, it is not advisable to try to dissolve in other buffers, such as Tris, because this may lead to nonspecific deposition of undissolved DAB particles on the sections.
3. Antibodies are prepared according to our published method *(7)*. Two distinct permutations are available; the antigen may either be attached to colloidal gold particles, or immunization may be by conventional means. If the latter, simpler approach is used, the titer is always lower, though still acceptable.

   We routinely raise antisera in young adult rabbits, rats, or guinea pigs. For each immunogen use one rabbit or two rats or two guinea pigs. It is preferable to raise antisera in animals that are the dominant individuals in any particular group. Accordingly, study the group of animals for a few minutes before selecting an animal for immunization because this will have a major impact on the degree of success in raising high quality antibodies.

4. Whereas the method outlined in Section 3.1.2. is the simplest immunization paradigm, it does not yield the best antisera in terms of titer. For optimal results, use the slightly more complex immunization paradigm *(7)* described in Section 3.1.3.
5. The quantities of immunogen quoted are based on the assumption that the animal being immunized is a rabbit. Smaller animals, such as rats and guinea pigs, need approximately half the amount of immunogen as needed for rabbits. In the case of the latter two species, it is not easy to inject the gold conjugate intravenously; instead administer the gold conjugate by ip injection.
6. It is difficult to take test bleeds from rats and guinea pigs, and we routinely choose not to take test bleeds from these animals. Instead, terminally exsanguinate these species under sodium pentobarbitone anesthesia (60 mg/kg, ip) 1 wk after the last of a series of four injections.
7. Many invertebrates, such as locusts, can be rapidly killed with an injection of fixative. Other tissues should be derived from animals that have been euthanized in an appropriate manner. Where practicable, fixative injection is a rapid method of euthanasia that simultaneously initiates fixation of the tissues. Check the suitability of this method with your local ethical regulatory body.
8. Rapid and sustained delivery of fixative immediately after anesthesia is essential for the fixation of amino acids. Failure to do so can result in anoxia-induced release and subsequent diffusion of amino acids away from the original source.
9. Mammalian tissues are normally best fixed by vascular perfusion via the heart. For large animals, larger bore cannulae can be fashioned from rigid plastic tubing.
10. The use of a large dose of anesthetic ensures that the process of anesthesia is very rapid, reducing the potential period of anoxia that might otherwise be encountered. If possible (e.g., in the case of rabbits, which have large accessible ear veins), it is preferable to give the anesthetic intravenously because this results in extremely rapid induction of anesthesia.
11. Preperfusion must be as brief as possible, commensurate with the removal of most of the blood (which will otherwise clot and ruin the perfusion); typically this takes about 15–20 s.
12. If desired, calcium channel blockers, such as cadmium chloride, cobalt chloride, and nickel chloride (each at 50 μ*M*), can be added to the preperfusion mixture. Their inclusion helps to reduce transmitter release, thereby maximizing the transmitter pool remaining in the neurons of interest. If calcium channel blockers are used the heart stops almost instantly, which may impair the overall quality of the perfusion. The tissue culture media must be buffered with HEPES rather than bicarbonate (and gassed with oxygen alone) to prevent precipitation of the blockers. Accordingly, use these calcium channel blockers only if you suspect that the amino acid of interest is being released during the perfusion process.
13. The size of the pieces of tissue is important because this determines the extent of further penetration of fixative into the tissues. Therefore, cut tissue pieces as small as possible, commensurate with other experimental constraints. Alterna-

tively, if appropriate, Vibratome sections of the tissues may be cut at this point, these sections then being postfixed as per Section 3.4.

14. The inclusion of the overnight fixation step at 4°C is essential. This is probably because fixation of amino acids by paraformaldehyde is a multistage process, proceeding via several rather unstable intermediate reaction products before the final linkage (a methylene bridge) is formed *(8)*.
15. Tissues that cannot be fixed by perfusion, such as the rabbit retina (which is avascular and thus fixes poorly by perfusion), invertebrate tissues and cultured cells, brain slices, and so forth, can be fixed by immersion. However, in such cases the tissue must be 1 mm or less in thickness to ensure both rapid and adequate fixation.
16. The tough sheath surrounding most ganglia of such animals as insects is a major barrier to the rapid penetration of fixative. To enhance fixative penetration, make small nicks in this sheath using a small knife fabricated from broken fragments of a razor blade attached to a handle fabricated from a Pasteur pipet, or using a keratomy knife (of the type used by eye surgeons).
17. A large excess volume of fixative is important to ensure an effective concentration of fixative is maintained. In some cases, addition of a small amount of detergent (e.g., 0.05%, v/v, Triton X-100) to the fixative as a wetting agent is recommended. This will assist the penetration of the fixative, especially in insects in which such structures as the tracheoles are hydrophobic and, thus, normally exclude the entry of aqueous liquids.
18. The precise conditions under which pre-embedding immunocytochemistry is performed are dictated by the types of tissues that are being examined. Antibodies are large molecules (150 kDa), thus their ability to penetrate into tissues is dependent upon the presence or absence of any permeability barriers. Sectioned tissues are relatively permeable to antibodies because the action of cutting the tissues slices open many cells—at least at the surface of the tissue—thereby permitting the easy entry of antibodies. Even so, some cellular compartments, such as synaptic terminals, may still not be easily accessible to the antibodies. Under such situations, further permeabilization is required. Such tissues as wholemounts of retina and invertebrate ganglia require specific permeabilization techniques because of the lack of cut surfaces. Specific requirements for labeling sections and wholemounts are detailed in the Notes that accompany the individual protocols.
19. Cryostat or freezing-microtome sections are very easy to immunolabel, because they are relatively thin, and the act of freezing also opens up the section as a result of formation of small ice crystals.
20. After tissues have been fixed with paraformaldehyde, they must be washed before immunocytochemistry can be performed. This wash step, which is carried out at 4°C, removes all unbound aldehyde molecules that might otherwise fix the antibodies to the tissues in a nonspecific manner. PBS is also used in all subsequent wash steps. Do not be tempted to use Tris-buffered saline (TBS) because this is an amine buffer; the amine can be fixed by any residual paraformaldehyde. In our experience, with some antibodies, use of TBS may result in labeling artifacts.

21. Tissues must be kept cool in order to minimize loss of amino acids because of breaking of the methylene bridges between the paraformaldehyde and the amino acids that can occur at higher temperatures *(9)*.
22. To minimize cellular destruction by ice crystals, the tissues must be cryoprotected prior to sectioning. The presence of high concentrations of sucrose results in the slow formation of small ice crystals, rather than the rapid formation of large ice crystals.
23. This permits reasonably rapid freezing, with minimal ice crystal damage to the tissue.
24. Slides are coated with an adhesive compound to ensure that the sections remain stuck to the slides. Coat (sub) the slides by dipping them into a solution of warm 0.2% (w/v) gelatin in distilled water containing 0.02% (w/v) chrome alum. Drain the excess liquid off and dry the slides in a warm (50°C) oven. The slides are then ready for use. Silanation is more complex procedure, but this coating retains sections strongly, and most importantly, does not yield nonspecific binding of antibodies (which may sometimes occur with gelatin-subbed slides).

    Silanation is a process in which the silane reagent is covalently linked to hydroxyl groups on the surface of the slide. The silane has a free amino group, allowing the subsequent binding of the dialdehyde, glutaraldehyde. Because the glutaraldehyde is a dialdehyde, it can also bind to amino groups in the section, linking the sections to the slide via the formation of Schiff bases. To silanate slides, prepare a 1% (v/v) solution of γ-aminopropyltriethoxysilane (Sigma, A3648) in distilled water. Adjust the pH of the solution to approx 3.5, using 1*M* HCl. Place slides into the solution overnight at 70°C (place a lid on the container to prevent evaporation). Thereafter, wash the slides in distilled water for 30 min, then dry them for a further 8 h at 100°C. Store slides at room temperature until needed. The silane solution can be reused many times. Immediately prior to use, activate the slides by placing them into a 1% (v/v) solution of glutaraldehyde for 3 h. Subsequently, wash the slides with distilled water and air dry them. The slides are then ready to use. It is also possible to purchase presilanated microscope slides (Sigma, S4651), although they are rather expensive.

    We routinely silanate glass microscope slides that have small wells 3–5 mm in diameter that have been formed by a paint coating. These are available commercially (not silanated) from Roboz (Rockville, MD). Alternatively, cut holes in strips of vinyl electrical insulation tape using a hole punch and apply this tape to microscope slides that have previously been silanated. This tape forms small wells that contain the drops of immunoreagents that are subsequently applied. The use of small wells conserves expensive immunoreagents and allows multiple dilutions of an antibody to be tested on different wells of a single slide. Sections of the tissues are cut and deposited onto the slides. The sections are allowed to air dry at room temperature for 20–30 min prior to commencement of immunolabeling.
25. Because small volumes of immunoreagents are applied to the sections, it is imperative to avoid evaporation. Therefore, place the slides in a humid chamber. Such a chamber is easily fabricated using a plastic box with plastic or glass strips

approx 7–10 mm thick glued in the base. The slides are then laid flat on these strips. Water is then added to the bottom of the box, making sure that the water level remains below the slides. A well-fitting lid further ensures that only minimal evaporation of the immunoreagents occurs.

26. Even if most of the unbound paraformaldehyde is washed away, it is still possible that residual aldehydes or other reactive groups remain in the tissue. These reactive groups can nonspecifically bind proteins, such as antibodies. To prevent this type of nonspecific binding, it is normal practise to block these binding sites by incubating the tissue in an excess of irrelevant protein, dissolved in PBS. The most commonly used protein is BSA. BSA works well under most circumstances; however, if high levels of background labeling are observed, then it is advisable to use a 10% (v/v) solution of serum (in PBS) taken from the same species that the secondary antibody is raised in (because the secondary antibody is unlikely to recognize "self" proteins).
27. Perhaps the most common reason for failure to achieve satisfactory/optimal immunolabeling in any given tissue is because an inappropriate primary antibody dilution has been selected. Most commercially available polyclonal antibodies to neurotransmitter amino acids are raised in rabbits, and typically have working dilutions of 1:500 to 1:5000 (**Beware:** Some manufacturers are rather imaginative in their quoted dilutions!).

    Most readers will readily accept the idea that if a reagent, such as an antibody, is present in insufficient quantities (so that only a few of the potential antibody binding sites in the tissue can be occupied by antibody molecules), then the ensuing immunolabeling will be weak. When faced with weak labeling or the possibility of such, a common response is to use much higher antibody concentrations. However, this too may have drawbacks. The most obvious drawback is that nonspecific background labeling increases with increased antibody concentration, especially where polyclonal antibodies are being used, because these preparations often contain contaminating antibodies against other molecules. The second and perhaps more serious problem is that increasing an antibody concentration too much may result in a paradoxical decrease in intensity of specific immunolabeling, especially where an antigen is present at high concentrations. This is referred to as the Bigbee effect *(9)*, and is thought to be a result of steric hindrance problems, with secondary detection reagents being unable to bind to the primary antibodies because the primary antibody molecules are packed in so densely over the sites where the antigen is located. The Bigbee effect is frequently encountered in amino acid immunocytochemistry. Thus, it is absolutely essential to determine, using a dilution series, which concentration of primary antibody gives the maximal signal-to-noise ratio for each and every tissue that is being investigated.
28. Longer incubations permit the use of lower dilutions of primary antibody, resulting usually in reduced nonspecific background labeling. The optimal dilution of primary antibody must be ascertained as detailed in Note 27.
29. Vibratome sections are usually immunolabeled as free floating sections rather than being mounted onto microscope slides. A fine paintbrush is used to transfer

sections between tubes containing each of the immunoreagents. Alternatively, if many sections are being labeled, use a tea strainer to drain off each successive solution. Sections cut using a Vibratome are less permeable to antibodies than cryostat sections, so incubation times must be extended. These increased incubation times bring with them the possibility of bacterial growth (*see* Note 30). To enhance antibody penetration, add 0.2–1.0% (v/v) Triton X-100 to the PBS-BSA used as the diluent for the primary and secondary antibodies and the SABHRP (if applicable).

30. To prevent bacterial contamination, add 0.05% (w/v) sodium azide (Sigma; **toxic**) as a bacteriostat to all antibody solutions. Do not add sodium azide to the SABHRP complex (if used) because azide inhibits the HRP enzyme. If a bacteriostat must be added to the SABHRP complex, then use 0.05% (v/v) thimerosal (**toxic**).
31. Wholemounts present their own particular problems. In general, treat wholemounts as if they are Vibratome sections. However, because they are generally thicker, and because they generally lack cut surfaces, antibody penetration is extremely slow and must be enhanced wherever possible. A typical penetration enhancement is to add Triton X-100 (to 1.0%, v/v) to the wash and diluent buffers. This detergent dissolves lipid membranes, and, when used in moderation, it aids antibody penetration; however, if used at too high a concentration, the tissue will disintegrate. Penetration of antibodies may also be enhanced by a variety of additional techniques. The most common technique is to sequentially dehydrate and rehydrate the specimen by passing through a graded series of alcohols. The alcohol solubilizes some membrane lipids while the shrinkage that also ensues opens up the extracellular spaces, allowing antibodies to permeate between cells. In general, this dehydration/rehydration technique should be used only if permeabilization by the means detailed above have failed, since cellular morphology can be distorted. Another method to enhance penetration is to use enzyme treatment. In particular, trypsin treatment is reported to reveal cryptic pools of amino acid transmitter in some nerve terminals. This action is probably a consequence of trypsin breaking open nerve terminals/synaptic vesicles that are otherwise inaccessible to immunoreagents, rather than the liberation of free amino acids during proteolysis, but the latter potential action must always be considered.
32. To enhance penetration of immunoreagents the ganglia should, where possible, be desheathed, since the collagenous connective tissue surrounding the ganglia presents a formidable penetration barrier. Where appropriate, nerves entering the ganglia of interest should be severed close to the ganglia to provide greater access for the immunoreagents. If practicable, small nicks may be made in the sheath (if still present) to mechanically desheath the ganglia. Preparations may be chemically desheathed using collagenase as follows: Dissolve 1 mg of collagenase (Sigma type VII, C 0773) in 1 mL of pH 7.0 sodium chloride solution (0.9%, w/v), containing in addition, 1 m*M* zinc chloride and 1 m*M* calcium chloride. Incubate the ganglia in this solution at 37°C, for 5–12 h. Thereafter, wash the ganglia in PBS-BSA for 1 h, then proceed to immunolabel the ganglia.

33. Antibody incubation times and conditions for wholemounts should be varied empirically depending on the dimensions of the wholemounts to be studied (i.e., the larger the wholemount, the longer the incubation times).
34. Immunolabeling of cultured cells grown on coverslips is a relatively simple process, especially if monolayers are being used. For detailed descriptions of how to grow cells on coverslips for subsequent immunocytochemical labeling, *see* ref. *10*. Because the cells do not have any cut surfaces, they must be permeabilized; this may be accomplished by adding Triton X-100 (to 0.2%, v/v) to all buffers. Alternatively, permeabilize the cells by immersing coverslips in cold methanol (4°C) for 10 min. Thereafter, rehydrate the cells with PBS. Triton X-100 may still be included in buffers if required.
35. The duration of each immunolabeling step is determined by the efficiency of the permeabilization steps. The primary antibody incubation step may need to be increased up to 24 h if permeabilization is poor.
36. If a rabbit primary antibody has been used, then a suitable secondary antibody directed against rabbit immunoglobulins should be selected; if a mouse primary has been used, select antimouse, and so forth. Secondary antibodies are commonly available with a fluorophore attached for use in immunofluorescence studies. Alternatively, they may be biotinylated, which permits subsequent binding of streptavidin-biotin-horseradish peroxidase complex, or they may have enzymes, such as horseradish peroxidase, attached directly.
37. The cleanest signals are generally obtained using antibodies labeled with fluorescine isothiocyanate (FITC) or carbocyanine dyes (Cy 3). If possible, avoid antibodies labeled with rhodamine (TRITC) or, to a lesser extent, Texas red, since in our hands at least, these tend to yield higher levels of background labelling.
38. For Vibratome sections, increase the incubation time to 3–4 h. For retina or other wholemounts, incubation times on the order of 8 h may be necessary. Likewise, for cultured cells, secondary antisera incubation times may need to be extended; this will have to be determined empirically.
39. Wash times need to be extended for Vibratome sections, wholemounts, and cultured cells. For Vibratome sections, wash in three changes of PBS over 1 h. For wholemounts, wash in three changes of PBS over 2–3 h, whereas for cultured cells, give three changes over 30 min.
40. Vibratome sections immunolabeled using fluorescently tagged secondary antibodies and cells grown on coverslips may be mounted and handled in the same way as for cryostat sections. Whole mounts may also be mounted on normal glass microscope slides unless they exceed 200 μm in thickness, in which case the wholemounts should be mounted in cavity slides or slides that have a thin washer or similar spacer element included to prevent the coverslip from squashing the preparation.
41. The main problems that may be encountered are either no labeling or excessively high background labeling. If no labeling is detected, one or more steps have failed and the origin of the failure should be elucidated in a stepwise manner. The problem is likely to be either one of inadequate retention of the target antigen or a defect in one of the immunoreagents.

The performance of the secondary detection reagents is easy to establish by immunoblotting. If a rabbit primary antibody is being used, apply a series of 1 μL droplets of 10-fold dilutions (1:100–1:100,000) of rabbit serum to nitrocellulose membranes and then use the secondary detection system to detect the bound rabbit immunoglobulins. If this fails, then the problem is in the secondary detection system. If not, the problem lies either with the primary antibody or the retention of the amino acid of interest. To determine if the problem is with the primary antibody, perform immunoblots as indicated in Note 48. If there is no problem with the primary antibody, then the problem lies in the retention of the molecule of interest, or the molecule of interest may not be present! High background labeling is usually a function of inappropriate dilutions of each reagent. This should be evaluated by modifying first the primary antibody concentration and then the secondary antibody concentration. If you are using an antibody raised against an amino acid conjugated to a protein, such as BSA, one potential problem is that this antibody may also recognize BSA. Accordingly, if BSA is used as a blocking agent, high levels of background labeling may arise. This can be cured by using another blocking agent, such as goat or sheep serum.

42. Biotinylated secondary antibodies are used as bridging reagents when greater signal intensity is required. Their binding is subsequently detected using streptavidin (a bacterial molecule that binds extremely avidly to biotin), linked either to a fluorophore, such as Texas red, or to an enzyme, such as horseradish peroxidase. Streptavidin is available linked to fluorophores, such as Texas red (*see* Notes 37 and 44), and to reporter molecules, such as horseradish peroxidase (streptavidin-biotin-horseradish peroxidase complex; SABHRP); in the latter case, labeling is ultimately revealed using a chromogenic substrate (3,3,diaminobenzidine; DAB) for the enzyme horseradish peroxidase.
43. These wash times should be increased for Vibratome sections and so forth, in line with the times indicated in Note 39.
44. Streptavidin is available linked to fluorophores, such as Texas red. It is usually purchased as a solution that can be used at a dilution of approx 1:200 in PBS-BSA. The solution containing the streptavidin-fluorophore conjugate is applied to the sections for 30 min. Thereafter, sections are washed and treated as if they had been labeled with a fluorescently tagged secondary antibody.
45. Streptavidin is available in the form of a streptavidin-biotin-horseradish peroxidase complex. This is available either as a preformed complex (e.g., from Amersham) or as kits containing two separate components that must be mixed immediately prior to their use (e.g., kits from Vector). We use the preformed complex because in our hands it is simple to use, gives excellent results, and is considerably cheaper. When diluting the stock solution, make sure that no sodium azide is present in the dilution buffer because azide inhibits the HRP enzyme.
46. In some circumstances (e.g., if high background staining is observed) it may be necessary to incubate tissues with the DAB peroxidase substrate solution, but omitting the hydrogen peroxide, for 10 min before adding the hydrogen peroxide

and incubating for a further 10 min. This preincubation with DAB in the absence of hydrogen peroxide permits the DAB to penetrate evenly into the tissues, ensuring even labeling throughout the depth of the tissue.

47. A 10-min incubation in the peroxidase substrate is also usually adequate for vibratome sections and for wholemounts. For cells, the length of incubation in the peroxidase substrate can be determined by periodically examining the cells under a microscope. The reaction may be stopped when a suitable signal has been attained.
48. To verify the specificity of antisera that you are using, immunoblotting (dot blotting) should be carried out using standard methods *(6)*. Prepare amino acid-paraformaldehyde-BSA conjugates in an identical manner to the amino acid-paraformaldehyde-thyroglobulin conjugates used to immunize the animals. Apply spots of each conjugate (1 μg) to nitrocellulose membranes (e.g., Sigma, N6020). Thereafter, incubate the nitrocellulose membranes for 15 min with PBS containing 1% (w/v) BSA (to block nonspecific binding) and then immunolabel the membranes with each antiserum at a range of dilutions (typically 1:500–1:50,000) for 2 h. Thereafter, wash the membranes with PBS and apply a biotinylated secondary antibody (diluted 1:500) for 2 h. Subsequently, wash the membranes and incubate with streptavidin-biotin-horseradish peroxidase complex (1:500) for 2 h. Reveal labeling using DAB as a chromogen.
49. The exact amount of conjugate needed to preabsorb the antibody will obviously vary with the titer of the antibody. As a general rule simply use a vast excess; typically 1–10 mg of conjugate/mL of diluted antibody.

References

1. Fonnum, F. (1984) Glutamate, a neurotransmitter in mammalian brain. *J. Neurochem.* **42,** 1–11.
2. Madl, J. E., Larson, A. A., and Beitz, A. J. (1986) Monoclonal antibody specific for carbodiimide-fixed glutamate: immunocytochemical localisation in the rat CNS. *J. Histochem. Cytochem.* **34,** 317–326.
3. Eranko, O. (1955) Distribution of adrenaline and noradrenaline in the adrenal medulla. *Nature* **175,** 88–91.
4. Geffard, M., Henrich-Rock, A.-M., Dulluc, J., and Seguela, P. (1985) Antisera against small neurotransmitter-like molecules. *Neurochem. Int.* **7,** 403–413.
5. Pow, D. V., Wright, L. L., and Vaney, D. I. (1995) The immunocytochemical detection of amino acid neurotransmitters in paraformaldehyde-fixed tissues. *J. Neurosci. Methods* **56,** 115–123.
6. Wright, L. L. (1995) A new type of GABAergic amacrine cell with a bistratified morphology. *Proc. Aust. Neurosci. Soc.* **6,** 198.
7. Pow, D. V. and Crook, D. K. (1993) Extremely high titre antisera against small neurotransmitter molecules: rapid production, characterisation and use in light- and electron-microscopic immunocytochemistry. *J. Neurosci. Methods* **48,** 51–63.

8. Hayat, M. A. (1989) *Principles and Techniques of Electron Microscopy.* Macmillan, London.
9. Bigbee, J. W., Kosek, J. C., and Eng, L. F. (1977) Effects of primary antiserum dilution on staining of "antigen-rich" tissues with the peroxidase-antiperoxidase technique. *J. Histochem. Cytochem.* **25,** 443–447.
10. Harlow, E. and Lane, D. (1988) *Antibodies: A Laboratory Manual.* Cold Spring Harbor Laboratory, Cold Spring Harbor, NY, pp. 178–179.

9

cGMP-Immunocytochemistry

Jan de Vente and Harry W. M. Steinbusch

1. Introduction

Immunocytochemistry (ICC) of small, water-soluble molecules requires some special precautions compared to ICC of large protein molecules. In general, protein antigens are relatively insensitive to the type of fixative used. Apparently, the epitopes are not altered in a decisive way by the fixative, immobilization is usually effective, and more than one epitope may be recognized by polyclonal antisera. In studying by ICC the location of small, hydrophilic molecules—such as neurotransmitters and cyclic nucleotides—choice of fixative plays a decisive role. First, fixation is necessary not only to preserve tissue morphology, but also to fix the molecules to the tissue matrix, otherwise these may be lost from the tissue. This is certainly the case for cyclic nucleotides *(1–4)*. Second, the fixative may covalently modify the hapten by fixing it to tissue protein. This has been shown to occur with serotonin *(5,6)* and also with small peptides *(7)*. Because the covalent modification of the hapten may be part of the antigenic determinant, it is a prerequisite that antisera are raised against conjugates prepared using a fixative that is also suitable for tissue fixation *(8–10)*.

Earlier studies on cyclic nucleotide immunocytochemistry used antibodies that had been developed for the radioimmunoassays of cAMP and cGMP *(11)*. These antibodies were raised against a 2'-succinylated cyclic nucleotide coupled to a carrier protein using the carbodiimide reaction. This procedure is not applicable to tissue fixation, although attempts in that direction have been made *(12)*. Immunocytochemistry using these antibodies was introduced by Wedner et al. *(13)*. Subsequently, it was shown that almost all cyclic nucleotides were lost from the tissue during tissue processing *(1–4)*, although some specific staining could be observed. It was suggested by Steiner *(4)* that these

From: *Methods in Molecular Biology, Vol. 72: Neurotransmitter Methods*
Edited by: R. C. Rayne Humana Press Inc., Totowa, NJ

antibodies might recognize cyclic nucleotides bound to the respective protein kinases. It also was reported that formaldehyde fixation improved the detection of cGMP *(14)*, which finds its basis in the studies that underlie the present chapter *(8–10)*. However, we demonstrated that these so-called Steiner antibodies exhibited an appreciably lower affinity to formaldehyde-fixed cGMP than antibodies raised specifically against the cGMP-formaldehyde-conjugate *(9)*.

cGMP is synthesized by the enzyme guanylyl cyclase (GNC). Several isoforms of this enzyme have been described *(15)*. The membrane-bound form is the so-called particulate form of the enzyme (pGNC); the cytosolic form is soluble (sGNC) *(16)*. pGNC may be coupled to receptors of a variety of peptide hormones depending on the tissue and the species *(15)*.

sGNC is activated by nitric oxide (NO; *17*). Recently, it has been demonstrated that NO has a regulatory function in a number of important physiological processes, e.g., vasodilation, immunomodulation, and neuromodulation *(18–20)*. NO, together with L-citrulline, is synthesized by the enzyme nitric oxide synthase from L-arginine and molecular oxygen *(21,22)*. Nitric oxide synthase has been found in almost all areas of the brain although important differences exist among various regions in the amount of enzyme present *(23–25)*. In the central nervous system (CNS), NO has important messenger functions *(19,26,27)*. Immunohistochemical studies using antibodies against the formaldehyde-fixed cGMP have presented evidence that the activation of sGNC may be one of the important functions of NO in the CNS *(10,28–31)* and peripheral *(32,33)* nervous systems.

This chapter gives a detailed account of the steps necessary to raise antibodies against formaldehyde-fixed cGMP, including criteria to be met in ascertaining the specificity of the antibodies, and outlines approaches to the study cGMP-metabolism in vivo or in vitro using cGMP-immunocytochemistry.

2. Materials

2.1. Preparation of cGMP-Formaldehyde Conjugates

1. Freshly depolymerized paraformaldehyde solution (8%, w/v): Add 4 g of powdered paraformaldehyde to 50 mL distilled water; add 3 drops of 1*N* NaOH. Heat gently at 65°C, stirring continuously until the solution has cleared completely.
2. Bovine thyroglobulin.
3. Bovine serum albumin (BSA).
4. Guanosine 3',5'-cyclic monophosphate, free acid (cGMP).
5. Phosphate buffer A: 0.2*M* sodium phosphate, pH 7.0.
6. Phosphate buffer B: 0.01*M* sodium phosphate, pH 7.0.
7. Ultrafiltration cell (Amicon, Beverly, MA; model 8050).
8. Diaflo ultrafilters XM50 (Amicon, cat. no. 14122).

2.2. Preparation of Formaldehyde-Treated Protein

Use the same materials as for Section 2.1., except cGMP, which is not required here.

2.3. Immunization Procedure

1. Freund's complete adjuvant.
2. Freund's incomplete adjuvant.
3. Saline: 0.9% (w/v) NaCl in distilled H_2O.

2.4. Antiserum Purification

Formaldehyde-treated protein: *See* Sections 2.2. and 3.2.

2.5. Specificity Testing of the cGMP-Antisera

2.5.1. Serum Specificity: Control Sera

1. Tris buffered saline (TBS): 50 m*M* Tris-HCl, pH 7.4, 0.9% (w/v) NaCl.
2. TBS-Triton X-100 (TBS-T): TBS containing 0.3% (v/v) TX-100.

2.5.2. Serum Specificity: The Gelatin Model System

1. Gelatin solution (Merck, DAB 8, Ph. Helv.; cat. no. 4072): 1% (w/v) in distilled water.
2. Phosphate buffer A: 0.2*M* sodium phosphate, pH 7.0.
3. Sucrose solution: 2*M* sucrose in 0.2*M* sodium phosphate, pH 7.0.
4. Fixative solution: 4% (w/v) freshly depolymerized paraformaldehyde in 0.1*M* phosphate buffer, pH 7.0, containing 1*M* sucrose.
5. TBS: *see* Section 2.5.1., item 1, except pH to 7.0.
6. Microscope slide coating solution: 0.5% (w/v) gelatin and 0.5% (w/v) chrome-alum in dH_2O (*see* Note 1).
7. Coated glass microscope slides: dipped in coating solution and air-dried (*see* Note 1).
8. Guanosine and guanosine mono-(disodium salt), di-(sodium salt), and triphosphate (sodium salt; GMP, GDP, and GTP).
9. Adenosine 3',5'-cyclic monophosphate, free acid (cAMP).
10. Adenosine and adenosine mono-(sodium salt), di-(sodium salt), and triphosphate (disodium salt; AMP, ADP, and ATP).
11. Tygon tubing, 3-mm id (Norton Performance Plastics Corp., Akron, Ohio).
12. Tissue-Tek OCT compound (Miles, Inc., Elkhart, IN; cat. no. 4583).
13. Cryomold (Miles, Inc.; cat. no. 4566).
14. 2-Methyl-butane (isopentane).
15. Liquid nitrogen.
16. Coverslips (Menzel, D-38116 Braunschweig, Germany; e.g., 24 × 50 mm).

2.5.3. The Nitrocellulose Model System

1. Nitrocellulose, 0.45 μm pore size (Bio-Rad, cat. no. 162-0115).
2. Donkey antirabbit antiserum (Jackson Immunoresearch Labs, West Grove, PA; cat. no. 711-165-152).

3. Rabbit peroxidase-antiperoxidase antiserum (DAKO, DK-2600, Glostrup, Denmark; cat. no. Z113).
4. Hydrogen peroxide (Merck; Perhydrol 30%, cat. no. 7209.250).
5. TBS, pH 7.4: *see* Section 2.5.1., item 1.
6. Tris buffer: 10 mM Tris-HCl, pH 7.4.
7. Nitrocellulose blocking solution: 1% gelatin in distilled H_2O.
8. DAB staining solution: 10 m*M* Tris, pH 7.4, containing 0.01% (v/v) hydrogen peroxide, 0.5 mg/mL diaminobenzidine (DAB).

2.6. Tissue Preparation for cGMP-Immunocytochemistry

2.6.1. Perfusion Fixation of Rats

1. Pentobarbital (Nembutal).
2. Saturated picric acid solution: Add 15 g of picric acid to 1 L of distilled H_2O, stir overnight, and leave the solution in contact with undissolved picric acid.
3. Hypodermic needle, 16 gage × 1½ in. (Sherwood Medical, St. Louis, MO; cat. no. 200045).
4. Calcium-free Krebs buffer: 121.1 m*M* NaCl, 1.87 m*M* KCl, 1.15 m*M* $MgCl_2$, 24.9 m*M* $NaHCO_3$, 11.0 m*M* glucose, pH 7.4, aerated with 5% CO_2 in O_2.
5. Perfusion fixative: 4% (w/v) freshly depolymerized formaldehyde in 0.1*M* sodium phosphate, pH 7.4, containing 0.2% (v/v) saturated picric acid solution.

2.6.2. Preparation of Vibratome Sections (Brain Tissue)

1. Vibratome (Technical Products Inc., St. Louis, MO).
2. Cyanoacrylate glue.
3. TBS, pH 7.4: *See* Section 2.5.1., item 1.

2.6.3. Preparation of Cryostat Sections (Brain Tissue)

1. Cryostat.
2. Cryoprotection buffer: 0.1*M* sodium phosphate, pH 7.4, containing 15–20% (w/v) sucrose.

2.6.4. Preparation of Brain Slices

1. Krebs buffer (with Ca^{2+}): 1.2 m*M* $CaCl_2$, 121.1 m*M* NaCl, 1.87 m*M* KCl, 1.15 m*M* $MgCl_2$, 24.9 m*M* $NaHCO_3$, 11.0 m*M* glucose, pH 7.4, aerated with 5% CO_2 in O_2.
2. Slice fixative solution A: 8% (w/v) freshly depolymerized formaldehyde in 0.1*M* sodium phosphate, pH 7.4.
3. Slice fixative solution B: 4% (w/v) freshly depolymerized formaldehyde in 0.1*M* sodium phosphate, pH 7.4, containing 15% (w/v) sucrose.
4. Phosphate-sucrose: 0.1*M* sodium phosphate, pH 7.4, containing 15% (w/v) sucrose.
5. McIlwain tissue chopper (Mickle Laboratory Engineering Co., Ltd.).
6. Filter paper (Schleicher & Schuell; cat. no. 595).
7. Stereo microscope (Zeiss, cat. no. 475002-9902, objective 475038).

8. Multiwell tissue culture chamber, 12 wells.
9. Temperature-controlled incubator with gas-inlet.
10. Vertical adjustment attachment (Harvard, cat. no. 50-2641).

2.7. cGMP-Immunocytochemistry

2.7.1. Immunofluorescence Method

1. TBS, pH 7.4: *See* Section 2.5.1., item 1.
2. TBS-T: *See* Section 2.5.1., item 2.
3. Aqueous mounting medium: 1 vol TBS plus 3 vol glycerol.
4. Goat antirabbit antiserum conjugated with fluorescein-isothiocyanate (Jackson Immunoresearch, cat. no. 111-096-003).
5. Donkey antirabbit antiserum conjugated with indocarbocyanine (CY3) (Jackson Immunoresearch, cat. no. 711-165-152).
6. Entellan (Merck; cat. no. 7961).

2.7.2. Free-Floating Section Method

1. TBS, pH 7.4: *See* Section 2.5.1., item 1.
2. TBS-T: *See* Section 2.5.1., item 2.
3. Tris buffer: *See* Section 2.5.3., item 6.
4. DAB staining solution: *See* item 8 in Section 2.5.3.
5. Coated microscope slides: *See* Section 2.5.2., item 7 and Note 1.
6. Microclear (JT Baker Chemicals; cat. no. 3905).
7. Microcover (JT Baker Chemicals; cat. no. 3906).

3. Methods

3.1. Preparation of a cGMP-Formaldehyde-Protein Conjugate

1. Prepare 5 mL of phosphate buffer A containing 20 mg of protein (bovine thyroglobulin or BSA) and 45.8 mg of cGMP (25 m*M*). Start the coupling reaction by adding 5 mL of freshly depolymerized paraformaldehyde. Stir this solution for at least 2 h at room temperature (*see* Note 2).
2. Purify the conjugate by repeated ultrafiltration in phosphate buffer B at 4°C using an Amicon stirred cell equipped with a XM50 filter (*see* Note 3).
3. Freeze-dry the conjugate and store at –20°C.

3.2. Preparation of Formaldehyde-Treated Carrier Protein

1. Prepare 5 mL of phosphate buffer A containing 20 mg of protein. Mix this solution with 5 mL of freshly depolymerized paraformaldehyde in distilled water. Stir this solution for at least 2 h at room temperature.
2. *See* Section 3.1. and perform steps 2 and 3.

3.3. Immunization Procedure

1. Suspend 0.5 mg of conjugate in 0.5 mL of saline and emulsify with 0.5 mL of complete Freund's adjuvant (*see* Note 4).

2. Collect a preimmune blood sample of the animal to be immunized in normal glass tubes. Allow the sample to stand for 30 min at 37°C. Afterward, place the clotted blood at 4°C overnight and isolate the serum on the following day. Store the serum in aliquots at –80°C.
3. Inject the conjugate suspension (1 mL, total vol) intradermally at 6–8 sites at the back of the animal.
4. At intervals of 3 wk, prepare a booster injection by suspending 0.5 mg of conjugate in 0.5 mL of saline and emulsify with 0.5 mL of incomplete Freund's adjuvant. Administer this booster injection intramuscularly at four sites.
5. Collect immune serum 7–10 d after each booster injection and isolate the serum as described under step 2. Store the serum in aliquots at –80°C. Repeat this procedure until no further increase in serum titer occurs (*see* Note 5).

3.4. Antiserum Purification

1. Thaw a sample of the stored antiserum. Incubate the sample for 30 min at 56°C (*see* Note 6).
2. Add 3 mg of the formaldehyde-treated carrier protein to each milliliter of antiserum (*see* Note 7).
3. Incubate for 30 min at 37°C and leave the sample overnight at 4°C.
4. Centrifuge for 15 min at 10,000*g* at 4°C.
5. Decant the serum and discard the pellet.
6. Freeze-dry serum aliquots of convenient size (e.g., 50 μL) and store them tightly capped at –20°C.

3.5. Specificity Testing of cGMP Antisera

3.5.1. Serum Specificity: Control Sera

1. Dilute preimmune sera and antisera in TBS-T (*see* Note 8).
2. Incubate preimmune sera and antisera in the presence of 1 mg/mL cGMP-formaldehyde-protein conjugate or 10 m*M* cGMP for 1 h at 37°C, followed by 18 h at 4°C (*see* Note 9).
3. Centrifuge for 15 min at 10,000*g* at 4°C.
4. Apply the preincubated antisera to nonbiological model sections and/or tissue sections (*see* Note 10).

3.5.2. Serum Specificity: The Gelatin Model System

1. Prepare serial dilutions in the range of 1 μ*M* to 10 m*M* of cGMP, GTP, GDP, GMP, guanosine, cAMP, ATP, ADP, AMP, and adenosine in sucrose solution.
2. Mix equal volumes of gelatin solution (*see* Note 11) with the solutions prepared in step 1 and keep warm (approx 50°C).
3. Cut Tygon tubing into pieces of approx 10 mm in length. Pipet aliquots (approx 100 μL) of the diluted solutions into the cut tubing.
4. Place the tubes in a cold (approx 4°C) environment to permit gelation.
5. Fix the nucleosides and nucleotides to the gelatin by adding fixative (three pieces of tubing into 1.5 mL). Leave the gels at 4°C overnight.

6. Wash 1 × 10 min with TBS followed by 1 × 10 min washing with TBS-T and 1 × 10 min with TBS. Remove the gels from the tubing.
7. Embed the gels in Tissue-Tek OCT compound using a cryomold.
8. Freeze the gels in isopentane prechilled with liquid N_2.
9. Cut sections (e.g., 10–16 μm) using a cryostat.
10. Thaw-mount sections on coated glass microscope slides (*see* Note 1). Store slides at –20°C until further use.
11. *See* Section 3.7. for cGMP-immunocytochemistry.

3.5.3. Serum Specificity: Nitrocellulose Model System

1. Prepare serial dilutions in the range of 1 μ*M* to 10 m*M* of the nucleotides and nucleosides mentioned in Section 3.5.2. in distilled water containing 1 mg/mL protein. This protein must be different from the carrier protein used for immunization (*see* Note 9).
2. Cut squares of nitrocellulose paper approx 10 × 10 cm; wear gloves when handling the nitrocellulose.
3. Spot 3 μL of the solutions (prepared as in step 1) in a row onto the nitrocellulose and allow to dry thoroughly.
4. Place 1 g of formaldehyde powder, together with a few drops of distilled water, into a vessel (e.g., a desiccator) that can be closed (*see* Note 12).
5. Place the spotted nitrocellulose squares into the vessel, close the lid, and heat at 75°C for 4 h.
6. Soak the nitrocellulose squares in the blocking solution for 1 h (*see* Note 13).
7. Wash the squares for 3 × 5 min in TBS.
8. Incubate the nitrocellulose squares with primary (rabbit) antiformaldehyde-fixed-cGMP antiserum at an appropriate dilution in TBS (*see* Note 14).
9. Wash the nitrocellulose squares for 1 × 15 min with TBS, 1 × 15 min with TBS-T, and 1 × 15 min with TBS.
10. Incubate with a secondary (donkey) antirabbit antiserum for 30 min, followed by the same washing procedure as described under step 9.
11. Incubate with a rabbit peroxidase-antiperoxidase antiserum for 30 min, followed by the same washing procedure as described under step 9.
12. Wash 1 × 5 min in Tris buffer, then add the DAB staining solution. Leave the squares while slowly shaking, until a deep brown staining is observed (5–20 min).
13. Wash at least 3 × 10 min with Tris buffer and air-dry the nitrocellulose.

3.6. Tissue Preparation for cGMP-Immunocytochemistry

3.6.1. Perfusion Fixation of Rats

1. Anesthetize the animal with an ip injection of (60 mg/kg) pentobarbital (Nembutal; *see* Note 15).
2. Open the chest of the animal and insert a cannula (a hypodermic needle, as specified in Section 2.6.1., item 3) into the left ventricle of the heart.
3. Start perfusion with 50 mL of cold, calcium-free Krebs buffer. Immediately after beginning the perfusion, open the right atrium to prevent a buildup of pressure.

Without delay, switch to perfusing with ice-cold perfusion fixative. Perfusion pressure should be 50–70 mm Hg using a total volume of 500 mL of fixative (*see* Note 16).

4. Dissect the tissue of interest.
5. Postfix the tissue for 2 h in the perfusion fixative.

3.6.2. Preparation of Vibratome Sections (Brain)

1. Place the postfixed tissue (Section 3.6.1., item 5) in ice-cold TBS.
2. Affix the tissue to the cutting block of the Vibratome using superglue.
3. Immerse the tissue in ice-cold TBS.
4. Cut sections of 30–50 µm.
5. Allow the sections to remain free-floating in ice-cold TBS (*see* Note 17).
6. Proceed for cGMP-immunocytochemistry as described in Section 3.7.

3.6.3. Preparation of Cryostat Sections (Brain)

1. Cryoprotect the postfixed tissue (Section 3.6.1., step 5) by overnight immersion in ice-cold cryoprotection solution.
2. Embed tissue in Tissue-Tek OCT compound and freeze with CO_2.
3. Cut tissue sections of 10–20 µm using a cryostat.
4. Thaw-mount sections as described in Section 3.5.2., step 10.
5. Store sections until use at –20°C.
6. Proceed for cGMP-immunocytochemistry as described in Section 3.7.

3.6.4. Preparation of Brain Slices

1. Decapitate the animal without anesthesia.
2. Remove the brain immediately and transfer it to ice-cold, Krebs buffer (*see* Note 18).
3. Dissect the area of interest.
4. Place the brain tissue on a filter paper on a precooled (4°C) stainless-steel chopping table (*see* Note 19).
5. Cut 300 µm transverse slices using a McIlwain tissue chopper (*see* Note 20).
6. Submerge the filter paper with the chopped tissue in ice-cold, aerated Krebs buffer.
7. Observe the slices using a dissecting microscope. Keep the slices submerged in ice-cold Krebs buffer and separate the slices from each other. Partition the slices into wells of a multiwell tissue culture chamber (12 wells), each containing 2 mL of Krebs buffer.
8. Slowly warm the slices to 36°C under an atmosphere of 5% CO_2 in O_2.
9. Incubate the slices, adding drugs as required (*see* Note 21).
10. Terminate incubations by adding 2 mL of ice-cold slice fixative A (*see* Note 22).
11. Keep slices on ice or in a cold room.
12. After 15 min, exchange slice fixative A for slice fixative B.
13. After 2 h exchange slice fixative B for phosphate-sucrose solution and incubate for another 30 min (*see* Note 23).

14. Align the slices in a plane, remove excess fluid, and freeze them in Tissue-Tek OCT compound using CO_2 (*see* Note 24).
15. Prepare tissue sections as described in Section 3.6.3, steps 3–6).

3.7. cGMP-Immunocytochemistry Using Immunofluorescence

1. Allow slides to come to room temperature and let them dry.
2. Wash the slides 3 × 10 min in TBS.
3. Apply anti-cGMP antisera in the appropriate dilution in TBS-T and incubate for 18 h at 4°C (*see* Note 14).
4. Wash slides with TBS 1 × 15 min, then wash with TBS-T 1 × 15 min. Wash again with TBS 1 × 15 min.
5. Apply secondary (goat or donkey) antirabbit antiserum conjugated with fluorescein-isothiocyanate or CY3 in the appropriate dilution in TBS-T and incubate for 1 h at room temperature.
6. Wash slides as detailed in step 4.
7. Coverslip the sections in aqueous mounting medium.
8. Drain from the slides as much superfluous mounting medium as possible and fasten coverslips using small drops of Entellan.
9. Store the slides at –20°C.

3.8. cGMP-Immunocytochemistry on Free-Floating Sections

1. Prepare sections as described in Section 3.6.2. Wash the sections 3 × 5 min with TBS.
2. Incubate the sections for 18 h at 4°C with the anti-cGMP antiserum diluted appropriately in TBS-T.
3. Wash slides with TBS 1 × 15 min, then wash with TBS-T 1 × 15 min. Wash again with TBS 1 × 15 min.
4. Incubate with a donkey antirabbit antiserum (diluted in TBS-T) for 2 h at room temperature.
5. Wash the sections as described under step 3.
6. Incubate the sections with rabbit peroxidase-coupled antiperoxidase antiserum (diluted in TBS-T) for 1 h at room temperature.
7. Wash the sections as described under step 3. Follow with an final wash of 1 × 10 min in Tris buffer.
8. Add approx 2 mL of DAB staining solution per five free-floating sections.
9. Check color development under the microscope.
10. Wash sections 3 × 10 min in Tris buffer.
11. Collect the sections on precoated glass slides. Keep the slides at room temperature until dry.
12. Dehydrate the sections through an ethanol series and cover the sections with a coverslip using a nonaqueous mounting medium, such as Entellan or Microcover (*see* Note 25).

4. Notes

1. Glass slides must be coated with a medium that ensures adherence of the sections to the slide. There are several options. Convenient and inexpensive is the coating

described in Section 2.5.2, item 6. However, it is also possible to use poly-L-lysine coated glass slides. Fetal tissue is sometimes more difficult to mount and in this case it is advisable to use slides that have been coated first with gelatin/chrome-alum and afterward with poly-L-lysine. Another method is to use polysiloxane (3-aminopropyl-triethoxysilane; Fluka) as a coating solution: Soak the slides for 20 min in a 2% (v/v) solution in acetone followed by two washes in acetone and two washes in distilled H_2O. Allow the slides to dry overnight at 37°C. A caveat: After mounting sections on polysiloxane-coated slides, extend the drying time to about 2 h (room temperature).

2. Coupling reagents other than formaldehyde are not suitable. A number of other cross-linkers were tested, including glutaraldehyde *(8)* (*see* also Notes 16 and 22), diethylpyrocarbonate, benzoquinone, malondialdehyde, N-chloromethyl maleimide, and acrolein. Only acrolein gave a stable incorporation of cGMP into the carrier protein. Acrolein is toxic and very unpleasant to work with, and is therefore unsuitable, unless special apparatus are developed. Acrolein has been used to prepare a cAMP-conjugate, that resulted in specific antibodies against acrolein-fixed cAMP *(34)*.

   The quantities of reagents stated in this section are dependent on the specific requirements. Recovery of the thyroglobulin will be better than 90%. The stated recipe yields sufficient conjugate to do the immunization of the animals and to perform some pre-absorption experiments (*see* Section 3.5.1.). Reaction times should not be shorter than stated, but may be extended to 16 or 18 h. Attention must be paid to the pH of the reaction mixture; a pH between 6.0 and 7.0 gives satisfactory results. At lower or higher pH less cGMP is incorporated into the protein *(14)*.

   The cGMP-formaldehyde-protein conjugate is not very stable. At elevated temperatures cGMP will dissociate from the protein *(8)*. However, at 4°C the conjugate is fairly stable and dissociation during the purification process will be negligible. Care must also be taken to perform incubations with the primary cGMP-antiserum in the cold.

   Unfortunately, no fixed mol/mol ratio of hapten vs carrier molecule will guarantee the (undefinable) necessary minimum level of hapten incorporation to ensure a good immune response. The number of haptens incorporated into the carrier protein is determined for a large part by the number of available functional groups in the carrier protein. The mol/mol ratio of formaldehyde-fixed cGMP with the bovine thyroglobulin is approx 75:1. However, excellent antibodies have been raised against conjugates with a mol/mol ratio below 10:1 *(11)*. The mol/mol ratio can be easily established using a radioactive tracer, provided the tracer is pure and no exchange of label occurs during the coupling reaction. Spectrophotometric analysis has also been used in establishing the mol/mol ratios of conjugates. An important condition is that no alterations are induced in the absorbance spectrum of the hapten when coupled to the carrier protein. For cGMP, both methods gave identical results *(8)*.

3. The choice of the filters is determined by the size of the conjugate. The cutoff value of the XM50 filter (M_r 50,000) is sufficient to warrant minimal loss. Instead

of repeated ultrafiltration one can also use classical dialysis against three changes of a large volume of 0.01*M* phosphate buffer, pH 7.0.

4. Good results are obtained using Freund's complete adjuvant as an immunostimulant and slow release vehicle. However, also other adjuvants may be used, such as Specol (Central Veterinary Institute, Lelystad, The Netherlands) or aluminum-gel (Imject-Alum, Pierce; cat. no. 77160). The important point is to prepare a stable emulsion. This can be done using a homogenizing system like a routine Potter system or a high-speed blade (polytron-type) system. Care must be taken for sufficient cooling of the homogenizing system. It is advisable to prepare twice the amount actually needed, because the emulsion is very sticky and a large portion adheres to the equipment and is lost.
5. The gelatin model system (Section 3.5.2.) is suitable to study the increase in titer of the antiserum. By means of limiting dilution, one may determine the maximal increase in immunofluorescence intensity and the dilution that gives 50% of the maximal increase in immunofluorescence at the lowest concentration of cGMP. When no further increase in the limiting dilution is observed, boosting is discontinued and the rabbit is bled.
6. The heat treatment inactivates the complement proteins present in the serum.
7. The polyclonal antiserum will contain a large number of antibodies that are directed against the conjugate-protein itself. To avoid the possibility of high background staining, these antibodies are removed by saturating them with the formaldehyde-treated carrier protein. The antibodies that are directed against the epitopes that include cGMP are for the greater part retained in the serum.
8. To enhance tissue penetration the dilution of the antisera is made in TBS containing a tissue sensitizer like Triton X-100 or Tween-30.
9. Sera are absorbed with a conjugate that contains formaldehyde-fixed cGMP, but that does not contain the same carrier protein used for immunization. After treating the antiserum with formaldehyde-treated carrier protein (Section 3.4.), it is not easy to ascertain that all antibodies directed only against the carrier protein have been removed. This may obstruct the interpretation of the results when the immunizing conjugate is used for preabsorption studies. Therefore, it is necessary to prepare a different conjugate for preabsorption purposes. Because the affinity of the cGMP-antiserum is far greater for formaldehyde-fixed cGMP than for free cGMP *(9)*, it is necessary to use a high concentration of free cGMP in preabsorption experiments.
10. Only a few control experiments are possible in immunocytochemistry. One is the application of a preimmune serum or a nonimmune serum, which should fail to give the desired staining reaction. Another control is an absorption or immunoinhibition test. As discussed under Note 9, it is desirable that this absorption is done with the antigen fixed to a protein that is different from the carrier protein used for immunization. This procedure attempts to mimic conditions encountered in the tissue (which, unfortunately, cannot be known with certainty). Nevertheless, the protein environment of cGMP in the tissue or cells after formaldehyde

fixation is an important factor that may be decisive for the successful application of a cGMP-antiserum.

The specificity of the antiserum should also be established in model systems. In these model systems the procedures for tissue processing should be closely followed *(35)*. The models presented in these protocols have proven their usefulness in testing the specificity of antisera *(5,6,8,9,35,36)*. An important tool in establishing the specificity of antisera is the biological model, in which changes in levels or quality of the antigen under study (e.g., by pharmacological means) should be reflected in the degree of immunohistochemical staining. A complication in this analysis is the assessment of the immunohistochemical staining itself. Quantification of the immunohistochemical signal is sometimes difficult (especially with immunofluorescence) and requires sophisticated instrumentation *(35,37)*. However, the combination of nonbiological models with biological model systems presents a powerful set of control experiments.

cGMP-antisera raised in different animals against formaldehyde-fixed cGMP may yield different staining patterns in biological tissue. So far only polyclonal antisera have been raised against formaldehyde-fixed cGMP. These antisera contain an unknown mixture of antibodies directed against cGMP together with a number of amino acids of the carrier protein *(38)*. That portion of the conjugate that interacts with a specific antibody is the antigenic determinant. Because cGMP is fixed to functional groups that have different amino acid environments, the antigenic determinants will differ. Animals will raise polyclonal antisera that will differ in their potential to recognize specific antigenic determinants. Unlike the case for some biomolecules (such as neurotransmitters) that are found in the cell within strictly localized and specialized structures, cGMP is free in the cell. Therefore, cGMP will become fixed to a variety of cellular proteins. In combination with the unknown potential of the polyclonal antibodies, this results in antisera that show differential staining of formaldehyde-fixed cGMP in biological tissue *(9)*. This finding may complicate the specificity testing of antisera against formaldehyde-fixed cGMP in biological tissue.

Complications certainly arise in the interpretation of the results obtained with antisera directed against formaldehyde-fixed cGMP. No definite conclusions can be made regarding structures that show no cGMP accumulation under a variety of stimuli: cGMP may be present but fixed to a tissue protein so that it presents an antigenic determinant that is unrecognized by a certain cGMP-antiserum.

11. The quality of the gelatin is important for preparing suitable gelatin models. Sometimes it is convenient to swell the gelatin in cold water for 2 h before heating gently to 50°C. For easy processing it is necessary to keep the gelatin solution warm until pipeting into the tubing.
12. It is convenient to use a desiccator that can be placed into a heating stove. Take care to let the desiccator cool down before opening under a fume-hood. Vapor fixation of cGMP has not been studied in detail. The stated procedure worked at the author's lab. It is likely that variations in time and temperature can be made.

13. For blocking the protein-binding capacity of the nitrocellulose one can use a variety of protein preparations. The working instructions supplied by nitrocellulose manufacturers usually will include some suggestions. However, gelatin is an inexpensive alternative to commercial formulations.
14. Antibody dilution has to be established empirically. The final dilution depends on the detection system used. Our rabbit anti-cGMP antisera are diluted 1:300 v/v in TBS-T in combination with a secondary goat or sheep antirabbit antiserum conjugated with fluorescein-isothiocyanate (or CY3) as a reporter molecule. Furthermore, using Vibratome sections antisera can, in general, be diluted 4–10 times farther than using cryostat sections.

    Several variations can be made in the detection system, e.g., using alkaline phosphatase as a reporter enzyme coupled to the tertiary antibody. A two-step reaction, in which the reporter enzyme is coupled to the secondary antibody, may also be used.
15. Perfusion fixation of experimental animals is a routine procedure for (neuro)anatomical studies. It has been known for some time, however, that anesthetic agents decrease the level of cGMP in the brain *(39)*. Therefore, using perfused animals to study the localization of cGMP in the brain has the disadvantage that the procedure is not optimal for maintaining cGMP levels *(8,10)*. Infusion of the nitric oxide donor compound sodium nitroprusside prior to the fixative increases cGMP intracellular to high levels in a region-specific manner; this can be readily demonstrated using cGMP-immunocytochemistry *(30)*. So far, no studies have appeared with other procedures to increase cGMP levels in specific structures in the rat brain in combination with perfusion fixation.
16. Instead of the fixative described, a Bouin fixation can also be used. This fixative consists of 25 mL distilled water saturated with picric acid, 25 mL 40% formaldehyde solution (Merck; cat. no. 4003), and 5 mL glacial acetic acid. It is not advisable to use glutaraldehyde fixation other than in low concentrations (<0.1% v/v), because glutaraldehyde has an adverse effect on cGMP fixation *(10)*.
17. Keep sections in ice-cold buffer because of the lability of the cGMP-formaldehyde-protein bond (*see* Note 2).
18. It is advisable to remove the brain as quickly as possible and to cool it immediately in order to slow down metabolic processes.
19. This is a process of trial and error, in which the filter paper should be damp enough to adhere to the stainless-steel table, but not so wet that the tissue does not stick to the filter paper during the chopping process.
20. The quality of the slice depends to a certain extent on the method used to prepare the slices *(40,41)*. Using the McIlwain tissue chopper it is possible to prepare 100-μm thick slices. These are tedious to work with. However, slices of 300-μm thickness can, with some care, be handled routinely. Furthermore, 300-μm presents a thickness that is acceptable in terms of oxygen supply to the deeper layers of the slice *(42,43)*.

    When doing pharmacological experiments using 300-μm thick tissue slices, one has to take into account the rate and degree of equilibration of the drugs throughout the slice. This has been studied so far only for a few compounds: atrial natriuretic factor (5 and 10 min) *(44)*, sodium nitroprusside (5 and 10 min), NMDA (1 min), kainic acid (1 min), several phosphodiesterase-inhibitors (30 min), hemoglobin as an NO-trapping compound, and the inhibitors of nitric oxide synthase N^G-nitro-L-arginine and N^G-methyl-L-arginine (both 15 and 30 min). Penetration of atrial natriuretic factor into slices has been studied by making transverse sections of brain slices at different time-points; the diffusion process was followed by visualizing the atrial natriuretic peptide-stimulated cGMP-production *(44)*. For the other compounds, equilibration was established by comparing the immunocytochemical results of sections originating from different depths of the same slice (unpublished).

When the effect of nitric oxide donors on cGMP-accumulation is studied, one must take into account the rate of NO-release from the donor compound. The diffusion rate of NO is quite high, because of the small size and lipophilicity of this molecule *(45)*. Sodium nitroprusside (SNP) is often used an NO-donor. This compound is inexpensive, which is an important asset. However, the byproduct after NO-release, $[Fe(CN)_5]^{2-}$, is not devoid of pharmacological activity *(46)*. In addition, SNP is light sensitive, although in most cases this property should not rule out use of this compound. In 10-min incubations with brain slices, SNP and the NO-donors S-nitroso-acetyl-penicillamine and streptozotocin were equally effective in raising cGMP levels (unpublished).

21. Incubation of tissue slices in vitro in Krebs buffer offers the opportunity to study the effects of drugs on cGMP metabolism. Phosphodiesterase activity is an important variable that must be controlled *(47)*. Unfortunately, this is a very complex task. The enzyme family of phosphodiesterases comprises at least eight different subfamilies, each with their specific substrate and inhibitor profiles *(48)*. Furthermore, there is a regional localization of these enzymes in several tissues *(49–51)*. Therefore, it is not possible to give a general solution to the problem of phosphodiesterase inhibition. Most studies so far have been done in the presence of isobutylmethylxanthine, which is a nonspecific phosphodiesterase inhibitor, or without any inhibitor at all. Nevertheless, it has been found that in the enteric nervous system of the dog, a combination of different phoshodiesterase inhibitors is required to obtain cGMP accumulation *(33)*. For this tissue a species difference is apparent *(32,33)*. It is an advantage of the in vitro incubation method that the presence of phosphodiesterase inhibitors can be easily controlled. However, for every application, the influence of different phosphodiesterase inhibitors has to be studied.

    The effects of agonists on cGMP accumulation is normally studied after preincubation with the slice for 30 min. Inhibitors of nitric oxide synthase or phosphodiesterase activity are normally present from the start of the incubation. The time-point at which the effect of the agonist is to be evaluated must be determined experimentally.

22. It is essential to use a formaldehyde-based fixative. The formaldehyde-based Bouin fixative (*see* Note 16) gives the same results. The coupling reaction of

formaldehyde with cGMP and protein components is not instantaneous *(8)* and does not run to completion *(3,8)*. Not all cGMP is fixed to the tissue protein. We found that 38 ± 6% of the original content of cGMP in cerebellar slices was fixed to the tissue protein after cutting and washing of the sections prior to application of the primary antibody. The other 62% of the cGMP was washed away during the washing procedure.

Several recent studies have appeared that discuss a messenger role for extracellular cGMP *(52–54)*. However, these studies used the technique of in vivo microdialysis to monitor cGMP levels. The significance of this efflux, however, remains to be demonstrated because it is conceivable that under the conditions of microdialysis, an artifactual efflux of cGMP occurs. A comparison of the efflux of cGMP from brain slices with the studies on the efflux of cGMP from cultured cells reveals important differences. Apparently, efflux of large amounts of cGMP from cultured cells under conditions of unphysiologically high cGMP levels (e.g., refs. *55–57*) is unrelated to the situation found during in vitro incubation of brain slices, where cGMP efflux was found to be 10% or less *(58)*.

23. It is necessary to wash out the fixative because it can give an artifact consisting of precipitated formaldehyde when the slices are frozen in CO_2. The 30 min washing time is close to the minimal time required. Washing time might be extended to at least 18 h without affecting cGMP-immunocytochemistry provided the slices are kept in the cold.
24. The pharmacological experiments yield a large number of slices that need to be embedded in Tissue-Tek OCT compound before sectioning. Because cutting of the slices one at a time would be too time-consuming, the following procedure has been adopted. A glass slide is covered with Parafilm and the slices are aligned next to each other. Excess liquid is removed carefully with pieces of filter paper. Next, the glass slide is pushed face down into liquid Tissue-Tek OCT compound on a cryostat chuck using a vertical adjustment attachment. In this position the slices are frozen into the Tissue-Tek using CO_2.
25. Dehydrate the sections by dipping for 3 min in each of the following (fresh) ethanol solutions: 70–70–96–96–100–100%. This is followed by two washes of either xylene or Microclear; slides are then cover slipped using Entellan or Microcover. Microclear and Microcover are good nontoxic alternatives to the more hazardous compounds, xylene and Entellan.

References

1. Cumming, R., Dickison, S., and Arbuthnott, G. (1980) Cyclic nucleotide losses during tissue processing for immunocytochemistry. *J. Histochem. Cytochem.* **28,** 54,55.
2. Ortez, R. A., Sikes, R. W., and Sperling, H. G. (1980) Immunohistochemical localization of cyclic GMP in goldfish retina. *J. Histochem. Cytochem.* **28,** 263–270.
3. Rall, T. W. and Lehne, R. A. (1982) Evidence for cross-linking of cyclic AMP to constituents of brain tissue by aldehyde fixatives: potential utility in histochemical procedures. *J. Cycl. Nucleotides Res.* **8,** 243–265.

4. Steiner, A. L., Ong, S., and Wedner, H. J. (1976) Cyclic nucleotide immunocytochemistry. *Adv. Cycl. Nucleotides Res.* **7,** 115–155.
5. Schipper, J. and Tilders, F. J. H. (1983) A new technique for studying specificity of immunocytochemical procedures: specificity of serotonin immunostaining. *J. Histochem. Cytochem.* **31,** 12–18.
6. Milstein, C., Wright, B., and Cuello, A. C. (1983) The discrepancy between the cross-reactivity of a monoclonal antibody to serotonin and its immunohistochemical specificity. *Mol. Immunol.* **20,** 113–123.
7. Berkenbosch, F., Schipper, J., and Tilders, F. J. H. (1986) Corticotropin-releasing factor immunostaining in the rat spinal cord and medulla oblongata: an unexpected form of cross-reactivity with substance P. *Brain Res.* **399,** 87–96.
8. De Vente, J., Steinbusch, H. W. M., and Schipper, J. (1987) A new approach to immunocytochemistry of 3',5'-cyclic guanosine monophosphate: preparation, specificity, and initial application of a new antiserum against formaldehyde-fixed 3',5'-cyclic guanosine monophosphate. *Neuroscience* **22,** 361–373.
9. De Vente, J., Schipper, J., and Steinbusch, H. W. M. (1989) Formaldehyde fixation of cGMP in distinct cellular pools and their recognition by different cGMP-antisera. An immunocytochemical study into the problem of serum specificity. *Histochemistry* **91,** 401–412.
10. De Vente, J. and Steinbusch, H. W. M. (1992) On the stimulation of soluble and particulate guanylate cyclase in the rat brain and the involvement of nitric oxide as studied by cGMP-immunocytochemistry. *Acta Histochem.* **92,** 13–38.
11. Steiner, A. L., Parker, C. W., and Kipnis, D. M. (1972) Radioimmunoassay for cyclic nucleotides. 1. Preparation of antibodies and iodinated cyclic nucleotides. *J. Biol. Chem.* **247,** 1106–1113.
12. Rosenberg, E. M., LaVallee, H., Weber, P., and Tucci, S. M. (1979) Studies on the specificity of immunohistochemical techniques for cyclic AMP and cyclic GMP. *J. Histochem. Cytochem.* **27,** 913–923.
13. Wedner, H. J., Hoffer, B. J., Battenberg, E., Steiner, A. L., Parker, C. W., and Bloom, F. E. (1972) A method for detecting intracellular cyclic adenosine monophosphate by immunofluorescence. *J. Histochem. Cytochem.* **20,** 293–299.
14. Chan-Palay, V. and Palay, S. L. (1979) Immunohistochemical localization of cyclic GMP: light and electron microscope evidence for involvement of neuroglia. *Proc. Natl. Acad. Sci. USA* **76,** 1485–1488.
15. Garbers, D. L. and Lowe, D. G. (1994) Guanylyl cyclase receptors. *J. Biol. Chem.* **269,** 30,741–30,744.
16. Waldman, S. A. and Murad, F. (1987) Cyclic GMP synthesis and function. *Pharmacol. Rev.* **39,** 163–196.
17. Murad, F. (1994) Regulation of cytosolic guanylyl cyclase by nitric oxide: the NO-cyclic GMP signal transduction system. *Adv. Pharmacol.* **26,** 19–33.
18. Moncada, S., Palmer, R. M. J., and Higgs, E. A. (1991) Nitric oxide: physiology, pathophysiology and pharmacology. *Pharmacol. Rev.* **43,** 109–142.
19. Garthwaite, J. (1991) Glutamate, nitric oxide and cell-cell signalling in the nervous system. *Trends Neurosci.* **14,** 60–67.

20. Knowles, R. G. and Moncada, S. (1994) Nitric oxide synthases in mammals. *Biochem. J.* **298,** 249–258.
21. Palmer, R. M. J., Ashton, D. S., and Moncada, S. (1988) Vascular endothelial cells synthesize nitric oxide from L-arginine. *Nature* **333,** 664–666.
22. Marletta, M. A. (1993) Nitric oxide synthase: structure and mechanism. *J. Biol. Chem.* **268,** 12231–12234.
23. Bredt, D. S., Hwang, P. M., and Snyder, S. H. (1990) Localization of nitric oxide synthase indicating a neural role for nitric oxide. *Nature* **347,** 768–770.
24. Vincent, S. R. and Kimura, H. (1992) Histochemical mapping of nitric oxide synthase in the rat brain. *Neuroscience* **46,** 755–784.
25. Rodrigo, J., Springall, D. R., Uttenthal, O., Bentura, M. L., Abadia-Molina, F., Riveros-Moreno, V., Martinez-Murillo, R., Polak, J. M., and Moncada, S. (1994) Localization of nitric oxide synthase in the adult rat brain. *Phil. Trans. R. Soc. Lond. B* **345,** 175–221.
26. Knowles, R. G., Palacios, M., Palmer, R. M. J., and Moncada, S. (1989) Formation of nitric oxide from L-arginine in the central nervous system: a transduction mechanism for stimulation of the soluble guanylate cyclase. *Proc. Natl. Acad. Sci. USA* **86,** 5159–5162.
27. Bredt, D. S. and Snyder, S. H. (1992) Nitric oxide, a novel neuronal messenger. *Neuron* **8,** 3–11.
28. De Vente, J., Bol., J. G. J. M., Berkelmans, H. S., Schipper, J., and Steinbusch, H. W. M. (1990) Immunocytochemistry of cyclic GMP in the cerebellum of the immature, adult, and aged rat: the involvement of nitric oxide. A micropharmacological study. *Eur. J. Neurosci.* **2,** 845–862.
29. De Vente, J. and Steinbusch, H. W. M. (1993) Immunocytochemistry of second messenger molecules: the study of formaldehyde-fixed cyclic GMP, in *Immunohistochemistry II* (Cuello, A. C., ed.), Wiley, New York, pp. 409–427.
30. Southam, E. and Garthwaite, J. (1993) The nitric oxide-cyclic GMP signalling pathway in rat brain. *Neuropharmacology* **32,** 1267–1277.
31. Koistinaho, J., Swanson, R. A., De Vente, J., and Sagar, S. M. (1993) NADPH-diaphorase (nitric oxide synthase)-reactive amacrine cells of rabbit retina: putative target cells and stimulation by light. *Neuroscience* **57,** 587–597.
32. Young, H. M., McConalogue, K., Furness, J. B., and De Vente, J. (1993) Nitric oxide targets in the guinea-pig intestine identified by induction of cyclic GMP immunoreactivity. *Neuroscience* **55,** 583–596.
33. Shuttleworth, C. W., Xue, C., Ward, S. M., De Vente, J., and Sanders, K. M. (1993) Immunohistochemical localization of 3',5'-cyclic guanosine monophosphate in the canine proximal colon: responses to nitric oxide and electrical stimulation of enteric inhibitory neurons. *Neuroscience* **56,** 513–522.
34. De Vente, J., Schipper, J., and Steinbusch, H. W. M. (1993) A new approach to the immunocytochemistry of cAMP. Initial characterization of antibodies against acrolein-fixed cAMP. *Histochemistry* **99,** 457–462.
35. Berkenbosch, F. and Tilders, F. J. H. (1987) A quantitative approach to cross-reaction problems in immunocytochemistry. *Neuroscience* **23,** 823–826.

36. Larsson, L. I. (1981) A novel immunocytochemical model-system for specificity and sensitivity screening of antisera against multiple antigens. *J. Histochem. Cytochem.* **29,** 408–410.
37. De Vente, J., Garssen, J., Tilders, F. J. H., Steinbusch, H. W. M., and Schipper, J. (1987) Single cell quantitative immunocytochemistry of cyclic GMP in the superior cervical ganglion of the rat. *Brain Res.* **411,** 120–128.
38. Atassi, M. Z. (1984) Antigenic structures of proteins. Their determination has revealed important aspects of immune recognition and generated strategies for synthetic mimicking of protein binding sites. *Eur. J. Biochem.* **145,** 1–20.
39. Kant, G. J., Muller, T. W., Lenox, R. H., and Meyerhoff, J. L. (1982) *In vivo* effects of pentobarbital and halothane anesthesia in levels of adenosine 3',5'-monophosphate and guanosine 3',5'-monophosphate in rat brain regions and pituitary. *Biochem. Pharmacol.* **29,** 1891–1896.
40. Garthwaite, J., Woodham, P. L., Collins, M. J., and Balasz, R. (1979) On the preparation of brain slices: morphology and cyclic nucleotides. *Brain Res.* **173,** 373–377.
41. Aitken, P. G., Breese, G. R., Dudek, F. F., Edwards, F., Espanol, M. T., Larkman, P. M., Lipton, P., Newman, G. C., Nowak, T. S., Panizzon, K. L., Raly-Susman, K. M., Reid, K. H., Rice, M. E., Sarvey, J. M., Schoepp, D. D., Segal, M., Taylor, C. P., Teyler, T. J., and Voulalas, P. J. (1995) Preparative methods for brain slices: a discussion. *J. Neurosci. Meth.* **59,** 139–149.
42. Lipinski, H. G. and Bingmann, D. (1986) pO_2–dependent distribution of potassium in hippocampal slices of the guinea pig. *Brain Res.* **380,** 267–275.
43. Lipinski, H. G. and Bingmann, D. (1987) Diffusion in slice preparations bathed in unstirred solutions. *Brain Res.* **437,** 26–34.
44. De Vente, J., Bol, J. G. J. M., and Steinbusch, H. W. M. (1989) cGMP-producing, atrial natriuretic factor-responding cells in the rat brain. An immunocytochemical study. *Eur. J. Neurosci.* **1,** 436–460.
45. Wood, J. and Garthwaite J. (1994) Models of the diffusional spread of nitric oxide: implications for neural nitric oxide signalling and its pharmacological properties. *Neuropharmacology* **33,** 1235–1244.
46. East, S. J., Batchelor, A. M., and Garthwaite, J. (1991) Selective blockade of N-methyl-D-aspartate receptor function by the nitric oxide donor, nitroprusside. *Eur. J. Pharmacol.* **209,** 119–121.
47. Sonnenburg, W. K. and Beavo, J. A. (1994) Cyclic GMP and regulation of cyclic nucleotide hydrolysis. *Adv. Pharmacol.* **26,** 87–114.
48. Beavo, J. A. and Reifsnyder, D. H. (1990) Primary sequence of cyclic nucleotide phosphodiesterase isozymes and the design of selective inhibitors. *Trends Neurosci.* **11,** 150–155.
49. Billingsley, M. L., Polli, J. W., Balaban, C. D., and Kincaid, R. L. (1990) Developmental expression of calmodulin-dependent cyclic nucleotide phosphodiesterase in rat brain. *Dev. Brain Res.* **53,** 253–263.
50. Furuyama, T., Iwahashi, Y., Tano, Y., Takagi, H., and Inagaki, S. (1994) Localization of 63–kDa calmodulin stimulated phosphodiesterase mRNA in the rat brain by *in situ* hybridization histochemistry. *Mol. Brain Res.* **26,** 331–336.

51. Polli, J. W. and Kincaid, R. L. (1994) Expression of a calmodulin-dependent phosphodiesterase isoform (PDE1B1) correlates with brain regions having extensive dopaminergic innervation. *J. Neurosci.* **14,** 1251–1261.
52. Vallebuona, F. and Raiteri, M. (1993) Monitoring of cyclic GMP during cerebellar microdialysis in freely-moving rats as an index of nitric oxide synthase activity. *Neuroscience* **57,** 577–585.
53. Luo, D., Leung, E., and Vincent, S. R. (1994) Nitric-oxide dependent efflux of cGMP in rat cerebellar cortex: an *in vivo* microdialysis study. *J. Neurosci.* **14,** 263–271.
54. Laitinen, J. T., Laitinen, K. S. M., Tuomisto, L., and Airaksinen, M. M. (1994) Differential regulation of cyclic GMP levels in the frontal cortex and the cerebellum of anesthetized rats by nitric oxide: an *in vivo* microdialysis study. *Brain Res.* **668,** 117–121.
55. Schini, V., Grant, N. J., Miller, R. C., and Takeda, K. (1989) Morphological characterization of cultured bovine aortic endothelial cells and the effects of atriopeptin II and sodium nitroprusside on cellular and extracellular accumulation of cyclic GMP. *Eur. J. Cell Biol.* **47,** 53–61.
56. Ørbo, A., Jaeger, R., and Sager, G. (1993) Effect of serum and cell density on transmembrane distribution of cAMP and cGMP in transformed (C4–I1) and non-transformed (WI-38) human cells. *Int. J. Cancer* **55,** 957–962.
57. Wu, X. B., Brüne, B., Von Appen, F., and Ullrich, V. (1993) Efflux of cyclic GMP from activated human platelets. *Mol. Pharmacol.* **43,** 564–568.
58. Tjörnhammar, M. L., Lazaridis, G., and Bartfai, T. (1986) Efflux of cyclic guanosine 3',5'-monophosphate from cerebellar slices stimulated by L-glutamate or high K^+ or N-methyl-N'-nitro-N-nitrosoguanidine. *Neurosci. Lett.* **68,** 95–99.

10

Microwave Antigen Retrieval in Formaldehyde-Fixed Human Brain Tissue

Suzanna J. Newman and Stephen M. Gentleman

1. Introduction

The use of formaldehyde as a reagent for fixing tissue was first identified over a century ago *(1)*, and because of its efficacy in a broad range of situations, at varying concentrations and with virtually all tissues, it has remained one of the most widely used fixatives. Formaldehyde is a reactive electrophilic species *(2,3)* that reacts readily with primary amines, amide groups, aromatic rings, and alcoholic hydroxyl groups, crosslinking various biological macromolecules. For histological purposes, this crosslinking is highly desirable in that it allows considerable morphological detail to be maintained, but such fixation invariably results in a corresponding time-dependent loss of antigenicity *(4)*. Consequently, early immunocytochemical protocols used fresh/frozen or alcohol-fixed tissue only *(5)*, retaining tissue antigenicity, but at the expense of morphology. Despite the effects of formalin fixation, however, the loss of antigenicity is neither complete nor irreversible. Indeed, formaldehyde-fixed tissue has been successfully used for the detection of many different tissue antigens, such as neurotransmitters *(6)*, neuropeptides *(7)*, cell surface markers *(8)*, and enzymes *(9)*. The suitability of formalin-fixed tissue for immunocytochemistry has been further enhanced with the development of sensitive indirect staining methods, such as PAP (peroxidase antiperoxidase) *(10)* and, more recently, avidin-biotin complex (ABC) methods *(11)*.

In addition to improvements in the detection methods, attempts to enhance antigenicity have led to the development of a variety of pretreatment steps, of which the use of proteolytic enzymes, such as pronase and trypsin *(12,13)*, have been particularly well documented. Tissue sections are typically incubated in low concentrations of proteolytic enzyme (0.1–0.001%)

From: *Methods in Molecular Biology, Vol. 72: Neurotransmitter Methods*
Edited by: R. C. Rayne Humana Press Inc., Totowa, NJ

for anything from a few minutes to several hours. This method has been shown to improve the staining of some antigens, but with longer incubations tissue morphology suffers.

In addition to the various enzymatic approaches, heat pretreatment has recently become an increasingly popular alternative technique for improving antigen availability. The majority of immunocytochemical protocols promote the use of low temperatures during processing and avoid heat because of potential problems of protein denaturation. Consequently, it is especially ironic that an increasing number of protocols now include heating tissue samples to over 90°C by autoclaving, conventional radiant heat, or microwaving *(14)*. Shi et al. were the first to describe the beneficial effects of microwaving tissue as a means of retrieving certain antigens from archival material *(15)*. Heating is believed to lead to partial protein hydrolysis, particularly in the presence of acidic or basic buffers, which has the effect of revealing previously undetectable antigenic sites. Studies involving the use of synthetic peptides and the effect of microwaving in cleaving specific peptide bonds have shown that the effects are confined to aspartyl residues *(16)*. These studies show that aspartyl bonds with phenylalanine are hydrolyzed within 4 min, whereas all other nonaspartyl-containing dipeptides remain approx 75% intact. This effect is not confined to microwaving, but occurs as the result of other forms of heating, supporting the belief that this form of antigen retrieval is through the effects of heat rather than microwaves as such. Many investigators have reported additional improvements in antigen retrieval if the tissue is heated in the presence of metallic salts or urea, which are believed to act by altering the hydrophobicity of the protein, thereby destabilizing the tertiary structure *(14)*. As with any pretreatment method, however, prolonged heating can also have deleterious effects on some tissue antigens.

To date, a number of different antigens have been described as benefiting from heat pretreatment. Although many of these are cell surface markers *(14)*, beneficial effects have also been described for neurotransmitters such as VIP and serotonin *(14,15)*. Additionally, comparisons between protease digestion and microwaving methods suggest that microwaving produces superior results *(17)*.

Despite the apparent advantages of using pretreatment methods to reveal previously undetectable epitopes, it should be understood that one of the main criticisms leveled at these techniques is the possible production of false positives *(18)*. Although these methods are not believed to significantly alter patterns of antibody binding *(19)*, some examples have been reported *(20)*. With respect to heat pretreatment, a number of large studies encompassing over 300 antibodies *(14,17)* reported no changes to the antibody-binding pattern, although overirradiation may increase background staining. Therefore, the staining pattern of each new antibody should be fully assessed in conjunction with the appropriate, pretreated control sera. Indeed, the assessment of control serum staining is essential if the resulting staining pattern is to be accurately validated and interpreted.

This chapter will describe in detail how one microwave pretreatment technique has been used for improving and retrieving tissue antigenicity. Three examples will be shown where microwaving has been used by the authors to successfully enhance neurotransmitter (e.g., neuropeptide Y, Fig. 1A,B), cell surface marker (e.g., major histocompatibility complex class II antigens-HLA-DR, Fig. 1C,D), and protein (β-amyloid, Fig. 1E,F) immunostaining of normal and pathological human tissue. The latter example demonstrates how microwaving can improve the antigenicity of archival human tissue stored in formaldehyde for over two decades. Section 3. describes the authors' standard immunocytochemical protocol for microwave pretreatment of 10-μm thick tissue sections of human brain. Although the following protocol has not been used by the authors on free-floating sections, it has been successfully applied to sections of formalin-fixed rat brain.

2. Materials

2.1. For Microwave Pretreatment

1. Domestic microwave (>650 W).
2. Citrate buffer: 10 m*M* citric acid, pH 6.0, in double-distilled water.
3. Plastic staining racks. Metal staining racks should not be used because they generate sparks.
4. Plastic staining dishes.

2.2. For Immunostaining

1. Plastic staining dishes.
2. Histoclear (National Diagnostics [Hessle, Hull, UK]; cat. no. HS200).
3. Graded alcohols: 98.8% ethyl alcohol and diluted IMS for 90 and 70%.
4. PBS/peroxide: 0.3% hydrogen peroxide (v/v) in PBS.
5. Phosphate-buffered saline (PBS): 150 m*M* NaCl, 7 m*M* Na_2HPO_4, 2 m*M* KH_2PO_4. Make fresh as required.
6. Normal serum (Vector [Peterborough, UK]; S-1000 for rabbit primary, S-2000 for mouse primary).
7. Primary diluent: PBS containing 0.3% (v/v) Triton X-100, 0.01% (w/v) sodium azide, and either normal goat serum (rabbit primary) or normal horse serum (mouse primary) at 2% (v/v).
8. Second and third layer diluent: 0.3% (v/v) Triton X-100 in PBS.
9. PAP pen (AGAR [Stanstead, Essex, UK]; cat. no. L41973).
10. Humid chambers (NUNC bio-assay dishes, Gibco, UK; cat. no. 2-40835A).
11. Filter paper, 24 cm in diameter.
12. Cotton stick applicators with cotton tips removed (Western Laboratory Supplies, Aldershot, UK).

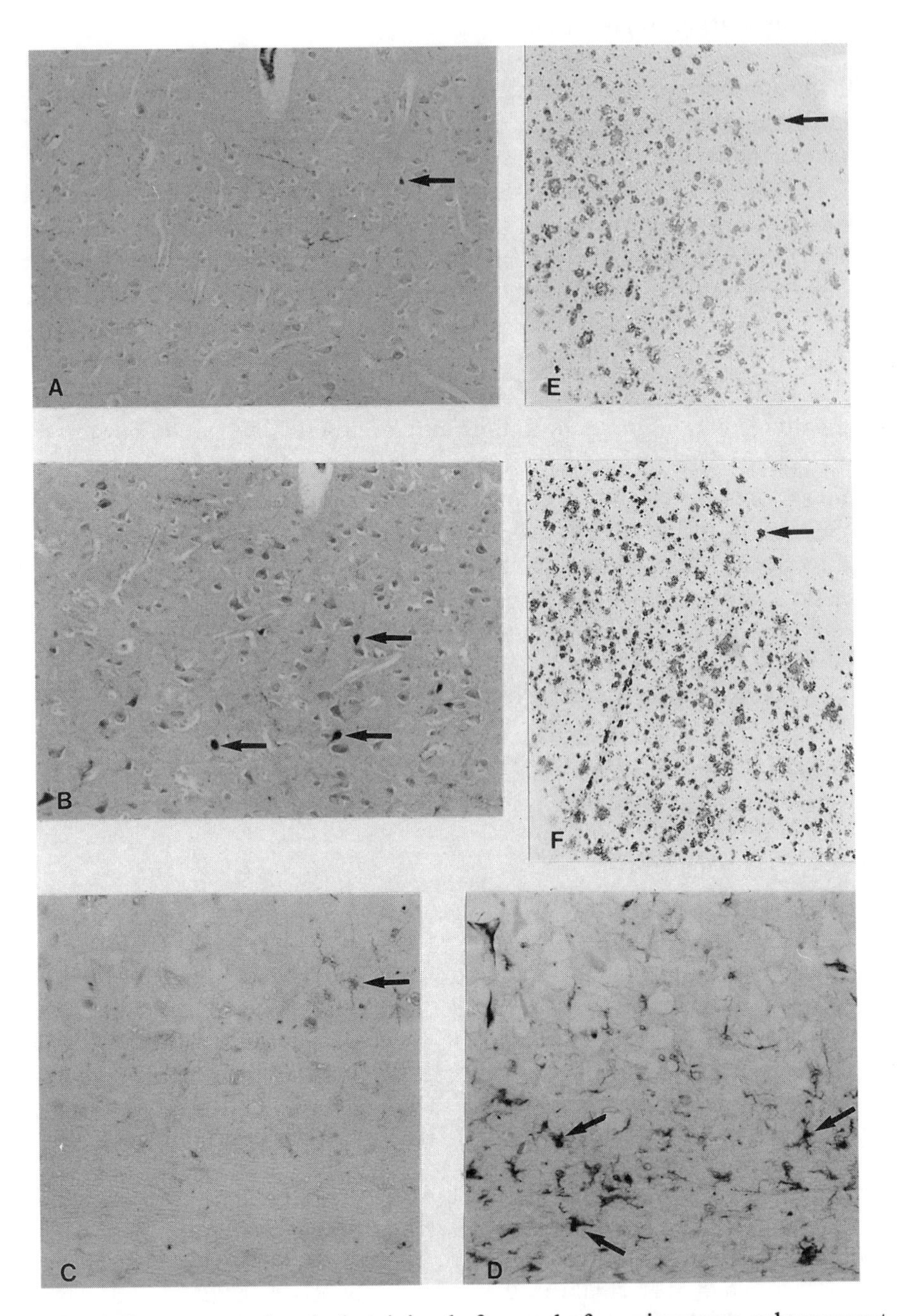

Fig. 1. Immunocytochemical staining before and after microwave enhancement. Neuropeptide Y-positive neurons (arrows) in normal human temporal cortex **(A,B)** ×120. HLA-DR positive microglia (arrows) in temporal cortex from a patient with Alzheimer's disease **(C,D)** ×200. β-Amyloid-positive plaques (arrows) in frontal cortex from a patient with Alzheimer's disease **(E,F)** ×40.

13. Biotinylated anti-IgG (Vector BA-1000 for rabbit primary, BA-2000 for mouse primary) Vectastain "Elite" kit (Vector PK-4000) (kit reagents should be prepared 30 min prior to use; *see* Section 3., step 16).
14. DAB stock solution: 25 mg/mL 3,3'-Diaminobenzidine (tetra-hydrochloride) in distilled water (Hazardous: *see* Notes 1 and 2).
15. DAB staining solution: To approximately 400 mL of PBS, add 4 × 1 mL aliquots of 3, 3'-DAB stock (for a 0.025% final concentration); add hydrogen peroxide to 0.03% (v/v, final concentration). Make fresh, just prior to step 19 in Section 3.
16. DPX mountant (BDH/Merck, Leicestershire, UK).
17. Coverslips (1.5 mm thick).

3. Methods

1. Prepare a humid chamber for the incubations in the immunocytochemistry procedure by placing a piece of moist filter paper in the base of a bio-assay tray. Then place four rows of cotton stick applicators in the chamber on which to rest the slides.
2. Place the slide-mounted sections in a slide rack and dewax the sections in Histoclear followed by rehydration in graded alcohols as follows: Histoclear for 15 min, Histoclear for 10 min, 98.8% EtOH for 10 min (×2), 90% IMS for 10 min, 70% IMS for 10 minutes, and dH_2O for 5 min.
3. Microwave approx 400 mL of citrate buffer (or sufficient to cover the slides) in a plastic staining dish for 4–5 min on full power until it simmers.
4. Place no more than six slides in a plastic staining rack (do not use metal!) (*see* Note 3). Microwave the slides at 70% power for 5 min (make sure the buffer continues to simmer).
5. After 5 min, check the level of the buffer to make sure that the slides are still covered (*see* Note 4). Top up if necessary with distilled water. Microwave for another 5 min.
6. After microwaving, leave the slides to cool for 15 min (*see* Note 5).
7. Rinse the slides in PBS (2 × 5 min).
8. Block endogenous peroxidase by incubating the sections for 30 min in PBS/ peroxide.
9. Rinse the slides in PBS (3 × 5 min).
10. Dry the area around each section with tissue paper, then carefully draw around the sections with the PAP pen (*see* Note 6).
11. Pipet either optimally diluted primary antibody (in primary diluent) or control serum onto the sections and place the slide into the humid chamber on the cotton applicator sticks.
12. Place the chamber in a refrigerator (at 4°C) overnight (*see* Note 7 for appropriate control steps).
13. Rinse the slides in PBS (3 × 5 min).
14. Incubate the sections with appropriate biotinylated anti-IgG at a concentration of 1:100 diluted in second layer diluent for 50 min at room temperature.
15. Rinse the slides in PBS (3 × 5 min).

16. Prepare third layer complex 30 min before use. Combine Vectastain Elite reagents A and B to give a final concentration of 1:200 with respect to both components; e.g., 5 μL A + 5 μL B + 990 μL of diluent.
17. Incubate sections in the third layer complex for 1 h at room temperature.
18. Rinse the slides in PBS (3 × 5 min).
19. Incubate the slides for 5–10 minutes in DAB staining solution (*see* Note 8).
20. Rinse the slides in PBS (3 × 5 min).
21. Dehydrate through an ascending series of alcohol washes (IMS—70, 90, and 98% ethyl alcohol—each for 5 min). Clear the slides in Histoclear (2 × 5 min) and permanently coverslip the sections with DPX nonaqueous mountant.

4. Notes

1. DAB is a suspected carcinogen and should only be handled in a fume cupboard. Lab coat, gloves, and safety glasses should be worn and waste DAB disposed of by an external contractor.
2. DAB solution may be aliquoted into 1 mL vol and frozen at –18°C. It will remain stable for 2–3 mo.
3. In order to ensure that the slides are equally affected by the heat treatment, it is important that only a few (five to six) slides are microwaved at a time. This is because the microwaves within the oven become unevenly distributed, leading to areas of low and high intensity. Those areas with a high intensity of microwaves are referred to as "hot spots." Tissue that is unevenly microwaved may produce variable or poor results. Microwave ovens with a turntable help to reduce this problem.
4. Always ensure that the tissue sections remain covered with buffer during the pretreatment step. Sections that are dry do not benefit from microwave irradiation since many proteins have to be hydrated in order to be susceptible to denaturation.
5. After microwaving, the majority of protocols advise that the sections be allowed to cool to room temperature in the buffer. Some investigators have noted that for certain antibodies, this part of the procedure is critical for successful immunostaining *(17)*, but for other antibodies this step appears to be less critical. Consequently, the effect of including this step of the protocol on staining should be assessed for each antibody.
6. The PAP pen produces a water-insoluble deposit that is used to prevent the different immunocytochemical reagents from draining off the section during incubations. The deposit is miscible with clearing agents, such as Histoclear and xylene, and will, therefore, be removed prior to slide coverslipping.
7. As with any technique, it is extremely important to include all appropriate controls. This is particularly the case with pretreatment methods, because this step can not only dramatically alter the staining pattern produced by the antibody of interest, but it can also affect the control normal serum staining pattern. Therefore, it is imperative that the effects of microwaving be assessed for each new antibody in conjunction with the relevant control serum. Ideally, additional sec-

tions should be microwaved and incubated with the preimmune serum from the animal used to raise the antibody, and diluted to the same concentration as the antibody of interest. Alternatively, if that serum is unavailable, use normal rabbit serum (for rabbit polyclonals), normal mouse serum (for mouse monoclonals), or preferably, if the isotype of the clone is known (e.g., IgG 1) then incubate the section with control isotyped ascites fluid. An additional control involves assessing the relative contribution (if any) of the second layer antibody to the staining pattern. This is achieved by omitting the primary antibody and using PBS instead; the rest of staining protocol remains the same.

8. Further enhancement of antigen labeling can be achieved by adding metal salts to the developing solution (e.g., 5% nickel chloride) that will result in a black end-product rather than brown, thereby increasing contrast *(21)*.

Acknowledgment

The authors would like to acknowledge Katherine Horton for providing the photographs of the bA4 immunostaining.

References

1. Blum, F. (1893) Der formaldehyde als hartungsmittel. *Z Wiss. Mikrosc.* **10,** 314.
2. Fox, C. H., Johnson, F. B., Whiting, J., and Roller, P. P. (1985) Formaldehyde fixation. *J. Histochem. Cytochem.* **33(8),** 845–853.
3. French, D. and Edsall, J. T. (1945) The reactions of formaldehyde with amino acids and proteins. *Adv. Protein Chem.* **2,** 278.
4. Leong, A. S.-Y. and Gilham, P. N. (1989) The effects of progressive formaldehyde fixation on the preservation of tissue antigens. *Pathology* **21,** 266–268.
5. Coons, A. H., Creech, H. J., and Jones, R. N. (1941) Immunological properties of an antibody containing a fluorescent group. *Proc. Soc. Exp. Biol. Med.* **47,** 200–202.
6. Gentleman, S. M., Falkai, P., Bogerts, B., Herrero, M. T., Polak, J. M., and Roberts, G. W. (1989) Distribution of galanin-like immunoreactivity in the human brain. *Brain Res.* **505,** 311–315.
7. Roberts, G. W. and Allen, Y. S. (1986) Immunocytochemistry of brain neuropeptides, in *Immunocytochemistry. Modern Methods and Applications* (Polak, J. M. and S. Van Noorden, eds.), John Wright and Sons, Bristol, UK, pp. 349–359.
8. Akiyama, H., Kawamata, T., Yamada, T., Tooyama, I., and McGeer, P. L. (1993) Expression of intercellular adhesion molecule (ICAM)-1 by a subset of astrocytes in Alzheimer disease and some other degenerative neurological disorders. *Acta Neuropathol.* **85,** 628–634.
9. Marshall, J. M., Jr. (1954) Distributions of chymotrypsin, procarboxypeptidase, deoxyribonuclease and ribonuclease in bovine pancreas. *Exp. Cell. Res.* **6,** 240–242.
10. Sternberger, L. A. (1979) *Immunocytochemistry* (2nd ed.), Wiley, New York.
11. Hsu, S.-M., Raine, L., and Fanger, H. (1981) Use of avidin-biotin-peroxidase complex (ABC) in immunoperoxidase techniques: a comparison between ABC and unlabelled peroxidase (PAP) procedures. *J. Histochem. Cytochem.* **29,** 577–580.

12. Huang, S. N., Minassian, H., and More, J. D. (1976) Application of immunofluorescent staining on paraffin sections improved by trypsin digestion. *Lab. Invest.* **35,** 383–390.
13. Finley, J. C. W. and Petrusz, P. (1982) The use of enzymes for improved localisation of tissue antigens with immunocytochemistry, in *Techniques in Immunocytochemistry*, vol. 1 (Bullock, G. R. and Petrusz, P., eds.), Academic, London, pp. 239–249.
14. Cattoretti, G., Pileri, S., Parravicini, C., Becker, M. H. G., Poggi, S., Bifulco, C., Key, G., D'amato, L., Sabbattini, E., Feudale, E., Reynolds, F., Gerdes, J., and Rilke F. (1993) Antigen unmasking on formalin-fixed paraffin-embedded tissue sections. *J. Pathol.* **171,** 83–98.
15. Shi, S.-R., Key, M. E., and Kalra, K. L. (1991) Antigen retrieval in formalin-fixed, paraffin-embedded tissues: an enhancement method for immunohistochemical staining based on microwave oven heating of tissue sections. *J. Histochem. Cytochem.* **39(6),** 741–748.
16. Chi-Yue, W., Shui-Tein, C., Shyh-Horng, S., and King-Tsung, W. (1992) Specific peptide-bond cleavage by microwave irradiation in weak acid solution. *J. Protein Chem.* **11(1),** 45–50.
17. Gown, A. M., de Wever, N., and Battifora H. (1993) Microwave-based antigenic unmasking. A revolutionary new technique for routine immunohistochemistry. *Appl. Immunohistochem.* **1(4),** 256–266.
18. Hayderman, E. (1979) Immunoperoxidase technique in histopathology: applications, methods and controls. *J. Clin. Pathol.* **32,** 971–978.
19. Horton, A. J. (1993) Microwave oven heating for antigen unmasking in routinely processed tissue sections. *J. Pathol.* **171,** 79–80.
20. Charalambous, C., Sing, N., and Isaacson, P. G. (1993) Immunohistochemical analysis of Hodgkin's disease using microwave heating. *J. Clin. Pathol.* **46(12),** 1085–1088.
21. Shu, S., Ju, G., and Fan, L. (1988) The glucose oxidase-DAB-nickel method in peroxidase histochemistry of the nervous system. *Neurosci. Lett.* **85,** 169–171.

11

Localization of Nitric Oxide Synthase Using NADPH Diaphorase Histochemistry

Maurice R. Elphick

1. Introduction

The gas nitric oxide (NO) was first identified as a putative neurotransmitter substance in the mammalian brain by Garthwaite et al. in 1988 *(1)*. In 1990, the enzyme responsible for synthesis of NO in the brain, NO synthase (NOS), was purified and antibodies to it were used to establish the distribution of NO-producing neurons in the rat brain *(2)*. NOS was also found to be responsible for an enzyme activity in the brain known as NADPH diaphorase that was first described by Thomas and Pearse in 1961 *(3–5)*.

NADPH diaphorase is an enzyme that catalyzes NADPH-dependent reduction of a tetrazolium salt, such as nitro blue tetrazolium (NBT), into an insoluble colored formazan. In theory, any NADPH-requiring enzyme could exhibit NADPH diaphorase activity. However, when NADPH diaphorase histochemistry was applied to aldehyde-fixed mammalian brain tissue, only a specific population of neurons stained *(6)*. These NADPH-diaphorase-positive neurons did not correspond to any of the known classical neurotransmitter systems and they remained a mystery until the discovery of NO-producing neurons in the brain.

NADPH diaphorase histochemistry is now the most widely used method for revealing the distribution of NOS in the nervous system. Using this approach, the anatomy of putative NO-producing neurons has been described in the nervous systems of several mammalian species, a variety of other vertebrate species, and invertebrates *(7–11)*. A major advantage of the NADPH diaphorase method over such alternatives as NOS immunocytochemistry and *in situ* hybridization is that the only specific reagents required are NBT and NADPH, both of which can be easily purchased. Moreover, to date anti-NOS antibodies have been raised only to forms of the enzyme obtained from mammals and

From: *Methods in Molecular Biology, Vol. 72: Neurotransmitter Methods*
Edited by: R. C. Rayne Humana Press Inc., Totowa, NJ

these immunoprobes may not recognize NOS in other animal types. Similarly, NOS cDNA sequences are only known for mammalian species *(12,13)* so *in situ* hybridization methods may be applicable only to mammals at present.

The major disadvantage of the NADPH diaphorase method is that it is not known whether this technique is always specific for NOS. In aldehyde-fixed mammalian nervous tissue, NADPH diaphorase activity can be attributed to NOS *(14)*, but in other animal types, where antibodies to native NOS proteins are not yet available, caution should be exercised when interpreting NADPH diaphorase staining. Other techniques should be used to verify the presence of NO-producing neurons in which NADPH diaphorase staining has been observed. For example, when the NADPH diaphorase staining technique was first applied to the insect brain, the most striking observation was intense activity associated with the olfactory processing centers, the antennal lobes *(10)*. Biochemical measurement of NOS activity in different parts of the insect brain also revealed high levels of NOS activity in the antennal lobes *(10)*, supporting the view that the observed NADPH diaphorase activity could be attributed to NOS.

In this chapter, the NADPH diaphorase staining technique will be described in detail. The methods here are based on those used by the author in investigating the anatomy of NO signaling in invertebrate nervous systems, in particular those of the locusts *Schistocerca gregaria* and *Locusta migratoria* and the mollusk *Lymnaea stagnalis (10,11)*.

Section 3.1. describes application of the NADPH diaphorase method to fixed, whole nervous tissue. This approach has not been applied to vertebrate central nervous systems because they are simply too large and multilayered for it to be of any use in revealing NOS anatomy. It has been used in studies of the mammalian peripheral autonomic nervous system *(15)*, but where the whole-mount staining method has proven especially useful is in the analysis of NOS anatomy in the smaller and simpler central nervous systems of invertebrates *(10)*.

Section 3.2. will describe preparation of cryostat-sectioned nervous tissue for NADPH diaphorase staining. This approach is the most widely used and has been applied to numerous vertebrate species and a few invertebrate species.

2. Materials

2.1. Both Methods

1. Phosphate-buffered saline (PBS): 0.1*M* sodium phosphate, pH 7.4, containing 0.9% (w/v) NaCl.
2. Paraformaldehyde fixative: 4% (w/v) in PBS. Heat gently at 65°C, stirring continuously until the solution has cleared completely.
3. Tris buffer: 50 m*M* Tris-HCl, pH 7.5.
4. Tris-T: 50 m*M* Tris-HCl, pH 7.5, containing 2% (v/v) Triton X-100.

5. NADPH diaphorase staining solution: 1 m*M* β-NADPH, 0.25 m*M* NBT in Tris buffer. Prepare fresh and filter just prior to the steps required (Section 3.1., step 4 and Section 3.2., step 7).
6. Filters, 0.2 μm (Sigma, St. Louis, MO; F-1387).

2.2. Staining of Whole Tissue Only

1. Cavity microscope slides.
2. Xylene-based mounting medium (e.g., BDH Fluoromount, cat. no. 36098 2B).
3. Xylene.
4. Ethanol.

2.3. Staining of Sectioned Tissue Only

1. Cryoprotection solution: 10% (w/v) sucrose in PBS (*see* Section 2.1., item 1).
2. Tissue-Tek OCT compound (Miles Inc., Elkhart, IN).
3. Slide coating solution: 0.5% (w/v) gelatin, 0.5% (w/v) chrome alum (chromium [III] potassium sulfate 12-hydrate) in distilled water.
4. Coated microscope slides (*see* Note 1).
5. Coverslips.
6. Cryostat.
7. Aqueous mounting medium (e.g., Immu-mount; Shandon, Pittsburgh, PA).
8. Nail varnish.

3. Methods

3.1. NADPH Diaphorase Staining of Whole Tissues

1. Incubate tissue with paraformaldehyde fixative for 4 h at 4°C (*see* Note 2).
2. Incubate tissue overnight at 4°C in Tris-T.
3. Wash tissue with Tris buffer (3 × 15 min) at room temperature.
4. Incubate tissue in freshly prepared, filtered NADPH diaphorase staining solution at room temperature in the dark until deep purple staining appears, usually about 60 min (*see* Note 3).
5. Wash tissue in Tris buffer (3 × 15 min).
6. Photograph at this stage by placing tissue on a cavity microscope slide under a coverslip with Tris buffer as a mounting medium (*see* Fig. 1A).
7. Produce permanent preparations by mounting in an aqueous-based medium, such as Immu-mount, or in a xylene-based mounting medium, such as Fluoromount, following dehydration through an ethanol series (i.e., 5 min each in 10, 20, 30, 40, 50, 60, 70, 80, and 90% ethanol in distilled water followed by three 5-min steps in absolute ethanol; *see* Note 4).

3.2. NADPH Diaphorase Staining of Tissue Sections

1. Incubate tissue with paraformaldehyde fixative for 4 h at 4°C (*see* Note 2).
2. Incubate tissue in cryoprotection solution overnight at 4°C.
3. Embed tissue in Tissue-Tek OCT compound and freeze (*see* Note 5).

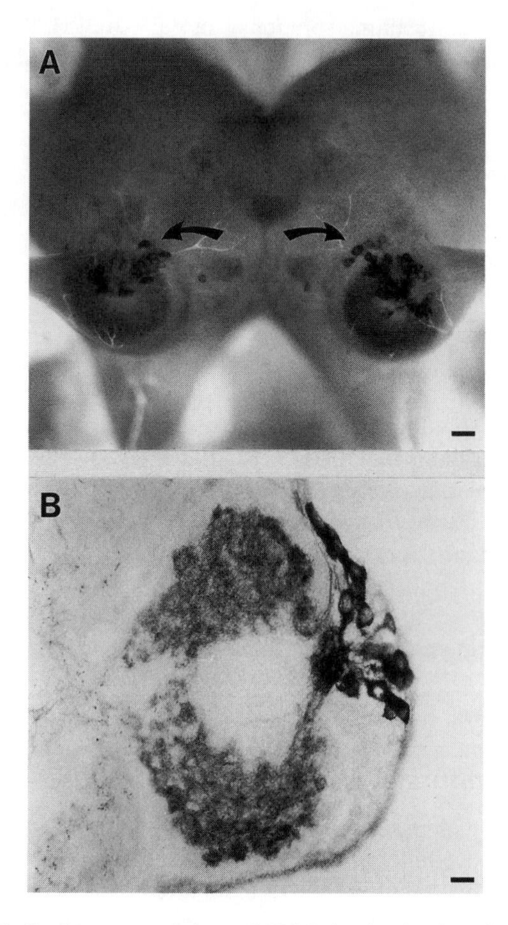

Fig. 1. NADPH diaphorase staining of NOS in the brain of the locust *Locusta migratoria.* **(A)** Whole-mount preparation showing staining of a population of neurons in the antennal lobes (arrows). **(B)** Parasagittal cryostat section of the locust brain showing stained neurons within antennal lobe. The staining seen here is very similar to that previously described in another locust species, *Schistocerca gregaria (10)*. Scale bars: (A) 80 µm, (B) 23 µm.

4. Cut sections of tissue (e.g., 20–40 µm) using a cryostat.
5. Collect sections on coated glass microscope slides. Keep at room temperature until dry.
6. Wash slides with Tris buffer 2 × 15 min and then wash 1 × 15 min with Tris-T.
7. Cover each slide with about 0.5 mL of freshly prepared filtered NADPH diaphorase staining solution. Leave slides in the dark at room temperature until deep purple staining is observed (usually about 30 min).

8. Wash slides in distilled water (3 × 15 min) and then cover sections with a coverslip using an aqueous-based mounting medium (*see* Note 4).
9. Leave slides to dry overnight, seal coverslip using nail varnish, and photograph (*see* Fig. 1B).

4. Notes

1. Microscope slides are coated with chrome alum-gelatin by immersion for 1 min in a solution of aqueous 0.5% (w/v) gelatin and 0.5% (w/v) chrome alum. Slides are then dried in an oven. The coating of slides with chrome alum-gelatin is required to provide a surface to which cryostat sections will adhere. If slides are not coated, sections may detach from slides during washing steps.
2. The specificity of the NADPH diaphorase staining technique for NOS is dependent on aldehyde fixation. For unknown reasons, the NADPH diaphorase activity of NOS is resistant to fixation, whereas the NADPH diaphorase activity of other NADPH-requiring enzymes in the brain is destroyed by aldehydes *(14)*. Fixation in 4% paraformaldehyde for 4 h at 4°C should provide sufficient fixation for small pieces of tissue (<0.5 cm^3). Longer fixation times or fixation at room temperature should be used if the staining observed appears nonspecific (i.e., if all of the tissue stains purple).
3. A modification of the whole-mount staining method may prove useful in revealing NOS anatomy in deeper regions of tissue (R. A. Colbert, personal communication). This technique involves preincubation of the tissue in 0.25 m*M* NBT/50 m*M* Tris-HCl, pH 7.5, at 4°C for 60 min before addition of β-NADPH (to 1 m*M*) and further incubation at 4°C. This ensures that the substrate of the NADPH diaphorase reaction (NBT) can penetrate throughout the tissue before reduction by H^+ ions derived from NADPH.
4. The purple formazan product of the NADPH diaphorase reaction is insoluble in water but can be soluble in alcohols. The author has found that dehydration through an ethanol series can remove some but not all of the formazan. For example, in locust brain sections NADPH diaphorase staining of the antennal lobes can still be clearly seen even after alcohol dehydration (*see* Fig. 1B), but in other parts of the brain, where staining is less intense, many of the fine features of staining are lost. Therefore, for sectioned material it is probably better to avoid alcohol dehydration and to use aqueous-based mounting media. With whole-mount preparations, however, xylene can be utilized as a clearing agent aiding visualization of staining within deeper regions of tissue. Therefore, although some formazan may be lost, alcohol dehydration prior to mounting in a xylene-based mounting medium is recommended for staining of whole tissue.
5. If it is important that the orientation of the sectioned tissue is known, then methods should be used whereby the tissue can be mounted in Tissue-Tek in a known orientation. The author uses two L-shaped brass molds and a small brass plate to form a cuboid cavity (0.75 cm^3) into which the tissue is placed in a known orientation. The cavity is then filled with Tissue-Tek before immersion in liquid N_2. This forms a frozen cuboid block that after removal from the mold (using a razor blade) can be attached to a cryostat chuck in a known orientation.

References

1. Garthwaite, J., Charles, S. L., and Chess-Williams, R. (1988) Endothelium-derived relaxing factor release on activation of NMDA receptors suggests role as intercellular messenger in the brain. *Nature* **336,** 385–388.
2. Bredt, D. S., Hwang, P. M., and Snyder, S. H. (1990) Localization of nitric oxide synthase indicating a neural role for nitric oxide. *Nature* **347,** 768–770.
3. Dawson, T. M., Bredt, D. S., Fotuhi, M., Hwang, P. M., and Snyder, S. H. (1991) Nitric oxide synthase and neuronal NADPH diaphorase are identical in brain and peripheral tissues. *Proc. Natl. Acad. Sci. USA* **88,** 7797–7801.
4. Hope, B. T., Michael, G. J., Knigge, K. M., and Vincent, S. R. (1991) Neuronal NADPH diaphorase is a nitric oxide synthase. *Proc. Natl. Acad. Sci. USA* **88,** 2811–2814.
5. Thomas, E. and Pearse, A. G. E. (1961) The fine localization of dehydrogenases in the nervous system. *Histochemie* **2,** 266–282.
6. Scherer-Singler, U., Vincent, S. R., Kimura, H., and McGeer, E. G. (1983) Demonstration of a unique population of neurons with NADPH-diaphorase histochemistry. *J. Neurosci. Methods* **9,** 229–234.
7. Mizukawa, K., Vincent, S. R., McGeer, P. L., and McGeer, E. G. (1990) Distribution of reduced-nicotinamide-adenine-dinucleotide-phosphate diaphorase-positive cells and fibers in the cat central nervous system. *J. Comp. Neurol.* **279,** 281–311.
8. Luebke, J. I., Weider, J. M., McCarley, R. W., and Greene, R. W. (1992) Distribution of NADPH-diaphorase positive somata in the brainstem of the monitor lizard *Varanus exanthematicus*. *Neurosci Lett.* **148,** 129–132.
9. Schober, A., Malz, C. R., Schober, W., and Meyer, D. L. (1994) NADPH-diaphorase in the central nervous system of the larval lamprey (*Lampetra planeri*). *J. Comp. Neurol.,* **345,** 94–104.
10. Elphick, M. R., Rayne, R. C., Riveros-Moreno, V., Moncada, S., and O'Shea, M. (1995) Nitric oxide synthesis in locust olfactory interneurons. *J. Exp. Biol.* **198,** 821–829.
11. Elphick, M. R., Kemenes, G., Staras, K., and O'Shea, M. (1995) Behavioral role for nitric oxide in chemosensory activation of feeding in a mollusc. *J. Neurosci.* **15,** 7653–7664.
12. Bredt, D. S., Glatt, C. E., Hwang, P. M., Fotuhi, M., Dawson, T. M., and Snyder, S. H. (1991) Nitric oxide synthase protein and mRNA are discretely localised in neuronal populations of the mammalian CNS together with NADPH diaphorase. *Neuron* **7,** 615–624.
13. Bredt, D. S., Hwang, P. M., Glatt, C., Lowenstein, C., Reed, R. R., and Snyder, S. H. (1991). Cloned and expressed nitric oxide synthase structurally resembles cytochrome P-450 reductase. *Nature* **352,** 714–718.
14. Matsumoto, T., Nakane, M., Pollock, J. S., Kuk, J. E., and Förstermann, U. (1993) A correlation between soluble brain nitric oxide synthase and NADPH diaphorase is only seen after exposure of the tissue to fixative. *Neurosci. Lett.* **155,** 61–64.
15. Grozdanovic, Z., Baumgarten, H. G., and Brüning, G. (1992) Histochemistry of NADPH diaphorase, a marker for neuronal nitric oxide synthase, in the peripheral autonomic nervous system of the mouse. *Neuroscience* **48,** 225–235.

12

In Situ Hybridization to Determine the Expression of Peptide Neurotransmitters

Niovi Santama

1. Introduction

The application of *in situ* hybridization for the detection of neuropeptide mRNA has made a very significant contribution to the understanding of fundamental mechanisms of gene expression in neural tissues. First, as a powerful tool for analyzing the pattern of spatial and temporal expression of neuropeptide genes, *in situ* hybridization has provided useful insights into the functional implications of mRNA distribution and localized neuropeptide synthesis. In addition, *in situ* hybridization allows the localization of different forms of alternatively spliced transcripts from a single neuropeptide gene to be determined with very high sensitivity and resolution, essentially at the single cell level. This feature is significant, because alternative pre-mRNA splicing has proven to be an almost ubiquitous mechanism for posttranscriptional regulation and information amplification in the nervous system.

The protocol described in this chapter makes use of single-stranded, synthetic DNA oligonucleotides (25–30 nucleotides) that are designed to be specific probes for each of the exons of a multiexon neuropeptide gene. The probes are nonisotopically 3' end-labeled with digoxygenin dUTP tailing. They are applied onto paraformaldehyde-fixed, paraffin-sectioned tissue, hybridized and washed at high stringency to ensure signal specificity. Hybrids are detected immunologically and the hybridization patterns for different exons are compared directly on consecutive serial sections. Using this protocol, the differential and mutually exclusive expression of the alternatively spliced transcripts of a neuropeptide gene may be clearly visualized at the single cell level. Although this protocol has been optimized for the invertebrate CNS (specifi-

From: *Methods in Molecular Biology, Vol. 72: Neurotransmitter Methods*
Edited by: R. C. Rayne Humana Press Inc., Totowa, NJ

cally for the pulmonate mollusc *Lymnaea stagnalis*) the basic steps can easily be adjusted to study neural tissue derived from other sources.

The significant advantages of the described protocol rest in the versatility of probe design, the increased tissue permeability and hybridization efficiency because of the small probe size and, finally, the superior tissue preservation as a result of the rapid freezing/dehydration and vapor-induced fixation. In addition, tissue preparation with the proposed protocol is compatible with immunocytochemistry. It is possible, therefore, to obtain serial consecutive paraffin sections from the same CNS and to combine *in situ* hybridization with immunocytochemistry for direct comparison of the signal from the two techniques in the same cells. For a more comprehensive and detailed review of *in situ* hybridization in general, the reader is referred to the excellent recent practical guide by Leitch et al. *(1)* and also to the seminar series by Harris and Wilkinson *(2)*.

2. Materials

To avoid degradation of target sequences by the ever-present RNases, double-distilled, sterile water, and sterile plasticware should be used for the preparation of all solutions. All glassware as well as the glass slides used for tissue section mounting should be baked in a 150°C oven overnight. It is recommended that all handling is carried out with gloves. These precautions are sufficient in most cases, but if RNase contamination appears to be a problem, treatment of water with DEPC and the inclusion of specific RNase inhibitors (such as RNasin from Stratagene, La Jolla, CA) will be required. All common chemicals and solutions are analytical grade.

2.1. Probe Preparation and Labeling

1. Custom-made synthetic oligonucleotides (usually 10–50 nucleotides), cleaved from the solid matrix, deprotected and dialyzed against distilled water (*see* Note 1).
2. Sephadex G25 (Pharmacia), prespun and packed in 1-mL disposable sterile plugged syringes.
3. TEN buffer: 10 m*M* Tris-HCl, pH 8.0, 1 m*M* EDTA, pH 8.0, 100 m*M* NaCl.
4. Digoxygenin-deoxy-uridine 5'-triphosphate stock solution at 25 n*M* in distilled H_2O (DIG-11-dUTP from Boehringer Mannheim, Mannheim, Germany).
5. Terminal deoxynucleotidyl transferase (25 U/µL from Boehringer Mannheim).
6. 5X labeling buffer: 1*M* potassium cacodylate, 125 m*M* Tris-HCl, pH 6.6, 1.25 mg/mL BSA (the buffer is supplied with the terminal transferase enzyme by the manufacturer).
7. Cobalt chloride solution: 25 m*M* $CoCl_2$ in distilled water (also supplied with item 5 by the manufacturer; *see* Note 2).
8. Hybridization buffer: 25% (w/v) deionized formamide, 3X standard sodium citrate (SSC; 1X SSC is 150 m*M* sodium chloride, 15 m*M* sodium citrate, pH 7.0),

0.1% (w/v) polyvinyl pyrrolidone, 0.1% (w/v) Ficoll, 1% (w/v) BSA, 500 μg/mL sheared herring (or salmon) sperm DNA, and 500 μg/mL *Escherichia coli* or yeast tRNA.

9. Sodium acetate solution: 3*M*, pH 5.2, in distilled water.
10. Ethanol: Absolute and 70% (v/v) in distilled H_2O.

2.2. Tissue Preparation and Prehybridization Treatment

A refrigerated vacuum chamber, incubation ovens at 37 and 60°C, a conventional microtome for paraffin sectioning, and a histological slide drying bench will be required.

1. Cryotubes, 2-mL capacity (Nunc).
2. Dissection medium: A HEPES-buffered saline medium or another medium of choice.
3. Freezing media: Pressurized freon aerosol and liquid nitrogen.
4. Embedding medium: Paraffin wax pellets, preferably the type blended with synthetic polymers/DMSO and congealing point at 56–58°C, obtainable from Sigma (St. Louis, MO).
5. Slide coating solution: 0.5% (w/v) chrome alum, 0.5% (w/v) gelatin in distilled water.
6. Chrome alum/gelatin precoated glass microscope slides (*see* Note 3).
7. Section mounting medium: 1% glycerin-albumin in distilled water, freshly prepared.
8. Xylene.
9. Methanol.
10. Phosphate-buffered saline (PBS): 140 m*M* NaCl, 2.7 m*M* KCl, 5.37 m*M* Na_2HPO_4, 1.76 m*M* KH_2PO_4, pH 7.2. To make 1 L, use: 8 g/L NaCl, 0.2 g/L KCl, 1.44 g/L Na_2HPO_4 (heptahydrate), 0.24 g/L KH_2PO_4 (anhydrous).
11. Pepsin solution: 0.1% (w/v) pepsin in 0.2*N* HCl, pH 2.0 (prepare fresh, just prior to use and prewarm at 37°C a few minutes before application; *see* Section 3.2., step 10).
12. Postfixative: 4% (w/v) paraformaldehyde, freshly made in PBS. Heat gently at 65°C, stirring continuously until the solution has cleared completely. Cool to room temperature.
13. Hydroxyl ammonium chloride: 1% (w/v) hydroxyl ammonium chloride, freshly made in PBS.
14. RNase A and DNase I (RNase-free, i.e., from Stratagene) for control preparations.

2.3. Hybridization and Immunological Detection of Hybrids

1. 10X SSC stock solution (*see* 1X SSC, Section 2.1., item 8).
2. Alkaline phosphatase (AP)-conjugated antidigoxygenin antiserum (Boehringer Mannheim).
3. Antibody dilution buffer: 100 m*M* Tris-HCl, 150 m*M* NaCl, pH 7.5, containing 1% (w/v) "blocking reagent" (from Boehringer Mannheim).

4. Alkaline phosphatase substrate: 165 µg/mL (w/v) x-phosphate (5-bromo-4-chloro-3-indolyl-phosphate, 4-toluidine salt) and 330 mg/mL (w/v) nitro blue tetrazolium chloride (NBT), mixed in 200 m*M* Tris-HCl, 10 m*M* $MgCl_2$, pH 9.2. Concentrated stock solutions of the two detection reagents (x-phosphate and NBT) can be prepared in 70% formamide, stored in frozen aliquots (–20°C), and mixed with the reaction buffer just prior to use.
5. Water-based mounting medium (*see* Note 4).

3. Methods

3.1. Probe Preparation and Labeling

1. Prior to probe labeling, purify the crude oligonucleotide by applying it, in a total volume of 100 µL, to the top of a prespun and packed 1-mL Sephadex G25 column that has been washed several times with sterile TEN buffer. Hold the spun column in a small Falcon tube containing a decapped Eppendorf tube at the bottom and centrifuge at 1600*g* for 3 min (*see* Note 5).
2. Collect the effluent, containing the oligonucleotide probe, in the decapped tube and then transfer to a fresh tube. Determine the concentration spectrophotometrically at 260 nm (*see* Note 6).
3. Label the oligonucleotides. Prepare the labeling reaction with distilled water to a final volume of 25 µL containing: 200 ng of oligonucleotide, 5 µL of 5X labeling buffer, 1 µL of $CoCl_2$ stock solution, 2.5 µL of DIG-11-dUTP, and 1 µL of terminal transferase. Permit the reaction to proceed for 15 min at 37°C (*see* Notes 7 and 8).
4. After labeling, make up the volume to 100 µL with water and precipitate the purified oligonucleotide by addition of 0.1 vol of sodium acetate solution and 2.5 vol of absolute ethanol. Keep at –20°C for at least 30 min and recover the probe by centrifugation at 15,000*g* for 15 min.
5. Discard the supernatant and rinse the pellet (which may be invisible) gently with 70% ethanol and air dry until all ethanol has evaporated.
6. Resuspend the pellet in hybridization buffer in a volume that will give a final concentration of 0.25 ng/µL (i.e., 800 µL). A convenient volume of 80 µL of this stock solution (equivalent to 20 ng of probe) can be directly applied to each slide (*see* Note 9).

3.2. Tissue Preparation and Prehybridization Treatment

1. A few minutes before starting the dissections, assemble a "quick-freezing set up." With a hot, fine injection needle puncture several holes at the top part of screw-cap cryotubes. Place these tubes at the top position of metal canes of the type used for storage in liquid nitrogen. Immerse the canes into a liquid nitrogen-filled Dewar flask so that the tubes are held above and cooled by the liquid, but are not immersed in it (*see* Note 10).
2. Dissect the central nervous system (CNS) (or other organs of interest) in dissection buffer. In the meantime, spray pressurized aerosol freon into the cryotubes of your quick-freezing setup, which will instantly liquefy on touching the precooled surface of the tube.

3. Drain excess liquid from the dissected tissue on waxed paper and immediately freeze the tissue by dropping it into the liquid freon in the cryotubes. Leave the tissue for 3–5 min, then screw the caps on and rapidly invert cryotubes so that all the liquid freon pours out of the prepunctured holes. Immediately immerse the capped tubes in liquid nitrogen. The tissue can be stored in liquid nitrogen practically indefinitely or can be used immediately.
4. For immediate use, transfer tubes inside the chamber of a refrigerated vacuum dryer (i.e., an Edwards Freeze-Dryer Modulyo), precooled at –50°C and dehydrate at about 0.06 mbar at –50°C for approx 24 h (*see* Note 11).
5. Quickly transfer the dehydrated tissue to a small Petri dish and place the dish on top of powdered paraformaldehyde in a tightly sealed jar, prewarmed at 60°C. Place the jar into a 60°C oven for 1.5 h (but not longer!) to fix the tissue in the paraformaldehyde vapors.
6. Immerse the fixed tissue in molten paraffin wax at 60°C for 1–2 h and then embed in fresh molten paraffin in bendable molds of convenient shape. Work rapidly over a hot surface (~60°C) and try to avoid trapping air bubbles in the molten paraffin. Arrange the tissue in a convenient orientation inside the mold to assist sectioning. Allow the paraffin to solidify, at least overnight, at room temperature.
7. Make 7–8 µm sections of the tissue with a conventional microtome. Arrange sectioned ribbons onto gelatin/chrome alum precoated slides and dry on a drying bench at 45–50°C for a few minutes until the section ribbon stretches well after the compression caused by the sectioning.
8. Dry the sections at 37°C for several hours or overnight.
9. Deparaffinize the sections by immersion into two changes of xylene for a total of 20 min, followed by a 2 min wash in 100% methanol.
10. Air dry the sections for a few minutes and digest with pepsin solution for 10 min at 37°C (*see* Note 12).
11. Postfix immediately with paraformaldehyde postfixative for 4 min. Rinse two times in PBS for 10 min in total.
12. Incubate slides in hydroxyl ammonium hydrochloride solution for 15 min and briefly wash in PBS for 5 min prior to hybridization.

3.3. Hybridization and Immunological Detection of Hybrids

1. Carry out hybridization overnight under RNase-free coverslips (avoid trapping air bubbles under the coverslip!) with 20 ng of probe per slide (in 80-µL hybridization buffer; *see* Section 3.1., item 6) at the optimum hybridization temperature (T_H), calculated for each probe (*see* Notes 13–15).
2. After hybridization, wash the sections for 20 min in 3X SSC at room temperature (during this first wash, coverslips will float away).
3. Wash the sections a second time: 20 min in 3X SSC at the hybridization temperature.
4. Perform a final wash for 5 min in 3X SSC at room temperature.
5. Incubate the sections in antibody dilution buffer, containing 1% blocking reagent, for 10 min.

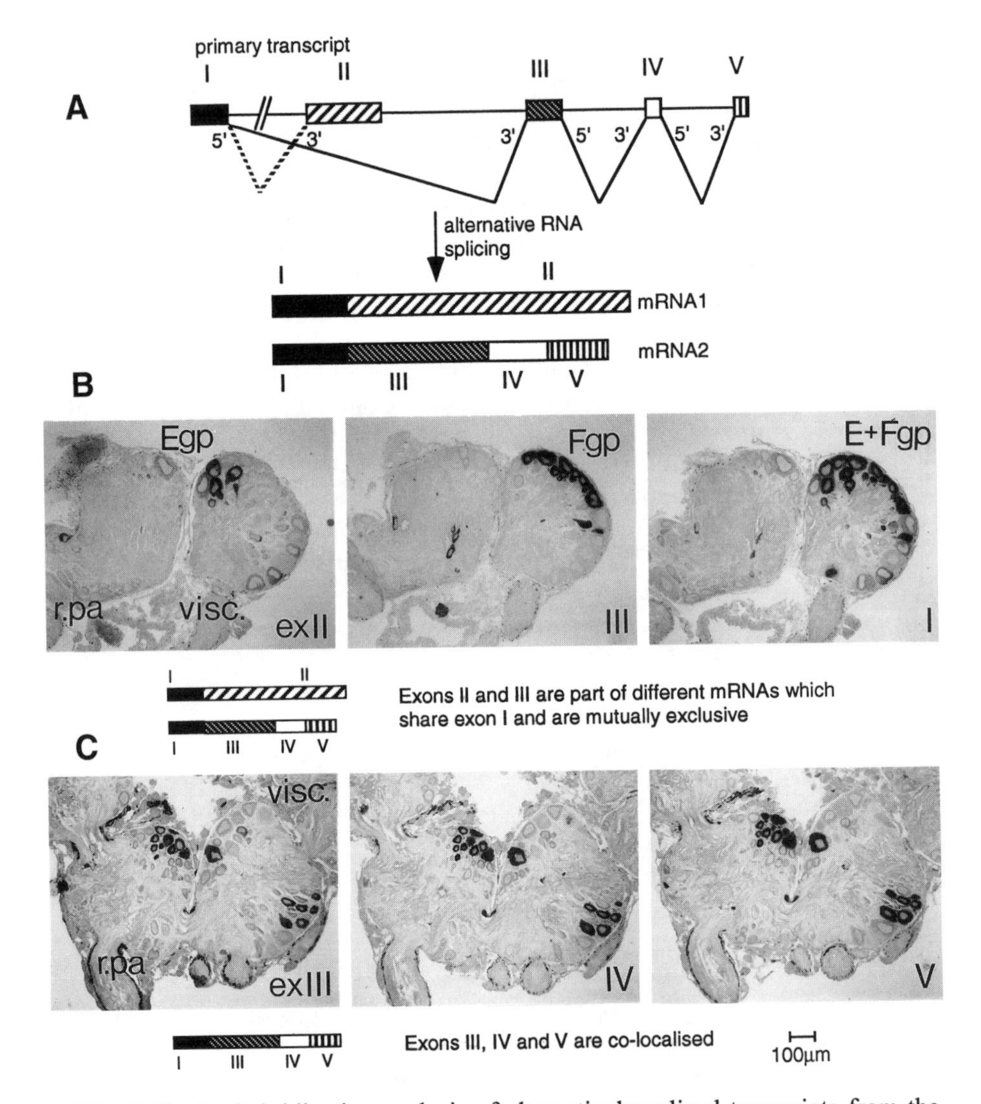

Fig. 1. *In situ* hybridization analysis of alternatively spliced transcripts from the FMRFamide neuropeptide locus in the pulmonate mollusc *Lymnaea stagnalis*, using a battery of oligonucleotides specific for each of the exons (*see also* ref. *6*). **(A)** The locus comprises five exons (I–V) and gives rise to two mRNAs, both sharing exon I, but containing either exon II or alternatively exons III-V *(3)*. **(B)** The set of consecutive serial sections from the CNS shown here has been probed with an oligonucleotide specific for exon II (left), exon III (middle), and exon I (right). It is clear that expression of exon II and III is mutually exclusive. In addition, because the two transcripts that con-

6. Apply alkaline phosphatase-conjugated antiDIG antiserum, diluted at 1:400 in the same buffer, for 1 h at room temperature.
7. Wash the sections in antibody dilution buffer, lacking the blocking reagent, for 15 min.
8. Visualize hybrids by the addition of the alkaline phosphatase substrate. Monitor the reaction (often over several hours) until a striking purple reaction product is obtained (*see* Notes 16 and 17). Care should be taken to prevent the sections drying out: Maintain sections well covered with the substrate solution during the developing process in a humid and light-proof chamber.
9. Briefly rinse the slides in PBS and mount in a water-based mountant. Observe and photograph (*see* Note 18).
10. Figure 1 shows results obtained using this protocol for the analysis of gene expression patterns of the locus encoding the FMRFamide-like family of neuropeptides in the brain of the pulmonate mollusc *Lymnaea*. In this case, alternative splicing generates two mutually exclusive types of transcript. Application to tissue sections of designer oligonucleotides, specific for each of the five exons, allowed—in combination with cDNA analysis *(3)*—the elucidation of transcript structure and the visualization and detailed mapping in the CNS of the two alternative mRNAs.

4. Notes

1. For the detection of neuropeptide precursor mRNA sequences, both RNA and DNA probes can be used. Single-stranded RNA probes (riboprobes), both antisense (mRNA-hybridizing) and sense (negative control) can easily be generated (and labelled at the same time) if the neuropeptide encoding sequence is subcloned into a suitable in vitro transcription vector (i.e., Bluescript). The high stability of RNA:RNA hybrids and the high sensitivity of detection is an advantage in this approach; however, the use of labile riboprobes may require careful handling (use of the RNase inhibitor RNasin from Stratagene in probe solutions and during hybridization may be a useful special precaution). DNA probes can be either cDNAs or PCR-generated (sequence amplification and probe labeling can, conveniently, be carried out simultaneously in this case also), but will require denaturation and possibly limited alkaline hydrolysis (to break probe into smaller fragments) prior to application. DNA oligonucleotides are especially useful both because they can be specifically designed and are thus the most appropriate for

---

tain, respectively, exons II and III both share exon I, the total number of exon I-expressing neurons is equal to the combined number of exon II- and exon III-containing neurons. (The identified E group and F group, Egp and Fgp, neurons in the visceral ganglion, visc. of *L. stagnalis*, are illustrated here). **(C)** Another set of consecutive serial sections hybridized with an oligonucleotide specific for exon III (left), exon IV (middle), and exon V (right), illustrates that the three exons are coexpressed and colocalized. (The right parietal, r.pa and visceral, visc., ganglia are shown). Scale bar: 100 µm.

the detection of particular exon-specific sequences in alternatively spliced messages and because of their small size (10–50 bp), which permits efficient penetration into tissues. Owing to the small probe size, even small nucleotide mismatches between probe and target can affect hybrid stability significantly. Therefore, while working out the conditions of hybridization (hybridization temperature and salt concentration of posthybridization washes), the appropriate balance between stable binding of the oligo probes and high stringency for good specificity has to be struck (*see* Note 13).

When designing exon-specific oligonucleotide probes, an area unique to each of the exons of interest obviously must be selected. To make sure that the proposed sequence is appropriate and in particular that there are no "clusters" of homology that would cause crosshybridization, perform a computer-aided sequence comparison to the entire neuropeptide gene sequence. Aim for a G-C to A-T ratio close to 1:1 and examine the proposed oligo sequence for regions of internal complementarity that may lead to the formation of hairpin loops or other secondary structures. As also mentioned in Note 15, to ensure specificity, it is preferable to design at least two different probes for each of the target sequences. These probes should reveal identical hybridization patterns.

For nonoligonucleotide probes, in vitro transcription (riboprobes), random-primed labeling, nick translation, and PCR-amplification (all for DNA probes) could be the methods of choice for probe labeling with either radioactive or non-radioactive labels. Unless target sequences are not abundant (neuropeptide mRNAs usually are abundant), the use of nonradioactively labeled probes is highly recommended because of:

a. Speed and ease of detection;
b. Superior resolution;
c. The long probe stability at –20°C (so that large batches of probe can be prepared and stored for many months with consistently good performance);
d. The convenience of safe handling.

2. Items 5–7 in Section 2.1. are supplied as a kit by Boehringer Mannheim.
3. Microscope slides must be precoated with 0.5% w/v gelatin, 0.5% w/v chrome alum to increase adherence of sections to the glass. This is a necessary step and if omitted, it will result in loss of material during further processing of slides. Prewarm about 200 mL of distilled water, add 1.5 g of gelatin, and warm further in a microwave oven until the gelatin dissolves completely. Allow to cool, add 1.5 g of chrome alum, and make up to 300 mL with water. Place in a glass trough and immerse microscope slides, arranged in racks, twice for 30 s in this solution. Dry slides for several hours at 37°C and store at 4°C. It is convenient to prepare a large batch (i.e., 50–100 slides) because slides can be maintained for up to 4 wk in the refrigerator.
4. The alkaline phosphatase product is soluble in organic solvent-based mounting media and will fade and disappear rapidly if such a medium is used! Instead, any commercial, water-based mountant, such as Immumount (Shandon) or Aquamount (BDH), is suitable.

5. An alternative, simple butanol extraction of oligonucleotides may be used instead of purification with Sephadex G25. In this case, dissolve the crude oligonucleotide in 1 mL of distilled water and extract two or three successive times with 400 μL of 1-butanol. Centrifuge for 3 min at 15,000*g*, discarding the liquid phase after each extraction. Rinse the pellet in 70% ethanol before redissolving in TE buffer (10 m*M* Tris-HCl, pH 7.4, 1 m*M* EDTA, pH 8.0).
6. To determine the oligonucleotide concentration, mix 1–5 μL of the oligo solution with 500 μL of water in a quartz cuvet and measure the absorbance at 260 nm, using water as a blank. To calculate the oligonucleotide concentration, first calculate $OD260/E_0$, where E_0 is the millimolar extinction coefficient, determined as the sum of the contributions of each base in the oligo, that is: T = 9.7, C = 9.2, G = 11.4, and A = 15.4. The ratio $OD260/E_0$ gives the average molarity of the oligo in millimolar (or 10^3 μ*M*; or in other words, the oligo concentration in pmol/μL). The concentration of the oligo in ng/μL is determined by multiplying the calculated molarity (in millimolar) by 330 (the average molecular weight of an ssDNA base) and by the number of nucleotides present in the oligo. (For example, for the 18-mer oligo TCGACCTGGATCCAAGGA, the $E_0 = [3 \times 9.7] + [5 \times 9.2] + [5 \times 11.4] + [5 \times 15.4] = 209.1$. If the measured OD260 is 10.65, then $OD260/E_0 = 0.0509$ m*M* or 50.9 pmol/μL and the concentration in ng/μL is $0.0509 \times 330 \times 18 = 302.3$ ng/μL).
7. The labeled probe from one such reaction (200 ng) is sufficient for hybridization on 10 slides.
8. For the end-labeling of oligonucleotides with terminal transferase, in addition to digoxygenin-11-dUTP, either biotin-11-dUTP (Sigma) or fluorescein-11-dUTP (Amersham International, Buckinghamshire, UK) can be used. A dideoxy-analog of digoxygenin-11-dUTP is also available (Boehringer Mannheim) and could be used for the 3' addition of a single-labeled UTP if the oligo U tail (resulting from the tailing with dUTP) appears to be a problem (i.e., nonspecific "sticking" to poly A mRNA tails). Although the choice of label is partly a matter of personal preference, the use of digoxygenin, a naturally occurring plant steroid, is highly recommended since the presence of endogenous biotin in some tissues can account for background noise. Acetylation of sections removes endogenous biotin if this proves to be a problem, and it also helps to prevent nonspecific, electrostatic probe binding by neutralizing positively charged molecules, such as proteins. In addition, the unavoidable bleaching of fluorescein during microscopic examination can be a disadvantage, especially when a detailed mapping analysis of the hybridization signal in the CNS is desired. If fluorescein-11-dUTP is used for probe labeling, remember to add an antifading agent, such as 1% (w/v) ethylenediamine or 0.1% (w/v) 1,4-diazabicyclo-[2.2.2]octane (DABCO), to the mounting medium.
9. It is advisable to check the incorporation of labeled nucleotides to the probe, especially since good batches of probes can be stored frozen for long-term use. Scintillation counting for radionucleotides *(4)* or standard dot-blot assays *(5)* for digoxygenin or biotin incorporation can be employed, and incorporation of

fluoronucleotides can be directly evaluated by examining 1–2 μL of probe, spotted on a glass slide, under an epifluorescence microscope.

10. This freezing and fixation protocol has been designed for (small) invertebrate CNS or organs, and may not be optimal for larger brains or thick tissues. In such cases, fix the material in freshly prepared 4% paraformaldehyde in PBS (perfuse if necessary) and then dehydrate through a graded ethanol series (50, 70, 80, 90%, and two times in 100% EtOH for 10 min each), prior to paraffin embedding. Alternatively, embed fixed material in cryoprotectant/mountant (e.g., OCT compound; Tissue Tek, Agar Scientific) and instantly freeze in liquid nitrogen-cooled isopentane, followed by liquid nitrogen. Cryosection material and allow sections to thaw and dry before fixing with 4% paraformaldehyde and using for *in situ* hybridization.
11. The length of dehydration depends on the size and thickness of the tissue and may have to be increased accordingly if the material is thick, e.g., >0.5 cm in diameter.
12. Protease treatment of the tissue that has been crosslinked by the paraformaldehyde fixation (usually with pepsin/HCl for paraffin-embedded tissue or alternatively with protease K or pronase E, all obtainable from Sigma) is necessary to increase target accessibility to the probe. Nucleic acid-bound proteins are digested with this treatment, unmasking the target, but a careful calibration of the type of protease and its concentration must be carried out so that a satisfactory compromise between good hybridization and good tissue preservation is achieved.
13. The importance of hybridization temperature cannot be overstated. If hybridization is carried out under nonstringent conditions, nonspecific signal is very likely to occur, regardless of probe specificity. This is particularly significant when the exon-specific pattern of expression of alternatively spliced transcripts is under investigation. The appropriate hybridization temperature using this protocol can be calculated as follows: T_H = ([2 × number of AT pairs] + [4 × number of GC pairs]°C – 15°C [decrease of hybrid stability caused by the inclusion of 25% formamide in hybridization buffer] – 5°C). The value of T_H should be in the range of 50–65°C for probes of 25–30 nucleotides in length.
14. Because hybridization is carried out at a high temperature for long periods, it is important to prevent drying out of sections. Well-sealed Perspex plastic boxes or empty plastic, Boehringer enzyme boxes, stuffed with tissues soaked in 3X SSC, and with a platform (i.e., made from cut plastic pipets taped to the bottom of the box) for placing slides horizontally, can be used as homemade hybridization chambers in a water bath.
15. The specificity of hybridization and exon specificity of each probe must be verified as follows:
    a. Treat sections with RNase A or DNase I for 30 min at 37°C, following the pepsin treatment and prior to postfixation. RNase A is applied at a final concentration of 200 ng/mL in 10 m*M* Tris-HCl, 1 m*M* EDTA, pH 7.5, and DNase I at a final concentration of 34 ng/mL in 50 m*M* sodium acetate, 10 m*M* $MgCl_2$, 2 m*M* $CaCl_2$, pH 6.5. Treatment with RNase A should eliminate all signal whereas DNase I treatment should have no effect.

b. Compare the hybridization patterns generated with the use of two different oligonucleotides, specific for different regions of the same target sequence. They should be identical.

c. Use an excess of unlabeled oligonucleotide as a competitor to verify a decrease or total elimination of the hybridization signal. This should not occur with a "competitor" oligonucleotide that is unrelated to the target sequence.

d. Compare the hybridization pattern and immunocytochemical pattern generated with the use of an oligonucleotide and an antibody specific for protein product of the target sequence, if available. Again, they should be identical.

16. It is good practice always to ensure that the substrate for the alkaline phosphatase reaction works before adding it to the hybridized tissue sections. Routinely, put aside the tube in which you had previously made the AP-anti digoxygenin antibody dilution. Drop a few microliters of the mixed substrate to the remaining droplets of antibody dilution and watch while a deep purple color develops within the next minute or so, indicating that the substrate is functional (and that you have correctly included all the components).
17. The enzymatic alkaline phosphatase color reaction for the visualization of hybrids will develop slowly, sometimes over several hours at room temperature. If this is the first time the hybridization is carried out, you should frequently monitor the progress of the reaction under low magnification at the microscope, but try to minimize prolonged exposure to light as much as possible. In practice, neuropeptide mRNAs are abundantly expressed and should be easily detectable. It is preferable to have a reaction that takes 2–3 h to develop rather than one that is very rapid or very slow (but still not long enough to maintain overnight). One could experiment with probe, antibody, enzyme substrate concentrations, and developing temperature (the reaction can also be carried out at 4°C) to obtain convenient development times. If the background is low (as it should be, if stringent hybridization temperature was used and washes were thorough) one could let the reaction develop over a long time in order to obtain a pronounced and striking colored product.
18. The purple colored precipitate, generated by the alkaline phosphatase detection reaction, is very photogenic in transmitted light microscopy. For increased resolution in photography, it is generally preferable to use small-grained, low ASA films, such as Kodak Ektar 125, Kodak Tmax 100, Agfa Ortho 25 (very high contrast), Fujicolor 100, and Fujichrome 100. For fluorescent photomicroscopy use high-speed films, such as Fujicolor/Fujichrome 400 or Kodak Ektar 1000.

References

1. Leitch, A. R., Schwarzacher, T., Jackson, D., and Leitch, I. J. (1994) *In Situ Hybridisation*. Royal Microscopical Society Microscopy Handbooks, 27. ßios Scientific Publishers, Oxford.
2. Harris, N. and Wilkinson, D. G. (1990) *In Situ Hybridisation: Application to Developmental Biology and Medicine*. Society for Experimental Biology, Seminar Series 40. Cambridge University Press, Cambridge.

3. Kellett, E., Saunders, S. E., Li, K. W., Staddon, J. W., Benjamin, P. R., and Burke, J. F. (1994) Genomic organisation of the FMRFamide gene in *Lymnaea*: multiple exons encoding novel neuropeptides. *J. Neurosci.* **14,** 6564–6570.
4. Sambrook, J., Fritsch, E. F., and Maniatis, T. (1989) *Molecular Cloning. A Laboratory Manual.* Cold Spring Harbor Laboratory, Cold Spring Harbor, NY.
5. Harlow, E. and Lane, D. (1988) *Antibodies. A Laboratory Manual.* Cold Spring Harbor Laboratory, Cold Spring Harbor, NY.
6. Santama, N., Benjamin, P. R., and Burke, J. F. (1995) Alternative RNA splicing generates diversity of neuropeptide expression in the brain of the snail *Lymnaea*: *in situ* analysis of mutually exclusive transcripts of the FMRFamide gene. *Eur. J. Neurosci.* **7,** 65–76.

13

Quantitative *In Situ* Hybridization Histochemistry

Andrew Levy

1. Introduction

The key advantage of *in situ* hybridization histochemistry with radiolabeled as opposed to nonradiolabeled probes is that relative differences in the amount of specific mRNA transcripts present in tissue sections can be accurately and reproducibly quantified. Precise quantification and localization of transcripts, as well as the enhanced sensitivity of ^{35}S-labeled probes, allows powerful qualitative as well as quantitative controls to be built into experiments. Interpretation of autoradiographs of tissue sections hybridized *in situ* depends on careful selection of control tissue and on the availability of a suitable computerized densitometer. Major limiting factors on the performance of *in situ* hybridization are the relative and particularly the absolute abundance of target transcripts. *In situ* hybridization on frozen sections produces higher signal intensity and lower background than *in situ* hybridization on wax-embedded sections, at the expense of tissue morphology (*see* Fig. 1).

2. Materials

Autoclaved, deionized water and stock solutions can be conveniently made up in the metal-capped, 500 mL glass bottles that manufacturers use to supply tissue culture medium, or in autoclavable polypropylene bottles. RNase contamination is minimized by adding known weights of water directly to reagents in weighed stock bottles rather than by using measuring cylinders and by using sterile, single-use plasticware, which can be assumed to be RNase-free. In addition, diethylpyrocarbonate (DEPC) water pretreatment is used to inactivate RNase (*see* Note 1); alternatively, essentially RNase-free water can be purchased (e.g., Sigma [St. Louis, MO], cat. no. 27,073-3 or W3500). Gloves should be worn throughout the procedure to prevent contamination by RNases from skin secretions.

From: *Methods in Molecular Biology, Vol. 72: Neurotransmitter Methods*
Edited by: R. C. Rayne Humana Press Inc., Totowa, NJ

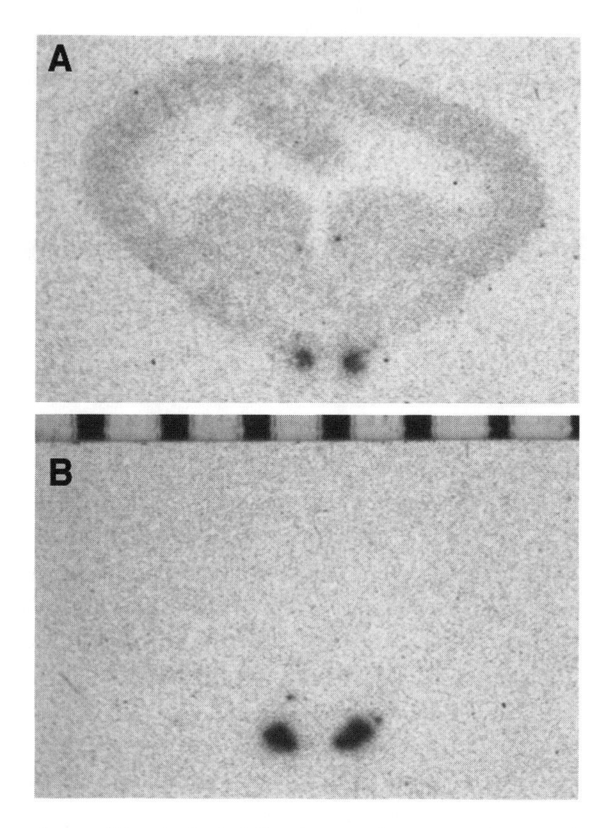

Fig. 1. Comparison of *in situ* hybridization on frozen sections and wax-embedded tissue. Macroscopic autoradiographs of 5 μm-thick wax-embedded **(A)** and 12 μm-thick frozen **(B)** coronal sections of rainbow trout brain hybridized to a probe complementary to melanin concentrating hormone (MCH2). Sections were treated similarly throughout (apart from the prehybridization protocols) and were exposed simultaneously to adjacent areas of the same sheet of Hyperfilm for 48 h. It can be seen that specific signal is higher and background binding lower (negligible) in the frozen section autoradiograph. The scale marks at the top of frame (B) are millimeter marks cast by the edge of a plastic rule when the image was captured using the program "NIH Image" running on the computer system described in the text.

2.1. Slide Preparation for Frozen Section Hybridization (see Note 2)

1. Standard microscope slides (e.g., 76 × 26 mm, Twin Frost: Raymond Lamb [London, UK], cat. no. E/25.3).
2. Stainless-steel slide racks (e.g., Raymond Lamb, cat. no. E/102 or E/102.6 for 25 or 48 slides, respectively).
3. TEEPOL HB7 detergent (Sigma, cat. no. T6152).

4. Gelatin/Chrome-Alum slide coating solution (*see* Note 3): Dissolve 2.5 g gelatin in 1 L of deionized, autoclaved water at 70°C and allow to cool before adding 0.25 g of chromium potassium sulfate.

2.2. Slide Preparation for Wax-Embedded Sections (see Note 2)

1. Silane slide coating solution: 2% (v/v) 3-aminopropyltriethoxysilane (Sigma, cat. no. A3648) in acetone. Prepare fresh just before use (*see* Section 3.2.).
2. Industrial Methylated Spirits (95% BDH [Leics, UK], cat. no. 30244).

2.3. Probe Labeling

1. Terminal transferase, buffer, and cobalt chloride solution (Boehringer Mannheim [Lewes, UK], cat. no. 220 582).
2. ^{35}S-αdATP (New England Nuclear [NEN, Stevenage, UK], cat. no. NEG-034H).
3. Nucleotide removal kit: QIAquick (Qiagen Ltd. [Dorking, UK], cat. no. 28304).
4. 100X TE: 1*M* Tris-HCl, pH 7.4, 100 m*M* EDTA.

2.4. Prehybridization of Frozen Sections

1. 10X Phosphate-buffered saline (10X PBS): Dissolve 90 g of NaCl, 1.65 g of KH_2PO_4, and 19.527 g of Na_2HPO_4 (dodecahydrate) in deionized water. Autoclave and store at room temperature, but discard if it develops a white precipitate. There is no need to adjust the pH.
2. PBS/formalin: Mix 400 mL of autoclaved, deionized water with 50 mL of 10X PBS and 50 mL of formalin (40% formaldehyde solution; BDH, cat. no. 10113). Measure 50 mL aliquots using sterile, single-use plasticware. This solution is stable at room temperature but should be checked for contamination prior to use.
3. Saline (*see* Note 4): 0.9% (w/v) NaCl in RNase-free water.
4. Acetylation buffer (*see* Note 5): 0.1*M* triethanolamine, 0.25% (v/v) acetic anhydride in saline. To make 500 mL, add 7 mL of triethanolamine solution to 500 mL of 0.9% saline. Immediately before use, add 1.25 mL of acetic anhydride by shaking vigorously for a few seconds.
5. Polypropylene Coplin jars (Raymond Lamb, cat. no. E/96.11).
6. Polypropylene 1 L bottles (BDH, cat. no. 215/0408/25; wide-mouthed, 215/0399/32).
7. Ethanol: 70, 80, and 95%, (w/w) with RNase-free water. Zero a polypropylene, 1 L bottle on a balance and add 100% ethanol. Multiply the weight of ethanol added by the following constants to give the weight of water to add: 0.543 for 70%, 0.317 for 80%, and 0.0677 for 95%. Remember to leave enough room for the water, particularly with the 70 and 80% solutions.
8. Chloroform.
9. Upright slide racks (Raymond Lamb, cat. no. E89.055).

2.5. Prehybridization of Wax-Embedded Sections

1. Xylene.
2. Proteinase K (Sigma, cat. no. P0390) stock solution: 10 mg/mL in water. Store at –20°C (*see* Note 6 and Section 3.5.).

3. $1M$ Tris-HCl, pH 7.5 (Sigma, cat. no. T7149).
4. 4% Paraformaldehyde: Add 4 g of paraformaldehyde in 100 mL of 1X PBS in a glass bottle. Heat for short bursts in a microwave oven, stirring frequently, until it dissolves. Cool and store at 4°C.
5. PBS/formaldehyde: 0.4% paraformaldehyde in PBS. Add 4 mL of 4% paraformaldehyde to 36 mL of 1X PBS (*see* Section 2.4., item 1). For this purpose, when stored at 4°C, this buffer remains stable for some months.
6. 2X SSC: Dilute 20X SSC (*see* Section 2.6., item 4) by 10X in water.

2.6. Hybridization

1. Plastic food boxes with glass or rigid plastic sheets separated by spacers (i.e., bottle lids).
2. NescoFilm (BDH, cat. no. 235/0414).
3. Pointed forceps (BDH, cat. no. 406/0070/00).
4. 20X SSC: $3M$ NaCl, 300 mM sodium citrate in water. Dissolve 175.3 g of NaCl and 88.2 g of sodium citrate in 1 L of deionized water. Add 1 mL of DEPC and disperse by vigorous shaking. Allow the solution to stand for ≥1 h, then autoclave. It is not necessary to pH this solution.
5. $5M$ DTT stock solution (*see* Note 7): Add exactly 1.05 mL of water to 1.93 g of DTT. Store aliquots (25–100 μL) at –20°C and discard after a single thaw.
6. Sheared, single-stranded salmon sperm DNA (Sigma, cat. no. D1626): Dissolve the DNA at 10 mg/mL in RNase-free water and sonicate it or pass it repeatedly through a 19-gage needle until its viscosity is greatly diminished, then store at –20°C in 1 mL aliquots. Before use, boil for 10 min and quench on ice.
7. Yeast tRNA solution: Add RNase-free water to make a 25 mg/mL solution. Store in 0.5 mL aliquots at –20°C. (Boehringer Mannheim, cat. no. 109 495).
8. 50X Denhardt's solution: 1.0% Ficoll, 1.0% polyvinylpyrrolidone, and 1.0% BSA in water. Make 100 mL by dissolving 1 g of each component in RNase-free water. Store in 1 mL aliquots at –20°C. May also be purchased premade (Sigma, cat. no. D2532).
9. 50% dextran sulfate: Add RNase-free water to the powder in a 50 mL tube to make 50 mL. Store at room temperature (Sigma, cat. no. D8906).
10. Hybridization buffer: 50% formamide, 4X SSC, 500 μg/mL sheared, single-stranded salmon testis DNA, 250 μg/mL yeast transfer RNA, 1X Denhardt's solution, and 10% dextran sulfate (mol wt 500,000). The buffer is stable at –20°C. To make 50 mL, mix the following components: 25 mL formamide (Sigma, cat. no. F7503; *see* Note 8), 10 mL 20X SSC, 2.5 mL 10 mg/mL sheared single-stranded salmon sperm DNA, 0.5 mL 25 mg/mL yeast tRNA, 1.0 mL 50X Denhardt's solution, 10 mL 50% dextran sulfate (mol wt 500,000), and 1 mL RNase-free water for a total of 50 mL.

2.7. Posthybridization Washes

1. 1X SSC: Dilute 20X SSC by 20-fold in water (*see* Section 2.6., item 4).
2. Polypropylene troughs (BDH, cat. no. 272/0078).
3. Warm air blower (e.g., hair drier).

2.8. Autoradiography

1. Cardboard sheets cut to fit X-ray cassettes.
2. Exposure cassettes (e.g., Sigma, cat. no. E9510).
3. Autoradiography film (Hyperfilm MP. Amersham [Arlington Heights, IL] cat. nos. RPN 6 [18 × 24 cm]), RPN 7 [30 × 40 cm], and RPN 8 [35 × 43 cm] are the most useful sizes).
4. Access to an automatic X-ray film developer or to developing chemicals and equipment.
5. K5 nuclear emulsion in gel form (Ilford Ltd. [Mobberley, UK], cat. no. 1355127).
6. Glycerol.
7. Glass dipping chamber.

2.9. Quantification of Autoradiographic Results

1. Radioactive standards (Amersham Microscales Code RPA 511 L [low] and RPA 504 L [high]).
2. Double-sided sticky tape.
3. Computerized image analysis system.

3. Methods

3.1. Slide Preparation for Frozen Section Hybridization (see Notes 2 and 3)

1. Transfer standard slides into stainless-steel slide racks (wearing gloves).
2. Stand racked slides in a large volume of deionized water (at least 5 L) containing 10–15 mL of TEEPOL HB7 for at least 1 h in a clean container such as an unused plastic bucket.
3. Wash extensively in clean running tap water for 30 min to remove all traces of detergent, then in at least five changes of deionized water.
4. Allow the slides to dry in a relatively dust-free environment (i.e., under a clean paper towel).
5. Gently dip racked slides into gelatin/chrome-alum coating solution for a second or two, drain, and draw off excess solution with a paper towel held against the underside of the bars of the racks.
6. Allow the slides to dry at room temperature between paper towels for ≥90 min, then redip.
7. Store dried slides in a dust-free environment at room temperature or at –20°C, ready for use.
8. Thaw frozen sections directly onto the slides, dry briefly over a warm plate at 37–40°C, and store at –20°C temporarily or at –80°C long term.

3.2. Slide Preparation for Wax-Embedded Sections

1. Wash slides in 95% Industrial Methylated Spirits twice for 5 min each and allow to dry.
2. Dip for 5 s into freshly prepared silane slide coating solution.

3. Wash twice in deionized water.
4. Air dry at room temperature and store at room temperature in a dust-free environment.

3.3. Probe 3' End-Labeling with ^{35}S-αdATP (see Note 9)

1. Mix the following in order in a screw top Eppendorf tube: 28 μL deionized, autoclaved water, 10 μL 5X terminal transferase buffer (supplied with the enzyme), 1 μL 5 μ*M* probe (*see* Note 10), 5 μL cobalt chloride solution (supplied with the enzyme), 5 μL ^{35}S-dATP, and 1 μL terminal transferase for a total volume of 50 μL.
2. Vortex and centrifuge very briefly to collect the reaction mixture at the bottom of the tube.
3. Incubate at 37°C for 1 h.
4. Elute labeled probe in 100 μL of water using a nucleotide removal column (*see* Note 11).
5. Add 1 μL of 100X TE and store at 4°C until use or at –20°C if TE is omitted.
6. Pipet 1 μL of probe into 5 mL of scintillant and count (*see* Note 12).

3.4. Prehybridizing Frozen Sections (see Note 13)

1. Allow slides to thaw face up on a sheet of aluminum foil for 10 min. Ensure that they are appropriately labeled in pencil.
2. Load the slides back to back in polypropylene Coplin jars (*see* Note 14).
3. Incubate at room temperature for 5 min in PBS/formalin.
4. Discard the solution by inverting the Coplin jar, and wash twice in PBS for 2 min (*see* Note 15).
5. Incubate in freshly made acetylation buffer for 10 min (*see* Section 2.4., item 4 and Note 5).
6. Pass the slides through a graded ethanol series, as follows: 70% ethanol for 1 min, 80% ethanol for 1 min, 95% ethanol for 2 min, 100% ethanol for 1 min, 100% chloroform for 5 min, 100% ethanol for 1 min, and 95% ethanol for 1 min.
7. Stand the slides upright on racks to dry.

3.5. Prehybridizing Wax-Embedded Sections (see Note 13)

1. Load the slides into polypropylene Coplin jars.
2. Incubate the slides in 100% xylene for 5 min, followed by two successive 5-minute incubations in 100% ethanol. Follow this with an incubation in 70% EtOH for 5 min.
3. Wash the slides in deionized, autoclaved water twice for 5 min each.
4. Incubate the slides in 2X SSC at 70°C for 10 min.
5. Rinse the slides in water for 2 min.
6. Incubate the slides in proteinase K solution for 1 h. For a single Coplin jar, add 2 mL of 1*M* Tris-HCl, pH 7.5, to 38 mL of water. Add 30 μL of Proteinase K stock solution immediately before use (*see* Note 16).
7. Rinse the slides in water for 2 min, then in PBS for 2 min.
8. Incubate the slides in 0.4% paraformaldehyde for 20 min at 4°C.
9. Rinse the slides in water twice for 2 min.

10. Stand the slides upright and remove excess water, but do not allow them to dry completely.

3.6. Hybridization (see Notes 17 and 18)

1. On a sheet of fresh aluminum foil, use a "no touch" technique to cut NescoFilm coverslips to the approximate size of the tissue sections using clean scissors and pointed forceps. The surface applied to the hybridization buffer should touch nothing once it has been peeled from the backing paper.
2. Complete the hybridization buffer by adding 1 μL of 5*M* DTT (*see* Note 18) and 10^5–10^6 counts of labeled probe/slide (45 μL). Pipet 45–50 μL of buffer onto each prehybridized slide.
3. Use the edge of the NescoFilm coverslip to ensure that the entire section is coated with buffer, then cover with the same piece of film, ensuring that no bubbles are trapped under it.
4. Cover the chamber and incubate overnight at 37°C.

3.7. Posthybridization Washing

1. Use a container (approx 250 mL) filled almost to the brim with 1X SSC to gently lift coverslips off using surface tension.
2. Rinse each slide briefly by moving it back and forth a few times in three sequential changes of 1X SSC (100 mL) to remove adherent buffer and store the slides in stainless-steel slide racks under 500 mL of fresh 1X SSC at room temperature until all of the coverslips have been removed.
3. Wash racked slides at 55°C in four changes of 15 min each of 1X SSC over a 1 h period (*see* Note 19).
4. Incubate in two changes of 500 mL 1X SSC at room temperature for 15 min each.
5. Dip briefly in water twice to remove salt deposits prior to drying.
6. Dry completely in a warm air stream for 10–20 min.

3.8. Autoradiography

1. Tape slides onto cardboard sheets cut to fit X-ray cassettes (*see* Note 20).
2. Include radioactive standards during exposure if the results are to be quantified (*see* Note 21).
3. Include qualitative and quantitative hybridized controls (*see* Note 22).
4. Place a sheet of autoradiography film over the slides and allow to expose at room temperature from 4 h to 4 wk, depending on the expected signal strength (*see* Note 23).
5. After suitable autoradiographs have been generated (which may entail more than one exposure) the slides can be dipped in nuclear emulsion if quantification of probe binding at the single cell level is required (*see* Note 24).

3.9. Quantification of Autoradiographs (see Note 25)

1. Turn on the light source and camera and allow to stabilize for 30 min.
2. Capture a sample of background density, four or five autoradiographs generated by the isotope standards, and a magnification standard (by capturing an image of a transparent plastic rule held against the light box) to calibrate the machine.

3. Set segmentation (*see* Note 26).
4. Collect data as cpm, or alternatively, as the product of cpm × image size (*see* Notes 27 and 28).
5. Use the same system to analyze microscopic autoradiographs at the single cell level (*see* Note 29).

4. Notes

1. Make a 0.1% (v/v) solution of diethylpyrocarbonate in distilled water. Shake vigorously for 30 s, allow to stand for at least 30 min, then autoclave. This cleaves remaining DEPC to CO_2 and water. DEPC is toxic and unstable even at 4°C. It should be handled in a fume cupboard and should not be tightly stoppered.
2. Slide coatings reduce nonspecific DNA binding to glass, but are primarily used to prevent tissue loss. Gelatin coating, although excellent for frozen sections, does not work for wax-embedded sections.
3. Chromium potassium sulfate (Chrome alum) is included as a preservative. If the slides are kept dry, or at –20°C, it can be omitted.
4. Saline (0.9%) can be made as a 20X solution and diluted as required, or obtained directly as 0.9% saline for irrigation or infusion from a pharmacy. From this source, it can be assumed to be RNase-free.
5. Stock acetic anhydride should be kept dry (over desiccant) at room temperature. If a syringe or pipet is used to measure the acetic anhydride, ensure that the graduations on the side of the barrel are not submerged because the ink tends to be soluble in this compound. Triethanolamine solution is viscous and best measured with a 10-mL syringe.
6. Stored at –20°C, proteinase K solution can be thawed and frozen a number of times without significantly diminishing in activity, although the solution may become slightly hazy.
7. DTT stock (5*M*) looks as if it will never dissolve, but it does if it is left alone for 30 min. We usually discard aliquots after a single thaw, although this is probably unnecessary.
8. Molecular biology grade formamide does not require any further treatment prior to use. Store at –20°C rather than room temperature.
9. Probes 24–48 bases in length are among the easiest to use. It is worth ensuring that they are not too "GC" rich and that there is no obvious scope for self hybridization (i.e., a run of "C"s at one end, and "G"s at the other). Published sequences are by convention sense; requests to probe manufacturers, therefore, need to be the complementary palindrome of the published sequence. End-labeling with ^{35}S αATP works well for almost all purposes. Cobalt chloride solution and labeling buffer are supplied with the terminal deoxyribonucleotidyl transferase (TdT).
10. Stock solution of purified probe is diluted in TE or distilled, autoclaved water to 5 μmol/L using the following calculation:

$$\text{fold dilution of stock} = (\text{concentration of probe } [\mu g/\mu L]) / (\text{length of probe} \times 0.0015) \quad (1)$$

11. Although labeled probe can be extracted with phenol-chloroform and chloroform, then precipitated with ethanol and salt, nucleotide removal columns such as the QIAquick nucleotide removal kit (*see* p. 14 of the manufacturer's handbook) are much quicker, more reliable and similar in price overall.
12. Eject the contents of the tip and then the tip into the scintillant to ensure accurate pipetting. Normally, a 3' end-labeling reaction with ^{35}S-dATP yields 40–120 × 10^6 counts (400,000 to just over 1,000,000 counts/μL when eluted as outlined in the methods).
13. Prehybridization protocols are designed to increase access of probe to the target while preserving tissue morphology. A balance has to be struck between the two processes because fixation (protein crosslinking) reduces probe penetration and, hence, hybridization efficiency. The procedure is carried out at room temperature.
14. Coplin jars are autoclaved when new, but can subsequently be reused without further treatment provided they are handled with gloves and kept inverted on a clean paper towel between use. Each holds 10 slides (back to back).
15. Ensure that the buffer is able to circulate between the slides by prying them apart with fine-tipped forceps. It helps if the PBS is poured directly over the ends of the slides.
16. This gives a final concentration of 7.5 μg/μL proteinase K in 50 m*M* Tris-HCl, pH 7.5.
17. Plastic food boxes with wet paper towels at the bottom provide a suitably humid microenvironment. The addition of a little $CuSO_4$ solution ensures sterility if required. Rest slides on glass or rigid plastic sheets separated by spacers (i.e., bottle lids).
18. A 45 μL aliquot of hybridization buffer is sufficient to cover a pair of adjacent sections of rat brain on one slide. Because the buffer is viscous and difficult to pipet, a safety margin of 10% extra should be made up since all attempts at quantification will be nullified if the same probe mix is not used for all of the sections to be compared.
19. Use a series of polypropylene troughs each containing approx 500 mL of 1X SSC. Prewarm for at least 1 h in a gently shaking water bath. Note that the solution tends not to reach temperature unless the water bath itself is set to 60 or 65°C.
20. Take care to keep the tape away from the sections, because some tapes are thick enough to hold the film away from the sections and result in a fuzzy image. Include a ^{14}C standard if the autoradiographs are to be quantified. It is ultimately counterproductive to mix up the slides from different experimental groups in an attempt to try to reduce errors caused by differences in sensitivity across single autoradiography films and minimize the effects of drift in the image analyzer and light source.
21. The response of autoradiography film to radioactive decay is nonlinear. At low levels film is relatively insensitive and at higher levels—once all of the silver ions have been converted to stable pairs of silver atoms—the latent image can develop no further. To take account of this, a series of radioactive standards are

opposed to the film along with the hybridized sections, and the optical density of the experimental autoradiographs compared to those produced by the standards.

Isotope standards can be made by spiking sonicated brain or other tissue with known amounts of ^{35}S. The technique requires a considerable amount of practice and luck before homogeneous, bubble- and ice crystal-free sections can be obtained, however, and even when successful, the method contaminates the cryostat and has to be repeated at intervals as the standards decay. Ready-made ^{14}C standards (energetically similar to ^{35}S) can be purchased from Amersham and although expensive, are ultimately very good value for the money. To mount them on slides, ignore the manufacturer's instructions in favor of using strong, double-sided sticky tape on an uncoated glass slide. Although each sheet of autoradiography film used to expose a single experiment should theoretically have a standard attached, it is usually best to calibrate the machine and define the minimum optical density to include in the analysis (segmentation) at the outset, and either keep the total number of slides to the number that can be exposed on a single sheet of autoradiography film ($\leq$70), or rearrange the slides and repeat the exposure and analysis from the beginning. Even though the analyzer and light source are likely to drift slightly during the process and the initial camera and computer settings are somewhat arbitrary, once defined, extremely accurate and reproducible comparisons can be made.

22. The use of poorly labeled probe or prolonged exposure times often produces considerable background that frequently varies in intensity depending on the nature of the tissue being examined. A section of rat brain, for example, might show more intense, but nevertheless still background binding to the hippocampus, which is cell body, and hence, nucleic-acid rich. Because of this, distinguishing specific binding from background in a relatively uniform tissue, such as the liver or heart, may be very difficult.

    Thus, whenever *in situ* hybridization is used, and particularly when *in situ* hybridization histochemistry is used to detect and quantify relatively rare targets, the inclusion of specific control sections is essential. In many cases, the quantitative and topographical information offered by *in situ* hybridization makes this straightforward, because hybridization should be confined to areas known to contain target transcripts and the amount of binding should change after physiological or pharmacological stimuli in vivo known to modify the transcription rate or mRNA stability. Pretreating sections with RNase or unlabeled probe is of very little value, because nonspecific as well as specific targets are removed or obscured in equal proportion. The use of nonsense probe is also nearly valueless, since its behavior in vitro may bear no relationship to that of the specific probe.
23. If no image is evident and the user is new to *in situ* hybridization, a number of troubleshooting steps should be followed. First, employ a positive methodological control. Use probe and tissue that will provide clear evidence of success (such as a probe complementary to growth hormone transcripts used in sections of anterior pituitary gland) and perform the entire procedure. It may also be of use to synthesize a probe against a different region of the target. If a probe that has

behaved impeccably for months or years begins to produce poor results despite apparently labeling normally, it is cost-effective and sanity-preserving to throw it away and have a new batch synthesized rather than try to find out what has gone wrong. Sometimes, using a number of different oligonucleotides against the same target can marginally increase sensitivity, although not often by as much as might be anticipated. If this method is used, the oligonucleotides should be labeled in separate reactions, then mixed.

Alternatively, decreasing hybridization temperature to ambient may be useful. Changing the wash temperatures or increasing the salt concentration is not useful in *in situ* hybridization. Neither is it useful to change the isotope to ^{32}P. If it is anticipated that the target is rare, it may well be impossible to detect it with oligodeoxynucleotides. In that case, switching to a ^{35}S-labeled riboprobe has a higher chance of success than any of the above maneuvers.

If no specific image is evident over background:

a. Ensure that the autoradiographic images are not overexposed (i.e., black, or near black) if they are to be assessed densitometrically.
b. Bear in mind that the specific activity of the probe rather than the total number of counts added per section is the critical parameter (i.e., whereas 10^5 counts of well-labeled probe/tissue section may produce specific binding, 10^6 counts of poorly labeled probe will not). Thus, for example, failing to dilute probe stock to 5 µm can impair results, even though the resulting labeled probe contains more counts overall.
c. Because the kinetics of *in situ* hybridization are not clear-cut, increasing the wash temperature or reducing the salt content of the wash solution does not work well. Hybridizing at 42 or 45°C may be useful, but above 50°C, considerable tissue loss should be expected.
d. Switch to a ^{35}S-labeled riboprobe.

24. Mix K5 nuclear emulsion in gel form with 2 vol of 0.5% aqueous glycerol, melt at 37°C, pour into a suitable dipping chamber at 37°C, and allow to stand for 10 min so that the bubbles rise to the surface and can be cleared by dipping a blank slide. Dipped slides should be allowed to gel for 30 min, then stored over desiccant at 4°C. Exposure time is two to three times longer than that required to expose autoradiography film. Note that if the emulsion is overheated (≥ approx 55°C), the background grain count will be high.
25. Densitometry is a surprisingly complex process. The image analysis system described below has proved to be by far the best, and also the least expensive of the several systems tried by the author over several years.
    a. Cameras: Although CCD cameras are more stable and less susceptible to shading artifact than tube cameras, they are also much less sensitive to low-light levels and are, therefore, difficult to use for analysis of magnified small images (≤1–2 mm) (*see* Fig. 1). A tube camera for phase contrast and differential interference microscopy (Hamamatsu Newvicon tube camera, cat. no. C2400-07) can be used with Vivitar extension tubes and a Micro-NIKKOR 55 mm lens (code 513851) (from local camera suppliers), or attached via a

suitable connector to the camera mount of a standard microscope. Care should be taken to prevent dust getting into the camera during change over from macro- to microanalysis.

b. Software: NIH Image, a superb, public domain (i.e., free) image analysis program written by Wayne Rasband at the National Institute of Mental Health, NIH, Bethesda, MD, is easy to use and provides accurate and reproducible results. The latest version of Image (1.58), including documentation, example images, and complete Pascal source code, is available by anonymous FTP from **zippy.nimh.nih.gov**, in the directory **/pub/nih-image**. There is a plain text README file (0README.txt) in the nih-image directory with more information. The NIH Image Web page is at **http://rsb.info.nih.gov/nih-image/**.

c. Computer Hardware: The program NIH Image V1.58, runs natively on the Power Macintosh. Older Apple Macintosh computers need to be upgraded with an image capture board (such as Data Translation's "QuickCapture", cat. no. DT2255-50 Hz). This is coupled to the camera controller with a 1.5 m cable from the same supplier (EP277-1.5 m cable with eight male BNC connectors, only one of which—video input—needs to be used). Data Translation supply copies of the software program "Image" free with the hardware. New "AV" Macintosh computers have built-in frame grabbers that can be used with a plug-in called "Plug-in Digitizer" available from **zippy.nimh.nih.gov** in the **/pub/nih-image/plug-ins** directory. The newer PowerMac 8500 computers have a higher quality and faster 24-bit digitizer that might well obviate the need for any additional hardware.

d. Light Source: Expensive stabilized light sources are available to trans-illuminate autoradiographs for image analysis. An X-ray box identical to those used in hospital radiology departments is quite satisfactory for most purposes (Wardray Products Ltd).

26. Since autoradiographic images invariably have a penumbra of decreasing optical density at their edges, a reproducible and unbiased mean optical density cannot be measured if the area to be included in the analysis is encompassed by eye. Thus, segmentation should be used. (This automatically selects the area to be analyzed by defining the minimum and maximum optical density to be included within an area of interest defined by the operator.) Once set, this removes all bias from identification of the edges of the images. For example, to manually define the optical density of this asterisk "*," it would have to be accurately drawn around. By using segmentation to define it, it would only be necessary to draw a circle around it for the computer to accurately distinguish it from the background white color of the page. The advantage in speed and accuracy is obvious, but the primary advantage of segmentation is that it allows the operator to set an optical density that will be used to define the edges of a series of gray objects reproducibly. This is essential if comparisons between them are to be made.

27. As the amount of target transcript within a single cell or cell cluster increases, the absolute amount of ^{35}S-labeled probe binding shows a concomitant rise. This

tends to increase not only the density of the autoradiographic image, but also the size of the image produced. The numerical difference between autoradiographs generated from low to high specific activity point sources can, therefore, be enhanced by multiplying the optical density and area of each image before comparing the values from different experimental groups. Segmentation (*see* Note 26) ensures that the data are collected completely objectively. Raw image data can be copied and pasted directly into statistical and graphical analysis programs on the Apple Macintosh computer for further analysis.

28. On average, differences in optical density between sequential sections mounted on the same slide (therefore treated, hybridized, and analyzed almost identically) are usually in the range of 10–20%. If standard errors appear to be too small to be consistent with this, it is likely that the film was overexposed and that many of the images could not further develop. Obtaining suitable exposure times to get all of the images within the measurable range may require more than one attempt, including rehybridization with different amounts of probe to bring the images into the range of the standards.
29. Surprisingly precise localization of the position of probe binding is possible even with ^{35}S-labeled probes, which have a decay particles path length of 17 μm in air. Grain counting by hand is gruelling and subject to bias and, unfortunately, morphological variability and the tendency for the number of individual grains to appear to approach unity as the whole field becomes black induces a number of computer grain counting programs to produce large and inconsistent errors unless the exposure is carefully controlled.

    An alternative method is to adjust the segmentation of unstained slides so that only the grains are defined (landmarks can be identified by racking the condenser down until a pseudo-phase appearance brings the unstained section into relief). Because grains are usually similar in size, a calibration curve can be drawn by hand by encompassing a series of different numbers of grains from one to 100 and plotting segmented area against grain number. This automatically takes into account the decrease in area when two grains overlap, which inevitably occurs as the grain density increases. In addition, in the event that two grains are exactly superimposed, the area will still tend to increase. Once the calibration curve has been drawn, the areas of interest can be encompassed, the segmented area measured, and the number of grains read off the graph.

14

Microbore Liquid Chromatography Analysis of Monoamine Transmitters

Sophie Sarre, Katrien Thorré, Ilse Smolders, and Yvette Michotte

1. Introduction

Ion-pair reversed-phase liquid chromatography (LC), as introduced by Johansson et al. in 1978 *(1)* is still the most widely used separation technique for analysis of the biogenic amines. It is usually coupled with electrochemical detection (ECD). The use of microbore LC was introduced by Scott and Kucera in 1976 *(2)*. These systems use columns with an internal diameter (id) of 1 mm or less. Microbore LC provides a significant advantage over conventional LC: The sample injected on to a microbore column is diluted by about 20-fold less than when injected on to a "normal" bore (e.g., 4.6 mm id) column. In this way, the compounds of interest become easier to detect. Microbore LC offers several other advantages:

1. The lower flow rates employed lead to a reduction in mobile phase consumption;
2. Microbore columns dissipate heat very rapidly and, therefore, are easily thermostatically controlled; and
3. Microbore columns have a higher mass sensitivity than their larger diameter counterparts (mass sensitivity is inversely proportional to the square of the column radius) and require smaller sample volumes to achieve the same concentration sensitivity.

These advantages do not come without some cost: Several modifications of conventional LC apparatus are necessary to permit use of microbore columns. In particular, when using microbore LC, it is essential to minimize system dead volume. Modifications of conventional LC apparatus necessary for microbore column work are described further in this chapter (*see* Note 1 and Fig. 1).

From: *Methods in Molecular Biology, Vol. 72: Neurotransmitter Methods*
Edited by: R. C. Rayne Humana Press Inc., Totowa, NJ

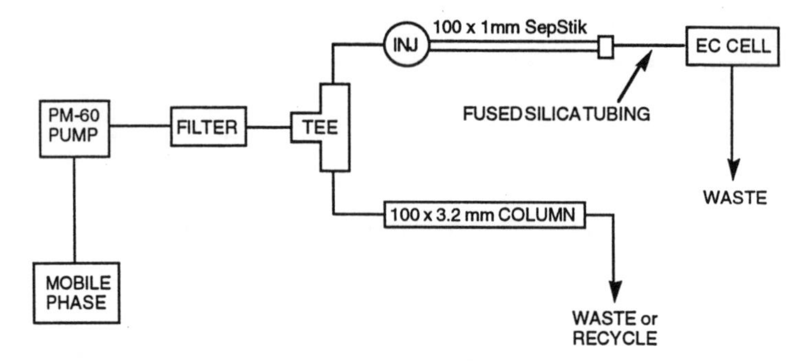

Fig. 1. Principle of the splitter kit for obtaining a microflow using a conventional LC pump system (courtesy of Bioanalytical Systems). *See* Note 1 for further description.

Since the introduction of in vivo brain microdialysis as sampling technique for neurotransmitters, there has been a revival in the use of microbore LC (protocols for in vivo microdialysis are presented and discussed in Chapter 19). Using microdialysis, low-molecular-weight compounds are sampled from the extracellular space where the turnover of neurotransmitters is very fast. Because only 10–20% of the sample is recovered, concentrations of the transmitters collected by microdialysis are quite low (nanomolar range). This means that routine detection of subpicogram amounts of neurotransmitters is required in very small volumes of dialysate. In many cases normal bore LC columns do not provide sufficient concentration sensitivity and, therefore, changes to the microdialysis parameters (elevated Ca^{2+} concentrations in the perfusion fluid, the use of uptake inhibitors, longer sampling times, and so forth) become necessary to enable detection of the transmitter of interest. By using microbore LC, however, such manipulation of experimental conditions can be avoided. Furthermore, because small volumes are injected onto microbore columns (usually only 10 µL), dialysates (usually 40 µL) can be split for analysis of other transmitters with different LC assays.

This chapter describes two microbore LC-based methods for measurement of monoamine neurotransmitters in rat brain dialysates: a method for the determination of the catecholamines (CA) noradrenaline (NAD), adrenaline (AD), and dopamine (DA) and a method for the determination of the indoleamine serotonin (5HT) and its major metabolite 5-hydroxyindoleacetic acid (5HIAA) *(3)*. For all compounds, a limit of detection of 1 pmol or less (in an injected volume of 10 µL) can be achieved. Basal serotonin can be measured in microdialysates without including a reuptake inhibitor in the microdialysis perfusion fluid. Both of these systems are routinely used in the authors' laboratory for the analysis of dialysates from striatum, hippocampus, and substantia nigra. Both methods are selective, robust, and can be automated.

2. Materials

2.1. Both Systems

1. Pulse-free LC pump or LC pump equipped with an efficient pulse-dampener.
2. Flow-splitter kit to give low flow rates across the microbore LC column (Bioanalytical Systems [Indianapolis, IN]; *see* Note 1 and Fig. 1).
3. High precision autosampler, equipped with cooling if possible (e.g., Kontron 465) or manual Rheodyne injector with low dispersion volume and PEEK stator (e.g., Rheodyne 9125).
4. Microbore column: 100 × 1 mm internal diameter.
5. Electrochemical detector (single or dual channel).
6. Strip chart recorder.
7. Integration computer program (e.g., Kontron MT2).
8. Antioxidant mixture: 0.01*M* HCl containing 0.1% (w/v) $Na_2S_2O_5$ (sodium disulfite) and 0.01% (w/v) Na_2EDTA (dihydrate). To prevent precipitation, add compounds only in the order given, as follows: In a 1 L flask, dissolve 10 mg of Na_2EDTA in approx 50 mL of purified water. Add 100 mg of $Na_2S_2O_5$, then about 50 mL of purified water. Add 1 mL of HCl (32%, w/w), and dilute further with purified water to a final volume of 1 L. Filter through a 0.2 µm membrane and store at room temperature. The solution is stable for 1 mo (*see* Note 2).
9. Ringer's solution: 147 m*M* NaCl, 1.1 or 2.2 m*M* $CaCl_2$, and 4 m*M* KCl. To make 250 mL, dissolve 2.15 g of NaCl, 0.04 or 0.08 g of $CaCl_2$ (dihydrate), and 0.075 g of KCl in fresh purified water and filter through a 0.2-µm membrane. This solution can be used for 1 wk (*see* Note 3 and Section 3.2.).

2.2. CA Analysis

1. Catecholamine stock solutions: 0.1 mg/mL in antioxidant mixture (*see* Section 2.1., item 8). Prepare separate stock solutions of NAD, AD, and DA by dissolving 5 mg of each in 50 mL of filtered antioxidant mixture. Stored at 4°C in brown glass bottles or glass bottles covered with aluminum foil, the solutions are stable for 2 mo.
2. Catecholamine standard solutions: 0.5, 1, and 2 pg/10 µL in Ringer's solution/antioxidant mixture. Prepare standard solutions daily by dilutions of each stock solution in a mixture of Ringer's solution and antioxidant mixture in a ratio of 4:1 for the analysis of samples from substantia nigra or hippocampus. For samples from the striatum, use a Ringer's to antioxidant ratio of 1:1 (*see* Note 4). First, prepare a between-stock solution (1 ng/10 µL) containing the three catecholamines by pipeting 10 µL of each stock solution in a flask and diluting to 10 mL with the Ringer's-antioxidant mixture. From this between-stock solution make three standards of 0.5, 1, and 2 pg/10 µL by diluting 5, 10, and 20 µL, respectively, to 10 mL with the Ringer's-antioxidant mixture.
3. CA mobile phase: 60 m*M* sodium acetate, 2 m*M* decanesulfonic acid, 0.1 m*M* Na_2EDTA, and acetonitrile (*see below* and Note 5). To make 1 L, dissolve 0.19 g of Na_2EDTA (dihydrate) in approx 50 mL of purified water, then add 8.2 g of

sodium acetate (trihydrate) and 0.5 g of 1-decanesulfonate. Dilute to 1 L with purified water and filter through a 0.2-µm membrane. (It is not necessary to pH this solution; the pH will be approx 5.8.) To 200 mL of this solution, add either 25 or 28 mL of acetonitrile to complete the preparation of the mobile phase (*see* Note 5).

4. Microbore column: 100 × 1 mm id with C8, 5 µm packing material (Sepstik or Unijet, Bioanalytical Systems).
5. Electrochemical detector: Single channel electrochemical detector with a glassy carbon working electrode (Antec, Leiden, The Netherlands). Set the operating potential to 450 mV vs a Ag/AgCl reference electrode (*see* Note 6). Use a 25-µm gasket to reduce the cell volume.
6. Vacuum degasser (*see* Note 7).

2.3. 5HT Analysis

1. Indoleamine stock solutions: 0.1 mg/mL in antioxidant mixture (*see* Section 2.1., item 8). Prepare separate stock solutions of 5HT, 5HTP, DOPAC, and 5HIAA by dissolving 5 mg of each in 50 mL of filtered antioxidant mixture. Stored at 4°C in brown bottles or bottles covered with aluminum foil, the solutions are stable for 2 mo.
2. Indoleamine standard solutions: 1, 2.5, and 5 pg/10 µL in Ringer's solution. Prepare standard solutions daily by dilutions of the stock solutions in Ringer's solution. First, prepare a between-stock solution (1 ng/10 µL) containing 5HT, 5HTP, and DOPAC by pipeting 10 µL of each stock solution into a flask and diluting to 10 mL with the Ringer's solution. From this between-stock solution make three standards of 1, 2.5, and 5 pg/10 µL by diluting 10, 25, and 50 µL, respectively, to 10 mL with the Ringer's solution.
3. 5HIAA standards: 100, 250, and 500 pg/10 µL in Ringer's solution (*see* Note 8). Prepare as for the other indoleamine standard solutions, but pipet 1 mL from the stock solution to make 10 mL of between-stock solution.
4. Citrate-acetate buffer: 20 m*M* citric acid, 0.1*M* sodium acetate, 0.1 m*M* Na_2EDTA, 0.1 m*M* octanesulfonate, 1 m*M* dibutylamine, pH 4.2. To prepare 1 L: Dissolve 37.2 mg Na_2EDTA (dihydrate) with about 50 mL purified water, then add 13.6 g of sodium acetate (trihydrate), 4.2 g of citric acid (monohydrate), 21.6 mg of octanesulfonate, and 170 µL of dibutylamine and dilute with purified water to 1 L. Adjust the pH of this solution to 4.2 with concentrated phosphoric acid and filter through a 0.2-µm membrane. This solution can be kept at 4°C for up to 3 wk.
5. 5HT mobile phase: 95% (v/v) citrate-acetate buffer, pH 4.2, and 5% (v/v) MeOH. Degas before use (*see* Note 7).
6. Microbore column: 100 × 1 mm id with a C18, 3 µm packing material (Sepstik or Unijet, Bioanalytical Systems).
7. Electrochemical detector: an instrument with dual glassy carbon working electrodes (W1 and W2), positioned in parallel (Bioanalytical Systems). Set the operating potential of W1 at 600 mV (the maximal oxidation potential of 5HT) and of

W2 at 525 mV (the half-maximal oxidation potential of 5HT). Use a 16-μm gasket to reduce the cell volume.

3. Methods

3.1. CA Analysis

1. After appropriate flushing of the new column (*see* Note 9), equilibrate the system with CA mobile phase, setting the flow rate at the pump to 0.7 mL/min. Check the flow rate through the microbore column by collecting during 1 min the eluting fluid from the detector cell outlet. The flow should be between 60 and 80 μL/min.
2. Make the connections with the electrochemical cell and switch the detector on with the potential at 450 mV and at a sensitivity that gives an acceptable baseline (in our laboratory this is 0.1 nA/V using the Antec detector).
3. Allow the system to equilibrate at least overnight (*see* Note 10).
4. Trace a baseline (*see* Note 11). If it is acceptable, inject a blank (e.g., Ringer's-antioxidant mixture; *see* Note 12). If no interfering peaks are observed, inject the three standards of the CA and establish a calibration curve (*see* Note 13).
5. Check the retention times for the standards. If all is well, begin analysis of the microdialysis samples (*see* Note 4).
6. Inject the dialysates immediately; there is no need for sample pretreatment because the dialysis procedure excludes large molecules from the samples.
7. Compare peak areas of dialysates and standards given by the computer printout and calculate the amount of the CA in one dialysate using the calibration curve (*see* Note 14).
8. Figure 2 shows chromatograms of a mixture of standards, a dialysate from the hippocampus, a dialysate from striatum, and a dialysate from the substantia nigra.

3.2. 5HT Analysis

1. After appropriate flushing of the new column (*see* Note 9), equilibrate the system with 5HT mobile phase at a pump flow rate of 0.8 mL/min. Check the flow rate through the column by collecting the eluting fluid from the outlet of the detector cell during 1 min. The flow should be between 60 and 80 μL/min.
2. Follow this four-step procedure to confirm that the 5HT peak measured is not contaminated with coeluting electroactive substances present in the dialysate:
   a. Step 1: Set W1 at 600 mV, range 0.2 nA/V; set W2 at 525 mV, range 0.2 nA/V. When a stable baseline is achieved (*see* Note 11), inject a blank and the standards containing 5HT, DOPAC, and 5HTP. Calculate the calibration curves for the different compounds. Also calculate the ratio of the peak height of 5HT obtained from W1 and W2. This value is characteristic for 5HT.
   b. Step 2: Using the same conditions as in step 1, inject the first three dialysis samples (*see* Note 4). If the ratio of the signals at W1 and W2 differ more than 20% from the mean ratio obtained in step 1, then the 5HT peak is contaminated and the dialysis experiment must be terminated. If the ratio is the same as in step 1, continue the dialysis experiment.

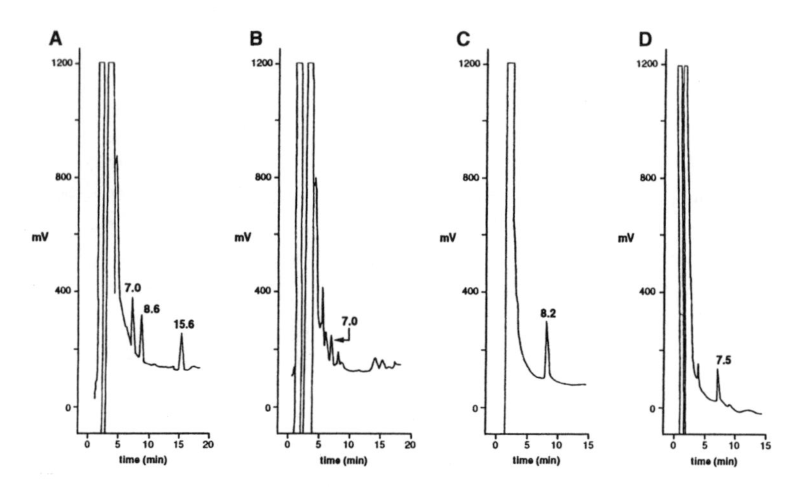

Fig. 2. Microbore LC assay for catecholamines in rat brain dialysates. All chromatograms are registered at 0.1 nA/V. Because the chromatograms in (C) and (D) were registered at different times and with different columns, elution times of DA are not identical. **(A)** Chromatogram of a standard CA solution containing 1 pg of each component. The mobile phase contains 25 mL of acetonitrile (*see* Note 5). Peaks are NAD (7 min), AD (8.6 min), and DA (15.6 min). **(B)** Chromatogram of a dialysate from hippocampus. The mobile phase contains 25 mL of acetonitrile. The peak eluting at 7 min is NAD. **(C)** Chromatogram of a dialysate from striatum. The mobile phase contains 28 mL of acetonitrile. The peak eluting at 8.2 min is DA. **(D)** Chromatogram of a dialysate from substantia nigra pars compacta. The mobile phase contains 28 mL acetonitrile. The peak eluting at 7.5 min is DA.

   c. Step 3: Set the range of W2 to 5 nA/V. (At 0.2 nA/V the current signal for 5HIAA exceeds 1 V, so that integration is faulty because the peak signal is cut off).
   d. Step 4: After the dialysis samples have been analyzed, inject the standards of 5HIAA. Establish the calibration curve (*see* Note 13).
3. Using the calibration curves, calculate the amount of 5HT and 5HIAA in the dialysis samples (*see* Note 15).
4. Figure 3 shows a single chromatogram of the separation of 5HTP, DOPAC, and 5HT in a standard mixture; a dual chromatogram (W1 and W2) of a 5HT standard; a dual chromatogram (W1 and W2) of a 5HIAA standard; and dual chromatograms of dialysates obtained from hippocampus and substantia nigra pars reticulata.

4. Notes

1. When working with microbore LC columns, low flow rates are used; typically the range is 50–100 µL/min. It is possible to create these flow rates with special-

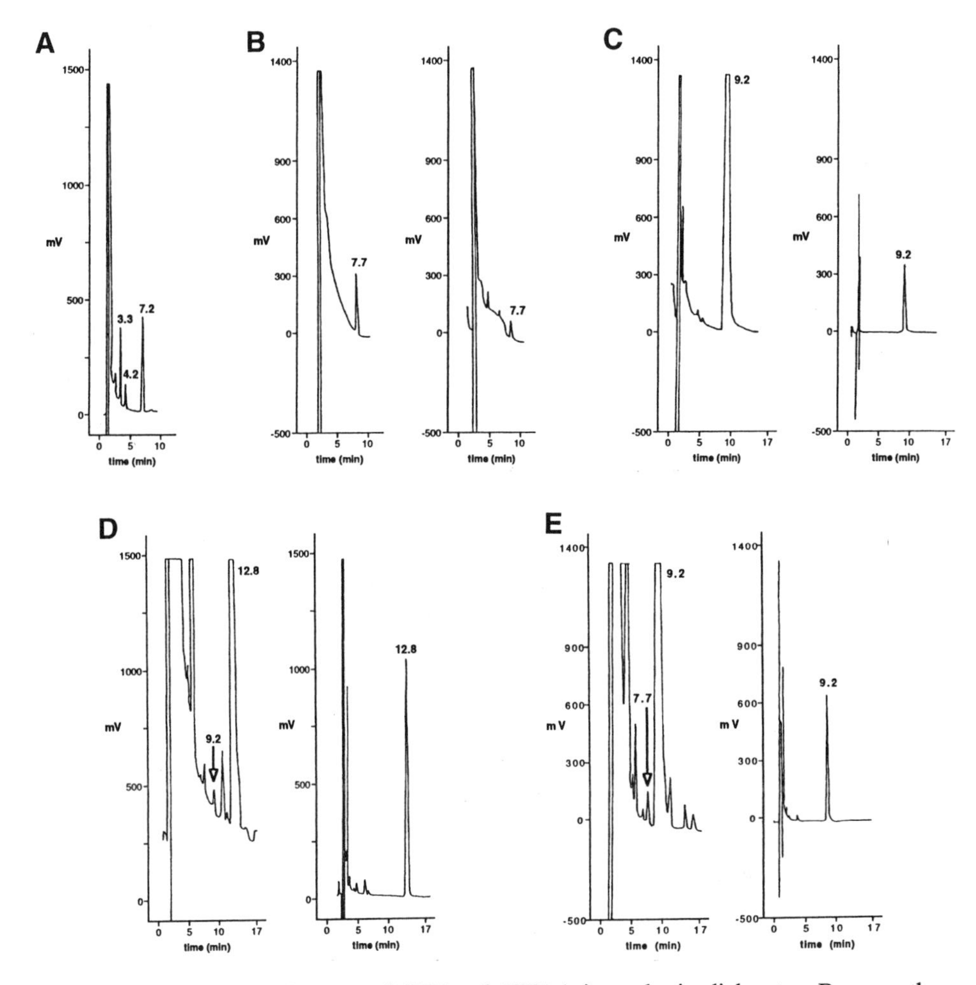

Fig. 3. Microbore LC assay of 5HT and 5HIAA in rat brain dialysates. Because the chromatograms shown in (D) were registered using a slightly slower than normal flow rate, elution times of 5HT and 5HIAA are longer than the corresponding times in (A), (B), (C), and (E). **(A)** Single chromatogram of a standard solution, containing 2.5 pg of each component, registered at W1 (625 mV and 0.2 nA/V). Peaks are 5HTP (3.3 min), DOPAC (4.2 min), and 5HT (7.2 min). **(B)** Dual chromatogram of a standard solution, containing 2.5 pg of 5HT, registered with W1 = 625 mV and 0.2 nA/V (left) and with W2 = 525 mV and 0.2 nA/V (right). The peak eluting at 7.7 min is 5HT. **(C)** Dual chromatogram of a standard solution, containing 250 pg of 5HIAA, registered with W1 = 625 mV and 0.2 nA/V (left) and W2 = 525 mV and 5 nA/V (right). The peak eluting at 9.2 min is 5HIAA. **(D)** Dual chromatogram of a dialysate obtained from hippocampus, registered with W1 = 600 mV and 0.2 nA/V (left) and with W2 = 525 mV and 5 nA/V (right). The peak eluting at 9.2 min is 5HT; the peak eluting at 12.8 min is 5HIAA. Note that at the higher detector sensitivity setting the response is off scale (left). **(E)** Dual chromatogram of a dialysate obtained from substantia nigra pars reticulata, registered with W1 = 625 mV and 0.2 nA/V (left) and with W2 = 500 mV and 5 nA/V (right). The peak eluting at 7.7 min is 5HT; the peak eluting at 9.2 min is 5HIAA. Note that at the higher detector sensitivity setting the response is off scale (left).

ized microbore syringe pumps. However, these pumps can be susceptible to air bubbles and have mechanical limitations. They are more difficult to purge and generally require more frequent maintenance. In the laboratory, we use conventional LC pumps and add a flow splitter, as shown in Fig. 1, to reduce the flow rate to the microbore LC column. In essence, the splitter consists of a tee that is placed between the pump and the injection valve. Approximately 90% of the mobile phase passing through the tee is diverted to a conventional LC column, to keep the back-pressure sufficiently high. The remaining 10% of the mobile phase flows via the injector to the microbore column. An inline filter is placed before the tee to prevent particles present in the mobile phase from reaching the microbore column.

There are a number of important considerations to follow in microbore LC. System pressure should be checked on a daily basis. A severe increase in pressure most likely indicates that the inline filter is dirty and needs replacing. Dirty inline filters can be reused after cleaning with 6*N* nitric acid.

It is essential when converting a conventional LC system to a microbore system that the dead volume of the system is kept to an absolute minimum. There are several ways to achieve this:

a. Use an injection valve with a low dispersion volume;
b. Connect the microbore column directly to the injection valve or, if this is not possible, use a fused silica connection tubing (id 50 μm);
c. Connect the microbore column directly to the electrochemical cell (EC) or, if this is impossible, use fused silica tubing to make the connection; and
d. In the EC cell, use a gasket that is thinner (15–25 μm) than that used conventionally.

The lifetime of a microbore column is generally shorter than that of a conventional column. The number of injections that can be tolerated by a microbore column under the conditions described here varies between 600 and 1200. Variations in retention times usually indicate column deterioration/blockage. Often the column lifetime may be extended by turning the column around so that the mobile phase flow is effectively reversed. Insufficient peak separation (poor resolution) is another sign of a column deterioration. In this case, the best solution is to change the column. In some applications, the use of a guard column to prolong the lifetime of the microbore column is possible. Routine (daily) determination of the flow rate through the microbore column allows quick diagnosis of column blockage, ultimately saving time and preventing loss of precious samples.

2. Several precautions must be taken when preparing mobile phases. To avoid clogging of the microbore column by free particles, **all** aqueous solutions (including those used for sample preparation and for analysis) must be filtered through a 0.2-μm membrane filter and an inlet filter must be placed before the microbore column. Always use solvents of the highest quality. For the electrochemical detection, the background current needs to be as low as possible and should not exceed 8 nA. This can be achieved by adding Na_2EDTA to the mobile phase and by using high quality solvents and water.

3. The Ca^{2+} concentration in the Ringer's solution used for perfusing the microdialysis probe can be varied according to the transmitter being studied. In normal circumstances, the concentration should mirror the concentration in the extracellular fluid; i.e., 1.1 m*M*. However, in some cases where the analysis is more demanding, the concentration may be increased to 2.2 m*M*. This stimulates the release of the transmitter slightly, so that monoamine concentrations in the dialysate are elevated. For example, this is done in the case of the determination of 5HT and NAD in hippocampus.
4. When working with an automated LC system, it is essential that stability of the monoamine samples be maintained during the duration of analysis (usually about 20 h). For the CA analysis, we observed that for larger amounts of DA in the dialysates (as is the case in samples from the striatum) the amount of antioxidant mixture (*see* Section 2.1., item 8) added to the samples must be increased. Therefore, for 40 µL dialysates from the striatum, we add 40 µL of antioxidant mixture, whereas for 40 µL samples of the substantia nigra or hippocampus, only 10 µL of antioxidant is added. To ensure stability of the analytes, dialysates are collected into a vial already containing the appropriate volume of antioxidant mixture (*see* Notes 14 and 15).
5. Two different mobile phases can be used for the CA analysis. In cases where NAD, AD, and DA are determined, only 25 mL of acetonitrile is added to the buffer. In cases where only DA is of interest, a faster elution can be achieved by adding 28 mL of acetonitrile to the buffer. (Note that we have not expressed the final ACN concentration as a percent. To express the ACN concentration in this way is dangerous because of the volume contraction that occurs when ACN [or any other organic solvent] is added to the aqueous buffer. Thus, addition of 25 mL of ACN to 200 mL of aqueous buffer is not equivalent to dilution of 200 mL of aqueous buffer to 225 mL with ACN.)

   In the chromatograms shown in Fig. 2, one or the other mobile phase is used according to the brain area studied. For example, in substantia nigra and striatum, usually only DA is measured, whereas in hippocampus, NAD is of primary interest. Also note that in the dialysates no AD is measured. This is normal. The presence of AD in the sample likely indicates that the blood–brain barrier is not intact and AD in the sample has originated from the periphery.
6. The potentials for both applications have been chosen after registration of a voltammogram. This is done by injecting a standard solution at increasing potentials. The peak heights are then plotted against the potential, and the working potential is chosen as the potential just before the voltammogram shows a plateau. This potential gives the best signal-to-noise ratio with the best selectivity. It is necessary to register a new voltammogram for each new application (change in mobile phase composition).
7. When using higher percentages of organic modifier in the mobile phase, degassing is very important. A vacuum degasser is preferred instead of helium degassing, because of the lower operating and maintenance costs and because vacuum does not affect the solvent composition. Degassing of the mobile phase generally

improves the baseline stability. Especially in gradient elution, noisy baselines and unreproducible retention times can be caused by the formation of gas bubbles in the pump heads. In electrochemical detection, the dissolved gasses in the mobile phase interfere with the oxidation-reduction reactions in the cell and result in baseline drift.

8. It is necessary to prepare the standards of 5HIAA separately from those of 5HT, 5HTP, and DOPAC, because the concentrations of the 5HIAA standards are 100-fold higher. (The concentrations of 5HIAA in the hippocampus are approx 100-fold higher than those of 5HT.) Impurities from the standards at these concentrations give coeluting peaks at the retention time of 5HT at 0.2 nA/V.
9. New columns must be treated by a flushing procedure to activate the C8 or C18 side chains that are bound to the silica packing material. Begin by flushing the column with 100% organic modifier (acetonitrile or methanol) for about 30 min at 0.7 or 0.8 mL/min (not too long because this can cause swelling and then blockage of the inline frit of the column). Then flush the column with a mixture of organic modifier plus filtered water (1:1) for the same amount of time, but at a lower flow rate, e.g., 0.4 mL/min, because this mixture increases the pressure of the system considerably. Then switch to filtered water again for 30 min at the normal flow rate. Now the mobile phase can be pumped through the system. If this procedure is not carried out properly, poor retention of the transmitters will result. This procedure is standard for all LC analyses.
10. Our LC-ECD systems are never switched off. When not in use, the mobile phase is simply continuously recycled and the electrochemical cell is left on. In this way, the cell is always equilibrated. Fresh mobile phase is prepared on at least a weekly basis to prevent bacterial contamination and consequent increased background current and noise. Refreshing the mobile phase also ensures that the organic modifier is present at the correct concentration: Over extended periods, the amount of organic modifier may decrease as a result of evaporation, resulting in longer retention times. When changing to a new mobile phase, always refill the reference electrode compartment with fresh mobile phase. This will improve baseline stability.
11. An acceptable baseline is one that is straight and with little noise.
    a. Irregular baselines may be caused by: leakage somewhere in the system (check all fittings) or irregular flow through the system because of clogging of the column (change the column; *see* Note 1) or blockage of fused silica tubing (replace it).
    b. Spikes in the baseline may be caused by: air bubbles inside the electrochemical cell (refill the cell with mobile phase), insufficient grounding of the system, or change of mobile phase (wait for complete equilibration!).

    Increased noise can be caused by a dirty working electrode. The increase in noise usually coincides with an increase of background current (this is the current of the system [mobile phase] when no sample is being injected). To solve this problem, clean the working electrode with water, then with methanol. If there is no improvement, polish the electrode. Polishing the working electrode must be reserved as a last resort, because this treatment ultimately shortens the lifetime of the electrode. Longer equilibration times are usually encountered after polishing. Polishing once every 6 mo is more than enough if the cell is being used for analysis of relatively clean dialysis samples.

12. Blank solutions (Ringer's or a mixture of Ringer's and antioxidant solution) may sometimes produce peaks that interfere with the peaks of compounds of interest. There may be several sources of such peaks. First, check the water quality. Inject water only, and if peaks appear, check the water purification system. Second, glassware should be rinsed only with purified water. Because working with dialysates is a clean procedure (no proteins), use of detergents is unnecessary. Detergent contamination can give blank problems as well as baseline stability problems because of the basic pH of many detergents. If interfering peaks still appear (this is possible), we recommend changing slightly the pH of the mobile phase (± 0.2 U).
13. Calibration curves are established by injecting a blank (e.g., dilution solvent for standards) and a minimum of three standards of the transmitters being studied (*see* Section 2.2., item 2 and Section 2.3., items 2 and 3). Using the concentrations of the standards and the corresponding peak areas, calculate a calibration curve with format $y = ax + b$ with linear regression analysis (y = peak area; x = concentration of transmitter in pg/10 µL; a = slope; b = intercept). The peak area recorded from the samples can be entered into the equation and x can be calculated. Taking into account the dilution factor (knowing the amount of antioxidant added to each vial), and the molecular weight of the transmitter allows the amounts in the dialysates to be expressed as picomole or femtomole per timed dialysis interval (usually 20 min; *see* Notes 14 and 15). Because of the poor stability of the catecholamines and the indoleamines, calibration curves should be constructed daily.

    The concentrations of standards used to construct the calibration curve depend on the expected basal values of the transmitters in the brain area being studied. If large increases of the transmitters are observed during analysis, one or two calibration points are added to cover the larger concentration range.
14. In the striatum, dialysis is carried out at 2 µL/min and samples are taken every 20 min using a 3-mm dialysis probe. Samples are collected into vials containing the appropriate volume of antioxidant solution (*see* Note 4). Typical baseline values are for NAD, 7.2 ± 1.5 fmol/20 min (mean ± SEM, n = 7) and for DA, 115.2 ± 13.1 fmol/20 min (mean ± SEM, n = 19). The limits of detection for NAD, AD, and DA are 0.6, 0.5, and 1.1 fmol injected (10 µL loop), respectively.
15. In the hippocampus, a 3-mm probe is placed and dialyzed at a flow rate of 1 µL/min with Ringer's solution containing 2.2 m*M* Ca^{2+} (*see* Note 3) and samples are taken every 20 min. Samples are collected into vials containing the appropriate volume of antioxidant solution (*see* Note 4). Typical baseline values under these conditions are for 5HT, 29.5 ± 5.1 fmol/20 min (mean ± SE*M*, n = 35) and for 5HIAA, 7.0 ± 0.7 pmol/20 min (mean ± SEM, n = 36). The limits of detection for 5HT, 5HTP, DOPAC, and 5HIAA are 0.5, 0.4, 0.5, and 0.7 fmol injected, respectively.

References

1. Johansson, I. N., Wahlund, K. G., and Schill, G. (1978) Reversed phase ion-pair chromatography of drugs and related organic compounds. *J. Chromatogr.* **149,** 281–296.
2. Scott, R. W. P. and Kucera, P. (1976) Some aspects of preparative-scale liquid chromatography. *J. Chromatogr.* **199,** 467–482.
3. Sarre, S., Marvin, C. A., Ebinger, G., and Michotte, Y. (1992) Microbore liquid chromatography with dual electrochemical detection for the determination of serotonin and 5-hydroxyphenylindoleacetic acid in rat brain dialysates. *J. Chromatogr. Biomed. Appl.* **582,** 29–34.

15

Microbore Liquid Chromatography Analysis of Amino Acid Transmitters

Ilse Smolders, Sophie Sarre, Guy Ebinger, and Yvette Michotte

1. Introduction

Microdialysis is a well-established sampling method for collecting neurotransmitters from the brains of freely moving rats (*1*; *see also* Chapter 19). To measure levels of amino acid transmitters in the microdialysates requires very sensitive methods of analysis owing to the subpicogram amounts of amino acids in small volumes of dialysates (in our experimental setup, the total volume is usually 40 µL). Often, conventional liquid chromatography (LC) systems are unable to provide the necessary sensitivity and, therefore, microbore LC has become a valuable analytical tool for the neurochemist *(2,3)*. The general principles, advantages, and disadvantages of microbore LC as an analytical method for measuring neurotransmitter levels in microdialysates are discussed further in Chapter 14. In this chapter, two different microbore LC methods—each optimized for analysis of different amino acid transmitters—will be described. Both methods are selective, robust, and can be automated.

Both LC systems depend on precolumn derivatization of primary amino acids with *o*-phthalaldehyde (OPA) in the presence of a thiol. These reactions are quantitative and produce isoindole derivatives that have electro-active and fluorescent characteristics, enabling use of highly sensitive electrochemical and/or fluorescence detection methods. In our laboratory we use these LC systems for routine determinations of extracellular levels of γ-aminobutyric acid (GABA), glutamate, and aspartate in rat striatum, substantia nigra, hippocampus, and cerebellum. The basal dialysate concentrations of GABA are in many brain areas at least 40 times lower than the basal dialysate concentrations of glutamate. Therefore, the assay for GABA is the most challenging. The respec-

From: *Methods in Molecular Biology, Vol. 72: Neurotransmitter Methods*
Edited by: R. C. Rayne Humana Press Inc., Totowa, NJ

tive methods are outlined briefly in Sections 1.1. and 1.2., and described in detail in Sections 2., 3., and 4.

1.1. LC Analysis of GABA in Brain Microdialysates

For analysis of the main inhibitory neurotransmitter, GABA, we have developed a separation method using isocratic elution and electrochemical detection. The best detection sensitivity for GABA is obtained after precolumn derivatization with OPA/*tert*-butylthiol. Because of the electrochemical characteristics of *tert*-butylthiol and the stability of the OPA/*tert*-butylthiol derivative, this thiol is the most suitable for GABA analysis. Use of *tert*-butylthiol necessitates a second derivatization step with iodoacetamide to scavenge excess thiol that otherwise would produce interfering peaks in the chromatogram.

Using this microbore LC method, the detection limit for GABA (taken as the amount corresponding to a signal-to-noise-ratio of 3) is 0.001 μ*M*; this is equivalent to 7 fmol injected onto the column. In comparison with conventional LC, this represents a fivefold increase in sensitivity, an important consideration when brain areas with a low GABA content are studied. Basal concentrations of GABA in rat striatum dialysates, for example, are 0.028 ± 0.005 μ*M*, illustrating the significant sensitivity margin afforded by our method.

1.2. LC Analysis of Glutamate and Aspartate in Brain Microdialysates

A second LC method, utilizing a gradient elution scheme and fluorescence detection, is employed in the determination of the main excitatory neurotransmitter, glutamate. This method may also be used to measure levels of the excitatory amino acid, aspartate.

Gradient elution is used in these assays to obtain short analysis times and consists of an isocratic step to elute the peaks of interest, followed by a wash-off step to elute other amino acids. Gradient elution in association with electrochemical detection suffers from baseline drift and requires long re-equilibration times, whereas the combination of gradient elution with fluorescence detection does not. Therefore, fluorescence detection is preferred. Again, OPA/thiol precolumn derivatization is used to produce isoindole compounds that facilitate detection. For derivatization of glutamate or aspartate for fluorescence detection, β-mercaptoethanol is an appropriate thiol and, in addition, requires only a single-step derivatization. The choice of β-mercaptoethanol is further recommended by the stability of the OPA/β-mercaptoethanol-glutamate and -aspartate derivatives.

Using this method of analysis, the detection limits (defined as above) for glutamate and aspartate are 0.008 μ*M*, or 67 fmol on the column. For comparison, basal concentrations of glutamate and aspartate in rat striatum dialysates are 1.085 ± 0.129 μ*M* and 0.080 ± 0.013 μ*M*, respectively.

2. Materials

2.1. Both Systems

1. 0.2-μm Membrane filters (*see* Note 1).
2. Sodium borate buffer: 0.01*M* sodium borate, in fresh, purified water, pH 9.0. Stored at –18°C, the buffer is stable for several months (*see* Note 2).
3. OPA solution: 15 m*M* *o*-phthalaldehyde. Dissolve 12 mg of OPA in 600 μL of HPLC-grade methanol. Add 5.4 mL of the sodium borate buffer (item 2). Mix well. Stored at 4°C, the solution is stable for 1 wk.
4. Ringer's solution: 148 m*M* NaCl, 2.3 m*M* $CaCl_2$, 4.0 m*M* KCl, pH 7.3. Dissolve the salts in fresh purified water, adjust the pH with NaOH, and filter through a 0.2-μm membrane.
5. Vacuum degasser (*see* Note 3).
6. Conventional LC gradient pump.
7. Flow splitter kit to provide low volumetric flow rates over the microbore column (Bioanalytical Systems, West Lafayette, IN; *see* Note 4).
8. Refrigerated (4°C), closed autosampler with 10 μL injection loop.
9. Microbore LC column: 5 μm C8 packing, 100 × 1 mm inner diameter.
10. Chart recorder.
11. Computer program for peak integration.

2.2. GABA Analysis

1. Mobile phase (GABA): 58% (v/v) 0.1*M* sodium acetate solution, pH 5.0 (containing 40 mg/L Na_2EDTA), 42% (v/v) acetonitrile. Mix and sonicate for 5 min (*see* Note 1).
2. GABA stock solution: 2.5 m*M* GABA in 0.1*M* HCl. Stored at –18°C the solution is stable for several months.
3. GABA dilution series: Dilute the GABA stock solution in fresh purified water, to obtain 0.010–0.100 μ*M* (*see* Note 2). Stored at –18°C, these are stable for at least 4 wk.
4. Working reagent A: Add 2.3 μL of *tert*-butylthiol to 1 mL of OPA solution and mix well. Protect this solution from light and air by using amber-colored borosilicate vials capped by a septum (thiols smell!). Prepare fresh every day.
5. Working reagent B: Dissolve 185 mg of iodoacetamide in 1 mL of methanol. Store in a borosilicate vial capped by a septum.
6. Electrochemical detector: single-channel amperometric detector with a glassy carbon working electrode. Use a 16-μm gasket to reduce the cell volume (*see* Note 5).

2.3. Glutamate and Aspartate Analysis

1. Mobile phase A: 90% (v/v) 0.1*M* sodium acetate, pH 6.0, 10% (v/v) acetonitrile. Mix and sonicate for 5 min (*see* Note 1).
2. Mobile phase B: 90% (v/v) acetonitrile, 10% (v/v) water. Mix and sonicate for 5 min (*see* Note 1).

3. Glutamate and aspartate stock solutions: 2.5 m*M* in 0.1*M* HCl. Stored at –18°C the solutions are stable for several months.
4. Glutamate and aspartate dilution series: 0.200–4.000 µ*M* for glutamate and 0.050–1.000 µ*M* for aspartate in fresh purified water (*see* Note 2). Stored at –18°C the solutions are stable for a few weeks.
5. Working reagent C: Add 2.3 µL of β-mercaptoethanol to 1 mL of OPA solution and mix well. Protect this solution from light and air by using amber-colored borosilicate vials capped by a septum (thiols smell!). Prepare fresh every day.
6. Fluorescence detector with a capillary flow cell.

3. Methods

3.1. GABA Analysis

1. Prepare the mobile phase (*see* Section 2.2., item 1).
2. Set the pump flow rate to 1 mL/min to produce a flow rate of ~90 µL/min through the microbore column (*see* Note 4).
3. Set up the detector. Assemble the electrochemical cell (*see* Note 5) and set the operating potential at +700 mV vs the Ag/AgCl reference electrode (*see* Note 6). Set the range at 2 nA/V and allow the system to equilibrate overnight (*see* Note 7).
4. After obtaining a stable baseline, prepare the reagents for derivatization (*see* Section 2.2., items 4 and 5).
5. Add 2 µL of working reagent A for every 10 µL of standard or microdialysate sample to form electroactive 1-alkylthio-2-alkylisoindoles. Allow the derivatization to proceed for 2 min at 4°C (*see* Notes 8 and 9).
6. Add 2 µL of working reagent B for every 10 µL of standard or sample to scavenge the excess of *tert*-butylthiol. The reaction time is 4 min at 4°C (total reaction time, steps 1 and 2, is 6 min).
7. Inject standards and samples immediately after the derivatization! An autosampler may be used to derivatize and inject automatically. If no autosampler is available, both steps can be performed manually.
8. Always inject first a derivatized blank (fresh Ringer's solution or fresh water) and a series of derivatized standards. Check the linearity, the sensitivity, and the reproducibility of the chromatograms (*see* Note 10).
9. Following this set-up procedure, analyze the dialysate samples.
10. Figure 1 shows example chromatograms of a blank, a 0.050-µ*M* GABA standard solution, and a dialysate obtained from rat striatum (*see* Note 11).

3.2. Glutamate and Aspartate Analysis

1. Prepare mobile phase solutions A and B (*see* Section 2.3., items 1 and 2).
2. Set the pump flow rate to 1 mL/min to produce a flow rate of ~90 µL/min through the microbore column (*see* Note 4).
3. Program the gradient conditions for the LC pump (*see* Note 12).
4. Set up the fluorescence detector so that the excitation wavelength is 350 nm and the emission wavelength is 450 nm (*see* Note 13).

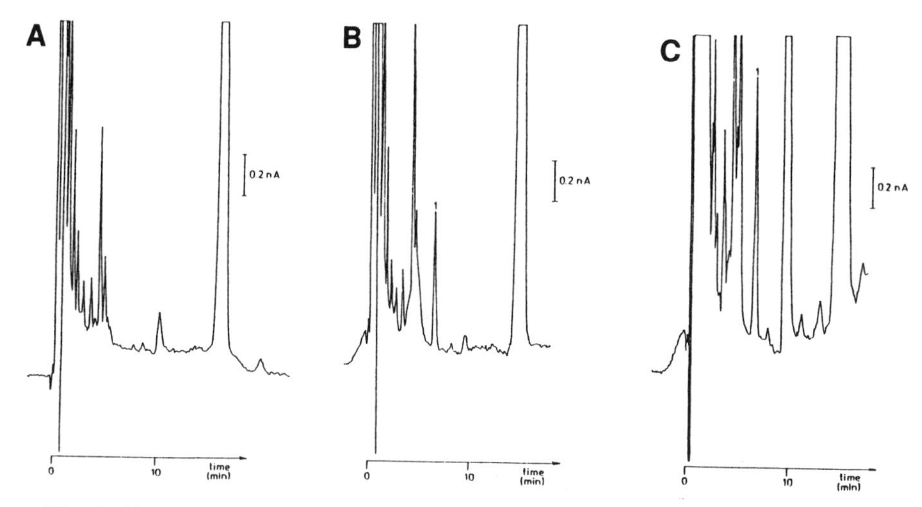

Fig. 1. Chromatograms of the LC assay for GABA by isocratic elution and electrochemical detection. **(A)** Blank injection. **(B)** 0.050 µ*M* GABA standard. **(C)** Dialysate obtained from rat striatum. Peak 1 (B,C) is GABA. The detector range was 2 nA, full scale. The large, late-eluting peak is because of the derivatizing reagent. The injection volume in each case was 10 µL.

5. To derivatize the samples, add 2 µL of working reagent C for every 10 µL of standard or microdialysate sample to form fluorescent 1-alkylthio-2-alkylisoindoles. The derivatization time is 2 min at 4°C (*see* Notes 8 and 9).
6. Inject standards and samples immediately after the derivatization!
7. After obtaining a stable baseline, load the gradient program file of the LC pump so that the gradient program runs with every injection you make. Inject a blank (fresh Ringer's solution or fresh water) and a series of standards before analyzing dialysate samples (*see* Note 10).
8. Figure 2 shows example chromatograms of a blank, a standard mixture containing 0.020 µ*M* aspartate and 1.000 µ*M* glutamate, and a dialysate obtained from rat striatum.

4. Notes

1. Several precautions must be taken when preparing mobile phases. Use only solvents of the highest quality. To prevent clogging of the microbore column by free particles, all aqueous solutions must be passed through a 0.2-µm membrane filter and an inlet filter must be placed before the microbore column. For the electrochemical detection, the background current must be as low as possible and may not exceed 8 nA. This can be achieved by adding Na_2EDTA to the mobile phase and by using high quality solvents and water.
2. It is very important that only fresh purified water is used for preparing all of the aqueous solutions (the standard solutions, the buffers, the blanks), and also for

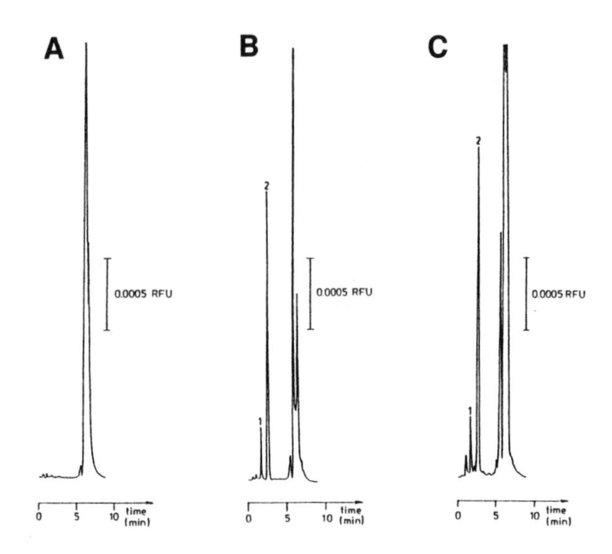

Fig. 2. (A) Blank injection. **(B)** Standard mixture containing 0.200 μ*M* aspartate (peak 1) and 1.000 μ*M* glutamate (peak 2). **(C)** Dialysate obtained from rat striatum. Peak 1 is aspartate and peak 2 is glutamate. The detector range was 0.01 Reference Fluorescence Units. The injection volume in each case was 10 μL.

flushing the syringes and needles for injection. Because of bacterial growth, aqueous solutions may contain free amino acids that contaminate standards and samples. Using membrane filtration, the micro-organisms—but not the produced amino acids—are removed. Do not allow disposable pipet tips and glassware to be exposed to dust and do not touch them with fingertips. The potential for contamination must not be underestimated when measuring subpicomole amounts of amino acids!

3. Degassing the mobile phases is obligatory. A vacuum degasser is preferred instead of helium degassing, because of the lower operating and maintenance costs, and because vacuum does not affect the solvent composition. Degassing of the mobile phase generally improves the baseline stability. Especially in gradient elution, noisy baselines and unreproducible retention times can be caused by the formation of gas bubbles in the pump heads. In electrochemical detection, the dissolved gasses in the mobile phase interfere with the oxidation-reduction reactions in the cell and result in baseline drift. In fluorescence detection, dissolved oxygen has a quenching effect; fluorescence response will be diminished, resulting in a loss of sensitivity.
4. The splitting technique is useful for both isocratic and gradient elution. The flow rate of the pump can be set to approx 1 mL/min, yielding a 10–12-fold lower, but constant flow rate over the microbore column because of the splitter. A typical LC setup employing a splitter to reduce the flow rate through the microbore column is depicted in Fig. 1 and described in Note 1 of Chapter 14.

5. The thin layer electrochemical cell consists of a glassy carbon working electrode, an Ag/AgCl reference electrode and an auxiliary electrode. Use a 16-μm thin layer gasket between the auxiliary electrode and the working electrode. Microbore analysis necessitates this thin gasket to reduce the cell volume. Store reference electrodes in a 3*M* NaCl solution. When assembling a reference electrode, avoid getting air bubbles in the reference electrode compartment. Air bubbles cause noisy baselines. Make sure that, after assembly, flow is continuous through the entire cell and there are no leaks.
6. The potential for this application has been chosen after registration of a voltammogram. This is done by injecting a standard solution at increasing potentials. The peak heights are then plotted against the potential and the working potential is chosen as the potential just before the voltammogram shows a plateau. This potential gives the best signal-to-noise ratio with the best selectivity. It is necessary to register a new voltammogram for each new application (change in mobile phase composition).
7. After equilibration, the baseline and the background current must be stable (*see* Note 11 of Chapter 14 regarding acceptable baselines). After a series of injections, do not switch the electrochemical detector off! When not in use, the mobile phase is simply continuously recycled and the electrochemical cell is left on. In this way, the cell is always equilibrated. Fresh mobile phase is prepared on at least a weekly basis to prevent bacterial contamination and consequent increased background current and noise. Refreshing mobile phase also ensures that the organic modifier is present at the correct concentration: Over extended periods, the amount of organic modifier may decrease because of evaporation, resulting in longer retention times. When changing to a new mobile phase, always refill the reference electrode compartment with fresh mobile phase. This will improve baseline stability.
8. The final pH of the derivatization mixture must be 9.0 or more for proper derivatization. Degradation of the derivatives is significant at pH 5.0–6.0. Our microdialysates, pH 7.3, need no pH adjustment; if the described volumes of working reagents are added to the samples, the pH of the final mixture is more than 9.0. However, if acid samples are to be analyzed, neutralization is required before adding the reagents (for instance, calculate the volume of a NaOH solution to be added to obtain a pH above 7.0; take the dilution factor into account when calculating the amino acid concentration).
9. It is important to maintain constant derivatization times and temperatures to obtain reproducible results. Automation with a refrigerated autosampler is very useful in this context.
10. Injection of a derivatized blank should not produce contaminant peaks in the chromatogram. Inject derivatized standard solutions of increasing concentrations and calculate the calibration curve $y = ax + b$ using linear regression analysis, where y is the peak area of the standard solutions and x the concentration. Linearity is verified by visual inspection of the calibration graph and calculation of the correlation coefficient that must exceed 0.9970. The detection limit of an amino

acid is the smallest concentration yielding a detector signal that is three times larger than the signal obtained for a blank solution. This detection limit must be at least 0.005 μ*M* or better. The reproducibility of the procedure is checked by replicate analyses (n = 6), repeatedly injecting the same standard. A standard deviation of the peak areas of <2% is considered satisfactory.

11. Late-eluting peaks can emerge in the GABA chromatogram. These peaks are produced by unidentified products of oxidation-reduction reactions. Such peaks are negligible if the derivatization is performed with freshly prepared reagents, the reagents are protected carefully from light and air, and the temperature is controlled (4°C).
12. The gradient program takes 9 min in total. For the first 4 min, pump 100% mobile phase A (0%B); this will elute isocratically the peaks of interest. Follow this with a wash-off step of late-eluting amino acids: At 4.1 min elapsed time, switch to 70%B. Program a linear gradient from 70 to 0%B between 4.1 and 6.5 min. Finish the program by pumping 0%B (100%A) for the last 2.5 min to allow the system to re-equilibrate. Enter in the program a "wait for autosampler injection signal" line; this way the gradient will not start while the autosampler is derivatizing the next sample, but it will start at the moment of injection.
13. Once the column has been adequately flushed (for a discussion, *see* Note 9 of Chapter 14) and equilibrated with mobile phase solution A, turn on the lamp of the fluorescence detector and allow it to warm up for about 15 min. This is more than sufficient as equilibration time. When not injecting samples, the fluorescence detector (unlike the electrochemical detector; *see* Note 7) is always switched off.

References

1. Benveniste, H. (1989) Brain microdialysis. *J. Neurochem.* **52,** 1667–1679.
2. Kissinger, P. T. and Shoup, R. E. (1990) Optimization of LC apparatus for determinations in neurochemistry with an emphasis on microdialysis samples. *J. Neurosci. Meth.* **34,** 3–10.
3. Smolders, I., Sarre, S., Ebinger, G., and Michotte, Y. (1995) The analysis of excitatory, inhibitory and other amino acids in rat brain microdialysates using microbore liquid chromatography. *J. Neurosci. Meth.* **57,** 47–53.

16

HPLC Methods to Isolate Peptide Neurotransmitters

G. Mark Holman and Timothy K. Hayes

1. Introduction

Coupled with solid-phase extraction methodology, reverse-phase high performance liquid chromatography (HPLC) has taken peptide isolation and purification from the very difficult to the almost routine. When combined with an efficient detection system (bioassay, ELISA, RIA, MALDI-MS), appropriate HPLC purification systems can provide peptides of sufficient purity for sequence analysis and structural characterization in a relatively short time. Despite these technical advances, the unique characteristics of the target peptide for any isolation campaign still will provide the investigator with a special challenge or two in adapting the individual methods discussed in this chapter. For those investigators interested in additional information and tips on peptide and protein separation and purification by HPLC, the authors would suggest obtaining the booklet listed in the references *(1)*.

To provide the selectivity required for ultimate purification of peptides from complex mixtures, our method relies on the use of several columns, each with different separation characteristics. Other workers have successfully employed alternative strategies using fewer columns, in which column selectivity is altered using ion-pair reagents, such as heptafluorobutyric acid, and/or by varying the organic solvents *(2)*. Both philosophies are equally valid. Regardless of the chosen strategy, it should be noted that dissection scissors can be very effective tools for simplifying purifications. Gäde and collaborators have traditionally used this approach. Most recently, they isolated and characterized two peptides from a portion of grasshopper brain. Four grasshoppers were required and the two peptides were isolated pure with a single HPLC run *(3)*.

From: Methods in Molecular Biology, Vol. 72: Neurotransmitter Methods
Edited by: R. C. Rayne Humana Press Inc., Totowa, NJ

The extraction/purification methods described here were developed for isolation of both small (pentamer) and somewhat larger (44 residue) peptides *(4,5)*. Two extraction/solid-phase fractionation systems are described. Each has been successfully used for peptide purifications from large (5 kg) whole-body insect extracts, but the methods are easily scaled down for smaller amounts of tissue and remain applicable for extracts of tissue from noninsect sources. Although the first two HPLC column systems were designed to process large extracts quickly, they are also easily scaled down for smaller extracts. In addition to our presentation of specific purification protocols, we describe, in Section 4., some sample-handling techniques to reduce peptide loss during purification, both from adsorption to glass and plastic surfaces and from oxidation during storage in water/acetonitrile/trifluoroacetic acid solutions.

2. Materials

2.1. Reagents

1. High purity water (glass-distilled or reverse osmosis).
2. Acetonitrile (MeCN), HPLC grade.
3. Methanol, reagent grade.
4. Acetic acid, reagent grade.
5. Trifluoroacetic acid (cat. no. T6508, Sigma, St. Louis, MO).
6. 2-(methylthio)ethanol (2-MTE) (cat. no. 22,642-4, Aldrich, Milwaukee, WI) (*see* Note 1).

2.2. Solutions (see Note 2)

1. 10% TFA: 10% (v/v) trifluoroacetic acid in water (*see* Note 3).
2. Methanol/water/acetic acid solution: µix the components 90:9:1.
3. 0.1% (v/v) TFA: Add 10 mL of 10% TFA to 990 mL of water.
4. 10% MeCN/0.1% TFA: Mix 100 mL of MeCN, 890 mL of water, and 10 mL of 10% TFA.
5. 30% MeCN/0.1% TFA: Mix 300 mL of MeCN, 690 mL of water, and 10 mL of 10% TFA.
6. 50% MeCN/0.1% TFA: Mix 500 mL of MeCN, 490 mL of water, and 10 mL of 10% TFA.
7. 80% MeCN/0.1% TFA: Mix 800 mL of MeCN, 190 mL of water, and 10 mL of 10% TFA.
8. 50% MeCN/0.085% TFA: Mix 500 mL of MeCN, 491.5 mL of water, and 8.5 mL of 10% TFA.
9. 95% MeCN/0.01% TFA: Mix 950 mL of MeCN, 49 mL of water, and 1 mL of 10% TFA.
10. 50% MeCN/0.01% TFA: Mix 500 mL of MeCN, 499 mL of water, and 1 mL of 10% TFA.

2.3. Solid-Phase Extraction (SPE) (see Note 4)

1. 35-cc SPE column: 35-cc syringe column containing 10 g of C-18 packing (Waters/Millipore, Milford, MA).
2. 60-cc SPE column: 60-cc syringe column containing 10 g of C-18 packing (Varian Sample Preparation Products, Harbor City, CA).
3. Filtration flask (1 or 2 L) with a rubber collar to fit SPE columns. Attach to house vacuum or water-driven aspirator column.

2.4. Equipment (see Note 5)

HPLC system: Two pumps; a gradient controller; septumless injector with a 2-mL sample loop; tunable UV detector with a 10-μL flow cell, set at 214 nm; and a chart recorder.

2.5. HPLC Columns (see Note 6)

1. DeltaPak C-18: 100 × 25 mm prepak and 10 × 25 mm guardpak. Particle size: 15 μm and 300 Å pores (Waters/Millipore; *see* Note 7).
2. DeltaPak C-4: 100 × 25 mm prepak and 10 × 25 mm guardpak. Particle size: 15 μm and 300 Å pores (Waters/Millipore; *see* Note 7).
3. Synchropak RP-1: 250 × 10 mm. Particle size: 6.5 μm and 300 Å pores (Phenomenex, Torrance, CA).
4. Vydac Phenyl: 250 × 10 mm. Particle size: 5 mm and 300 Å pores (Phenomenex).
5. Deltabond C-8: 250 × 4.6 mm. Particle size: 5 mm and 300 Å pores (Keystone Scientific, Bellefonte, PA).
6. Vydac Phenyl (Diphenyl): 250 × 4.6 mm. Particle size: 5 mm and 300 Å pores (Phenomenex).
7. Waters I-125 Proteinpak: 300 × 7.8 mm (Waters/Millipore).

2.6. Miscellaneous Items (see Note 8)

1. Screw-cap polyethylene tubes, 50-mL.
2. Screw-cap polyethylene scintillation vials, 20-mL.
3. 12 × 75 mm and 13 × 100 mm polypropylene tubes with caps.
4. Polypropylene microfuge (Eppendorf) tubes.
5. Refrigerated centrifuge, including heads to support centrifuge bottles of capacity appropriate to the scale of the extracts.
6. Nalgene polypropylene copolymer centrifuge tubes/bottles.
7. Tissue homogenizer: Polytron or glass mortar-pestle type, depending on quantity and type of tissue.
8. Whatman (Maidstone, UK) No. 2 filter paper or equivalent.
9. Nylon filters (40-mm diameter), 5 and 1.2 μm (Micron Separations, Westboro, MA), used in conjunction with a 47-mm magnetic filter funnel (Gelman Sciences, Ann Arbor, MI) and 1000-mL vacuum filtration flask.
10. Rotary evaporator equipped with heated water bath.
11. Speed-Vac apparatus (*see* Note 9).

3. Methods

3.1. Extraction and Pre-HPLC Treatment (see Note 10)

3.1.1. 10% TFA Extraction (see Note 11)

1. Homogenize tissue at ambient temperature in 10% TFA using 10 mL of solution/g of tissue (*see* Note 12).
2. Transfer the homogenate to appropriate centrifuge bottles/tubes, cap, and centrifuge at 4°C for 30 min at 10,000*g*.
3. Prepare a 35-cc SPE column (*see* Section 2.3., item 1) by passing 50 mL of 80% MeCN/0.1% TFA followed by 50 mL of 0.1% TFA through the column. Attach the SPE column to a filtration flask and apply a vacuum to assist solvent flow without exceeding 10 mL/min ("vacuum assist"; *see* Notes 4 and 13).
4. Filter the supernatant through Whatman #2 filter paper then apply, with vacuum assist, 40 g-Eq of extract to the prerinsed 35-cc SPE column (*see* Notes 13 and 14).
5. Rinse the inner walls of the column and flush using 50 mL of 0.1% TFA. Save the eluate until activity is demonstrated (*see* step 8), then discard.
6. Elute bound peptides stepwise, passing consecutive 50-mL aliquots of each solvent through the 35-cc SPE column (solvents 4, 5, 6, and 7 in Section 2.2.), collecting each fraction in a 50-mL, screw-cap tube:
   a. 10% MeCN/0.1% TFA.
   b. 30% MeCN/0.1% TFA.
   c. 50% MeCN/0.1% TFA.
   d. 80% MeCN/0.1% TFA.
7. Add to each 50-mL fraction 5 μL of 2-MTE, cap, shake well, and store at 4°C (*see* Note 1).
8. Remove a portion of each collected fraction; evaporate the solvent and assay to determine which fraction(s) contain peptide(s) of interest prior to further purification of the sample(s) by HPLC (*see* Note 15).

3.1.2. Methanol-Water-Acetic Acid Extraction (see Note 16)

1. Homogenize tissue at ambient temperature in MeOH/water/acetic acid solution using 10 mL of solution/g of tissue (*see* Note 12).
2. Transfer the homogenate to appropriate centrifuge bottles, cap, and centrifuge at 4°C for 30 min at 10,000*g*.
3. Filter the supernatant through 5- and 1.2-mm nylon filters (stacked, with the 5-mm filter on top), collecting into a 1- or 2-L vacuum flask. Vacuum assist solvent flow without exceeding 10 mL/min (*see* Note 13).
4. Pass (with vacuum assist) the filtered supernatant through a 60-cc SPE column (see Section 2.3., item 2). Wet the column with methanol just before use (no TFA is necessary; *see* Note 17).
5. Transfer sufficient SPE-purified supernatant to fill 80% of the centrifuge tube/bottle. Add sufficient water and 10% TFA to fill the bottle and achieve a final TFA concentration of 0.1%, v/v (e.g., 200 mL of supernatant, 47.5 mL of water, 2.5 mL of 10% TFA; *see* Note 17).

6. Centrifuge at 4°C for 30 min at 10,000*g*.
7. Filter the supernatant through a fresh set of 5- and 1.2-mm nylon filters (stacked, with the 5-mm filter on top). Use vacuum assist as described in step 3 (*see* Note 13).
8. Transfer 40 g-Eq of the supernatant into a round-bottomed flask and remove the methanol by rotary evaporation (*see* Notes 14 and 18).
9. Prepare a 35-cc SPE column by passing 50 mL of 80% MeCN/0.1% TFA followed by 50 mL of 0.1% TFA through the column, with vacuum assist, at a flow rate of ≤10 mL/min (*see* Note 13).
10. Transfer the MeOH-free eluate from the flask onto the prerinsed 35-cc SPE column. Rinse the flask with 50 mL of 0.1% TFA and apply the rinse to the SPE column. Use vacuum assist at a flow rate of ≤ 10 mL/min (*see* Note 13).
11. Elute peptides from the SPE column by stepwise elution as described in Section 3.1.1., step 6. Add to each fraction 5 μL of 2-MTE, cap, shake well, and store at 4°C (*see* Note 1).
12. Remove a portion of each collected fraction, evaporate the solvent, and assay each to determine which fraction(s) contain peptide(s) of interest prior to further purification of the sample(s) by HPLC (*see* Note 15).

3.2. HPLC Purification (see Note 19)

3.2.1. Loading the Samples (see Note 20)

1. Equilibrate with 0.1% TFA at the recommended flow rate for the column (*see* Note 21).
2. Introduce the sample via pump B such that 1 part sample is diluted with 5 parts of 0.1% TFA (*see* Notes 20 and 21).
3. After loading the sample, rinse the sample tube with an aliquot of the solvent in which the sample was dissolved (i.e., having the same %B) and pump the rinse onto the column. Rinse the sample tube a second time with the same solvent and load the second rinse onto the column (*see* Note 22).
4. Turn off both pumps and divert the flow to waste. Clear the B pump-lead by pumping about 10 mL of 50% MeCN/0.085% TFA through the B pump (to waste), then reduce the flow rate to zero (*see* Note 23).
5. Turn on pump A and pump 0.1% TFA to waste until no 50% MeCN/0.085% TFA is present in the eluent (10 mL is usually sufficient). Reduce the flow rate to zero and close the flush valve (*see* Note 23).
6. Use the solvent programmer to establish the initial solvent running conditions, then initiate the run according to the specifications for respective purification steps (*see* Sections 3.2.2., 3.2.3., and 3.2.4.).

3.2.2. Purification System 1: Columns 1 and 2

1. Set up the LC system as follows:
   a. Pump A: 0.1% TFA.
   b. Pump B: 50% MeCN/0.085% TFA.
   c. Flow rate: 7.5 mL/min.

   d. Detector: UV, 214 nm; output range, 2.0 aufs (absorbance units to full scale; the least sensitive setting).
   e. Program: Linear gradient from 100% A to 100% B over 120 min.
2. Having loaded the sample onto column 1 as described in Section 3.2.1., initiate the gradient program (*see* Notes 20–22).
3. At 10 min into the program, begin collecting 2 min fractions (15 mL each) into 20-mL scintillation vials.
4. At the end of the run, add 2 μL of 2-MTE to each fraction, cap, and shake well. Store at 4°C until assay (*see* Note 1).
5. Assay and select the appropriate fraction(s) for further purification (*see* Note 15).
6. Load the appropriate fraction onto column 2 using the method outlined in Section 3.2.1.
7. Repeat the procedures described in steps 1–4.
8. Assay and select the appropriate fraction(s) for further purification (*see* Note 15).

3.2.3. Purification System 2: Columns 3 and 4

1. Set up the LC system as follows:
   a. Pump A: 0.1% TFA.
   b. Pump B: 50% MeCN/0.085% TFA.
   c. Flow rate: 2 mL/min.
   d. Detector: UV, 214 nm; output range, 2.0 aufs (the least sensitive setting).
   e. Program: Linear gradient from 100% A to 100% B over 120 min.
2. Having loaded the sample onto column 3 as described in Section 3.2.1., initiate the gradient program (*see* Notes 20–22).
3. Collect 2 min (4-mL) fractions in 13 × 100-mm tubes throughout the entire run.
4. At the end of the run, add 1 μL of 2-MTE to each fraction, cap, and shake well. Store at 4°C until assay (*see* Note 1).
5. Assay and select the appropriate fraction(s) for further purification (*see* Note 15).
6. Load the appropriate fraction onto column 4 using the method outlined in Section 3.2.1.
7. Repeat the procedures described in steps 1–4.
8. Assay and select the appropriate fraction(s) for further purification (*see* Note 15).

3.2.4. Purification System 3: Columns 5 and 6

1. Set up the LC system as follows:
   a. Pump A: 0.1% TFA.
   b. Pump B: 50% MeCN/0.085% TFA.
   c. Flow rate: 1.5 mL/min.
   d. Detector: UV, 214 nm; output range, 1.0–0.5 aufs depending on estimate of peak height from previous run.
   e. Program: Linear gradient from 100% A to 100% B over 120 min.
2. Having loaded the sample onto column 5 as described in Section 3.2.1., initiate the gradient program (*see* Notes 20–22).
3. Collect 2 min (3-mL) fractions in 12 × 75-mm tubes. Alternatively, individual peaks may be collected as they appear.

4. At the end of the run, add 1 μL of 2-MTE to each fraction, cap, and shake well. Store at 4°C until assay (*see* Note 1).
5. Assay and select the appropriate fraction(s) for further purification (*see* Note 15).
6. Load the appropriate fraction onto column 6 using the method outlined in Section 3.2.1.
7. Repeat the procedures described in steps 1–4.
8. Assay and select the appropriate fraction(s) for further purification (*see* Note 15).
9. Using a Speed-Vac apparatus (with the heater off) evaporate to dryness the appropriate fraction(s) or peak(s) collected from the column 6 run. Flush with inert gas, cap, and store at 4°C.

3.2.5. Purification System 4: Column 7

1. Set up the LC system as follows:
   a. Pump A: 95% MeCN/0.01% TFA.
   b. Pump B: 50% MeCN/0.01% TFA.
   c. Flow rate: 2 mL/min.
   d. Detector: UV, 214 nm; output range, 0.2 or 0.1 aufs.
   e. Program: Isocratic at 100% A for 4 min, then a linear gradient to 100% B over 80 min (total run time, 84 min).
2. Dissolve completely the residue from the appropriate column 6 fraction(s) in 0.4 mL of 80% MeCN/0.1% TFA, then add in sequence—with thorough mixing between additions—three 0.4-mL aliquots of MeCN (*see* Note 24).
3. Load the 1.6 mL of solution using an injector equipped with 2-mL loop. Inject the sample and immediately begin the solvent program.
4. Return the injector to the load position (loop bypass) at 4 min into program.
5. Collect peaks as they appear throughout the run. Use a fraction collector, collecting 2 min fractions, but keep a finger on the "rack advance" button to collect individual peaks in single tubes.
6. Immediately remove the solvent from the desired fractions by vacuum centrifugation. Flush the tubes with inert gas, cap, and store at 4°C (*see* Note 25).

4. Notes

1. A few words about 2-(methylthio)ethanol (2-MTE) are appropriate: This antioxidant, recommended to us by David Schooley (University of Nevada/Reno), prevents oxidation of methionine residues but does not reduce disulfide linkages. At the concentrations used in these procedures, 2-MTE does not interfere with chromatographic separations or bioassays. It is to be used when samples will be held in solution while awaiting subsequent purification; it is not added to the final isolates because these are immediately taken to dryness. One hundred microliters of 2-MTE per liter of extract (0.01%, v/v) is sufficient to prevent peptide oxidations. Slightly higher concentrations (up to 0.033%, v/v) are used in some cases (e.g., in Sections 3.2.2., 3.2.3., and 3.2.4.) to simplify pipetting. 2-MTE is labeled as a harmful liquid and is combustible, so aliquoting should be done in a fume hood. Others working in the laboratory will be appreciative because 2-MTE is correctly labeled **stench**.

2. The MeCN-containing solutions listed in Section 2.2. should be prepared the day before expected use, if possible, to assure complete equilibration to room temperature. When water and MeCN are initially mixed, an endothermic process ensues and the mixture becomes quite cold. Ten minutes of sonication or helium sparging just prior to use will greatly reduce bubble formation during chromatography, which can result in uneven baselines at moderate to high detector sensitivities. Because solvent contraction occurs when MeCN and water are mixed, interchangeable use of v + v and "make to final volume" results in solvent concentration variability. We arbitrarily choose to use vol + vol when making MeCN-water-TFA solutions.
3. The 10% TFA stock solution is used to make working solutions for solid-phase extractions and HPLC. The 10% TFA solution can be stored at room temperature and is much safer and easier to handle than neat TFA because it does not rapidly volatilize when exposed to air. However, solution making should still be confined to a fume hood.
4. Solid-phase extraction (SPE) columns/cartridges are available in sizes ranging from <100 mg to 10 g. The size chosen depends on the total mass of material in the extract being purified. For instance, a small extract of 50–500 mg tissue could efficiently be fractionated on a 200-mg cartridge. A wide array of SPE columns, containing diverse types of reverse-phase and ion-exchange packings, are available from various manufacturers. We do not endorse the particular brands of SPE columns; columns from other suppliers may be superior to those we have used in specific applications.
5. The brand of the HPLC system does not matter; many manufacturers produce excellent apparatus. A few key specifications are important, however, to perform the methods presented here. If large diameter columns (25-mm) are to be used, the pumps should support flow rates of 9 mL/min and the UV detector cell must be able to withstand back pressures generated by these high flow rates (refer to the appropriate operating manual or contact the manufacturer's technical representative). A solvent selection manifold, at least on the B pump, is helpful for sample loading (otherwise, the B pump solvent lead and filter must be cleaned after sample loading prior to introduction of the B solvent; *see* Section 3.2.1.).
6. Specific brands are listed only because these are columns of which we happen to have direct experience. Similar columns obtained from other sources may work less well, as well, or better for a given application. The Vydac diphenyl column offers a unique selectivity in comparison to any other phenyl reverse-phase columns we have tried. A Biosep Sec-S2000 column, 300 × 7.8 mm (Phenomenex) can be substituted for the Waters/Millipore I-125 column. We would encourage all chromatographers to become curious and follow their instincts. The Waters I-125 Protein column normal-phase system was discovered in our laboratory in the late 1970s when attempts to use that column in the aqueous molecular sieve configuration (for which it was designed) failed to achieve the results we desired *(6)*. Only because of curiosity was it tried in the normal-phase system, which subsequently became the major breakthrough in our purifications of insect neuropeptides *(7)*.

7. A cartridge holder, referred to as the Radial Compression Module (RCM), is required for the DeltaPak columns (columns 1 and 2 in Section 2.5.). For smaller extracts, 100 × 8 mm cartridges containing the same packings as columns 1 and 2 can be substituted for the 100 × 25 mm cartridges, although a different cartridge holder is required. When using the smaller cartridges, reduce the flow rate to 1.5–2 mL/min.
8. Peptides can and do stick to plastic surfaces. This is especially true for small amounts of highly purified peptides in solutions containing low concentrations of MeCN. Adsorption is enhanced during refrigerated storage. To avoid the loss of precious materials, the authors outline a rinse procedure that has proved successful in reducing adsorption losses (*see* Note 22).
9. It is very important to protect Speed-Vac vacuum pumps from the corrosive effects of solvent components, particularly TFA. For the past 15 years we have been using high quality/heavy duty motor oil (such as Shell Rotella T 30w) in our piston and rotary vacuum pumps. No pump failures resulting from corrosion have occurred. The high detergent/antioxidant package in these oils absorbs water and TFA is neutralized. We have never observed two phases (water-oil) when the pump oil is drained and have not observed a degradation of vacuum.
10. Homogenization/extraction is the first of the series of fractionations that ultimately result in a peptide of a purity suitable for mass spectral and/or automated sequence analysis. A solvent or mixture of solvents in which the peptide is soluble and stable is used to extract the peptide from the natural source (tissue, or even whole animals). Homogenization disrupts the physical structure of the source tissue and centrifugation allows discarding of insoluble components. The supernatant of the homogenate contains virtually everything in the tissue, both inorganic and organic, that is soluble in the extraction solvent. The solid-phase extraction procedures outlined in Sections 3.1.1. and 3.1.2. fractionate the various components present based on polarity. The most polar components (salts, sugars, most amino acids) are not retained by the solid-phase matrix and pass directly through the column with 0.1% TFA. At the other end of the spectrum are the very hydrophobic components (certain lipids, pigments) that bind so tightly to the column packing that they cannot be eluted. In between these two extremes are the peptides, most of which will be eluted from the column matrix with 10% MeCN/0.1% TFA (very small peptides) to 50% MeCN/0.1% TFA. By increasing (stepwise) the amount of MeCN in the solvent, the peptides can be separated into several fractions based on polarity, whereas highly polar nonpeptide materials and extremely hydrophobic compounds are eliminated from the extract. The removal from the extract of the compounds that bind irreversibly to the SPE matrix is quite important since the first HPLC column contains a similar packing and would be ruined rather quickly by passing non-SPE purified extracts through it. The initial HPLC column used in our method costs about 37 times as much as the SPE-column!
11. The 10% TFA extraction method is excellent for small extractions but can become quite expensive when large extracts are anticipated. The acidic methanol extrac-

tion (Section 3.1.2.) is somewhat less expensive to use. Supernatants from 10% TFA extractions should be passed through the solid-phase column as soon after extraction/centrifugation as possible to reduce time of exposure to the high TFA concentration. If supernatants of 10% TFA extracts are stored under refrigeration, some precipitation may occur, especially with insect whole-body extracts. Extracts containing precipitate should be brought to room temperature, sonicated for 5 min, and refiltered before the first SPE step.

12. The authors do not cool the extraction solution during homogenization. The correct choice of the homogenization equipment (Polytron for almost all samples, glass tissue grinder for small to very small samples) and the solvents utilized for extraction assure a quick homogenization of the tissue such that temperature increase is not a problem. After the initial centrifugation some investigators may want to re-extract the pellet, especially with small tissue samples. The authors would suggest pooling the second supernatant with the initial one before the solid phase purification step.
13. A flow rate of about 10 mL/min through the SPE columns seems to be near optimum. At higher flow rates compounds may be pulled too quickly through the matrix for proper adsorption to occur, whereas lower flow rates waste time. With vacuum assist, the flow rate can easily exceed 10 mL/min but can be controlled with an adjustable clamp attached to a section of rubber tubing attached to the main vacuum line.
14. The authors use the term gram-equivalents (g-Eq) to define the quantity of an extract present in a fraction or a specified volume of solvent. For instance, if 500 g of tissue was homogenized in 5000 mL of 10% TFA, centrifuged, and filtered as per instructions and the final volume of filtered supernatant was 5500 mL, then each 11 mL would contain the soluble components from 1 g of tissue. Forty g-Eq would be present in 440 mL filtrate (40 g-Eq × 11 mL/g-Eq = 440 mL). In the final purification steps, a 3-mL fraction may contain all of "Peptide X" present in the 500 g tissue extract. In that case, 500 g-Eq would be present in 3 mL (or 3000 μL if preferred) and 1 g-Eq would consist of 0.006 μL (6 mL). These calculations are useful for determining how much of the extract can be applied to a SPE column or HPLC column without overload. Initial assays for activity (biological, biochemical, chemical) are quantified in terms of how many g-Eq are necessary to achieve a specific response in the assay. After each HPLC purification step, an assay based on g-Eq is used to locate the activity of interest and to determine if losses have occurred since the previous assay.

    In the example of the 500-g tissue extract above, multiple SPE runs are required to process the entire 500 g-Eq. Separate stations can be set up to run multiple, parallel SPE columns or the 40 g-Eq aliquots can be processed in series on the same SPE column (after the 80% MeCN/0.1% TFA elution, 50 mL of 0.1% TFA is passed through the column followed by another 40 g-Eq of the extract). A compromise between the series/parallel strategy is usually the most efficient. In the example above, the authors would set up four SPE columns and use each for three consecutive 41.6 g-Eq runs (a total of 12 runs). The lipid-pigment component that binds irreversibly to the SPE packing is usually dark and remains near the top of the column. This dark band increases in length with subsequent runs, until finally the whole column is a dark brown color. Three or four 40 g-Eq runs usually leave a dark band that extends downward for about one-half of the column length. At this point the SPE column should be discarded. The total quantity of irreversibly binding components in an extract will vary depending on the source tissue, so a SPE column may remain viable for far more than three or four runs, or may be spent after one run.

15. In general, sufficient material to generate a strong positive response in the chosen assay system should be set aside for testing. At the same time, one does not want to expend more material than necessary. The authors normally commit no more than 1% of each sample for assay. If 10% or more of the sample is utilized, then the sample is pretty well spent before final purification and little remains for identification. Aliquots reserved for assay are taken to dryness in the Speed-Vac, then dissolved in the appropriate solution (usually a buffered saline) just prior to assay.
16. If the acidic methanol system (described in Section 3.1.2.) produces an unsatisfactory extraction, we would suggest extracting with 50% MeCN/0.1% TFA, then diluting 1:10 (v/v) in 0.1% TFA before solid-phase purification. Unlike the 10% TFA extracts, supernatants from acidic methanol and 50% MeCN/0.1% TFA extracts can be stored at 4°C after addition of 2-MTE (to prevent sample oxidation; *see* Note 1) prior to solid phase extraction.
17. The 60-cc SPE column is used to quickly remove the extremely hydrophobic lipids extracted into the methanol during homogenization. The number of g-Eq that can be passed through the 60-cc SPE column before it is spent varies with the tissue extract. A slightly yellow, greasy-looking area first appears at the top of the column then slowly extends downward as more methanolic extract passes through the column. When this band approaches the bottom, discard the column and change to a new one. The peptides are not retained by the column.

    The 4 + 1 part dilution of supernatant and water/TFA causes additional lipids to precipitate. Refrigerated centrifugation causes that precipitate to stick to the walls and bottom of the centrifuge tube/bottle. Filtration (*see* Section 3.1.2., step 7) should be performed immediately after centrifugation while the supernatant is still cold.
18. Remove only the methanol plus a bit of water to ensure that all of the methanol is gone. The droplet size of condensed methanol is much smaller than for water. The methanol has been removed when the small (MeOH) droplets dripping from the condenser coil become much larger (water) droplets.
19. The methods described in this section all make use of TFA as an ion-pair reagent, and for most peptides this choice will be more than adequate. A complete discussion of buffer systems for reversed-phase HPLC is beyond the scope of this chapter, but a few alternatives are worth mentioning. An ion-pair reagent other than TFA can be used if samples are sensitive to low pH; for example, a phosphate buffer at pH 6.0 (vs phosphate-buffered MeCN as the B solvent) may be

employed. Care must be taken in using phosphate buffers to avoid precipitation at high organic modifier concentration (e.g., do not exceed 50% MeCN). A volatile buffer mixture of TFA and TEA (triethylamine) can be used for preparing buffers ranging in pH between 4.0 and 6.5. It is well to recall that all silica-based HPLC packings are eroded to some extent by pH over about 7.0. For those who desire reverse-phase separations operating at higher pH, the authors would suggest a column containing an organic polymer based packing.

Another strategy for fractionating compounds is to use a low pH ion-pair reagent (TFA, HFBA) for the first run, then make a second run of the active first-run fraction through the same column with pH modification only (i.e., no ion pair reagent). HCl can be used in this context at 6 m*M* in solvent A vs 50% MeCN/6 m*M* HCl in solvent B. This strategy was used by the authors during purification of a large whole-body mosquito extract. The peptide of interest, a 44-residue CRF-related peptide, eluted in a 2-min fraction from purification system 2 (*see* Section 3.2.3.). When this fraction was rerun on the same system, except using solvents containing 6 m*M* HCl instead of 0.1% TFA, the UV-absorbing material was spread over 20 min (10 fractions) but a single 2-min fraction contained the CRF-related peptide.

20. With the exception of the Waters ProteinPak column, loading is performed by pumping the samples onto the column via pump B so that the sample is diluted with 0.1% TFA (1 part sample plus 5 parts 0.1% TFA). Sample dilution is achieved by setting the flow rates for the pumps so that the flow rate of pump A (the output of which is 0.1% TFA) is five times the flow rate of pump B. This procedure has several advantages. First, it ensures that the solvent composition of the sample is very near to the initial solvent conditions of the solvent gradient program. Second, the procedure obviates the need to freeze-dry samples. This means that the organic solvent component of the samples is not reduced/removed while peptides are held in potentially "sticky" plastic storage/collection tubes (to which peptides are likely to adsorb in a largely aqueous environment).
21. Below are listed the recommended flow rates to be used in loading samples on the HPLC columns:
    a. For columns 1 and 2 (Section 3.2.2.), equilibrate with 0.1% TFA at 7.5 mL/min (100% pump A). When the column is equilibrated, start the B pump at 1.5 mL/min to introduce the sample. This gives a total flow rate of 9 mL/min.
    b. For columns 3 and 4 (Section 3.2.3.), equilibrate with 0.1% TFA at 2.5 mL/min (100% pump A). When the column is equilibrated, start the B pump at 0.5 mL/min to introduce the sample, giving a total flow rate of 3.0 mL/min.
    c. For columns 5 and 6 (Section 3.2.4.), equilibrate with 0.1% TFA at 1.5 mL/min (100% pump A). When the column is equilibrated, start the B pump at 0.3 mL/min to introduce the sample, giving a total flow rate of 1.8 mL/min. Alternatively, fivefold dilution with 0.1% TFA can be performed in a 12 × 75 mm tube. Apply the sample to the column via the injector by repeated injection while the system is running at initial solvent conditions. After the final injection, return the injector to its "load" position (loop bypass), equilibrate for 5 min, then initiate the gradient program.

22. There is no particular rule regarding a minimum solvent volume for rinsing sample tubes between each step, but following are some specific recommendations. When loading samples (volume approx 50 mL) onto column 1, rinse the sample tube with 2 × 5 mL of solvent. For loading columns 2 and 3 (sample volumes approx 15 mL), rinse with 2 × 4 mL; for columns 4 and 5 (sample volumes approx 4 mL), rinse with 2 × 2 mL. For loading column 6 (sample volume approx 3 mL), rinse with 2 × 1 mL. Rinsing the sample tube prior to initiating the column 7 step is usually unnecessary.

    The composition of the rinse solvent will be dictated by your determination of the solvent composition of the sample. Estimate the percentage of solvents A and B (for columns 1–6, solvent A is always 0.1% TFA and solvent B is always 50% MeCN/0.085% TFA) using your knowledge of the gradient conditions and the time of elution of the fraction in the previous HPLC step. Make up a rinse mixture accordingly from solvents A and B.
23. If the B pump does not have a solvent-select manifold, then the line from the solvent reservoir must be cleaned after each sample loading procedure. It would be somewhat easier to use a separate piece of tubing, running from the inlet manifold to the sample tube during sample loading, then unplugging this tubing and replacing it with the line from the solvent B reservoir. If one is so fortunate as to have a third pump available, that pump could be dedicated to sample loading (the loading pump can be cleaned during the HPLC run).
24. Virtually every peptide the authors have handled over the past 20 years has been freely soluble in 80% MeCN (0.01% TFA). Most peptides, however, do not dissolve very well in 95% MeCN (0.01% TFA), which is the initial condition "A" solvent for column 7. Because it is always best for the sample to be dissolved in the initial conditions solvent, the difficulty of solubilizing peptides in 95% MeCN presents a problem. This difficulty may be overcome by first dissolving the peptide in 80% MeCN (containing either 0.1 or 0.01% TFA), then diluting to 95% MeCN with the stepwise addition of three equal aliquots of neat MeCN, mixing between additions. Once in solution at 80% MeCN, the peptides will normally remain dissolved at least for a short time as the MeCN concentration is increased to 95%. This time should be sufficient to load and inject the sample. The authors would not recommend holding this solution for more than a few minutes because precipitation could occur, in which case the sample should be taken to dryness by vacuum centrifugation and the dried peptide redissolved and rediluted.
25. Prior to automated sequencing or mass spectrometry, one may wish to concentrate the peptide in a somewhat smaller tube. Transfer the sample to a 1500 or 500 μL Eppendorf polypropylene tube and combine with a rinse of the 12 × 75 mm tube (rinse with 100 μL of 80% MeCN/0.01% TFA; it is advisable to rinse the tube several times if practicable). Immediately remove the solvent by vacuum centrifugation. We advise using only the "transparent" (uncolored) Eppendorf tubes to handle peptide solutions. Do not use tubes of any kind with "O" rings or septum closures because they have been shown to contain compounds that can interfere with mass spectral analysis. If a microbore HPLC system is available,

the dried peptide from column 7 can be passed through a microbore column at a low flow rate (25–50 mL/min), collected in a 500-mL Eppendorf tube, and the solvent evaporated. This HPLC step will remove hydrophobic contaminants that may have been extracted from the 12 × 75 mm tube.

References

1. *The Handbook of Analysis and Purification of Peptides and Proteins by Reversed-Phase HPLC.* Vydac (The Separations Group), Hesperia, CA.
2. Kataoka, H., Troetschler, R. G., Li., J. P., Kramer, S. J., Carney, R. L., and Schooley, D. A. (1989) Isolation and identification of a diuretic hormone from the tobacco hornworm, *Manduca sexta*. *Proc. Natl. Acad. Sci. USA* **86,** 2976–2980.
3. Gäde, G. and Kellner, R. (1995) Isolation and primary structure of a novel adipokinetic hormone from the pyrgomorphid grasshopper, *Phymateus leprosus*. *Reg. Peptides* **57,** 247–252.
4. Hayes, T. K., Holman, G. M., Pannabecker, T. L., Wright, M. S., Strey, A. A., Nachman, R. J., Hoel, D. F., Olson, J. K., and Beyenbach, K. W. (1994) Culekinin depolarizing peptide: a mosquito leukokinin-like peptide that influences insect Malpighian tubule ion transport. *Reg. Peptides* **52,** 235–248.
5. Clottens, F. L., Holman, G. M., Coast, G. M., Totty, N. F., Hayes, T. K., Mallet, A. I., Wright, M. S., and Bull, D. L. (1994) Isolation and characterization of a diuretic peptide common to the house fly and stable fly. *Peptides* **15(6),** 971–979.
6. Holman, G. M., Cook, B. J., and Nachman, R. J. (1986) Isolation, primary structure and synthesis of two neuropeptides from *Leucophaea maderae*: members of a new family of cephalomyotropins. *Comp. Biochem. Physiol.* **84C,** 205–211.
7. Holman, G. M., Nachman, R. J., and Wright, M. S. (1990) A strategy for the isolation and structural characterization of certain insect myotropic peptides that modify the spontaneous contractions of the isolated cockroach hindgut, in: *Chromatography and Isolation of Insect Hormones and Pheromones* (MacCaffery, A. R. and Wilson, I. D., eds.), Plenum, New York, pp. 195–204.

17

Neuropeptide Expression Patterns as Determined by Matrix-Assisted Laser Desorption Ionization Mass Spectrometry

Ka Wan Li and Wijnand P. M. Geraerts

1. Introduction

The diversity of neuropeptide transmitters is fairly large, as compared to other, "classical" neurotransmitters, such as acetylcholine and glutamate. Currently, immunocytochemistry and *in situ* hybridization are the methods of choice to map neuropeptide expression patterns in the nervous system. A disadvantage of these methods, however, is that they give limited clues to the structural identities of processed and biologically active peptides contained in the cells. Moreover, these techniques do not reveal previously undescribed cotransmitters.

Recently, matrix-assisted laser desorption ionization-mass spectrometry (MALDI-MS) has been exploited to determine the complete profile of peptide messengers contained in a small sample of nervous tissues and even single neurons. This method is fast, sensitive, and provides highly specific and semiquantitative data of the peptides contained in complex mixtures of biological molecules. Other mass spectrometric methods, such as fast atom bombardment and electrospray mass spectrometry, are less sensitive and are also less suitable for the analysis of crude biological samples.

MALDI is a soft ionization technique for introducing large and delicate molecules, such as peptides and proteins, into the mass spectrometer for analysis. The methodology as it is known today was pioneered by Hillenkamp and coworkers *(1)*. The essence of the technique is that the analyte molecules are embedded in an excess amount of a suitable matrix. Then, a laser beam (usually a UV laser) is irradiated on the matrix, which subsequently transfers the absorbed energy to the sample, causing the molecules to desorb and ionize into

From: *Methods in Molecular Biology, Vol. 72: Neurotransmitter Methods*
Edited by: R. C. Rayne Humana Press Inc., Totowa, NJ

the gas phase. The desorbed ionized molecules are pulled by an ion accelerator to a fixed kinetic energy, into a long field-free drift tube. Because the velocity of the molecules depends on their masses (specifically, mass-to-charge ratios), the difference between the starting time and the arrival time of an individual ion at the detector can be used to calculate an ion's mass.

The technology of MALDI-MS has been considerably improved in recent years. One factor that has led to much better resolution is the addition of a reflectron. This is an ion mirror with the electrode voltage arranged in such a way that it largely corrects for the initial velocity distribution of a packet of particular ion species. The correction increases the mass resolution, which is especially useful in case a mixture of analytes is assayed.

MALDI-MS routinely achieves sensitivities at the low fmol (10^{-15} mol) level, which is comparable to that of antibody-based detection assays. A low amol (10^{-18} mol) detection level has also been reported, but is not applicable to direct peptide profiling of single cells and nervous tissue *(2)*. In general, the method as outlined in Section 3. has been optimized for the analysis of peptidergic cells (containing peptidergic analytes as the predominant molecular species in the cells). It is worth noting that all living cells contain a great variety of organic and inorganic molecules, including a bulk of phospholipids that may have masses corresponding to those of small neuropeptides. Fortunately, the method described here selectively sorts out peptides and proteins for ionization and subsequent mass measurement. This process is possibly matrix-dependent.

2. Materials

2.1. Apparatus

1. Mass spectrometer: There are several commercially available MALDI-MS apparatus. These have been recently reviewed in the section "product review" of *Analytical Chemistry* (497 A, 1995). The analysis of the relatively complex peptide profiles of nervous tissue requires a high resolution mass spectrometer, i.e., an apparatus equipped with a reflectron (or in the linear mode with a capacity for delayed ion extraction).
2. A zoom microscope and CCD camera for sample viewing facilitates the localization of the matrix crystals for subsequent laser irradiation.

2.2. Reagents

1. Matrix for MALDI-MS: 2,5-dihydroxybenzoic acid dissolved in distilled water (10 g/L) containing 0.1% (v/v) trifluoroacetic acid.
2. 0.1*M* acetic acid.
3. 0.1% (v/v) trifluoroacetic acid in water containing 60% (v/v) acetonitrile.
4. C18 solid-phase extraction column (e.g., Sep-Pak; Millipore, Bedford, MA).

3. Methods

3.1. Single Cell Analysis

This method is applicable to the giant identifiable molluskan neurons *(3,4)*, dissociated (vertebrate) peptidergic (neuroendocrine) cells, and possibly also (dissociated) peptidergic cancer cells.

1. Identify the cell under a binocular microscope.
2. Using gentle suction, pull the cell into a glass electrode having a tip diameter roughly equivalent to that of the cell. Take care to minimize the uptake of buffer solution bathing the cells.
3. Viewing the procedure under the microscope, blow the cell onto the metal target of the mass spectrometer.
4. Apply 0.5–1 μL of matrix (*see* Note 1) to the cell.
5. Using a fine pair of forceps, break the cell and mix the contents into the matrix solution.
6. Dry the preparation under a gentle stream of warm air. This usually takes a few minutes. A ring of small matrix crystals should form around the outer rim of the preparation; these crystals contain the neuropeptides (*see* Notes 2–4).
7. Insert the metal target containing the sample into the mass spectrometer and orientate to an appropriate position for laser irradiation. This is facilitated by viewing of the sample under a zoom microscope and a CCD camera.
8. Focus a UV laser beam on the matrix crystals and record the mass spectra. Usually 30–50 single shot mass spectra from a single sample should be summed and averaged to give a good signal-to-noise ratio (*see* Note 5).
9. A typical single cell mass spectrum is shown in Fig. 1.

3.2. Nervous Tissue Analysis

1. Obtain small pieces of nerve (~0.5 mm) or small clusters of neurons and either follow directly the procedure described in Section 3.1. (first, *see* Note 6) or extract and prepurify the analytes of interest, following the procedure in steps 2 and 3 (*see* Note 7).
2. Boil the tissue of interest in 50-fold excess of 0.1*M* acetic acid for 5 min and follow with homogenization and centrifugation.
3. Load the supernatant into a small C18 solid-phase extraction column and elute bound material with aqueous 7.0 m*M* trifluoroacetic acid containing 60% (v/v) acetonitrile. Use the smallest possible volume of eluent to minimize dilution of the sample.
4. Mix 1 μL of this prepurified sample with 1 μL matrix solution on the metal target, dry, and perform mass spectrometry as described in Section 3.1.

4. Notes

1. Although matrices other than 2,5-dihydroxybenzoic acid may be used for single cell analysis, in our hands 2,5-dihydroxybenzoic acid gives the best results in

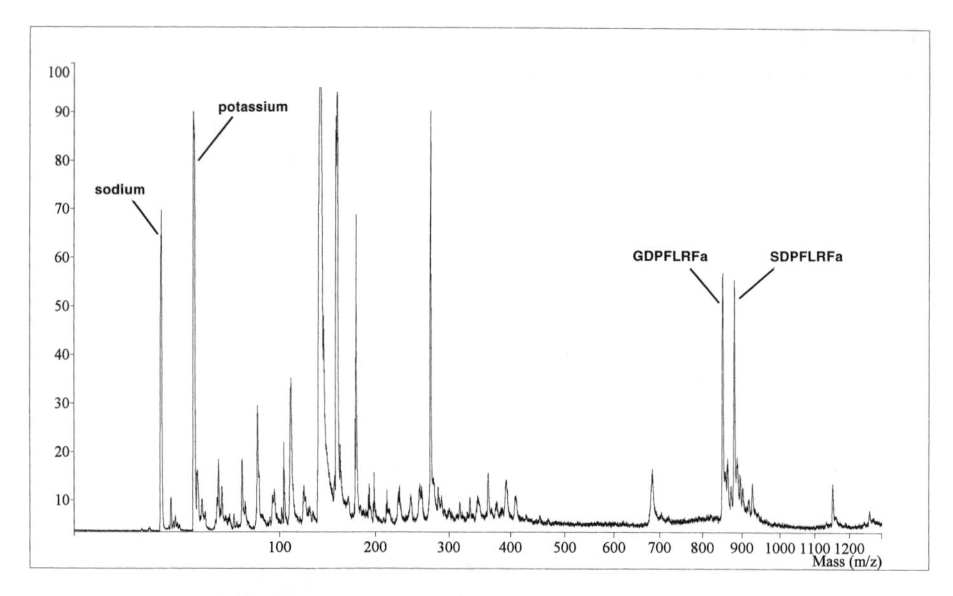

Fig. 1. Direct mass spectrometry of a single neuron from the B-group neurons in the right parietal ganglion of the mollusc *Lymnaea stagnalis*. Previous studies *(6)* using immunocytochemical and *in situ* hybridization techniques revealed that the mRNA encoding the SDPFLRFa/GDPFLRFa peptide precursor and the peptides derived from it are expressed in the B-group neurons. The MALDI-MS single cell analysis shows the presence of two prominent molecular species with masses corresponding to those of the peptides SDPFLRFa and GDPFLRFa. The sodium and potassium ions are also shown. The mass range <400 Daltons is usually dominated by many small molecules, which possibly are derived from the matrix solution. The presence of analytes with masses <400 Daltons, therefore, may be masked by matrix peaks, in which case the use of other types of mass spectrometry (e.g., electrospray ionization mass spectrometry) should be considered. *x*-axis, m/z is mass to charge ratio; *y*-axis, intensity in arbitrary units.

terms of higher sensitivity and increased resolution. In some cases, the addition of 10–30% (v/v) acetonitrile to the matrix solution may improve the mass spectrometric analysis.

2. Other organic molecules, such as phospholipids, membrane bound proteins, and other insoluble particles, presumably reside in the center of the preparation, and because the central region contains no matrix crystal these "unwanted" analytes will not be ionized and thus, not be detected.
3. Among the mass spectrometric techniques, MALDI-MS is the least susceptible to the presence of salts and detergents in the sample. However, high levels of contaminant may lead to poor crystalization of the matrix and consequently, a

poor mass spectrum. In these cases a prepurification step as described in Section 3.2. is strongly recommended.

4. If the tissue component is very complex, or if the peptide of interest is only a minor component of the sample, a fractionation by means of (micro-) reversed-phase liquid chromatography should be carried out. The collected fractions may then be assayed by mass spectrometry.
5. For (semi-)quantitative analysis the analytes contained in different crystals within a preparation should be sampled. Generally, over 150 single shots generated from different locations of the preparation should be averaged. An internal standard that is structurally similar to the analyte may be included in the sample to give a better quantitative result *(5)*.
6. After the preparation is dried, the insoluble piece of tissue in the center of the preparation should be removed.
7. Very often a tissue preparation is considerably bigger than the volume of the matrix solution (i.e., 1 µL), and in addition, may contain large amounts of unwanted materials that could be detrimental for mass spectrometry of the peptides. In this case, the sample should be extracted and prepurified as described in Section 3.2.

References

1. Karas, M. and Hillenkamp, F. (1988) Laser desorption ionization of proteins with molecular mass exceeding 10,000 Daltons. *Anal. Chem.* **60,** 2299–2301.
2. Jespersen, S., Niessen, W. M. A., Tjaden, U. R., van der Greef, J., Litborn, E., Lindberg, U., and Roeraade, J. (1994) Attomole detection of proteins by matrix-assisted laser desorption/ionization mass spectrometry with the use of picolitre vials. *Rapid Comm. Mass Spectrom.* **8,** 581–584.
3. Jiménez, C. R., van Veelen, P. A., Li, K. W., Wildering, W. C., Geraerts, W. P. M., Tjaden, U. R., and van der Greef, J. (1994) Neuropeptide expression and processing as revealed by direct matrix-assisted laser desorption-ionization mass spectrometry of single neurons. *J. Neurochem.* **62,** 404–407.
4. Li, K. W., Hoek, R. M., Smith, F., Jiménez, C. R., van der Schors, R. C., van Veelen, P. A., Chen, S., van der Greef, J., Parish, D. C., Benjamin, P. R., and Geraerts, W. P. M. (1994) Direct peptide profiling by mass spectrometry of single identified neurons reveals complex neuropeptide processing pattern. *J. Biol. Chem.* **269,** 30,288–30,292.
5. Nelson, R. W. and McLean, M. A. (1994) Quantitative determination of proteins by matrix-assisted laser desorption/ionization time-of-fight mass spectrometry. *Anal. Chem.* **66,** 1408–1415.
6. Bright, K., Kellett, E., Saunders, S. E., Brierley, M., Burke, J. F., and Benjamin, P. R. (1993) Mutually exclusive expression of alternatively spliced FMRFamide transcripts in identified neuronal systems of the snail *Lymnaea*. *J. Neurosci.* **13,** 2719–2729.

18

GC/MS Determination of Biogenic Amines in Insect Neurons

David G. Watson, Richard A. Baines, John M. Midgley, and Jonathan P. Bacon

1. Introduction

The biogenic amines, which include dopamine, noradrenaline, adrenaline, and octopamine, are an important class of signaling molecule utilized by the nervous systems of almost all multicellular animals. In the insect central nervous system (CNS), the presence of some of these amines (dopamine and octopamine) has been inferred by a variety of sensitive, although somewhat nonspecific techniques, including immunohistochemical and histofluorimetric studies *(1)*, radioenzymatic assays *(2)*, and high-performance liquid chromatography coupled to electrochemical detection (HPLC-ECD) *(3)*. This chapter describes the detection of biogenic amines in the insect CNS using gas chromatography-mass spectroscopy (GC-MS). This technique offers a significant advantage in the detection of such biomolecules over similarly sensitive techniques, such as HPLC-ECD, because GC-MS relies not only on chromatographic retention time but also on molecular mass to ascertain identity.

The ability to identify molecules by their molecular mass also offers the advantage of distinguishing among compounds that are modified by the inclusion of isotopes, such as deuterium atoms. These molecules, although not differing in their volatility from their endogenous equivalents, differ in molecular mass and are therefore separable from endogenous nondeuterated molecules (a feature not provided by the use HPLC-ECD). We have exploited this feature to demonstrate conclusively that deuterated *p*-tyramine is a physiological precursor of *p*-octopamine (the invertebrate equivalent to noradrenaline) in the insect CNS *(4)*. In addition, the facility afforded by GC-MS of adding deuterated internal standards to biological matrices when measurement of endogenous

From: *Methods in Molecular Biology, Vol. 72: Neurotransmitter Methods*
Edited by: R. C. Rayne Humana Press Inc., Totowa, NJ

analytes is carried out enables any losses of the analyte incurred to be precisely compensated for in extraction, derivatization, and chromatography.

The underlying principle of GC-MS involves the initial separation of molecules according to their volatility using, in our case, a fused silica capillary column that is wall-coated with a methylsilicone polymer that serves as the liquid stationary phase. The mobile phase is usually helium and molecules move through the column at different rates depending on their volatility, which in many cases decreases roughly in proportion to molecular weight. This is particularly true if a nonpolar stationary phase is used. Such columns may be commonly up to 60 m long but for most purposes 12-m columns will suffice. (A 12-m column provides about 60,000 theoretical plates of separating power compared to about 8000–10,000 plates for an average HPLC column.) As molecules exit the column they are bombarded by an ionizing beam of radiation provided by a rhenium filament. Resultant collisions either remove or add electrons from the molecules to create, respectively, positively charged ions (the +1 ions predominating) or negatively charged ions with a charge of –1. These ions are focused electrostatically and separated on the basis of their charge-to-mass ratio by the quadrupole rods of the mass spectrometer. The exquisite sensitivity and selectivity of the mass detection associated with GC-MS allows low picogram (10^{-12} g) amounts of a molecule to be quantified.

This chapter describes the use of this sensitive technique to measure rigorously biogenic amine levels in individually isolated insect identified neurons, and we present data from the locust protocerebral medulla 4 (PM4) neurons as a specific example *(5)*. We also describe the use of GC-MS to monitor the biosynthesis of biogenic amines. The example we present is the synthesis of *p*-octopamine, in populations of the dorsal unpaired median (DUM) neurons of the locust, by following the fate of one of its precursors in deuterated form, $[^2H_6]$ *p*-tyramine *(4)*.

2. Materials

All chemicals, unless otherwise specified, may be obtained from Sigma-Aldrich Chemical Co. (Dorset, UK).

2.1. Synthesis of Deuterated p-Tyramine

1. 4-Methoxybenzaldehyde.
2. Deuterated nitromethane.
3. Round-bottomed flasks: 1 two-necked, 100-mL; 2 single-necked, 100-mL.
4. Mercury thermometer, –20 to 110°C.
5. KOH solution: 2.4 g of KOH in 24 mL of deuterated methanol ($CH_3O[^2H_1]$).
6. Pressure equalizing funnel, 100-mL (Fisher Scientific, Loughborough, UK).
7. Magnetic stirrer and 1-cm stirrer bead.

8. Dilute HCl: 0.1*M* HCl in distilled H_2O.
9. Two sintered glass funnels.
10. Two separating funnels, 500-mL.
11. Chloroform.
12. Rotary evaporator.
13. Two Quickfit conical flasks with ground-glass stoppers, 250-mL.
14. Anhydrous sodium sulfate.
15. Dry diethyl ether.
16. Soxhelet extraction apparatus and thimble.
17. Lithium aluminum deuteride (LiAl[$^2H_4$]) solution: 1.05 g of LiAl[$^2H_4$] in 33 mL of dry diethyl ether; prepare in a Soxhelet extraction thimble attached to a round-bottomed flask.
18. Kieselguhr: 30 g, supported on a sintered glass funnel (Fluka, Dorset, UK).
19. Dry HCl gas.
20. Deuterium bromide solution: 47% [$^2H_1$]Br, w/w, in deuterium oxide.
21. Nitrogen gas.
22. Methanol.

2.2. Removal of Identified Neurons from the Locust CNS

1. Sylgard (Dow Corning) covered dissection dish.
2. Finely-sharpened forceps (e.g., Fine Science Tools, Haverhill, Suffolk, UK).
3. Locust saline: 200 m*M* NaCl, 3 m*M* KCl, 9 m*M* $CaCl_2$, 5 m*M* HEPES, pH 7.2.
4. Glass micropipet (internal tip diameter approx 80 µm), fire polished and prerinsed just prior to use with bovine serum albumin (BSA) solution.
5. BSA solution: 1 mg/mL BSA in distilled H_2O.
6. Lucifer yellow dye: 4% (w/v) Lucifer yellow in aqueous solution.
7. Ice-cold 0.1*M* HCl.

2.3. Incubation of Neurons in Deuterated Precursor Molecules

1. Plastic culture dishes, 35-mm diameter (sterile).
2. 96-Well tissue culture plates (sterile).
3. L-15 cell cuture medium (Gibco-BRL).
4. Deuterated *p*-tyramine (synthesis described in Section 3.1.).
5. Clear Perspex box.
6. Eppendorf vials (1.5-mL)
7. Dilute HCl: 0.1*M* HCl in dH_2O.

2.4. Analysis of Biogenic Amines

1. Screw-cap sample tubes, 3.5 mL, with aluminum-lined cap (BDH-Merck, Cutterworth, UK).
2. Deuterated amine standards: 0.25 ng/µL, e.g., [$^2H_3$] *m*-octopamine (*see* Note 1).
3. Acetonitrile, HPLC grade.
4. Ascorbic acid solution: 10 mg/mL ascorbic acid in methanol.
5. Potassium phosphate buffer: 1*M* potassium phosphate, pH 7.8, in distilled H_2O.

6. 3,5-Ditrifluoromethylbenzoyl chloride.
7. Ethyl acetate, HPLC grade.
8. Ammonium hydroxide solution: 10*M* ammonium hydroxide in distilled H_2O.
9. Anhydrous sodium sulfate.
10. OFN nitrogen gas (British Oxygen Company, Guildford, UK).
11. N,O-bis(trimethylsilyl) acetamide.
12. Gas chromatography column: 12 m × 0.2 mm id × 0.33 µm film (Hewlett-Packard HP-1).

3. Methods

3.1. Synthesis of Deuterated p-Tyramine (see Note 2)

1. Dissolve 5.6 g of 4-methoxybenzaldehyde and 2.6 g of deuterated nitromethane in 8.2 mL of $CH_3O[^2H_1]$ in a two-necked, 100-mL, round-bottomed flask and cool to –10°C in an ice/salt bath.
2. Attach a pressure-equalizing funnel containing KOH solution and insert a thermometer through the other neck of the flask. Begin to stir the solution vigorously using a magnetic stirrer.
3. Add the KOH solution dropwise to the stirring solution, ensuring that the temperature is maintained below 0°C. After all of the KOH solution has been added, continue stirring for 30 min.
4. Pour this reaction mixture into 150 mL of ice-cold, dilute hydrochloric acid to yield a yellow precipitate of $[^2H_1]$ 4-methoxy-β-nitrostyrene.
5. Collect the precipitate by filtration, then dissolve it in 200 mL of chloroform. Transfer the solution to a separating funnel and allow to stand for 1 h while a layer of water separates out.
6. Collect the lower, organic layer and dry by shaking in a conical flask with 20 g of anhydrous sodium sulfate. Allow the anhydrous sodium sulfate to settle and decant the dried solution into another flask.
7. Evaporate the chloroform to yield yellow crystals (*see* Note 3).
8. Place 1.2 g of $[^2H_1]$ 4-methoxy-β-nitrostyrene in a Soxhelet extraction thimble and attach to the Soxhelet extraction apparatus a round-bottomed flask containing the lithium aluminum deuteride solution. Reflux the solution so that the nitrostyrene is slowly added to the solution in the round-bottomed flask. Continue refluxing for 19 h.
9. After 19 h, cool the reaction mixture in ice and cautiously (dropwise) add 25 mL of ice-cold water.
10. Filter the mixture through a bed of Kieselguhr. Separate the ether layer using a separating funnel and further extract the aqueous layer four times with 50 mL of ether.
11. Dry the combined ether extracts with anhydrous sodium sulfate as described in step 6.
12. Pass dry HCl gas into the solution, thus precipitating $[^2H_3]$ 4-methoxyphenylethylamine hydrochloride. Recover the solid by filtration (*see* Note 4).

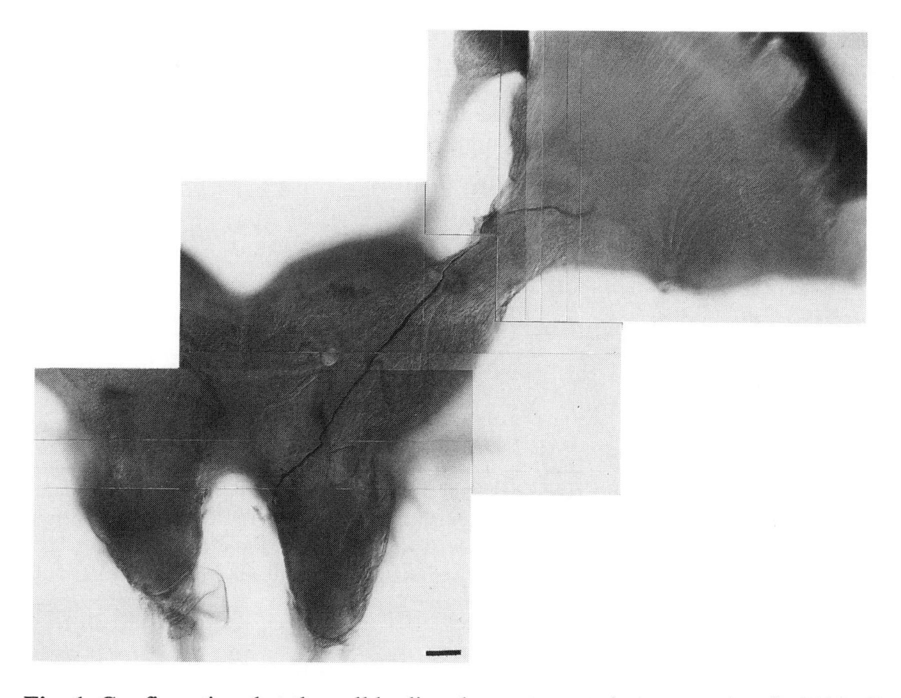

Fig. 1. Confirmation that the cell bodies shown to contain octopamine (in Table 1) belong to PM4 neurons. The photomontage shows the major projections of a Lucifer yellow-stained PM4 neuron revealed by anti-Lucifer yellow immunohistochemistry. The cell body is missing because it was removed for biogenic amine analysis. *See* Note 9 for additional commentary. Scale: 100 μm.

13. Dissolve the 4-methoxyphenylethylamine hydrochloride from step 12 in 3.7 mL of 47% deuterium bromide and reflux the solution under a nitrogen atmosphere for 2 h.
14. Remove most of the solvent by rotary evaporation to yield a paste that can be dissolved in approx 5 mL of methanol and then recrystallized by slow addition of 30 mL of ether (*see* Note 5).

3.2. Removal of PM4 Neurons from the Locust CNS (see Note 6)

1. Anesthetize the locust on ice, remove the brain from the head capsule, and pin out onto Sylgard.
2. Under locust saline, desheath the medio-ventral surface of the brain using finely sharpened forceps.
3. Expose the large PM4 cell bodies (50–60 μm in diameter) by gently teasing them apart with a saline jet from a mouth-held glass micropipet (*see* Note 7).
4. When a cell body of the appropriate size and position is made visible, impale it with a microelectrode and perform electrophysiological recordings to determine its identity (*see* Note 8). If desired, dye-mark physiologically identified cells with Lucifer yellow to ensure that the correct cells are removed (*see* Note 9 and Fig. 1).

Table 1
PM4 Cell Bodies Contain Substantial Quantities of *p*-Octopamine[a]

Sample contents	*p*-Octopamine, pg/sample	*p*-Octopamine, pg/blank	*p*-Octopamine, pg/cell
Four PM4 cell bodies, dye-marked	102.10	1.80	25.08
Four PM4 cell bodies, dye-marked	101.28	18.93	20.59
Seven PM4 cell bodies, not dye-marked	197.11	8.51	26.94
Seven PM4 cell bodies, not dye-marked	227.79	3.55	32.04
Eight control cell bodies	ND[b]	ND	ND

[a]The octopamine content per cell is calculated by subtracting the background level detected in a matched blank tube from the level detected in the sample. The resulting value is then divided by the number of cells in the sample. The mean octopamine content for these four samples is 27.07 pg/cell (0.143 pmol/cell). Dye marked cells were labeled with Lucifer yellow.
[b]ND: not detectable.

5. Remove the cell bodies by gentle suction through the fire-polished micropipet (*see* Note 10).
6. Collect approx 5–10 cell bodies in a 5-µL drop of saline and then wash the cells in five sequential 5-µL drops of saline, transferring the cell bodies from drop to drop using the micropipet (*see* Note 11).
7. After the fifth wash, take up the cells in approx 5 µL of saline and eject them into an Eppendorf tube containing 200 µL of ice-cold 0.1*M* HCl. As a blank control, take 5 µL of saline from the last drop in which the isolated cells were washed and add this to a separate tube of HCl.
8. Freeze samples and blanks in liquid nitrogen and store at –70°C until further analysis.
9. As a crucial control, remove equivalent numbers of neuronal cell bodies (preferably from the same general CNS region) known not to synthesize the biogenic amine under investigation. Process as described above (*see* Note 12 and Table 1 for results).

3.3. Incubation of DUM Neurons in Deuterated Precursor Molecules (see Note 13 and Fig. 2)

1. On a plastic culture plate, collect approx 50 isolated DUM cell bodies in a drop of saline using the method described in Section 3.2. but without individual physiological identification (i.e., omitting Section 3.2., step 4; *see* Note 13).
2. Transfer all the cells to 100 µL of L-15 medium in a well of a sterile 96-well plate. Add 100 µL of L-15 medium alone to several adjacent wells to serve as blanks.

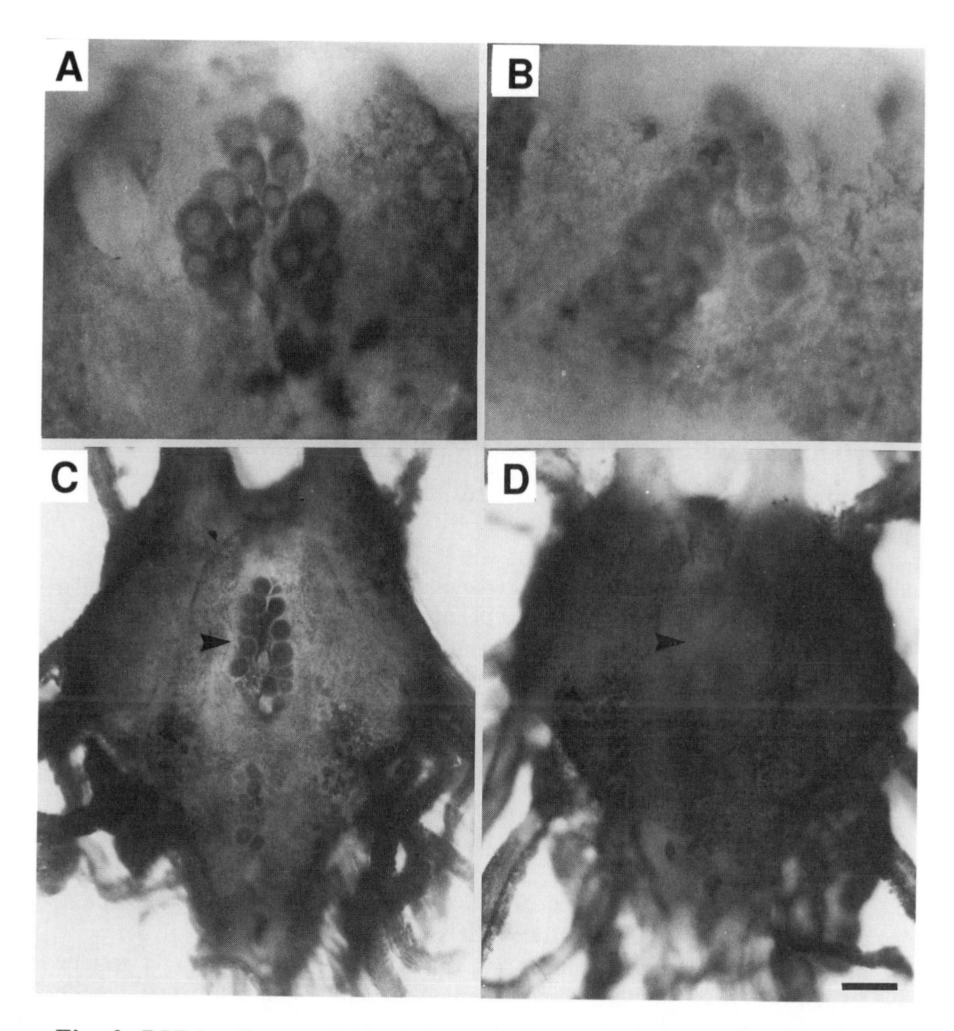

Fig. 2. DUM cells contain both tyramine-like and octopamine-like antigens. **(A)** Antibody to octopamine stains a number of cell bodies and their processes in the metathoracic ganglion. Among these positive cells are the anteriorly situated DUM neurons. Scale: 40 µm. **(B)** The DUM cells similarly stain, albeit more weakly, with antibody directed against *p*-tyramine. Scale: 40 µm. **(C)** DUM cells can be preferentially stained using neutral red (large arrow head; *see* Note 19 and ref. *11*). Other, smaller cells also stain with neutral red (small arrowhead). Scale: 100 µm. **(D)** A neutral red-stained metathoracic ganglion in which the DUM neurons have been selectively removed prior to the staining procedure. Note that although the DUM cell bodies are missing (large arrowhead), the other neutral red staining cells (small arrowhead) are still present, indicating the selectivity of our removal process. Scale: 100 µm.

3. Add a measured amount of deuterated precursor to the cells and to the corresponding blanks. Other control samples should comprise DUM cells in L-15 medium without deuterated precursor (*see* Note 14).
4. Place the multiwell dish in a humidified, sealed container (e.g., a plastic box containing moist paper towels) and incubate at 25°C.
5. After 24 h, use a mouth-held glass micropipet under a dissecting microscope to carefully remove the culture medium without disturbing the cells.
6. Without disturbing the cells, wash each well once with 100 μL of saline, then add 100 μL of 0.1 *M* HCl. After 5 min, transfer the contents of the well to Eppendorf vials (*see* Note 15).
7. Store the samples at –70°C until required for analysis.
8. *See* Fig. 3, Table 2, and Note 16 for results from GC-MS analysis.

3.4. Analysis of Biogenic Amines by GC-MS

All procedures can be conducted in 3.5-mL aluminum-lined screw-cap sample tubes.

1. To the lysate of cells in 0.1*M* HCl (*see* Section 3.3., steps 6 and 7), add 5 ng of deuterated *m*-octopamine *(8)* in 20 μL of acetonitrile (or 5 ng of deuterated *p*-octopamine *(8)* if *p*-octopamine is being measured) and 50 μL of ascorbic acid solution (*see* Note 1).
2. Add 1 mL of potassium phosphate buffer, mix, and then add 2 μL of 3,5-ditrifluoromethylbenzoyl chloride and shake the mixture vigorously for 3 min.
3. Extract the aqueous buffer layer with 2 mL of ethyl acetate and remove the organic layer to another 3.5-mL sample tube.
4. Shake the organic layer with 0.5 mL of ammonium hydroxide solution for 1 min.
5. Allow the two layers to separate. Remove the organic layer and dry it by passing through anhydrous sodium sulfate (*see* Note 17).
6. Evaporate the ethyl acetate under a stream of nitrogen and take up the residue in 50 μL of N,O-bis(trimethylsilyl) acetamide. Heat the solution for 30 min at 60°C, allow to cool, and then add 50 μL of ethyl acetate.
7. Inject a portion of the sample (2 μL) into the GC-MS system (*see* Note 18).
8. Figure 3 shows a typical selected ion trace obtained by GC-MS monitoring of a derivatized extract from isolated identified neurons (see the caption for further details).

4. Notes

1. These are freshly prepared from 1 mg/mL methanolic stock solutions of the deuterated amines, diluting the stock solution serially, as follows: 1:40 (50 μL diluted to 2 mL), 1:10 (200 μL diluted to 2 mL), and 1:10 (200 μL diluted to 2 mL) with acetonitrile. This achieves a total dilution of 1:4000 to yield a 0.25 ng/μL solution.

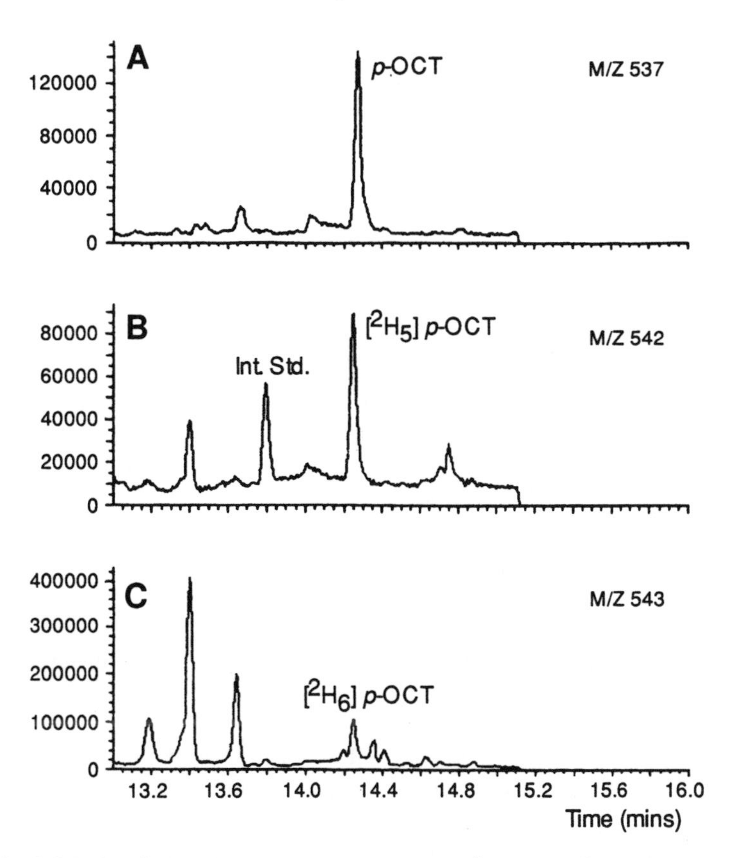

Fig. 3. GC-MS selected ion-monitoring trace of [$^2H_5$] and [$^2H_6$] *p*-octopamine (as their DTFMB-TMS derivatives) from 50 DUM cells incubated with [$^2H_6$] *p*-tyramine. **(A)** Detection of endogenous *p*-octopamine (mass-to-charge ration 537); **(B,C)** Deuterated [$^2H_5$] *p*-octopamine (mass-to-charge ratio 542) and [$^2H_6$] *p*-octopamine (mass-to-charge ration 543), respectively. These deuterated products must have been synthesized by the DUM cells from [$^2H_6$] *p*-tyramine, because they are not present in the controls. Int. Std.: Internal standard ([$^2H_3$] *m*-octopamine) added at the time of derivatization to allow quantification of amines recovered. The *y*-axis of the figure represents instrument response in arbitrary units. M/Z: Mass-to-charge ratio for the particular ions being monitored, which in each case corresponds to the molecular weights of the derivatives of the species present.

2. The preparation of other deuterated biogenic amine precursors, such as deuterated *m*-tyramine and deuterated dopamine, can be carried out along similar lines to the procedures described for *p*-tyramine. Some examples of the preparation of other deuterated biogenic amines may be found in ref. *8*.
3. The expected yield of [$^2H_1$] 4-methoxy-β-nitrostyrene in this preparation is approx 3.3 g. Sufficient deuterated *p*-tyramine can be synthesized by taking only

Table 2
Biogenic Amine Biosynthesis in Isolated DUM Cells from *Locusta migratoria*[a]

Amine	Amount, ng/cell body
[$^2H_5$/$^2H_6$] *p*-Octopamine	0.04 ± 0.01
[$^2H_5$/$^2H_6$] Dopamine	0.008 ± 0.004

[a] [$^2H_6$]p-tyramine (12 μg) incubated with 50 DUM cell bodies for 24 h, yields significant amounts of deuterated *p*-octopamine, but negligible amounts of deuterated dopamine. Values = mean ± SE for five separate experiments.

a portion of the product (1.2 g), as noted in Section 3.1., step 8. This is simply to economize on the use of expensive deuterated reagents in subsequent steps.

4. The expected yield is approx 0.54 g.
5. The expected yield is approx 0.46 g.
6. Protocerebral medulla 4 (PM4) neurons are a pair of cells located in the medial deutocerebral region of the locust brain; these cells are immunoreactive to octopamine antisera *(5)*. Antigenicity of neurons to antibodies specific to particular biogenic amines is not, by itself, sufficient evidence of transmitter content. More rigorous methods of analysis, e.g., GC-MS, are necessary to identify transmitter content.

   Although the protocol in Section 3.2. describes specifically removal of PM4 cell bodies from the locust brain, the method is appropriate for removal of identified neurons from other invertebrate preparations. Note also that a number of locust brains will be required to obtain the recommended number of PM4 cell bodies to perform the GC/MS determination.
7. We find it convenient to hold the micropipet in a silcone bung (from a Microcap; Laser Laboratory Systems, Sarisbury Green, Southampton, UK) in the wide end of a Pasteur pipet. The other end of the Pasteur pipet is attached to a mouthpiece using flexible silicon tubing.
8. Microelectrode impalement under the dissecting microscope allows unequivocal physiological identification of the cell prior to its removal; these methods are fully described in ref. *9*. In the case of PM4 neurons, a particularly characteristic feature is the occurrence of antidromic action potentials in response to high-intensity light stimuli *(5)*.
9. To confirm that the cell bodies analyzed were from PM4 neurons, Lucifer yellow (a fluorescent dye) was injected (using 5 nA hyperpolarizing pulses of 100 ms duration at 5 Hz for 5–10 min) into the cell bodies prior to their removal. After removal of the cell bodies, the brain was fixed for 1 h in 4% (w/v) paraformaldehyde in PBS before processing with anti-Lucifer yellow antiserum (Molecular Probes, Leiden, The Netherlands). In all such preparations, processes in the protocerebrum and optic lobe, with the characteristic anatomical features of PM4, were stained (Fig. 1). We find that the presence of Lucifer yellow in the cell body

does not affect subsequent biogenic amine analysis, and recommend this technique to demonstrate unequivocally the identity of removed cells.

10. It is desirable to remove the soma together with a short length of axon to prevent rupture of the cell body. Also, prerinsing the pipet with BSA prevents the cells from sticking to the glass micropipet. A sample of BSA should be analyzed for biogenic amine content as a control.
11. Washing is important to remove extracellular amines that are present in animal CNS.
12. Each PM4 cell body was found to contain, by GC-MS, approx 25 pg of *p*-octopamine. Neighboring cells, which do not stain with octopamine antisera, analyzed simultaneously as controls contain no detectable octopamine (*see* Table 1).
13. The locust CNS contains approx 100 DUM neurons, distributed dorsomedially in the ganglia of the ventral nerve cord. These neurons contain and release *p*-octopamine *(10)*. It is possible, therefore, to remove and maintain in vitro an approximately homogeneous population of DUM cell bodies to monitor the biosynthesis of *p*-octopamine (Fig. 2). Because of the stereotyped location and large size of these cells, physiological confirmation of their identity prior to removal is unnecessary.
14. For octopamine biosynthesis we used 12 µg of [$^2H_6$] *p*-tyramine in a minimal volume of water.
15. Homogenization and centrifugation of the samples are unnecessary because the cells lyse on contact with the HCl. Cell debris that may remain in the samples does not interfere with the GC-MS analysis.
16. To follow the biosynthetic route of *p*-octopamine, we exposed isolated groups of DUM neurons to deuterated [$^2H_6$] *p*-tyramine for 24 h *(4)*. This procedure yielded significant quantities of [$^2H_5$] and [$^2H_6$] *p*-octopamine (approx 0.04 ± 0.01 ng/cell body). Nondeuterated endogenous *p*-octopamine was also present (Fig. 3; Table 2). Controls showed no detectable amounts of deuterated *p*-octopamine *(4)*. The mixture of deuterated ([$^2H_5$] and [$^2H_6$] *p*-octopamine) products is consistent with the [$^2H_6$] *p*-tyramine containing approx 20% of [$^2H_5$] *p*-tyramine and/or the possibility that the hydroxylation of the ring or side chain of the precursor may involve the loss of a hydrogen atom or a deuterium atom. We cannot readily distinguish between these two possibilities. There is only slight evidence for the formation of deuterated dopamine from [$^2H_6$] *p*-tyramine; this would be seen as a peak at approx 14.75 min.

    These results demonstrate that incubation of isolated cells with deuterated precursors provides a means of elucidating neurotransmitter biosynthetic pathways. A range of deuterated precursors, including deuterated amino acids, could be produced for more extensive studies.
17. Insert a small cotton wool plug into a Pasteur pipet, fill to approx 3 cm with anhydrous sodium sulfate, and transfer the sample to the Pasteur pipet. Then force the solution through the bed of anhydrous sodium sulfate with a Pasteur pipet bulb.
18. The conditions for GC/MS are as follows. Injector temperature is 250°C; oven temperature is initially 100°C (1 min) and is then increased to 300°C at 10°C/min.

The transfer line temperature is 280°C. The mass spectrometer is operated in the negative ion chemical ionization mode with methane as the reagent gas, introduced to give a source pressure of approx 1.2 mbar. The instrument is regularly tuned using the ions at mass-to-charge ratio 452, 595, and 633 of the perfluorotributylamine (PFTBA) calibrant. The mass spectrometer must be kept clean, oxygen must be carefully excluded from the system, and the electron multiplier must be changed every 6 mo to produce the optimal performance required for this type of work.

The mass-to-charge ratios for typical ions that are monitored for amines in the selected ion mode are: 449 (*m*- and *p*-tyramine), 455 ([$^2H_6$] *m*-tyramine and [$^2H_6$] *m*-tyramine), 537 (*p*-octopamine and dopamine), and 542 and 543 ([$^2H_5$] and [$^2H_6$] *p*-octopamine and [$^2H_5$] and [$^2H_6$] dopamine).

19. Neutral red staining follows the method described in ref. *11*. Briefly, desheathed ganglia were immersed in Neutral red (0.01 mg/mL in saline) for 30–60 min. This procedure stains the ganglia uniformly and a subsequent destaining period in normal saline leaves the DUM cells stained bright red against a pink background. This technique has only been used here to demonstrate the fact that the DUM cells can be successfully removed from the ganglion without prior physiological or anatomical identification.

References

1. Klemm, N. and Axelsson, S. (1973) Detection of dopamine, noradrenaline and 5-hydroxytryptamine in the cerebral ganglion of the desert locust, *Schistocerca gregaria* Forsk (Insecta: *Orthoptera*). *Brain Res.* **57,** 289–298.
2. Evans, P. (1978) Octopamine distribution in the insect nervous system. *J. Neurochem.* **30,** 1009–1013.
3. Downer, R. G. H., Bailey, B. A., and Martin, R. J. (1985) Estimation of biogenic amines in biological tissues, in *Neurobiology* (Gilles, R. and Balthazart, J., eds.), Springer Verlag, Berlin, pp. 248–262.
4. Baines, R. A., Zhou, P., Midgley, J. M., Bacon, J. P., and Watson, D. G. (1996) *p*-Tyramine is a precursor for both *p*-octopamine and dopamine in the insect nervous system: determination by GC/MS of deuterium incorporation. *J. Neurochem.*, submitted for publication.
5. Stern, S., Thompson, K. S. J., Zhou, P., Watson, D. G., Midgley, J. M., Gewecke, M., and Bacon, J. P. (1995) Octopaminergic neurons in the locust brain: morphological, biochemical and electrophysiological characterisation of potential visual neuromodulators. *J. Comp. Physiol.* **177,** 611–625.
6. Shafi, N., Midgley, J. M., Watson, D. G., Smail, G. A., Strang, R., and Macfarlane, R. G. (1989) Analysis of biogenic amines in the brain of the American cockroach *Periplaneta americana* by GC-NICIMS. *J. Chromatogr. Biomed. Appl.* **490,** 9–19.
7. Watson, D. G., Zhou, P., Midgley, J. M., Milligan, C. D., and Kaiser, K. (1993) The determination of biogenic amines in four strains of the fruit fly, *Drosophila melanogaster. J. Pharm. Biomed. Anal.* **11,** 1145–1149.

8. Couch, M. W., Gabrielsen, B. M., and Midgley, J. M. (1983) The synthesis of deuterated hydroxyphenylethanolamines and their metabolites. *J. Label. Comp. Radiopharma.* **20,** 933–949.
9. Thompson, K. S. J. and Bacon, J. P. (1991) The vasopressin-like immunoreactive (VPLI) neurons of the locust, *Locusta migratoria*: II. Physiology. *J. Comp. Physiol.* **168,** 619–630.
10. Evans, P. D. (1980) Biogenic amines in the insect nervous system. *Adv. Insect Physiol.* **15,** 317–473.
11. Evans, P. D. and O'Shea, M. (1978) The identification of an octopaminergic neurone and the modulation of a myogenic rhythm in the locust. *J. Exp. Biol.* **73,** 235–260.

19

Monitoring Amino Acid Neurotransmitter Release in the Brain by In Vivo Microdialysis

Mark Zuiderwijk and Wim E. J. M. Ghijsen

1. Introduction

Microdialysis is a frequently used technique to collect continuously in vivo various endogenous chemical substances from the extracellular space of discrete brain regions. In comparison to other sampling methods (such as the push-pull technique), microdialysis causes only minor tissue trauma. This feature makes microdialysis the technique of choice when studying in vivo extracellular transmitter levels in relationship to diverse physiological and pathophysiological processes in several species, including humans *(1,2)*. Because of limitations of the technique, microdialysis methods are unsuitable for analysis of rapid, transient changes in extracellular transmitter levels. The microdialysis technique is, however, well suited to detect slow alterations in neurotransmitter concentrations, as occurring, for example, during ischemia *(3)*.

The microdialysis technique requires that a small-diameter (<300 μm) dialysis tube be stereotaxically implanted in a defined brain area. Perfusion of the dialysis tube with Ringer solutions enables diffusion of small molecules down their concentration gradient from the brain extracellular fluid and into the tube. The substances in the collected dialysates can be identified and measured by various analytical methods, either on-line or after storage *(4)*. Extracellular transmitter levels may be measured either under basal conditions or after evoked release using, for example, high KCl concentrations in the perfusate. Because some classes of neurotransmitters, (such as the amino acids and catecholamines) are rapidly cleared from the extracellular space by specific reuptake mechanisms *(5)*, it is often of interest to examine dialysate levels of these transmitters after perfusion with substances that will block these reuptake mechanisms.

From: *Methods in Molecular Biology, Vol. 72: Neurotransmitter Methods*
Edited by: R. C. Rayne Humana Press Inc., Totowa, NJ

In this chapter, we provide a method to construct concentric microdialysis probes and describe the use of such probes in monitoring extracellular levels of the amino acids glutamate and γ-amino butyric acid (GABA) in the rat hippocampus. This brain region is intensively studied with respect to changes in synaptic transmission by both glutamate and GABA in several processes, such as long-term potentiation *(6)*, epilepsy *(7)*, and ischemia *(3)*. Furthermore, we describe a method to stimulate in vivo transmitter release from hippocampus cells by perfusing high KCl concentrations through the probe. We also describe use of the glutamate carrier blocking agent L-*trans*-pyrrolidine-2,4-dicarboxylate (L-*trans*-PDC), and the GABA carrier blocker N-(4,4-diphenyl-3-butenyl)-3-piperidinecarboxylic acid (SK&F 89976-A), to perturb basal and K^+-evoked amino acid levels *(8)*. A method to measure the amino acid transmitter levels in microdialysates is presented in Chapter 15.

2. Materials

2.1. Construction of Microdialysis Probes

1. Artificial kidney dialysis membrane: OD 0.23 mm (Asahi, AM160-Nova).
2. Fused silica tubing: ID 75 μm, OD 150 μm (Composite Metal Services).
3. Polyethylene tubing: ID 0.38 mm, OD 1.09 mm (Portex).
4. Stainless steel hypodermic needles: 25 gage × 5/8 in. (orange) and 30 gage × 1/2 in. (white; Microlance).
5. Araldit 2-component epoxy glue (AW2102, Ciba-Geigy).
6. Pattex 2-component Super-mix glue (Henkel).
7. Pattex Super-gel adhesive (Henkel).
8. Razor blades (Schick, platinum).
9. Dumont #5 forceps (Fine Science Tools).
10. Tube cutter for 1/16 in. tubing (Chrompack).
11. Binocular dissecting-type microscope, capable of 40× magnification.

2.2. Implantation of Microdialysis Probes

All surgical tools (items 2–5) are from Fine Science Tools.

1. Stereotaxic equipment (Kopf).
2. Scalpel handle (no. 3) with surgical blades (no. 10).
3. Forceps: serrated and curved with 0.7 mm wide tips.
4. Fine forceps: curved, type Dumont #7.
5. Vascular clamps: curved, 35 mm length.
6. Bone-scraper: curved, 9.5 cm length.
7. Cotton buds (Q-tips).
8. Electric drill with 0.65 and 1.30 mm bits (Minicraft).
9. Coagulation apparatus with angled cutting electrode (Aesculap).
10. Hair clipper.

11. Lidocaine gel, 5% (Astra Pharmaceutica).
12. Alcohol (70%) and acetone.
13. Skull-screws (1 mm diameter, 3 mm length) and screwdriver.
14. Pipet tips (Gilson, 2–200 μL).
15. Dental acrylic cement (Simplex Rapid).
16. Electrical heating mattress.
17. Binocular microscope: *See* Section 2.1., item 11.

2.3. Microdialysis Experiments

1. Winged infusion sets: 25 gage × 3/4 in. (0.5 × 19 mm; Terumo Europe NV, Leuven, Belgium).
2. Ether.
3. Syringe-type infusion pump (e.g., Harvard Apparatus, type 22) with 1.0-mL glass syringes.
4. Multiposition valve with zero dead volume and fitting screws with ferrules suitable to fit 1/16 in. tubing (Valco Instruments).
5. Automated sample changer (Gilson, model 221).
6. Stainless-steel septum-piercing needle: 9.5 cm length, for item 5 (Gilson, type 460952).
7. Vials: 7 × 41 mm (Gilson, type 450830).
8. Caps: for item 7 (Gilson, type 450831).
9. Recirculation cooler.
10. Fused silica tubing: *See* Section 2.1., item 2.
11. Polyethylene tubing: *See* Section 2.1., item 3.
12. Tygon pump tubing: ID 0.6 mm, OD 2 mm (Ismatec).
13. PEEK tubing (Upchurch, type 1531).
14. Hypodermic needles: 25 gage × 5/8 in. (orange) and 23 gage × 1 in. (blue; Microlance).
15. Pattex super-mix: *See* Section 2.1., item 6.

2.4. Stock Solutions

Prepare all solutions using Milli-Q (Millipore) purified water. Filter solutions 1–5 through a 0.45-μm nitrocellulose filter (Schleicher & Schuell) and store at 4°C. Store solutions 6 and 7 at –20°C.

1. Normal K^+-stock solution: 60 m*M* KCl and 3.0*M* NaCl.
2. High K^+-stock solution: 2.0 *M* KCl and 1.06*M* NaCl.
3. $CaCl_2$-stock solution: 1.0*M* $CaCl_2$.
4. HEPES-stock solution: 0.5*M* HEPES.
5. NaOH-stock solution: 1.0*M* NaOH.
6. SK&F-stock solution: 20 m*M* SK&F 89976-A (Smith, Kline & French).
7. L-*trans*-PDC-stock solution: 100 m*M* L-*trans*-PDC (Tocris Neuramin) in 100 m*M* NaOH.

2.5. Solutions for Microdialysis Experiments

Prepare solutions 1–5 fresh for each experiment.

1. Artificial cerebrospinal fluid (aCSF): 3 m*M* KCl, 150 m*M* NaCl, 2 m*M* $CaCl_2$, 5 m*M* HEPES. Prepare using stock solutions 1, 3, and 4, and adjust pH to 7.4 with 1.0*M* NaOH (stock solution 5).
2. High K^+-aCSF: 100 m*M* KCl, 53 m*M* NaCl, 2 m*M* $CaCl_2$, 5 m*M* HEPES. Prepare using stock solutions 2, 3, and 4, and adjust pH to 7.4 with 1.0*M* NaOH. Note that the osmolarity is kept equal to aCSF by replacing NaCl for KCl.
3. SK&F 89976-A/L-*trans*-PDC: Dilute SK&F-stock and PDC-stock in aCSF to either 0.1 m*M* or 0.5 m*M* SK&F 89976-A, and either 0.2 m*M* or 1.0 m*M* L-*trans*-PDC.
4. High K^+-SK&F 89976-A/L-*trans*-PDC: Dilute SK&F-stock and PDC-stock in high K^+-aCSF to either 0.1 or 0.5 m*M* SK&F 89976-A, and either 0.2 or 1.0 m*M* L-*trans*-PDC.
5. Chloral hydrate solution: 8% (w/v) chloral hydrate in 0.9% (w/v) NaCl. Stored at 4°C, the solution is stable for 2 wk.
6. Urethane solution: 25% (w/v) urethane in 0.9% (w/v) NaCl and 2% (v/v) heparin (5 IE/mL).
7. TCA/homoserine solution: 10% (w/v) trichloroacetic acid and 5.0 μ*M* L-homoserine. Store at 4°C for up to 2 mo.

3. Methods

3.1. Construction of Concentric Microdialysis Probes (see Note 1)

For construction of the probes, *see also* Fig. 1.

1. Using a razor blade, cut pieces of fused silica (FS) tubing, polyethylene (PE) tubing, and dialysis membrane to lengths of 16, 15, and 15 mm, respectively.
2. Insert a 30 gage hypodermic needle into the 15 mm PE tube to a depth of 7 mm, bend the tube, and push the needle through the tube wall.
3. Thread the 16 mm FS tube through the needle using a forceps. Start from the needle's opening at the sharp end and insert the FS tube toward the white plastic end of the needle until about 10 mm of the FS tube still protrudes from the sharp end of the needle.
4. Keeping hold of the FS tube, remove the needle. Make sure that only 2 mm of the FS tube protrudes through the wall of the PE tube.
5. Glue the FS tube to the PE tube with a small application of Pattex supergel. After the supergel has set for 1 h, cover the joint with a small drop of Pattex supermix to strengthen the attachment. Allow the adhesives to set for a further 1 h.
6. Using a tube cutter, cut off the distal 7 mm of a 25-gage hypodermic needle (i.e., the sharp end of the needle).
7. Fit the cut 7 mm part with its sharp end over the inlet side of the FS tube.

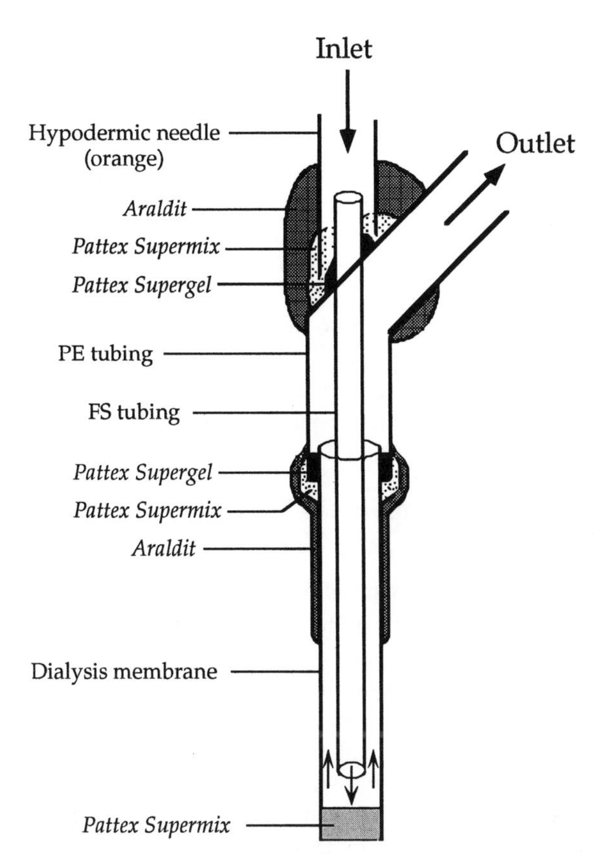

Fig. 1. Schematic drawing of the concentric microdialysis probe. For clarity of presentation, the dimensions depicted are not in scale. The length of the unsealed microdialysis membrane is 2.0 mm. The outer diameter of the probe is 230 µm.

8. Glue the needle to the PE tube with Pattex supermix. Allow the adhesive to set for 1 h, then strengthen this connection with an application of Araldit and allow this to set for a further 1 h.
9. Using forceps, place the dialysis membrane around the outlet side of the FS tube, feeding the membrane toward the PE tube until 2 mm of the membrane protrudes into (inside) the PE tube.
10. Now, carefully trim the dialysis membrane with a razor blade (under a binocular microscope) so that it is about 0.5 mm longer than the FS tube (the exact length is not critical yet).
11. Pull the membrane out of the PE tube and seal the "bottom" end of the membrane with a small dot of Pattex supermix (do not allow the glue to touch the FS tube). Allow the glue to dry for 1 h.

12. Trim the end of the glued tip using a razor blade (under the microscope) until an approx 200 μm (not less) "plug" of glue remains.
13. Again feed the membrane around the FS tube into the PE tube, until a gap of about 200 μm separates the glue plug at the membrane tip from the outlet end of the FS tube.
14. Glue the upper end of the membrane to the PE tube with Pattex supergel and, after 1 h, apply Pattex supermix.
15. Seal (under binocular microscope) the membrane with Araldit, beginning at the membrane/PE tube junction, leaving a 2.0 mm length of membrane exposed at the tip (measured from the Araldit to the glue plug at the membrane tip).
16. Store the probes dry, dust-free, and at room temperature. Probes may be stored in these conditions for several weeks prior to use.
17. Prior to use for in vivo microdialysis, test the probes in vitro to determine recovery of amino acids (*see* Note 2).

3.2. Implantation of Microdialysis Probes into the Rat Hippocampus (see Note 3)

1. Fill a 1.0-mL glass syringe with aCSF and mount the syringe in the infusion pump. Connect a microdialysis probe and start perfusion at a rate of 2.0 μL/min (*see* Note 4).
2. Anesthetize a male Wistar rat (200–350 g) with chloral hydrate (40 mg/100 g body weight, ip).
3. Shave the top of the rat's head and mount the rat in the stereotaxic frame (tooth bar 5.0 mm above interaural line; *see* ref. *9*) with lidocaine applied to the ear bars. Cover the eyes with a moistened tissue to avoid desiccation. Use a heating mattress thermostatted at 37°C to maintain the rat's body temperature.
4. Using a scalpel, make a firm cranio-caudal incision of the skin, in the middle on top of the rat's head. Start cranial, between the eyes, and cut to caudal until between the ears. Take care that the peri-ost is being cut also (*see* Note 5).
5. Using a bone-scraper, scrape the peri-ost and the skin from the skull in mediolateral direction toward the lateral upstanding edges on both sides of the skull (where the jaw-muscles are attached).
6. Grasp the skin and peri-ost together, with serrated forceps, and pinch them between vascular clamps (two on each side). Place the vascular clamps aside the animal's head, so that the skull remains free from skin and peri-ost.
7. Using a coagulation apparatus, thoroughly cauterize all blood vessels on top of the skull until the bleeding has stopped. Then carefully swab with alcohol and acetone (using cotton buds), respectively, until the skull is clean and dry (*see* Note 5).
8. Using an electric drill, drill one hole (0.65 mm diameter) at 2.4 mm A/P and 1.4 mm M/L on the left side of the skull and another hole on the right side of the skull, near the bregma, completely through the skull. Take care that the underlying dura mater is not damaged. Enlarge the left-side hole with a 1.30-mm bit (*see* Fig. 2).

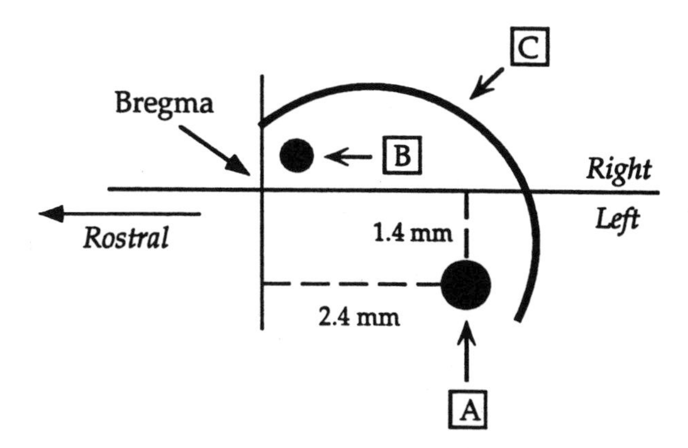

Fig. 2. Positioning of bore holes on the rat skull. *See* Section 3.2. for complete instructions. **(A)** Hole for microdialysis probe; **(B)** Hole for skull-screw; **(C)** Rimmed section (half circle) of 2–200 μL pipet tip.

9. Turn a skull-screw in the right-side hole, through the skull (not into the underlying brain).
10. Using fine forceps, carefully pluck on the dura mater underneath the left-side hole (under binocular microscope; *see* Note 6). Put a small piece of paper tissue in the hole to stop bleeding, and remove afterward.
11. Disconnect the microdialysis probe from the perfusion set-up (leaving the probe filled with aCSF) and cover the inlet and outlet of the probe with tape to avoid desiccation. Then mount the probe in a stereotaxic arm so that the dialysis membrane is exactly vertically orientated.
12. Lower the microdialysis probe through the left-side hole in the skull (under microscope) until its tip just touches the cortex surface. Then lower the probe gently and slowly for another 3.4 mm plus an extra 0.2 mm to compensate the length of the glued membrane tip (*see* Note 6).
13. Cut the rimmed section (wide end) of a 2–200 μL pipet tip with a razor blade and split it into two half circles.
14. Place one rimmed half circle around the skull-screw and the implanted probe (**do not** touch the probe; *see* Fig. 2).
15. Fix the half circle, together with the probe and the skull-screw, to the skull by filling the half circle with dental acrylic cement. Allow the cement to dry for about 5 min. Then carefully remove the stereotaxic arm from the microdialysis probe and build up a firm construction of dental acrylic cement around the probe. Allow the cement to dry for another 5 min.
16. Remove the rat from the stereotaxic frame and put the animal in its cage for about 1 h on a heating mattress thermostatted at 37°C.
17. Allow rats to recover for 1 d before performing microdialysis experiments (*see* Note 7).

3.3. In Vivo Microdialysis Experiments

1. Microdialysis experiments are carried out 20–24 h after implantation (*see* Note 7).
2. Anesthetize rats briefly with ether and infuse freshly prepared urethane (0.15 g/100 g body weight, iv) with a winged infusion set via the lateral tail vein (*see* Note 8).
3. Mount the rats in the stereotaxic frame to enable free breathing. Use a heating mattress to keep the rat warm (37°C).
4. Fill respective glass syringes with aCSF and high K^+-aCSF and mount the syringes in the infusion pump.
5. Connect the outlet of the multiposition valve to the microdialysis probe inlet and begin perfusion with aCSF at 2.0 μL/min. Connect the outlet of the probe to the automated sample changer (*see* Note 4 for connections).
6. Set up vials containing 4 μL TCA/homoserine in the automated sample changer. Connect the changer to a recirculation cooler that maintains the temperature at 2°C.
7. After perfusing for 1.5 h without collecting samples (*see* Note 9), begin to collect samples in the vials over 20 min intervals (40 μL/vial).
8. After 2 h of sample collection (6 × 20 min samples), switch the multiposition valve to begin perfusion with high K^+-aCSF. Perfuse with this solution for 1 h (collecting 3 × 20 min samples), then switch back to aCSF and collect four additional, consecutive 20 min samples (80 min).
9. At the end of the experiment, sacrifice the rat by decapitation.
10. Vortex mix the collected samples and store them at –20°C until analysis by high performance liquid chromatography (HPLC) (*see* Note 10).
11. To study effects of uptake carrier blockers on basal and K^+-evoked extracellular amino acid levels, follow the same procedure but use SK&F 89976-A/L-*trans*-PDC and high K^+-SK&F 89976-A/L-*trans*-PDC instead of aCSF and high K^+-aCSF for perfusion.
12. Figure 3 shows extracellular levels of GABA and glutamate in rat hippocampus in vivo, as determined by microdialysis and HPLC (*see* Note 11).

4. Notes

1. The major advantage of the concentric dialysis probe (as compared to U-shaped or transverse probes) is its small diameter. Because of this, the rat brain suffers minimal implantation damage and it is possible to achieve better spatial resolution of sampling of discrete brain regions. Furthermore, the vertical implantation of this probe allows concomitant measurement of electrical transmission in the dialyzed field by coimplanted electrodes according to well-defined stereotaxic coordinates.
2. A rough estimate of the actual extracellular glutamate and GABA concentrations in rat hippocampus can be derived by correction for the in vitro recoveries of the microdialysis probe, determined using solutions with known amino acid concentrations. The recovery depends on several parameters, such as membrane length, perfusion rate, composition of perfusion fluid, membrane type, and temperature *(10)*.

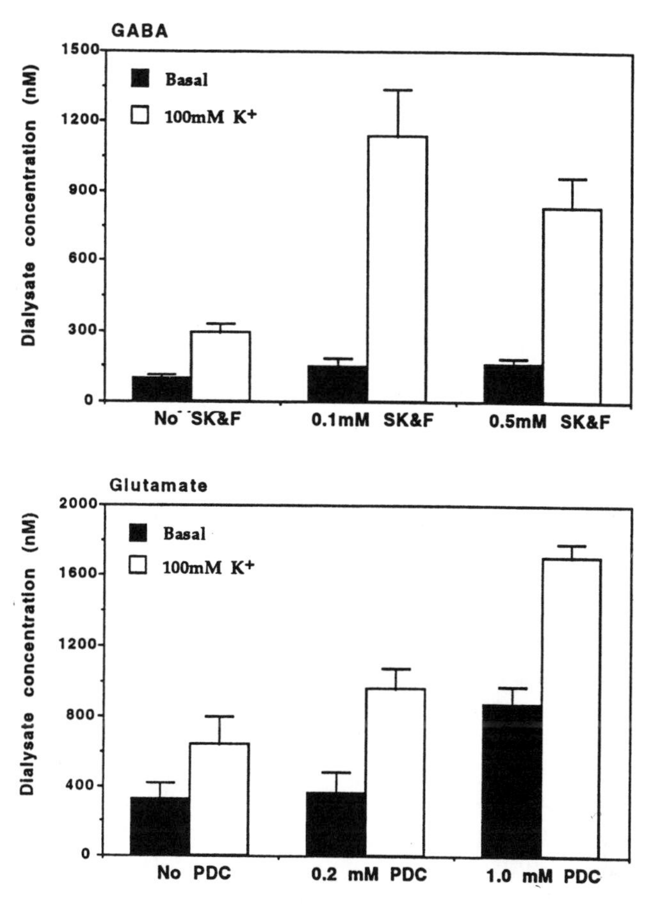

Fig. 3. Extracellular levels of GABA and glutamate in rat hippocampus in vivo, as determined by microdialysis and HPLC. Basal and K^+-evoked levels of GABA **(top)** and glutamate **(bottom)** and the effects of SK&F 89976-A and L-*trans*-PDC on the respective transmitter levels are shown. Dialysates were collected over 20 min in each case. Basal values represent means ± SEM of six dialysates collected prior to K^+-elevation. Because K^+ addition induced a transient rise in extracellular amino acid levels, K^+-elevated levels represent peak transmitter concentrations measured during 1 h perfusion (i.e., the highest of three 20 min samples) with 100 m*M* KCl. *See* Section 3.3. for details of the microdialysis procedure and Note 11 for further discussion.

In our experiments, 8.9 ± 0.4% of GABA and 7.6 ± 0.1% of glutamate were recovered. Similarly, not all of the KCl present in the 100 m*M* KCl perfusion fluid will pass the probe-membrane to depolarize hippocampus tissue. The K^+-recovery determined under our experimental conditions was 28.6 ± 2.0%.

3. We implanted one rat every day because our in vivo microdialysis set-up allowed the measurement of only one animal at a time. The implantation procedure, as described in Section 3.2., takes about 1 h.
4. The connections for the microdialysis set-up are as follows:
   a. To the inlet of the microdialysis probe: The glass syringes (which are mounted in the infusion pump) are connected to 23-gage hypodermic needles. These needles are connected to PEEK tubing, of about 10 cm length, via approx 3 cm of Tygon tubing. The PEEK tubing tightly fits into the inlets of the multiposition valve using fitting screws and ferrules. The outlet of the multiposition valve is connected to FS tubing of about 20 cm length. A strong connection is made by inserting the FS tube into approx 3 cm of PEEK tubing and fitting the PEEK tubing into the outlet of the valve. The outlet end of the FS tube is inserted into a 25-gage hypodermic needle until the FS tube just protrudes from the sharp end of the needle. The FS is glued to the needle at the sharp end with a small droplet of Pattex supermix (the glue easily flows into the space in between the wall of the needle and the FS tube). The needle's sharp end is inserted into 2.5 cm of PE tubing that forms the connection to the inlet of the microdialysis probe.
   b. From the outlet of the microdialysis probe: The outlet of the probe is connected to the sharp end of a 25-gage needle. An FS tube of about 60 cm length is inserted into the needle and glued as described above. The outlet end of the FS tube is inserted into a septum piercing needle and glued as described above. The piercing needle is mounted in an automated sample changer.
5. It is important that the incision of the skin on top of the rat's head does not touch the muscles caudal to the ears, since this can cause unnecessary bleeding and muscle damage. Furthermore, it is important that the skull is clean and dry before the microdialysis probe is implanted; otherwise the probe cannot be firmly fixed with dental acrylic cement.
6. Before lowering the microdialysis probe into the brain, the dura mater must be removed. By plucking with fine forceps, the dura tears open by its own tension. Take care that large blood vessels, sometimes present here (the vessels are clearly visible as thick red strings in the dura mater), are not damaged. Always check the opening by pricking with the tip of the forceps into the brain tissue; if the cortex surface indents this means that the dura is not entirely removed yet. When the dura is properly removed, there is no noticeable resistance as the microdialysis probe is lowered into the cortex tissue. To minimize implantation damage, lower the probe very slowly, over a period of about 5 min.
7. We have compared results of microdialysis experiments performed at 1.5 h with those 20–24 h after probe implantation. Elevation of transmitter levels after K^+-depolarization was greater in perfusates obtained 20–24 h after probe implantation. In addition, transmitter levels were less variable in perfusates taken 20–24 h after implantation. Both effects were more pronounced for glutamate than for GABA.

8. Apply the urethane slowly and in small amounts (about 0.1 mL), because large amounts of urethane at a time can cause the rat to go into shock. Remove ether from the animals immediately after the initial application of urethane.
9. To ensure stable transmitter levels in the collected perfusates, a period of at least 1.5 h perfusion without collecting samples is necessary.
10. Before HPLC analysis, centrifuge stored samples (40 μL) for 4 min (12,000*g*) and transfer 30 μL of the supernatant to new vials for analysis. Our samples were analyzed using a method of precolumn derivatization with *o*-pthalaldehyde followed by reversed-phase HPLC and fluorimetric detection *(11)*.

    The detection limits of our HPLC analysis are 20 fmol for GABA and 10 fmol for glutamate. In rat hippocampus, this analysis sensitivity allows measurement of 5 min samples instead of 20 min, and 1 mm dialysis membrane length instead of 2 mm, as observed in recent experiments in our laboratory.

    A similar method to measure amino acid transmitter levels in microdialysates, using microbore HPLC, is presented in detail in Chapter 15. Please note that the Smolders et al. method makes use of different precolumn sample derivatization procedures than used here and, therefore, step 6 of Section 3.3. will need to be altered accordingly should the Smolders et al. method be chosen.
11. In the experiments presented in Fig. 3, the hippocampus was perfused by both SK&F 89976-A and L-*trans*-PDC, and glutamate and GABA levels were determined in the same sample. In separate experiments, no crosseffects of SK&F 89976-A on glutamate levels, or L-*trans*-PDC effects on GABA levels were found *(8)*. Neither SK&F 89976-A nor L-*trans*-PDC interfere with the chromatograms. The effect of the higher concentration L-*trans*-PDC on basal glutamate levels (Fig. 3) can possibly be explained by action of this inhibitor as a substrate for the glutamate uptake carrier as well *(12)*.

    To determine the contribution of the vesicular release pools to the K^+-induced increases in amino acid levels, Ca^{2+}-dependency of the K^+-induced release was determined. About 50% of the K^+-evoked release of both glutamate and GABA was Ca^{2+}-dependent. In contrast, K^+-induced increases in the extracellular levels of the putative amino acid transmitters aspartate and taurine was completely Ca^{2+}-independent *(8)*.

Acknowledgments

We would like to thank B. H. C. Westerink and coworkers (State University Groningen) for providing us the method to construct concentric microdialysis probes. We appreciate H. Sandman (T. N. O. Zeist) for packing HPLC columns. Financial support for microdialysis and HPLC equipment was provided by the Netherlands Organization for Scientific Research (N. W. O.).

References

1. Ungerstedt, U. (1991) Introduction to intracerebral microdialysis, in: *Microdialysis in the Neurosciences* (Robinson, T. E. and Justice, J. B., Jr., eds.), Elsevier Science, B. V., Amsterdam, pp. 2–22.

2. During, M. J., Ryder, K. M., and Spencer, D. D. (1995) Hippocampal GABA transporter function in temporal-lobe epilepsy. *Nature* **376,** 174–177.
3. Globus, M. Y.-T., Busto, R., Martinez, E., Valdés, I., Dietrich, W. D., and Ginsberg, M. D. (1991) Comparative effect of transient global ischemia on extracellular levels of glutamate, glycine and γ-aminobutyric acid in vulnerable and non-vulnerable brain regions in the rat. *J. Neurochem.* **57,** 470–478.
4. Westerink, B. H. C., Damsma, G., Rollema, H., De Vries, J. B., and Horn, A. S. (1987) Scope and limitations of *in vivo* brain dialysis: a comparison of its application to various neurotransmitter systems. *Life Sci.* **41,** 1763–1776.
5. Attwell, D. and Mobbs, P. (1994) Neurotransmitter transporters. *C. Op. Neurobiol.* **4,** 353–359.
6. Bliss, T. V. P. and Collinridge, G. L. (1993) A synaptic model of memory: long-term potentiation in the hippocampus. *Nature* **361,** 31–39.
7. Kamphuis, W. and Lopes da Silva, F. H. (1990) The kindling model of epilepsy: the role of GABAergic inhibition. *Neurosci. Res. Comm.* **6,** 1–10.
8. Zuiderwijk, M., Veenstra, E., Lopes da Silva, F. H., and Ghijsen, W. E. J. M. (1996) Effects of uptake carrier blockers SK&F 89976-A and L-*trans*-PDC on *in vivo* release of amino acids in rat hippocampus. *Eur. J. Pharmacol.* **307,** 275–282.
9. Pellegrino, L. K., Pellegrino, A. S., and Cushman, A. J. (1979) *A Stereotaxic Atlas of the Rat Brain.* Plenum, New York, NY.
10. Benveniste, H. and Hüttemeier, P. C. (1990) Microdialysis—theory and application. *Prog. Neurobiol.* **35,** 195–215.
11. Verhage, M., Besselsen, E., Lopes da Silva, F. H., and Ghijsen, W. E. J. M. (1989) Ca^{2+}-dependent regulation of presynaptic stimulus-secretion coupling. *J. Neurochem.* **53,** 1188–1194.
12. Griffiths, R., Dunlop, J., Gorman, A., Senior, J., and Grieve, A. (1994) L-*trans*-pyrrolidine-2,4,-dicarboxylates behave as transportable, competitive inhibitors of the high-affinity glutamate transporters. *Biochem. Pharmacol.* **4,** 267–274.

20

In Vivo Detection of Neurotransmitters with Fast Cyclic Voltammetry

Julian Millar

1. Introduction

Electrochemical techniques detect materials by oxidizing them in solution at a positively polarized "working" electrode. The oxidized material gives up electrons that are collected by the working electrode and generate a current flow through it. The detection of this current is the basis of the measurement method. Several compounds of biological interest are oxidizable under these conditions, and are, therefore, said to be "electroactive." These include the amine neurotransmitters dopamine, noradrenaline (norepinephrine), adrenaline, and serotonin. Other transmitters, such as acetylcholine, γ-amino-butyric acid (GABA), glycine, and glutamate, are not electroactive and are, therefore, not detectable using electrochemical techniques. The neurotransmitter gas nitric oxide is also electroactive and can be detected electrochemically, as can several other compounds of interest in the brain, including ascorbic acid, uric acid, and several metabolites of the amine transmitters.

Electrochemical techniques have been used in analytical chemistry for more than 50 yr, but there was little interest in or understanding of them by biologists until the 1970s. A breakthrough occurred in 1973 when Ralph Adams and his colleagues *(1)* showed that it was possible to implant a working electrode in a rat brain and detect electroactive materials in vivo. This heralded a new chapter in electrochemistry because it now appeared to be possible to study the release of neurotransmitters in intact animals without the use of complicated radioactive labeling techniques.

A significant step forward was made in 1978 when Gonon et al. *(2)* introduced the use of working electrodes based on carbon fibers, instead of the carbon paste electrodes initially used by Kissenger et al. *(1)*. These carbon fiber electrodes

From: *Methods in Molecular Biology, Vol. 72: Neurotransmitter Methods*
Edited by: R. C. Rayne Humana Press Inc., Totowa, NJ

were made from glass pipets from which a 0.5 mm length of 8 or 12 μm diameter carbon fiber protruded from the end. This electrode gave a considerable improvement in terms of reproducibility and reliability over the carbon paste electrodes and its development led to a surge of interest in in vivo voltammetric techniques. This interest was reflected by the appearance in the middle 1980s of a large number of publications that made use of voltammetric methods.

However, there were still certain limitations in the technology. The relatively long length of carbon fiber exposed at the tip of the Gonon electrode meant that voltammetric signals could only be obtained from relatively large groups or "pools" of neurons, and not single cells. The electrochemical techniques favored at the time—such as differential pulse voltammetry or differential normal pulse voltammetry—limited the rate of sampling to about once or twice per minute. This slow sample rate, combined with the long tip length, meant that most of the electrochemical signals that could be detected were changes in extracellular levels of transmitter metabolites (e.g., DOPAC, 5-HIAA) rather than the transmitters themselves. There was a considerable amount of confusion about whether amine transmitters after release were contained wholly within the synaptic cleft and then taken back up from within the cleft without ever leaking out into the general extracellular space, or whether significant synaptic "overspill" occurred.

In 1979 Armstrong-James and Millar *(3)* introduced a carbon fiber electrode with a much shorter tip length. The exposed carbon tip was only 10–50 μm, instead of the 500 μm or more used in the Gonon electrode. This shorter electrode was initially designed as an extracellular voltage recording electrode, suitable as a replacement for tungsten electrodes for studying extracellular spike potentials. However, Armstrong-James et al. *(4)* found that the electrodes could be used for a new form of linear sweep voltammetry, initially called "High Speed Cyclic Voltammetry," but that has subsequently become known as "Fast Cyclic Voltammetry (FCV)." With FCV, catecholamine and indoleamine neurotransmitters could be detected in vitro at submicromolar levels at sample rates of up to 16 or more samples per second. In 1984 Millar et al. showed that FCV could be used to detect dopamine released from nerve terminals in the striatum of a rat after electrical stimulation of dopaminergic axons *(5)*. Interest in FCV then increased steadily, in particular when the method was found to be suitable for monitoring transmitter release in brain slices (*see* review by Stamford; *6*; and Chapters 1–3).

1.1. What Exactly Is Fast Cyclic Voltammetry?

Voltammetry is simply the name given to electrochemical techniques where a voltage is applied to an electrode and the resulting current is measured. FCV is a form of "linear sweep" voltammetry. In linear sweep voltammetry the

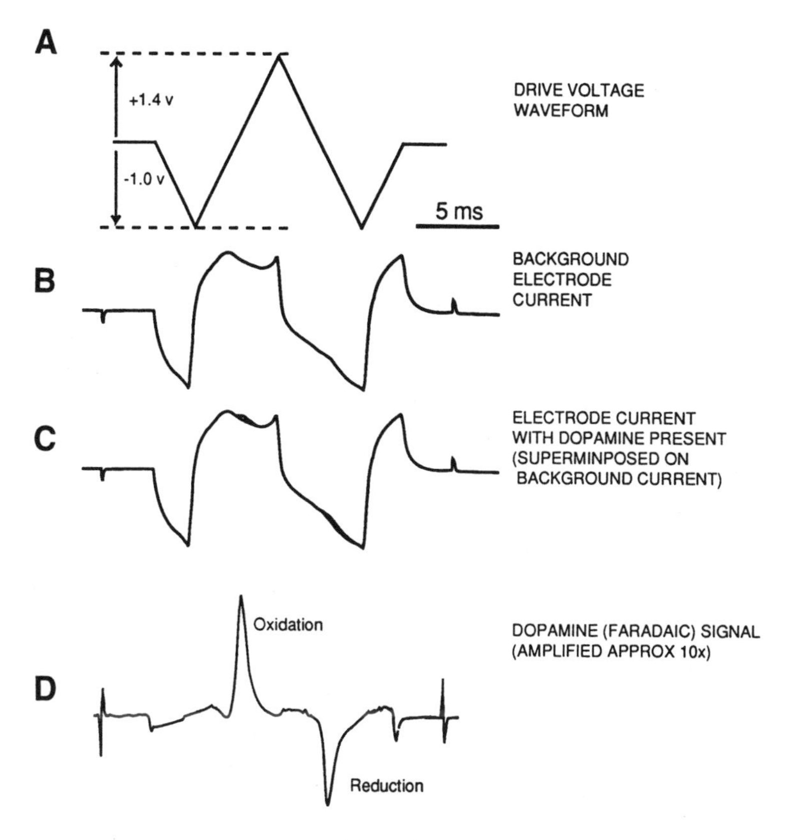

Fig. 1. The voltammetric scan in FCV. The traces shown in panels **(A–D)** are described and discussed in Section 1.1.

applied voltage is a linear ramp. In FCV, as in most voltammetric methods, a three-electrode system is used, comprising a working electrode, a reference electrode, and an auxiliary electrode. The three electrodes are all placed in electrical contact via an aqueous electrolyte solution. The electrolyte may be in a beaker, for calibration of the electrode, or it may be the extracellular fluid in the brain or in a brain slice. The FCV amplifier subjects the working electrode to a diphasic or triphasic set of anodic and cathodic voltage sweeps in a period of a few milliseconds. This is a voltammetric scan. Because the scan only lasts a few milliseconds, it can be repeated many times per second if required. The usual triphasic ramp pattern for FCV is shown in Fig. 1A. This pattern of positive and negative-going voltages generates a pattern of currents through the working electrode, even in simple saline solution in the absence of any electroactive (i.e., oxidizable) material. This "background current" is caused by the complex impedance of the electrode/electrolyte interface. (This interface has the approximate characteristics of a voltage-dependent capacitor in parallel with a resistor and diode). The amplitude of the background current may be two or more orders of magnitude greater than the oxidation current. A typical background current for a carbon fiber electrode is shown in Fig. 1B. The amplitude and shape of the background current is a function of the temperature, pH, and conductivity of the solution. The resolution of the FCV analysis depends on the stability of the background current.

When electroactive materials are added to the solution (e.g., by release from neurons) extra current flow is generated through the electrode at certain specific regions on the waveform. When the positive-going ramp exceeds a certain voltage level, oxidizable molecules at the surface of the electrode oxidize and transfer electrons to the working electrode. This electron influx is known as the 'faradaic' current. The faradaic current is, over a limited range, proportional to the concentration of the oxidizable material in the solution. Figure 1C shows two superimposed scans: one similar to that shown in Fig. 1B in phosphate-buffered saline (PBS), and one immediately after dopamine had been added to the solution to produce a final concentration of 1 μM. Figure 1D shows the difference between the scans in 1B and 1C, amplified approx 10-fold. Note the well-defined oxidation peak and also the reduction peak. If trace 1D is compared with trace 1A it can be seen that the faradaic oxidation current occurs over the range 300–900 mV, with a peak at about 600 mV. There is also a rereduction faradaic current with a peak at about –200 mV, which has approximately the same size and area as the oxidation current. The height of the oxidation peak (at a certain voltage) can be detected on each scan and fed out to a chart recorder. The chart recorder hard copy is usually the main way of logging FCV results; the detailed information from FCV scans, such as shown in Fig. 1D, are normally only used for compound identification.

The negative sweep of the voltage after the positive part of the scan enables the oxidized dopamine to be rereduced back to its original form, as is shown in Fig. 1C,D. This rereduction cycle is a unique feature of FCV and has several benefits. First, it prevents buildup of oxidized material on the surface of the electrode. The products of electro-oxidation can polymerize to form insoluble end-products that coat the working electrode and thereby slow down or stop the access of new substrate to the surface. This process results in a progressive loss of electrochemical sensitivity and is called electrode "poisoning." The rereduction cycle in FCV means that poisoning is very rarely a problem in FCV experiments. A second benefit of rereduction is that, even if it does not adhere to the electrode surface, the oxidized material may be toxic to the surrounding cells. Diffusion of a toxic material, even in small quantities, can harm the surrounding neurons even in an acute experiment, and clearly will do much more damage in experiments with chronically implanted electrodes.

1.2. Advantages of FCV

With conventional linear sweep voltammetry, the voltage at the working electrode is increased relatively slowly, for example, at 0.1–1 V/s. If there is a single electrochemical reactant in solution, there is little current flow before the oxidation potential for the reactant is reached. (The electrode interface with the solution has a primarily capacitative impedance, and the slow rate of change of potential means that little current can flow.) Once the potential at the working electrode has reached a sufficiently positive level, the reactant present at the surface of the electrode loses electrons to the electrode and becomes oxidized. Because the voltage change is slow relative to the oxidation process, this oxidation is complete within a few millivolts. Thus, relative to the timecourse of the ramp, the current increases in an almost stepwise fashion, and the concentration of reactant at the electrode surface drops equally rapidly to zero. After this initial step, the amplitude of the current (i.e., the rate of reaction) depends on the rate of transfer of new reactant to the surface. This is normally by diffusion from the bulk solution. As the region around the electrode becomes more and more depleted of reactant, new reactant has to diffuse farther and farther from the bulk solution, and so the curve decays. Because the electrochemical current depends on diffusion of new reactant, we can say the reaction is diffusion controlled.

In contrast, using FCV, the voltage at the working electrode is changed at a much faster rate, up to nearly 1000 V/s. This means that the reactant may start oxidizing at one potential, but the potential has increased significantly before the reaction reaches completion. In fact, because the potential at the working electrode is not allowed to dwell at its positive peak, but is immediately driven back down to a negative potential, the concentration of reactant at the surface may never reach zero. The brief time allowed for the scan means that only the reactant present at the working electrode surface before the sweep is started is able to take part in the reaction; there is no time for exchange with reactant in the bulk solution. The magnitude of the electrochemical current is, therefore, controlled not by diffusion of reactant to the surface, but by the rate of reaction of the reactant already present at the surface. Electrochemists call this a reaction rate-controlled process rather than a diffusion-controlled process.

The practical importance of this for the application of FCV is that the voltage at which a peak electrochemical signal appears in an FCV is usually much greater than the voltage at which the compound in question starts to oxidize. Thus, for example, dopamine might be oxidizable at a steady potential of 200 mV, but it will show a peak on an FCV scan of perhaps 600 mV. Similar "lags" in peak current occur with all compounds. Because of this rate-control factor in the oxidation, the rate of rise of the voltage is of crucial importance in an FCV scan. For stability and reproducibility, the dV/dt value must be absolutely fixed from scan to scan. In fact, one of the main keynotes in the design of an FCV amplifier is in the production of a very stable and reproducible ramp voltage.

1.3. Overview of Methods Presented

The sections that follow describe how to make the carbon fiber electrodes used as the working electrodes in FCV, how to set up an FCV amplifier, and how to test and calibrate the carbon fiber electrodes before use. These techniques are suitable for the detection and measurement of dopamine, noradrenaline, and serotonin (5-HT) in the whole brain or in brain slices.

2. Materials

2.1. Equipment (see Notes 1 and 2)

See Note 13 for a list of equipment suppliers.

1. Voltammetry amplifier.
2. Storage oscilloscope.
3. Y-T chart recorder.
4. Microelectrode micromanipulator.
5. Reference electrode.
6. Auxiliary electrode.
7. Carbon fiber working electrode (constructed as described in Section 3.).

2.2. Construction of Carbon Fiber Electrodes

See Note 13 for a list of equipment suppliers.

1. A "tow" of carbon fibers (Goodfellow Metals, Cambridge, UK; catalog no. C005725).
2. A flat-topped light box (as used for viewing photographic transparencies).
3. Electrode glass (2 mm od is recommended).
4. Two glass test tubes.
5. Modeling clay ("Plasticine" brand is best, although "Blu-Tack" can be used).
6. At least two pairs of no. 5 fine jeweller's forceps, available from FST Inc. Alternatively, RS Components type 549-628 extra fine tweezers may be used.
7. Acetone (industrial grade is sufficient).
8. A microelectrode puller.
9. A hot air blower (available from RS Components).
10. Fine dissecting scissors (available from FST).
11. A micromanipulator (the same as item 4 in Section 2.1.).
12. Standard histology microscope capable of 100–200× magnification.

13. Plastic-coated, miniature stranded signal wire (thin enough so that the insulated wire fits smoothly into the electrode glass). RS type 357-334 is suitable.
14. Conducting paint (silver-doped paint; e.g., RS type 555-156).
15. Diamond disk bevelling wheel (Narishige Scientific Instruments, Japan).
16. Tungsten microelectrode (*see* Note 13).
17. Step-up transformer (1:10).
18. Glass slides.
19. Phosphate-buffered saline (PBS): Made up from ampules or sachets, type P3813 supplied by Sigma Chemical Co. (St. Louis, MO) (10 m*M* sodium phosphate, 138 m*M* sodium chloride, 2.7 m*M* potassium chloride, pH 7.4).

3. Methods

3.1. Construction of an Electrode Blank

1. Place the tow of fibers on the light box (*see* Note 3). Cut from the tow a hank of fibers of about 15 cm long.
2. Place this hank in a fold of paper or a test-tube so that a bundle of fibers 2–3 cm long protrudes from the end. Place the paper fold or test-tube flat on the light box so that individual fibers may be teased from the protruding bundle and withdrawn using plastic-tipped forceps (*see* Note 3).
3. Using a blob of Plasticine, fix a second test tube on the light box so that it is nearly, but not quite horizontal (the mouth tilted very slightly upward).
4. Place a standard electrode glass tube (10 cm long by 2 mm od; this is the electrode "blank") into the test tube so that it protrudes from the end of the test tube by 1 cm or so (*see* Note 4).
5. Squirt a few milliliters of acetone into the test-tube. The acetone should immediately run up inside the electrode glass tube by capillary action. If air bubbles form in the electrode blank, remove it from the test tube, shake to remove the acetone, and gently replace in the tube. If the blank is replaced with care the acetone should fill the blank smoothly without bubbles.
6. Tease one carbon fiber from the hank. Using the plastic-tipped forceps, grasp the fiber about 1 cm back from its end and push it carefully into the acetone-filled electrode blank (*see* Note 5).
7. Once the end of the fiber is a few millimeters into the electrode blank, release the forceps and grip the fiber again about 1 cm further back. Again, push the fiber into the acetone-filled electrode blank. Repeat this operation until the fiber has been inserted all the way into the blank.
8. Cut off the excess carbon fiber with fine dissecting scissors, leaving 1 mm or so protruding from the top of the electrode blank.
9. Carefully remove the glass blank complete with fiber from the test-tube. Immediately blot one end with tissue to remove the acetone. The fiber will now stick to the inner wall of the glass tube by surface tension.
10. It is now necessary to "pull" the glass blank into a micropipet with the carbon fiber inside and to fit a connecting wire. Proceed to Section 3.2.

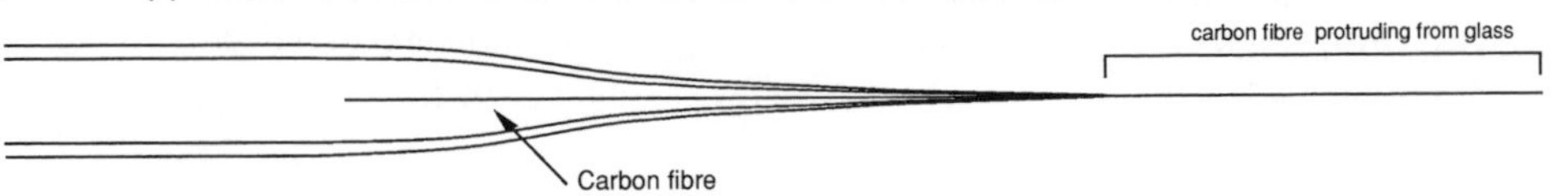

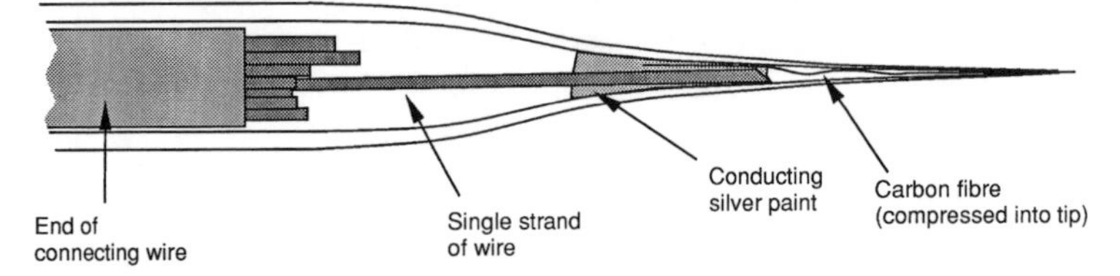

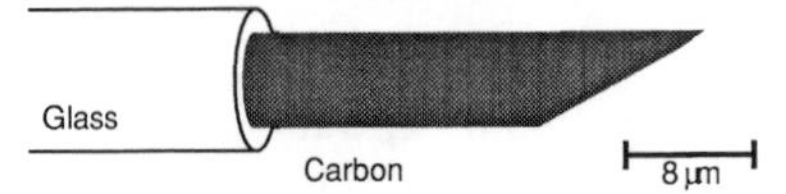

Fig. 2. Stages in the construction of a carbon fiber microelectrode. The steps involved are detailed in Sections 3.2. and 3.3. The beveling procedure is discussed in Note 8.

3.2. Pulling the Glass Micropipet

1. Hold the tube for a few seconds in the warm air stream of a blower or small hair dryer to dry the glass blank completely.
2. Place the carbon fiber-filled glass blank in the electrode puller and pull it into a microelectrode. The glass should melt and contract around the fiber, finally breaking to leave an open pipet in one end of the puller and a pipet with a carbon fiber protruding from the glass at the other end (*see* Note 6 and Fig. 2A).
3. Before removing the electrode from the puller, use a pair of fine scissors to cut the protruding carbon fiber (which may be several centimeters long) back to a cm or so beyond the end of the glass at the tip of the micropipet. This is easily done without magnification.
4. Transfer the pipet to a micromanipulator so that the tip can be viewed under a microscope (100× or 200× magnification). Visually check that the glass at the tip is not cracked, that it forms a tight seal around the carbon fiber, and also that a

sufficient length of carbon fiber is present in the stem of the pipet to make a contact when a connecting wire is inserted. (If not, start again from Section 3.1.).

5. Strip back 1–2 cm of insulation from a piece of miniature-stranded signal wire and cut all but one strand off, leaving a single strand of wire of about 2 cm in length.
6. Dip the end of this strand into a solution of conducting silver paint and immediately insert the wire into the stem end of the electrode. Push the wire inward, toward the electrode tip, so that the conducting paint-coated end of the strand makes contact with the carbon fiber and presses it against the glass (*see* Fig. 2B).
7. Apply a drop of cyanoacrylate glue to the stem end of the electrode to secure the wire.
8. Proceed to Section 3.3., choosing either to cut (Section 3.3.1.) or spark-etch (Section 3.3.2.) the carbon fiber tip.

3.3. Trimming the End of the Carbon Fiber

3.3.1. Cutting

1. Mount a pair of fine forceps on a micromanipulator so that the forcep tips can be placed in the field of view of a microscope.
2. Anchor one arm of the forceps firmly on the manipulator and arrange a lever or other mechanism to close the forceps smoothly by pressing on the other arm (*see* Note 7).
3. Place the microelectrode in a second manipulator or fix it with some plasticine on a microscope slide and place on the stage of the microscope.
4. Advance the tip of the microelectrode (with the excess fiber protruding from the end) until it lies within the jaws of the forceps.
5. Close the forceps; this usually produces a surprisingly clean "snap" cut of the fiber.
6. Inspect the cut end of the fiber. If it appears jagged (for example as in Fig. 2C), smooth the fiber tip using a diamond disk beveling wheel (*see* Note 8 and Fig. 2D). If the fiber tip is smooth, proceed to Section 3.4. (*see* Note 9).

3.3.2. Spark-Etching

1. Place the electrode on a micromanipulator under a microscope.
2. Attach the electrode-connecting wire to one side of the output of a 1:10 step-up transformer. Drive the input to this transformer from a physiological stimulator that can give pulses of up to 100 V output. (This enables the final voltage on the electrode to be up to 1000 V).
3. Connect the other side of the transformer to a tungsten microelectrode.
4. Mount the tungsten electrode perpendicular to the carbon fiber electrode, placing the tip 100 μm from the point where the carbon fiber leaves the glass. Position the tungsten electrode close to the carbon fiber tip without quite touching it.
5. Set the stimulator to deliver pulses of 50 or 100 μs duration at a frequency of 20 pulses/s and slowly increase the voltage. At some point (usually 400–600 V), a small spark will appear between the tip of the tungsten electrode and the carbon

fiber. The spark will rapidly eat away the carbon. Etch completely across the fiber so that a 100 μm tip is left protruding from the glass.

6. Now slowly advance the carbon fiber electrode so that the carbon fiber tip is etched progressively back toward the glass. Stop when the final length of protruding carbon fiber is about 50 μm.
7. Bevel the electrode by the method described in Note 8 to remove the spark-eroded surface (*see* Note 10 and Fig. 2). The beveled electrode should have a final length of carbon protruding 30–50 μm beyond the glass.
8. Store the finished electrode (*see* Note 9) or proceed directly to Section 3.4.

3.4. Conditioning Electrodes

1. Place the working, reference, and auxiliary electrodes into a beaker of PBS. Set up the monitor oscilloscope so that the drive waveform and electrode current can be observed during the conditioning process.
2. Apply a normal FCV waveform (+1.4 V, –1.0 V) at 16 scans/s for 5 min.
3. Now reduce the scan repetition rate to 2/s and observe the electrode current closely. If the electrode is properly conditioned the electrode current (the "background current") will be stable, and will not fluctuate from scan to scan. If current fluctuation is observed, the cycle must be repeated, several times if necessary, until the scan current is stable (*see* Notes 11, 12, and Fig. 3).
4. Store the conditioned electrode in PBS until ready for use (*see* Note 9).

4. Notes

1. A list of equipment manufacturers is given in Note 13. For FCV recording you need at minimum an FCV amplifier, a chart recorder, and an oscilloscope. A commercial instrument designed specifically for FCV is the Millar Voltammeter (P. D. Systems Ltd., London, UK). This instrument has the drive ramp "hard-wired" in and so is relatively inflexible. However, it is optimized for FCV studies and has a very stable ramp voltage. Several computer-controlled instruments are now appearing where the ramp (drive) waveform is generated by software. Among them are the Ivec-10 from Medical Systems Corp., and a modified Axopatch amplifier from Axon Instruments.

   Because the drive potential in FCV occurs in milliseconds, it is standard procedure to monitor both the drive potential and the electrode current on an oscilloscope. A digital rather than an analog oscilloscope is mandatory, because the scan repetition rate is too low to see properly with a nonstorage oscilloscope. The oscilloscope should be able to store a reference signal onscreen for comparison with the live signal. Almost all researchers using Millar Voltammeters use one of the Nicolet range of digital storage oscilloscopes because the sensitivity and flexibility of these instruments make them ideal FCV monitors. If an Ivec or Axon system is used, the computer monitor is normally used to monitor the drive voltage and electrode current, and so an oscilloscope is a useful, but not essential, extra.

   Because the current as a result of the oxidation of a particular compound occurs at a fixed place on the background signal, a sample-and-hold (more accurately, a track-and-hold) circuit is normally used to monitor the current amplitude at that point on each scan. The output of this amplifier is updated at the rate of the scan (usually 2 or 4 scans/s), and so can be output to a chart recorder as a record of the oxidation current changes for a particular compound. Any *y-t* chart recorder with a full-scale response time of 1 s or better will do for this task.

The micromanipulator (positioner) is required to hold the carbon fiber electrode. A large number of positioners are sold for different kinds of microelectrode work. Positioners suitable for extracellular recording are also suitable for FCV. Together with the positioner(s), you will also need various clamps, stands, and so forth. The hardware requirements for FCV are fundamentally the same as for extracellular single unit recording, and the basic equipment for the latter will also serve for the former.

2. The reference electrode **must** be a high-stability silver-silver chloride type. Any noise or drift in the potential registered by this electrode will be transferred to the voltammetry signal, so it is important to use a good quality reference electrode. The WPI (World Precision Instruments) "Dri-Ref" series of reference electrodes can be strongly recommended for FCV. The author has also had good results with the sintered Ag/AgCl electrodes supplied by Clark Electromedical (UK). It is not recommended that researchers use homemade reference electrodes, because they are rarely as stable as the commercial product. The nature of the auxiliary electrode is far less crucial, since potentials generated at this electrode are not transferred to the voltammetry signal. The auxiliary electrode may be made of any noncorroding metal. Most workers prefer to use either a stainless steel or platinum wire. Finally, although carbon fiber microelectrodes are available commercially (*see* Note 13), it is much better to make your own, because you then have control over the tip size and character. Making the electrodes is quite straightforward provided you have the right equipment; most people can learn the technique in a day or so.
3. Most people can see a single fiber (7 μm diameter) when it is lit from behind on a light box. If necessary, a simple magnifying lens can be used. The light box should have as large an area as possible; 0.5 × 0.5 m is more than adequate. It should be placed on a bench where there are no drafts, because the fibers are very easily disturbed by air currents.

   Handle the fibers using a pair of no. 5 jeweller's forceps to which small plastic collars have been fitted over the tips. The plastic collars can be made by stripping short (2–5 mm) lengths of plastic insulation tubing from thin low-voltage signal wires. Carbon fibers are extremely brittle, and untipped forceps can easily break them during handling.
4. Do not use glass with a filament inside for easier filling (Ω glass)—use standard tubing.
5. The acetone lubricates the glass and stops the fiber sticking to the wall because of static electricity. Ethanol can be used instead, but it does not evaporate so quickly when it is removed.
6. The high tensile strength of the carbon fiber prevents it breaking at the point where the glass fractures.) You will almost certainly have to experiment with the

pull heat and solenoid settings before you get the right combination. Common problems at this stage are a bent end to the electrode where the glass has become distorted proximal to the tip or a poor seal between the glass and the carbon. If such problems are observed, discard the electrode and start again!

7. The difficulty here is that human fingers have too much physiological tremor to be able to hold the forceps steadily enough to cut the fiber with the desired accuracy. The forceps need to be constrained in some way and closed by remote control. There are various ways to achieve this; for example, a hole can be drilled through the body of the forceps and a nut and bolt threaded through. Rotation of the nut on can be used to provide the required fine control of closure. Alternatively, the remote control cable of a camera (the type where a small bolt is pushed out of the end of a flexible cable) can be used to press on one arm of the forceps with the necessary delicacy. The method to be used will depend on the ingenuity of the researcher's workshop staff as much as anything!
8. With practice, the tip can be cut by this method to a precision of about ±5 μm. If the end of the carbon remains jagged after the cut, it may be smoothed using a diamond disk beveling wheel (the sort normally used to bevel glass micropipets—Narishige supply an instrument for this purpose). Mount the electrode in a micromanipulator at an angle of about 20° to the horizontal. Use a dissecting microscope to observe the wheel. Advance the electrode gently until it can be seen to just touch the rotating surface. Allow 20–40 s contact with the wheel; this produces a clear bevel that can be observed under the microscope (*see* Fig. 2).
9. Store finished carbon fiber electrodes by pressing the stem glass on to a blob of modeling clay stuck on a microscope slide. This holds the electrode tip in the air and the slide serves as a holder when the electrode is tested for a good electrochemical background current (*see* Section 3.4.). Electrodes may be stored dry indefinitely. Once conditioned, however, an electrode should be stored in PBS until ready for use. If allowed to dry out, an electrode should be reconditioned briefly before use. In any case, immediately before use, connect the electrode to the voltammeter and carry out scans at 4 scans/s continuously for a few minutes or until the background current has stabilized. If the background current will not stabilize despite repeated conditioning, there may be a problem with the electrical pickup (*see* Note 12). Consult with colleagues who do extracellular single-unit recording. The rules of grounding and layout that apply to extracellular recording can also be applied to FCV.
10. Although beveling is an "optional extra" for cutting, it is mandatory for electrodes that have been spark-etched. The spark-eroded surface can be electrically noisy and must be removed by beveling before the electrode is used (*see* Note 8).
11. This repetitive cycling is believed to take any amide, hydroxyl, or other groups that are present on the carbon surface through a repetitive series of oxidation and reduction cycles until a steady state is reached. However, because of our ignorance of the surface chemistry of the carbon fibers, this hypothesis is really only an educated guess. What has been empirically demonstrated, however, is that an electrode that gives a mainly capacitative current during the positive part of the

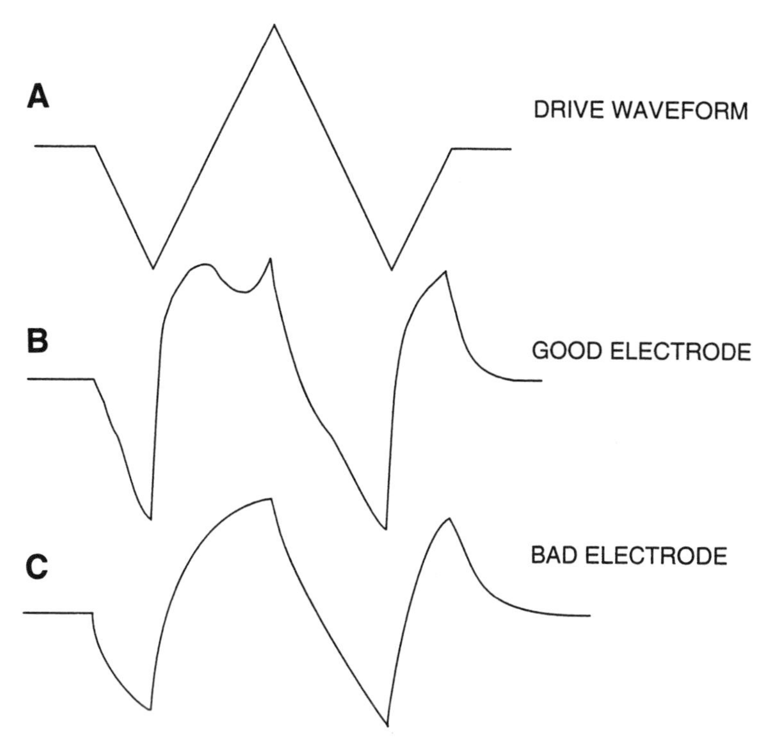

Fig. 3. Comparison of FCV waveforms. *See* Section 3.4. and Note 11 for a discussion of the salient features of each.

scan is associated with increased stability and a high signal-to-noise ratio for voltammetric recording. This is shown by a flat plateau-like shape to the current during the positive scan. For most electrodes, the waveform can be brought closer to this shape by conditioning.

After conditioning, a good electrode should have a noise level of <1 nA when observed on a chart recorder printout of the current scan (i.e., a current pulse of 10 nA should have a signal-to-noise ratio of >10:1). An example of a good background current is shown in Fig. 3B, and for comparison, Fig. 3C shows an unacceptable background current from a poor electrode. Note the differences in shape between the currents on the positive part of the scan waveform (the poor electrode has a much more rounded shape).

Since there is an unavoidable variation in the precise size and characteristics of electrodes, it is preferable for each electrode to be calibrated using solutions of the compound(s) to be investigated in a range of known concentrations. In general, it is only feasible to study changes in one electroactive transmitter in any one experimental protocol; signals from two or more compounds may overlap and/or interact in other ways.

When used for in vivo electrochemistry, a new electrode should be used for each experiment. This is because glial cells or neuronal fragments can collect on the surface of the carbon, thus forming a diffusion barrier for the electroactive compounds. These layers of cell debris may or may not be visible under the microscope, but a drop in sensitivity in vitro is a sure sign that an electrode is contaminated. Electrodes can be cleaned by overnight immersion in diluted nitric acid, followed by repeated washing with distilled water and reconditioning. However, with a little practice, the electrodes are so easy to make that it is no problem to make a fresh one for each experiment.

12. Stability of FCV signals:
    a. Drift: Without doubt, one of the biggest nuisances in FCV recording is a slow drift in the size and shape of the background current. This drift can be seen as a slow DC shift in the sample-and-hold chart recorder output. Drift can be caused by many factors. A poor reference electrode is often the culprit; if drift is excessive, the first thing to check is the reference electrode. You can do this with an oscilloscope or DVM (digital voltmmeter) if you have two reference electrodes. With both electrodes in a beaker of saline, connect one to the ground terminal of the DVM or oscilloscope, and the other to an input channel. Observe the voltage. Does it drift more than a few mV/s? If so, one or other of the reference electrodes is faulty. Clean them (see manufacturer's instructions) and replace if necessary.

       If the drift persists with a known good reference electrode, try more drastic conditioning of the carbon fiber electrode. A scan to +2.0 V will produce a certain amount of hydrolysis at the peak positive potential (seen as a sharp peak of extra current). This extra current produces oxygen that will sometimes clean off any debris attached to the electrode. But be warned, prolonged positive scans can erode or destroy the carbon fiber tip.
    b. Mains pickup: Even more annoying than drift can be the problem of mains pickup. This manifests itself as an "aliasing" of the electrochemical sample-and-hold signal. When this happens, the chart recorder trace oscillates at some frequency (usually between 0.1 and 5 Hz) which is a "beat" between the mains frequency (50 or 60 Hz), and a derivative of the crystal oscillator timing circuit in the voltammeter. The cause of this pickup is usually magnetic coupling between the carbon fiber electrode and either a mains lead, or some ground loop in the mains earth (ground) circuit. Magnetic pickup of this sort is not eliminated by a Faraday Cage: The cage will block electric fields but not magnetic ones! The only way to minimize this kind of problem is to first place the preparation (brain slice or whatever) on a heavy mild steel slab (at least 1 cm thick, but the thicker the better). Connect the slab to the ground terminal of the voltammeter preamplifier (and to no other ground point). Anything else near the electrode that needs to be grounded should be connected to a multiway "star" connection point as close as possible to the preamplifier ground. Now remove all mains leads physically as far away as possible from the slab. The mild steel should soak up any stray fields near the elec-

trode and channel through the steel. Using a star configuration and grounding via the first stage amplifier should minimize ground loops. Finally, if pickup persists, try to shorten the working electrode lead, if possible, mounting the electrode on the preamplifier itself. A shorter lead reduces the coupling between the electrode input and any interfering fields that are still present.

13. Equipment manufacturers/suppliers:
    a. Suppliers of microelectrode glass and reference electrodes:
        i. Clark Electromedical Instruments, PO Box 8, Pangbourne, Reading RG8 7HU, UK. Fax: +44 (0)1734 845 374.
        ii. FHC Inc., 69 Pleasant Street, PO Box 337, Brunswick, ME 04011, USA. Fax: (207) 729-1603.
        iii. W. Krannich, 37079 Göttingen, Elliehäuser Weg 17, Germany. Fax: +49 (640) 966-7673.
    b. Manufacturers of microelectrode pullers:
        i. Narishige Scientific Instruments, 9-28 Kasuya 4-Chome, Setagaya-Ku, Tokyo, Japan. Fax: +81 813 3308 2005 (in the USA, Fax: (516) 759-6138).
        ii. Campden Instruments, King Street, Silby, Loughborough LE12 7LZ, UK. Fax: +44 (0)1509 816 097.
        iii. Mecanex, Vuarpilliére 29, 1260 Nyon, France. Fax: +33 41 22 61 52.
        iv. Stoelting Co., 620 Wheat Lane, Wood Vale, IL 60191, USA. Fax: (708) 860-9775.
        v. Sutter Instruments Co., 40 Leveroni Court, Novato, CA 94949, USA. Fax: (415) 883-0572.
        vi. World Precision Instruments, International Trade Center, 175 Sarasota Boulevard, Sarasota, FL 34240-9258, USA. Fax: (813) 371-1003.
    c. Manufacturers of electrochemical amplifiers:
        i. Axon Instruments, 1101 Chess Drive, Foster City, CA 94404, USA. Fax: (415) 571-9500.
        ii. Medical Systems Corp., 1 Plaza Road, Greenvale, NY 11548, USA. Fax: (515) 621-8503.
        iii. P.D. Systems Ltd., Estate House, Pool Close, West Molesey, Surrey KT8 ORN, UK. Fax: +44 (0)181 979-4648.
    d. Supplier of carbon fiber: Goodfellow Cambridge Ltd., Cambridge Science Park, Cambridge CB4 4DJ, UK. Fax: +44 (0)1223 420 639.
    e. Suppliers of carbon fiber microelectrodes:
        i. A-M Systems Inc., 11627-A Airport Road, Everett, WA 98204, USA. Fax: (206) 353-1123.
        ii. Diamond General Corp., 3965 Research Park Drive, Ann Arbor, MI 48108, USA. Fax: (313) 747-9703.
        iii. Radiometer Analytical, 72 Rue D'Alsace, 69627 Villeurbanne, France. Fax: +33 78 68 88 12.
        iv. MBP Electrodes, Department of Physiology, Queen Mary College, Mile End Road, London E1 4NS, UK. Fax: +44 (0)181 983-0467.

f. Oscilloscopes: Nicolet Test Instruments Division, 5225 Verona Road, Madison, WI 53711, USA. Phone: (800) 387-3385.
g. Jeweller's forceps and dissecting scissors: FST Inc., 202-277 Mountain Highway, North Vancouver, BC V7J 3P2, Canada. Fax: (604) 987-3299.
h. Electrical wire, conducting paint, hot air blowers and other sundries: RS Components Ltd., P.O. Box 99 Corby, Northants NN17 9RS, UK. Fax: +44 (0)1536 405678.

References

1. Kissinger, P. T., Hart, J. B., and Adams, R. N. (1973) Voltammetry in brain tissue—a new neurophysiological measurement. *Brain Res.* **55,** 209–213.
2. Gonon, F., Cespuglio, R., Ponchon, J.-L., Buda, M., Jouvet, M., Adams, R. N., and Pujol, J.-F. (1978) Mesure électrochimiquecontinue de la libération de dopamine réalisée dans le néostriatum du rat. *Compte Rendus* **286,** 1203–1206.
3. Armstrong-James, M. and Millar, J. (1979) Carbon fiber microelectrodes. *J. Neurosci. Meth.* **1,** 279–287.
4. Armstrong-James, M., Millar, J., and Kruk, Z. L. (1980) Quantification of noradrenaline iontophoresis. *Nature* **288,** 181–183.
5. Millar, J., Stamford, J. A., Kruk, Z. L., and Wightman, R. M. (1985) Electrochemical, pharmacological, and electrophysiological evidence of rapid dopamine release and removal in the rat caudate nucleus following electrical stimulation of the median forebrain bundle. *Eur. J. Pharm.* **109,** 341–348.
6. Stamford, J. A. (1990) Fast cyclic voltammetry-measuring transmitter release in real-time. *J. Neurosci. Meth.* **34,** 67–72.

Index

The Aloha Quilt

A Quilter's Holiday

The Lost Quilter

The Quilter's Kitchen

The Winding Ways Quilt

The New Year's Quilt

The Quilter's Homecoming

Circle of Quilters

The Christmas Quilt

The Sugar Camp Quilt

The Master Quilter

The Quilter's Legacy

The Runaway Quilt

The Cross-Country Quilters

Round Robin

The Quilter's Apprentice

Canary Girls

A Novel

JENNIFER CHIAVERINI

wm
WILLIAM MORROW
An Imprint of HarperCollins*Publishers*

This is a work of fiction. Names, characters, places, and incidents are products of the author's imagination or are used fictitiously and are not to be construed as real. Any resemblance to actual events, locales, organizations, or persons, living or dead, is entirely coincidental.

HarperCollins books may be purchased for educational, business, or sales promotional use. For information, please email the Special Markets Department at SPsales@harpercollins.com.

FIRST EDITION

Designed by Kyle O'Brien
Title page art © Daniela Iga/Shutterstock

Library of Congress Cataloging-in-Publication Data

Names: Chiaverini, Jennifer, author.
Title: Canary girls : a novel / Jennifer Chiaverini.
Description: First edition. | New York : William Morrow, 2023. | Includes bibliographical references.
Identifiers: LCCN 2022027046 | ISBN 9780063080744 (hardcover) | ISBN 9780063080768 (ebook)
Subjects: LCSH: World War, 1914–1918—Great Britain—Fiction. | World War, 1914–1918—Women—Fiction. | Women soccer players—Fiction. | LCGFT: Historical fiction. | Novels.
Classification: LCC PS3553.H473 C36 2023 | DDC 813/.54—dc23/eng/20220609
LC record available at https://lccn.loc.gov/2022027046

ISBN 978-0-06-308074-4

23 24 25 26 27 LBC 5 4 3 2 1

TO MARTY, NICK, AND MICHAEL,

WITH ALL MY LOVE

Canary Girls

1

AUGUST-DECEMBER 1914

LUCY

Lucy rested the heavy sack of vegetables and paper-wrapped meat on her hip, reluctant to set it down at her feet beside her suitcase despite the ache in her arms. She didn't really fear that some hungry villain would dart across the train platform and snatch away her hard-won provisions the moment she relaxed her guard, but with two sons and a footballer husband to feed, she dared not take that chance. London had been in a state of anxious turmoil when she and the boys had departed for Surrey four days ago, and, glancing about Paddington station as they waited to change trains, Lucy could not tell whether things had settled in their absence or had grown more desperate.

Less than a fortnight before, Germany had ignored the British government's midnight ultimatum to withdraw its troops from Belgium, plunging Great Britain and Germany into a state of war. Excited, boisterous crowds had filled the streets of London, shouting and cheering and waving their hats in the air, and at Trafalgar Square, two rival demonstrations had broken out on either side of Nelson's Pillar, one for the war and one against. A vast throng had assembled outside Buckingham Palace, singing "God Save the King" with tremendous sincerity and fervor. The song had dissolved into roars of approval when King George, wearing the uniform of an Admiral of the Fleet, had appeared on the balcony overlooking the forecourt, where the Queen, the Prince of Wales, and Princess Mary had soon joined him.

Lucy and her husband had been at their tan brick, ivy-covered home in Clapham Common at the time, but they had heard frequent shouts of "War! War!" outside their windows, and learned the rest from the papers and neighbors who had ventured out. Lucy had hardly known what to think. In school, she had been taught that another major war in Europe would be highly improbable in the future because modern weapons were so dreadful that they served as a deterrent rather than a threat. But highly improbable was not impossible, and now Great Britain was at war.

The next morning, it seemed that nearly every housewife in England had been seized by an irresistible impulse to fill her cupboards for an imminent siege. Unaware of the rising panic, Lucy had gone out to do her marketing as usual and had been startled to discover other women bustling about, empty baskets dangling from their elbows, strain evident in their pinched mouths and furrowed brows. Long queues had formed at the doors of several of her favorite shops, and hand-lettered signs had appeared in some front windows announcing that the store had sold out and closed early. After waiting in a queue for two hours and finally gaining admittance only to discover little more than a few bits and bobs left over, Lucy made her meager purchases and set off for home, uneasy. She could manage tea and supper that day, but what about tomorrow and the day after that?

"I hear delivery vans are being ransacked on the streets," Lucy's neighbor Gloria told her later that afternoon as their children played together in the private, enclosed garden all the homes on their block shared. "My sister in Bayswater says her neighbor was on her way home from the shops, arms full up with parcels, when three women, complete strangers, accosted her and accused her of hoarding."

"Hoarding?"

"Exactly! How can it be hoarding if she can carry it all in her own two arms? Anyway, these so-called ladies, bold as brass, helped themselves to my sister's neighbor's groceries. The shops may be empty, they declared, but their families wouldn't starve for others' greed. Then they scurried off and left her to carry the scraps home. Didn't give her a penny for what they took, either."

"That's robbery," said Lucy, aghast. "Where were the police?"

Gloria shrugged. "Guarding the shops or keeping watch for German spies, I should think."

Lucy could only shake her head, speechless. Such madness in their own city. And it was only the first full day of the war.

Unsettled, she returned her gaze to her sons, who laughed and shouted as they passed a football back and forth with the other children. Jamie, the eldest, was slim, black-haired, and fair-skinned like herself, while Simon was ruddy and sturdily built, his square jaw and thick, sandy hair so like his father's. How guiltily grateful she felt knowing that at eight and six years of age, her sons were too young to go to war. At thirty-two, Daniel, though strong and fit and brave, was just old enough that he would not be expected to volunteer. Though her husband loved King and Country as much as any Englishman, Lucy trusted that he would not be tempted to enlist. How could he abandon his thriving architecture firm with his career on the rise? And how could he leave Tottenham Hotspur without their star center forward, so close to the start of the season and with his inevitable retirement drawing ever nearer? Only a few days before, as Lucy had massaged the aches from his hamstrings after a grueling practice, he had confided that he knew his best days on the pitch were behind him. As beloved as he was by fans and teammates alike, eventually he would be replaced by a younger, stronger, swifter man.

But only on the pitch. No one could ever take Daniel's place in their family or her heart.

Lucy had known Daniel all her life, or at least, she could not remember a time before knowing him. In Brookfield, the village in rural southwest Surrey where they both had been born and raised, Daniel had been Lucy's elder brother's friend first. Both boys were three years older than herself, three years wiser and infinitely bolder, although

Daniel had always been kinder and more patient than Edwin. It was Daniel who had taught Lucy how to swim in the shallows of the deep, rushing brook that had given their village its name, while Edwin had splashed and shouted with the other boys, ignoring her or perhaps having truly forgotten she was there. Daniel never teased her the way Edwin did, mocking her shyness, jeering when she blushed, tugging her braid sharply whenever she forgot to keep a respectful distance. Daniel was her defender. After one mild rebuke from him, Edwin would roll his eyes and let her be.

Eventually, as the years went by, Edwin outgrew his bullying ways—a fortunate thing indeed, since after university he would become headmaster of the village school. Meanwhile, the other girls discovered what Lucy had known all along: Daniel was the handsomest, cleverest, and most wonderful boy in Brookfield, perhaps in all of Surrey. What silent torment Lucy suffered when her friends sighed over his warm brown eyes and swooned at his smile! She had no choice but to feign indifference rather than reveal how much and how hopelessly she adored him. In this pretense her customary shyness served her well, and her friends—and Daniel himself—had been none the wiser.

In those days, she could not have imagined that Daniel might one day return her affection. He had always been kind to her, but he was kind to everyone. Sometimes he walked Lucy home from school, carrying her books, but he would have done the same for anyone younger and smaller, and her house was on his way home. When Daniel danced with her on festival days, that, obviously, was because she was an excellent dancer; even boys who equated shyness with dullness enjoyed dancing with Lucy. But none of those other partners lingered very long after the music faded. Invariably, their attention would drift from her bashful, nearly inaudible conversation to light upon another girl, one who could charm and flirt with aplomb. Sometimes, though, when Daniel held her gaze and smiled, Lucy could almost believe she was as charming as those bright-eyed, laughing girls. At a dance or after church or wherever their paths happened to cross, Daniel's face would light up when he spotted her, and they would fall effortlessly into conversation like the nearly lifelong friends they were, sharing observations and confidences and private jokes, with no need to impress each other or anyone who might be watching.

One day in her sixteenth autumn, Lucy was posting a letter to Daniel at university when she was struck by the dismaying notion that perhaps the fondness between them existed only because Daniel thought of her as a younger sister—nothing less, but nothing more. How lamentable if it were true! Even as a child, Lucy had never thought of Daniel as a brother. She already had two: Edwin, of course, and George, the eldest, who was studying medicine in Edinburgh, preparing to join their father's practice.

Lucy loved her brothers dearly, but two sufficed.

Many months later, near the end of the spring term, Lucy was walking in the village with friends when the conversation turned to the absent young men who, they hoped, would soon return to Brookfield for the summer. Naturally Daniel's name came up, and as several of her companions sighed longingly and others began scheming how best to catch his eye, Lucy felt warmth rising in her cheeks. Her friends were all such pretty, laughing, good-hearted girls. Surely one of them would win Daniel's heart before the autumn—

"Oh, would you stop, all of you?" exclaimed Nettie, Lucy's best friend. "Can't you see you're making Lucy absolutely miserable?"

"Nettie, no," Lucy murmured, but it was too late. All eyes were on her, some sympathetic and understanding, others astonished.

"Lucy and Daniel?" said Kathleen in wonder, pitch rising with each word. "Are you in love?"

"Of course not," Lucy replied, shaking her head, forcing a laugh. "Don't be silly."

"You've adored him for ages," countered Nettie, amused. "Deny it if you like. Go on, tell me I'm a liar, if I am."

"A person can be *wrong*," Lucy said carefully, wishing, as ever, that she did not so easily blush, "without being a liar."

Nettie laughed. "Well, I'm neither, at least not now and not about this."

"The rest of us never stood a chance," another girl lamented, smiling. "Lucy has always been Daniel's favorite."

"No, I haven't," said Lucy, startled. "Have I? What do you mean?"

A ripple of laughter rose from the circle of friends. "Oh, Lucy," said Nettie fondly. "How could you not know?"

"Lucy and Daniel," mused Kathleen, nodding. "Well, of course. It's obvious once someone says it aloud."

Lucy begged them not to make a habit of *that*, for Brookfield was a rather small village and gossip spread swiftly. Oh, but how wonderful the phrase sounded, even when her friends teased: Lucy and Daniel.

Whether her friends ignored her pleas and echoed the phrase until it eventually made its way around to Daniel, or whether he came up with the idea entirely on his own, by late summer, he and Lucy had shared their first kiss. Three years later, they were married. Lucy followed Daniel to London, where he was playing for Chelsea and working as a junior architect. In the nine years that had passed since then, he had become a partner in his own architecture firm, the captain of Tottenham Hotspur, an Olympic champion who led England to two gold medals, and a devoted father to the two most wonderful boys in England.

And now that war had come, Lucy would not give up her beloved husband, not even for King and Country.

When Daniel returned home from the office later that afternoon, Lucy raced to the door and flung her arms around him, pressing her cheek against his chest, overcome by a wave of terrible gratitude that she would be spared the grief and loss that would inevitably afflict so many other wives.

The fierceness of her embrace surprised her husband. "What's all this?" Daniel asked, kissing the top of her head. "I'm only a few minutes late. Surely dinner isn't spoiled."

"You know that's not why I'm upset," said Lucy, her voice muffled against his lapel. "Still, it's kind of you to pretend to misunderstand, so *I* may pretend I'm not a coward."

"You're no coward," said Daniel, cupping her chin and lifting her face toward his. "England is at war with a formidable enemy. Concern and, yes, even fear, are perfectly reasonable responses."

He might feel differently when he learned how their fellow Londoners' fear and concern had affected his dinner. Collecting herself, Lucy took his hand and led him to the dining room. "Prepare to be underwhelmed," she warned him as Jamie and Simon darted in, hands and faces freshly scrubbed. The family settled around the table and Lucy served thin slices of pork in onion gravy, with roasted turnips and bread and butter. At first all Jamie and Simon wanted to talk about was the war—the recruiting posters they had seen plastered up and down the street, the older lads in the neighborhood who had rushed off to enlist, and the tragic unfairness that boys their age weren't allowed.

Lucy knew most of the lads her sons had mentioned, if only through her acquaintance with their mothers. "Seems rather rash to enlist so soon," she said, keeping her voice even as she spooned the last of the turnips onto the boys' plates and the pork onto Daniel's. "It's early days yet. One could not possibly know what one is getting oneself into."

"Billy Warren says it'll be over by Christmas," said Jamie. "He says he's got to win valor on the battlefields of France before then or it'll be too late."

"If he don't go now, he'll miss the war," Simon chimed in.

"If he *doesn't* go now," Lucy corrected. "Which sounds like a splendid idea. If it'll be over so quickly, it hardly seems worth the trouble."

"But Mum," said Jamie, brow furrowing, "'Your King and Country Need You.'"

"Not *her*," said Simon. "She's a mum."

"I'm only saying what's on the posters."

Noting Lucy's deepening frown, Daniel quickly changed the subject. Soon the boys were caught up in an eager discussion of football, the team's pre-season training, and the traditional cricket match between Tottenham Hotspur and Chelsea coming up on Saturday. When Lucy thanked Daniel with a smile, his warm brown eyes shone with affection.

Only after the boys were asleep and they were preparing for bed themselves did Lucy tell Daniel about her misfortune at the market and the frightening robbery of Gloria's sister's neighbor. "I'm terribly afraid I'll find nothing but bare shelves when I assail the shops again tomorrow," she said, trying to make a joke of it. "I know everyone says the war will be over by Christmas, but we can't go hungry until then."

"Not everyone says it'll be over by Christmas," Daniel warned, deftly removing one of her hairpins, and then another, until her dark locks tumbled free from their upswept coil. Facing her, he ran his hands through her hair, spreading it like a silken cloak upon her shoulders.

His touch sent a delicious, warm frisson of pleasure down Lucy's spine, but, with an effort, she kept her focus on the immediate crisis. "Whether it's over by Christmas or Easter or Sunday next, between now and then, we have to eat. What am I to do?"

"What indeed?" Daniel took her hand, sat down on the edge of the bed, and pulled her onto his lap. "There's always Brookfield."

She rested her hands lightly on his shoulders. "Go all the way to Surrey for groceries?"

"Why not? Our families would be delighted to see Jamie and Simon, and what could be better for the boys than a week in the country? As for groceries, I can't imagine they had a run on the shops in Brookfield."

"No, I suppose not." Surely their sensible friends and former neighbors, reassured by the abundance of local farms and their own kitchen gardens, would have stubbornly refused to panic despite the chilling declaration of war.

"You might not even need to visit the shops," Daniel continued, smiling as he kissed her cheek. "My mother would gladly fill your basket with as much veg and cheese and sausages from her own cellar as you could carry."

"I'd need a second basket for my mother's jams," said Lucy, warming to the idea. "And I could stop by Brandt's bakery for a loaf of rye and some of your favorite raisin buns."

"You're an angel," Daniel declared, kissing her. Then his smile faded. "I only hope . . ."

"What?"

"Well, Dieter Brandt—he's German, you know. I hope no one gives him any trouble on that account."

Lucy was so surprised she laughed. "Why should they? He and Mrs. Brandt have lived in Brookfield for ages, since long before you and I were born. Their own children were born there, too, and their grandchildren. Surely by now the Brandts are as English as we are."

"I agree with you, darling, but not everyone will. With war fever sweeping the country . . ." Daniel shook his head.

When he left the thought unfinished, Lucy ventured, "Shall we go tomorrow, then?"

"You and the boys shall." Daniel touched her gently on the nose. "Sadly, I can't get away."

"Of course." Lucy nodded briskly to conceal her disappointment. "The cricket."

"And the Henderson offices," he reminded her. "I'm meeting with the foreman and the owner on Tuesday. If all goes well, we'll break ground next week."

"Very well, then." Lucy sat up straight and squared her shoulders, as best she could, seated on his lap. "The boys and I shall undertake this mission ourselves."

"That's my girl." Daniel cupped her chin in his hand and kissed her.

The next morning after Daniel left for the office, Lucy packed a bag for herself and another for the boys, locked up the house, and set out for the train station. The boys trailed along at her heels, fairly bounding with excitement as they planned their week's adventures. By early afternoon they had arrived in Guildford, where Daniel's father met them and carried them the last few miles to Brookfield in the horse-drawn wagon he used for his carpentry business.

Lucy's mother met them at the front gate of the Evans family home, a two-story brick-and-timber Edwardian residence just off the town square. The medical practice—once their father's, then his and George's both, and now George's alone—took up the front half of the ground floor, with the kitchen in the back. Mum shared the spacious living quarters above with George, his wife, Eleanor, and their two daughters.

It was a lovely homecoming, and the days flew past in a whirlwind of outings in the countryside with the boys and joyful reunions with friends and family, occasionally interrupted by grave discussions of the country's preparations for war. Recruiting posters sprung up like crocuses in the spring on signposts and fences throughout the village, and every day brought news of more local men who had enlisted.

One night at dinner, George confessed that he had considered joining the Royal Navy as a medical officer, but had reluctantly decided against it. "Even if the war *is* over by Christmas, four months is simply too long to leave Brookfield without its doctor. Unless, of course, I found a substitute—"

"We discussed this," protested Eleanor. "We agreed you should not go. If this hypothetical substitute has no practice of his own, let *him* enlist and go to France."

"I never expected to go to France, dear," said George mildly, peering at her over his glasses as he took her hand. "I supposed I might serve in a hospital for wounded soldiers here in England."

"We're going to need more hospitals than what we've got," said Edwin, frowning as he ran a hand through his hair, as dark as Lucy's, without a thread of the silver-gray that had begun to color their eldest brother's head and beard. "I hope the government has a scheme to build more. Say, Lucy, that sounds like ideal war work for your Dan. Building hospitals. He'd be a very busy fellow, I should think."

Lucy sipped water to clear the lump forming in her throat. "If Daniel's only alternative is the infantry," she said, setting down her glass, "I'll be sure to suggest building hospitals, but I should prefer he not enlist at all."

"Goodness," her mother exclaimed. "Daniel in the infantry. George in the navy. What nonsense! You three are married men, and you're not as young as you once were."

"No need to be unkind, Mother," said Edwin, feigning injury. "I'm only thirty-two."

"Precisely. Lord Kitchener called for one hundred thousand volunteers aged nineteen to thirty. That means young, vigorous lads, not men approaching middle age."

Edwin's wife laughed. "Edwin, George, and Daniel—especially Daniel—aren't as decrepit as you suggest. They're in the prime of life."

"Quite so, Alice," said George. "In my particular case, I daresay my experience as a physician—"

"No, dear," said Eleanor firmly. "You'll find another way to serve, one that doesn't require you to abandon your responsibilities here."

George offered Edwin a helpless shrug. "What can I say? She loves me too much to let me go."

"Yes, I see." Edwin turned to Alice. "Darling, *you* aren't trying very hard to persuade *me* not to enlist."

"Why should I?" said Alice. "Someone has to go. Someone has to defend poor Belgium."

"But why me? Aside from my belief that Britain is honor-bound to stand by her allies, and the fact that I'm loath to ask another man to fight in my place while I cower fearfully at home."

"Because you love your country, of course, and you want to defend England from a German invasion. As a teacher you're a natural leader of men, and—" Alice smiled dreamily. "You would look absolutely splendid in an officer's uniform."

As the others laughed, Lucy managed a smile. She knew that Alice teased Edwin only because he surely had no intention of enlisting. As for George, it was all too true that the army would soon need skilled physicians most desperately. And despite Edwin's remarks, she could not imagine the army recruiting droves of architects, and how thankful indeed she was for that.

Glancing to the far end of the table, she was relieved to find her sons conversing in hushed, eager tones with their cousins, cheerfully ignoring their elders. Good. There was no reason for them to fear, even for a moment, that their father or uncles might suddenly march off to war.

On the last day of their visit, Lucy packed up the provisions her mother and mother-in-law had shared from their own pantries and a few other staples purchased at the shops. In the afternoon she stopped by Brandt's bakery for the treats she had promised Daniel, and was taken aback by subtle changes to the storefront. A large Union Jack flew above the entrance, and in the display window, a large sign announcing "Good British Breads & Buns" had replaced the trays of pumpernickel, *Brötchen*, and *Stollen*, and a platter of *Hörnchen* was labeled "Croissants." Inside, Lucy found one of the Brandt daughters arranging loaves, rolls, and sweets in one of the glass cases while another swept the floor, both of them wearing identical lace-trimmed white aprons over their muslin summer shirtwaists and long skirts. The young women glanced up from their work and cheerfully greeted Lucy by name, although their smiles faded slightly when the only other customer left without purchasing anything.

At that moment, Mr. Brandt emerged from the kitchen carrying a tray of fragrant, perfectly browned loaves of rye bread in his thickly muscled arms, his face flushed from the heat of the oven, his thinning blond curls damp with perspiration beneath his white toque. "Ah! Miss Evans," he exclaimed in his rich, accented baritone, setting the tray down on the counter. "Welcome, welcome."

"It's 'Mrs. Dempsey,' Papa," chided the elder daughter, throwing Lucy an apologetic smile.

"Of course, of course." Mr. Brandt chuckled and waved a hand as if to shoo away his mistake. "What would you like today, my dear? This good Suffolk rye bread is fresh from the oven. I know it's one of Mr. Dempsey's favorites."

"And mine too," said Lucy, bemused. She had never heard the baker refer to the bread as Suffolk rye before, and as far as she knew, rye from Suffolk wasn't inherently superior to grain grown and milled elsewhere. Why mention it at all, except to distinguish it from German rye?

She selected a loaf of rye and a half dozen raisin buns, resisting the temptation to ask what had become of the German baked goods that had always been so popular there. As Mr. Brandt boxed up her selections, he inquired after Daniel and the boys, and she asked about his family in turn. "Our Herman and Otto have enlisted in the East Surrey Regiment," Mr. Brandt said, his smile both proud and strained. "They were among the first boys from Brookfield to volunteer."

"How very brave of them," said Lucy, hiding her dismay. Of course the Brandt boys would hasten to prove their loyalty, for their family's sake as much as their own. Only a few days before, the government had announced that all "enemy aliens," including longtime residents like Mr. and Mrs. Brandt, were required to register at their nearest police station. Alice thought the new measure was a necessary precaution to foil German spies, while Edwin predicted that it would burden good, law-abiding immigrants with suspicion and restrictions. "You and Mrs. Brandt must be very proud."

"*Ja*, we must be, but we miss our sons very much already." Mr. Brandt mopped his brow with a handkerchief and shrugged. "But everyone says the war will be over by Christmas, so they will not be away such a very long time."

"Surely not," Lucy agreed, wishing she believed it.

A day later, Lucy and her sons were back in London, the tasty bread and buns neatly tucked into Jamie's knapsack on top of several jars of jam and a peck of apples the boys had harvested that morning from their grandmother's orchard. Simon carried a somewhat lighter load in his pack, and the remaining supplies filled Lucy's bags. Along with their luggage, it was quite a lot to carry through Paddington station, but Daniel would meet them at their last stop and help them the rest of the way home.

As they made their way through the concourse, Lucy was surprised to see so many men in uniform already, less surprised by the recruiting posters that had multiplied on walls and posts since she had last passed through. They reached their departure platform with time to spare, and Lucy reminded the boys to keep close to her and stay away from the edge while they waited. Keeping one eye on the clock and another on her sons, Lucy allowed her thoughts to wander, from parting conversations with her mother and brothers, to her eager longing to reunite with Daniel at home, the Brandts' plight, and what she might make for dinner that evening.

"Mum, look!" Simon shouted. "That's not allowed!"

Startled from her reverie, Lucy turned and glimpsed Simon's wide eyes and outstretched arm and Jamie beside him gazing in disbelief. Once assured that her sons were safe and sound, she followed the line of Simon's pointing finger and discovered nearly four dozen men descending the platform to the rails below. Lucy's first thought as she watched them pick their way across both sets of tracks was that they were railroad workers, but they wore no uniforms, just ordinary suits and hats. Upon reaching the other side, the men climbed onto the opposite platform, which had been virtually empty until then, a curious thing for that time of day. There sat a special train bound for Falmouth, according to the sign, but the blinds were drawn and the men made no attempt to board. Instead they assembled on the platform near the parlor car, removed their hats, and began to sing, deeply and with intense passion: "*Deutschland, Deutschland über alles, über alles in der Welt.*"

Stunned, for a moment Lucy could only stare in disbelief as the German national anthem echoed off the stone walls.

The British passengers exchanged looks of astonishment and wild indignation. "What infernal cheek!" a gray-haired woman to Lucy's right scolded, wagging a finger at the Germans.

"Well, I'm blowed," a young man in a straw hat exclaimed, hands clenching into fists as he approached the edge of the platform. "'Germany over All' sung in London—and with a war on!"

Somewhere behind Lucy, a rich baritone voice sang, "God save our gracious King, long live our noble King, God save the King!"

All around, other voices took up the song. "Send him victorious, happy and glorious—"

At the sound of her sons' treble voices, Lucy also quickly joined in. "Long to reign over us, God save the King!"

The two groups sang and shouted their anthems across the tracks at each other, louder and more fiercely with each verse. Then, as suddenly as they appeared, the German men replaced their hats and swiftly dispersed, some disappearing down one corridor, others up one staircase or another. None dared cross the tracks to confront the Britons glaring at them—but perhaps that was because a shrill whistle had announced the arrival of the train.

Lucy's hands trembled as she beckoned to her boys and ushered them aboard. "I can't believe it," Jamie fumed as they settled into their seats. "How dare they sing their national song here! We're at war, and they're in our country!"

"Very disrespectful of them, and imprudent as well," said Lucy, stowing their bags, glancing about to make sure they had all of their belongings. "They could have been set upon by an angry mob of patriotic Englishmen."

Jamie's face lit up. "Do you really think so?"

"No, of course not," Lucy quickly amended. "Not in the middle of Paddington station."

Simon's brow furrowed in serious contemplation. "Mum, do you s'pose they were German spies?"

When Jamie guffawed, Lucy silenced him with a look. "Germans, yes. Spies, no," she said. "Spies would have to be terribly bad at their jobs to reveal themselves so dramatically, don't you think?"

Simon nodded thoughtfully. Lucy smiled and straightened his cap, but the scene had unsettled her. She supposed she understood why a German abroad in wartime might be moved to burst spontaneously into his national anthem, but this incident must have been coordinated ahead of time. Why enemy aliens would so imprudently declare their loyalties to Germany was one question. Another puzzle was why they had chosen Paddington station for their defiant serenade. Why not Waterloo station? Why not Trafalgar Square?

She pondered the questions as the train carried them on the last leg of their journey, but she was no wiser by the time they reached Clapham Junction. Lucy and her boys quickly disembarked and reunited at last with Daniel on the platform. As they walked home, the boys described the strange event, talking over each other in their eagerness to astound their father.

"I believe I know what that was all about," said Daniel as he unlocked their front door and the boys raced inside ahead of them. "The Austrian and German ambassadors were expelled from Great Britain. Count Mensdorff was expected to leave London by train today for Portsmouth, and then sail to Genoa."

Lucy and Daniel followed the boys inside, carried the bags into the kitchen, and set them on the table. "I'm surprised there was no military guard to keep anyone from getting near his train." Shuddering, Lucy removed her hat and hung it on the back of a chair. "It was such a strange thing to witness. It made the war feel so close. I dread to think how much closer it will come before it's all over."

Daniel embraced her from behind and kissed the back of her neck. "Don't worry about that now, darling," he murmured, his voice low and comforting. "We're all together again, safe and sound."

Sighing softly, she leaned back against his strong chest and relaxed into his arms. She wanted him to promise her that the war would never cross their threshold, but Daniel never made promises he could not keep, nor would she find any comfort in empty reassurances.

In the days that followed, Lucy was relieved to discover that order had been restored to London's markets and food was reassuringly plentiful. Jamie and Simon prepared for the new school term, or rather, Lucy prepared them for it. After a worrisome few days during which Daniel's client considered postponing construction until after the war, the crew broke ground for the Henderson offices. To Lucy's relief, Daniel completed pre-season training with no serious injuries—more aches and pains and fatigue than he cared to admit, but nothing that would keep him off the pitch.

Yet ordinary life unfolded amid a rising war fever and its inescapable consequences. Several football clubs found their training regimens disrupted when their pitches were commandeered by the War Office for military training. Territorial battalions marched and drilled on Everton's pitch at Goodison Park. Manchester City's grounds were turned into a stable for more than three hundred cavalry horses.

"It could be worse," said Daniel, as Lucy rubbed liniment on his lower back after a particularly grueling practice. "Politicians and others are questioning whether it's appropriate to carry on with football at all in these circumstances. Some teams have had so many players enlist that they've scratched their fixtures for the season."

Lucy felt her breath catch in her throat. "Not Tottenham, surely."

"No, not Tottenham," he replied, frowning slightly. "Not yet."

But evidently not enough men were enlisting, for throughout London, recruiting posters seemed to increase daily in number and variety. Particularly disquieting for Lucy were the specific appeals for sportsmen to enlist, and for women to urge their sons and husbands to answer the call to arms. The press enthusiastically described training exercises and dignitaries' reviews of the troops at military encampments, often focusing on certain Pals battalions—groups of men from particular towns, schools, or professions who had enlisted together with the understanding, backed by the promise of Secretary of State for War Lord Kitchener, that they would serve together. By the end of September, Kitchener's scheme had produced the Stockbrokers' Battalion from the City of London, the Grimsby Chums comprised of former students of Wintringham Secondary School, and more than fifty battalions from towns and cities throughout Britain.

To Lucy that seemed like a great many soldiers indeed, but the call to arms persisted, with heightened urgency. Rumors that conscription might be on the way compelled many young men to enlist while they could still choose a favorite Pals battalion. The women of Britain found their own way to contribute to the war effort when Queen Mary and Lord Kitchener called upon them to knit three hundred thousand pairs of socks for the army by November. Dutifully, Lucy purchased skeins of yarn and took up her needles. Working a few hours in the evenings and in spare moments throughout the day, she could produce a new pair of socks every second day. She and her friend Gloria occasionally knitted together in the garden while their children played, quietly discussing the rumors that not all was going well for the British Expeditionary Force in France. Sometimes rumors were all they had. Thanks to overzealous censors, when it came to reports from the

Western Front, the newspapers were often too vague or understated to merit reading.

What the press did make clear was that it would take more than warm, dry socks to sustain the troops and the dependents they had left behind. On Saturday, 22 August, several professional football clubs organized a series of matches to benefit the Prince of Wales's National Relief Fund. Tottenham Hotspur would play Arsenal at White Hart Lane, and Lucy had promised to take Jamie and Simon to watch their father play. She hoped the charitable effort would quiet the grumbling from certain quarters about athletes playing games on manicured pitches while braver, more patriotic men bled and died on the muddy battlefields of France and Belgium. But her hopes plummeted when she and the boys arrived for Daniel's match only to discover a small crowd of mostly older gentlemen demonstrating near the entrance, brandishing signs demanding that football be suspended for the duration. Jamie and Simon stared in astonishment as the gentlemen called out to fans as they approached the turnstile, urging them to go home rather than share in the teams' disgrace.

"That's Daniel Dempsey's wife," one of the men suddenly shouted. Lucy jumped as many pairs of eyes turned her way. "Say there, Mrs. Dempsey! Aren't you ashamed of your husband, running around on a pitch while our country is engaged in a life-and-death struggle? Wouldn't you rather see him a hero in uniform?"

Mortified, Lucy lifted her chin, seized her sons' hands, and marched past the protestors without acknowledging them, breathing a sigh of relief only after she and the boys had passed through the gate. Inside, a large and enthusiastic crowd had assembled, mostly working-class men who evidently did not object to sport in wartime. And why should they? It seemed to Lucy that football provided a welcome distraction from the war, an essential release of tension that sustained morale.

"Why do those men hate football?" Simon asked as they made their way to their seats.

"I don't believe they *hate* football," Lucy replied, "but they see it as . . . an extravagance during wartime."

Those men were not alone, she knew, and their numbers were increasing.

Not even the extraordinary sums footballers raised for the National Relief Fund and other charities could shift the tide of public opinion in the teams' favor. One by one, prominent gentlemen from the military, government, and business publicly condemned the playing of games during a national crisis. Lucy dreaded to see Daniel brooding over their denunciations. "This is the worst controversy the sport has faced since the conflict over paying players first arose," he said, studying a particularly irksome pamphlet.

"Surely it's not as bad as that," protested Lucy. Thirty years before, a few disgruntled teams had accused certain rivals of paying their players, giving them an unfair advantage. Although paying players was not prohibited, opponents contended that it would harm the sport, forsaking the ideals of sportsmanship and the true spirit of the game in favor of the false idols of financial incentive and profit. Outrage and hostility had fomented through the years until the Football Association had finally split into two factions, professional and amateur. Although in recent years the sides had warily reconciled, some amateur abhorrence for professionalism lingered, despite three decades of evidence that paying players had not ruined the game. Nor had any footballer become wealthy on his earnings. Most professional players earned only a few pounds a week, hardly enough to support oneself or raise a family. Most players earned a living from another trade or profession, trained and played matches in their off hours, and appreciated the extra measure of comfort and security their football income provided. Only a very few footballers were paid enough to forgo other work if they wished, and Daniel was not one of those.

In September, as the fledgling controversy worsened, the Football Association issued a statement declaring that the season would go on as scheduled, but any player who wished to enlist would be released from his contract without penalty, and those who did not enlist would engage in military drill and rifle training as a team. Recruitment posters would be displayed prominently at each team's grounds, and dignitaries would be invited to address audiences during the interval to urge all players and spectators who were physically fit and otherwise qualified to join the army at once.

Yet even those efforts did not quell the grumbling. In October, Lucy and her sons encountered more protestors outside St. James' Park when Tottenham played Newcastle United, men in black wool coats and fine top hats holding signs with slogans such as "Your Country Needs You" and "Are You Forgetting There's a War On?" and "Be Ready to Defend Your Home and Women from the German Huns!" As the season progressed, Lucy and the other Tottenham footballers' wives noted White Hart Lane's dwindling crowds, as recruitment drives claimed some spectators and negative publicity drove away others.

"It's not fair that our husbands must endure such unjust criticism while men in other professions—indeed, in other sports—escape scrutiny," said Minnie Bailey, thoroughly vexed, keeping her voice low so the spectators seated nearby would not overhear. She had been Lucy's closest friend among the wives ever since they had traveled to Stockholm together to attend England's final against Denmark in the 1912 Olympic Games. Minnie's then fiancé, Horace, had been England's reserve goalkeeper; although he had not played, he still shared in all the honor and praise and adoration England's fans had bestowed upon the team.

What a difference two years made, Lucy thought ruefully. Britain's footballers, only recently a source of tremendous national pride, had become the objects of scorn and derision.

"Minnie's right," another wife chimed in. "I don't hear any lords or legislators demanding that golfers, cricketers, and polo players cease playing for the duration and trade in their team kit for a soldier's khaki."

"And yet footballers are condemned, when they've enlisted in greater numbers and raised more relief funds than any other sportsmen," said Lucy.

"It's class prejudice, that's what it is," said Elsie May, sighing. "Football is the people's game, the pastime of the working classes rather than the gentry."

Lucy and the others agreed there was nothing for it but to bear the indignity with grace and support their husbands through it. All would be forgotten after the war ended and life returned to normal.

But as the weeks passed and the controversy smoldered, two distressing truths became evident to Lucy: The Football Association was steadily losing its case in the court of public opinion, and the war was very unlikely to be over by Christmas.

In late November, Daniel told Lucy that a group of Heart of Midlothian players had enlisted together in a new battalion of the Royal Scots being raised by Sir George McCrae, a former MP for East Edinburgh. "A few days before, the *Edinburgh Evening News* had suggested that the club should be renamed 'The White Feathers of Midlothian,'" he said, grimacing.

Lucy was aghast. "They ran off to enlist because of some childish name-calling?"

"What better way to silence their critics? The Hearts' entire first and reserve team joined up, along with members of the board and the staff and nearly five hundred fans. Now players from other Scottish teams are racing to join them before McCrae's Battalion is entirely filled." Daniel shook his head and grinned. "What a terrifying sight that would be—hundreds of fierce Scotsman footballers and fans hurling themselves at the unsuspecting enemy. I've faced them on the pitch. I wouldn't want to confront them on the battlefield."

"I could almost pity the Germans," said Lucy. Then her thoughts darted to the nearly seven hundred civilians slaughtered by the Kaiser's army in the village of Dinant, the countless numbers the Germans had killed elsewhere in Belgium and in France, the young British soldiers who had already lost their lives—and she decided that the Germans deserved every moment of terror, in equal measure to all they had inflicted upon innocents.

"Heart of Midlothian are leading the Scottish League," Daniel said, as if lost in thought, "and yet they're willing to cast aside their shot at the cup in order to enlist. No dodgy reporter can fault their courage now."

Uneasy, Lucy made no reply. Daniel's admiration for the Scottish footballers filled the room with a heavy, haunting presence, and she was afraid to say something that might persuade him to join them.

In early December, Tottenham Hotspur received a letter from the Football Association inviting the club to send delegates to a meeting at the FA offices in Russell Square. "The owner and manager will attend, and I've been asked to represent our players," Daniel told Lucy.

"Sounds like quite an honor," she said, bracing herself, for a certain intensity in his expression warned her that he had more to say.

"Lucy, darling." Daniel took her hand, kissed it, and pressed it to his heart. "McCrae's Battalion may have been first onto the pitch, but we English footballers play to the whistle. The purpose of the meeting is to discuss whether to form an English Footballers' Battalion for Kitchener's Army."

Lucy's heart plummeted. Daniel was trying to break the news gently, but she knew him too well.

If a Footballers' Battalion was formed, when the whistle blew, Daniel would take the field among them.

2

JANUARY-MARCH 1915

APRIL

In her four years at Alderlea, April had never seen Lady Rylance as flustered and indignant as she was that frosty January morning, moments after her husband's startling announcement.

Not even on Boxing Day a fortnight ago had the mistress been this upset. It was meant to have been the merriest day of the festive season for the household staff, but Harrison Rylance, the eldest son and heir, had spoiled their holiday by stealing away to accept a commission with the Derbyshire Yeomanry. Her duties complete, her half day off barely begun, April was in her shared attic bedroom dressing for the traditional Housemaids v. Footmen Boxing Day football match when the mistress had shrieked so horribly that she might have witnessed a murder in the drawing room, or a maid clumsily overturning an ashtray on the best Persian rug.

A moment later, the staff had been summoned to attend to Harrison's distraught parents and bewildered younger siblings as they came to terms with his patriotic disobedience. April had served brandy and tea and biscuits in the drawing room, silently mourning the cancelled football match, which was always great fun, and the lost Boxing Day afternoon off, which was all but sacrosanct. Eventually Lord Rylance had convinced his wife that their son had done a splendid thing by pledging to serve King and Country and to defend England's honor. "I would accept a commission myself, if not for this blasted leg," Lord Rylance had grumbled, slapping his thigh above the knee, which he had injured years before in a fall from a horse. "I might still, if His Majesty commands."

"Heaven forfend," Lady Rylance had exclaimed, laying a long, slim, beringed hand over her heart. She had embraced her eldest, then held him at arm's length, scolding him with one breath and praising him with the next. How thoughtless he was to have deceived them, but how honorable and courageous he was to answer the call to arms! They were exceedingly proud to have an officer in the family, but if Harrison had only told them his intentions, his father could have arranged a more advantageous commission.

"If I had told you," Harrison had protested, laughing, "you would have tried to stop me."

"Could I have done?" his mother replied, her smile wavering. Then she had pressed a handkerchief to her mouth and fled the room.

While the scene unfolded, April had been obliged to keep her eyes down and expression placid, revealing not the slightest flicker of astonishment, even though she had never seen the mistress so upset. Not that she spent much time in Lady Rylance's company. April occasionally served at table, but most of her work—washing, cleaning, fetching, scrubbing—was done out of the family's sight, beginning before dawn while the family slept until long after they had retired for the night.

April's mum had not told her what domestic service would be like, perhaps because she did not herself know. A word of warning wouldn't have changed April's mind anyway. The Tipton family desperately needed their eldest children to earn wages, and the position at Alderlea was simply too great an opportunity to let slip by. It had only come April's way because the housekeeper, Elizabeth Wilson, was April's mum's cousin. They had been good chums when they were girls, and had kept in touch through letters after their lives had taken them in dramatically different directions. Mrs. Wilson had achieved a position of respect and authority in one of the most gracious country houses in Derbyshire, while April's mum had followed a well-trod path into marriage, motherhood, and increasingly reduced circumstances.

"This could be your chance at a better life," April's mum had reminded her as she set off for the station, adjusting April's hat, picking a stray thread off her coat sleeve. "Be pleasant, be quiet, work hard, and do everything Cousin Bitsy tells you to do, and you could make something of yourself. You might even rise to become a cook or a housekeeper yourself someday."

"I will," April had replied solemnly, giving her mum a peck on the cheek. Though she was only fifteen, she was terribly eager to go, not because the work sounded particularly exciting, but because it was either domestic service or the mills, and she didn't want the mills. The only other respectable alternative was to stay home and mind the little ones while her elder sister accepted the housemaid position, her mum took in more piecework, and her dad drank his pay away.

It was a long journey by train to Derbyshire, and after that a jostling ride in a farmer's hay wagon to reach the estate. April was weary and disheveled when she finally knocked on the back door of Alderlea, a glorious three-story Elizabethan mansion of buff sandstone and gray slate set amid vast, rolling lawns at the edge of a pretty wood. The tall, golden-haired housemaid who answered her knock looked only a few years older than herself. "Yes?" she asked, friendly enough. "What d'you need?"

"I'm April Tipton," she said. "I'm here to see Mrs. Elizabeth Wilson, please."

"Oh, you must be the new girl." The housemaid opened the door wider and beckoned her inside. "Congratulations or condolences, as the case may be."

"Erm, congratulations, I hope. I don't have the job yet. And thanks."

"Don't mention it." She tossed April a grin over her shoulder as she led her down the hallway to the housekeeper's room. "Here we are at the dragon's lair," she said in a conspiratorial whisper as they halted outside the door. "Good luck."

With that, the maid swept off, leaving April alone in the passageway amid the clatter of dishes and enticing aromas drifting from a kitchen somewhere nearby. Collecting herself, she took a deep breath, knocked on the door, and entered when commanded to do so. Inside, a woman in her middle forties sat at a desk studying a ledger and receipts through wire-rimmed glasses attached to a fine silver chain around her neck. When she glanced up inquiringly, April bobbed a curtsey. "Good day, ma'am. I'm April Tipton," she said, handing over an envelope with her references and a letter from her mum. "Hannah Tipton's daughter."

"Yes, of course." The housekeeper rose to accept the papers, and April tried not to fidget while she read them. Elizabeth Wilson had her cousin's gray-streaked chestnut hair, apple cheeks, and cleft chin, but otherwise April saw no family resemblance. She wore a navy blue wool suit, a crisp white blouse, and perfectly polished, sensible heeled boots, finer clothes than April had ever seen her mum wear. April's mum often smiled, though sometimes tiredly, and she hunched her shoulders, a habit developed from years bent over her sewing. Studying her mum's cousin while she read, April noticed that her expression was stern, her gaze coolly appraising, her posture straight and imposing. April reminded herself that her mum was very fond of the housekeeper. She must be a good sort, even if the blonde maid had called her a dragon.

The housekeeper invited April to sit, queried her about her schooling and prior work experience, and explained the rules of the household. First and foremost, April was never to call her Cousin Bitsy, not even when they were alone; it was always to be Mrs. Wilson. As the newest maid, April would rank below everyone in the house except the scullery girl, and she must show all due respect. Her workday would begin at five o'clock sharp, when she would rise, wash, put on her day uniform, and light the kitchen fire. Next she would tend to the rooms on the ground floor, cleaning the grates, dusting, polishing, sweeping the hall and the stairs, washing the front steps, and polishing the brass door fixtures. Then she would light fires in the family's dressing rooms and carry hot water to their bedchambers. After breakfast, she would clear and air the bedchambers, turn the mattresses, and make up the beds, then clean other rooms such as the library and the study, lighting fires where necessary. All this must be accomplished by luncheon, when April must be prepared to wait at table if the footmen were occupied with other duties. In the early afternoon, after changing out of her morning work clothes into her more formal uniform of black dress, white apron, and white cap, she would mend and darn, answer the door and the telephone, and announce callers. As evening approached, she would prepare the dining room for dinner and serve at table if needed; afterward, she would bring tea to the drawing room for the family and any guests, and light fires in the bedchambers. In addition to these daily responsibilities, every week April would assist with the laundry and ironing; clean the carpets, wallpaper, and windows; and wash down the paintwork.

"That sounds like rather a lot," April managed to say.

"You'll have Sunday afternoons off."

"Even so," April said without thinking, quickly adding, "That is, thank you, ma'am."

Mrs. Wilson's expression softened a trifle. "You won't be responsible for everything yourself, my dear. You'll share the work with three other housemaids. You met Mary on your way in. You'll meet the others later." Mrs. Wilson rose. "Pending approval from her ladyship, of course."

April quickly stood and tried to smooth the wrinkles from her dress. "I'm to meet Lady Rylance? Now?"

"Don't panic, child. Of course now. If she doesn't like the look of you, you'll have time to catch the train back to Carlisle this evening." Mrs. Wilson studied April for a moment, then sighed. "Just keep your gaze down modestly, clasp your hands in front like so—" She demonstrated. "And don't say anything unless her ladyship addresses you. You'll be all right. Come along, now."

April swallowed hard and obeyed, wishing she had time to wash the dust of travel from her hands and face and arrange her flaxen hair, stick-straight and butter yellow, into a tidier bun. As she trailed after the housekeeper down corridors and up two flights of back stairs, she tried to tuck stray wisps under her cap, but without a mirror, she might have been making a worse mess of it.

All too soon Mrs. Wilson led her into a splendid sitting room, with a marble fireplace at one end, a gleaming piano at the other, and all sorts of lovely gilded and brocaded furnishings in between, warm wood and fabric in shades of sage green and lavender. It seemed that every level surface boasted a crystal or porcelain vase bursting with fragrant flowers. Tall windows adorned with heavy lavender velvet curtains let in sunlight and breathtaking views of the lush forest and rolling hills.

It occurred to April that this single room was larger and more abundantly furnished than her family's entire home. And Alderlea had many more such rooms. How could even four housemaids clean the entire mansion in a single day?

In the center of the room, three women dressed in elegant frocks sat chatting and sipping tea around a low table laden with a tea service and trays of tiny sandwiches and pastries. One woman perched gracefully in a lavender brocade armchair, the two others on the floral, claw-legged sofa opposite her. The woman in the armchair appeared to be in her late forties, slender and fair-skinned, with glossy black hair and an aquiline nose. Her dark eyebrows arched at the sight of Mrs. Wilson entering with April. "Yes, Wilson?" she inquired. "What is it?"

"If you please, ma'am, we've taken on a new housemaid." With a turn of her wrist, Mrs. Wilson indicated April, who bobbed a curtsey. "I'm well acquainted with her mother, and she brings two characters, one from a former schoolteacher and another from her minister. They both attest that she is a hardworking, reliable girl."

April felt the lady's eyes weighing and measuring her. "Is she an experienced maid? She looks quite young."

"She's fifteen, ma'am, the third eldest of eight. This would be her first position, but she has been helping her mother keep house and mind her siblings since she was quite young."

"Hmm." Sighing, Lady Rylance turned to her elegant companions.

"It's so difficult to find good servants these days. One has to bring in novices and train them oneself." The other ladies murmured in commiseration. Turning back to Mrs. Wilson, Lady Rylance asked, "What is she called?"

"April, ma'am. April Tipton."

"April?" Lady Rylance's nose wrinkled in distaste. "Oh, no. That will never do. April is too frivolous for a maid. We shall call her Mary."

"We already have a Mary, ma'am."

"Do we indeed?" Lady Rylance put her head to one side, thinking. "Oh, yes, the lanky girl, saucy grin. Well, let's call this one Ann. Unless we already have an Ann?"

"No, ma'am."

"Ann it is, then." April's new mistress bathed her in a radiant smile. "Welcome to Alderlea, Ann. Work hard, be punctual, and we'll get along splendidly."

"Yes, ma'am. Thank you, ma'am," April murmured, heart thudding as she curtsied again, but the mistress and her friends had resumed their conversation and took no notice. Ann the housemaid was quickly forgotten.

Mrs. Wilson inclined her head deferentially to the unwitting ladies and silently left the room. A bit stunned, April followed. "You made a good impression," Mrs. Wilson remarked quietly when they were alone in the hallway.

April murmured a reply, bewilderment giving way to annoyance. She *liked* her name. It was fresh and bright and lively like spring itself, like *April* herself. Lady Rylance might not think it suitable for a girl of her station, but no one had any right to change it.

But if April wanted to keep her new job, she supposed she would have to go along with it.

Mrs. Wilson led April back downstairs to the housekeeper's chamber and instructed her regarding her wages and room and board. Afterward, she summoned the same housemaid who had answered April's knock. "Mary will take you to the laundry and give you your uniforms," Mrs. Wilson said. "After that, she'll show you to your bed in the maids' quarters so you can settle in. You can begin work bright and early tomorrow morning."

With that, Mrs. Wilson dismissed them.

"Come on, then, Ann," said Mary, beckoning April to follow her down the passageway. "Let's get you sorted. What did you think of Lady Rylance?"

"She changed my name," said April woefully. "I'm not really Ann."

The other maid shrugged. "That's all right. I'm not really Mary. My name's Marjorie."

"Let me guess." April stuck her nose in the air, and in a plummy voice intoned, "That's too frivolous a name for a maid."

Marjorie laughed, delighted. "You have her exactly! But listen—" She halted for emphasis, and April did too. "Don't ever let Mrs. Wilson or the butler—or anyone but me, really—hear you mocking the family or you'll be sacked without a character. And without a character, you'll never get another job in service, although there are worse fates than that, surely."

"Thanks for the warning," said April. Apparently she had already found her first friend and ally at Alderlea.

"As for your new name, don't let it bother you," Marjorie said, continuing toward the laundry, which, judging by the smell of bleach and starch on the air, could not be far. "Mrs. Wilson and the butler will call you Ann, but you won't see her ladyship very often, and when you do, she likely won't address you at all."

That much, like most of the advice and warnings Marjorie had shared with her in the four years since that awkward first day, had proven true. April almost never saw Lady Rylance or her husband except brief glimpses with the length of a corridor between them. Nor was April seen. Unlike ladies' maids, who dressed their mistresses and arranged their hair and often became companions and confidantes, girls like April and Marjorie were meant to be invisible.

Yet April saw enough of her mistress in four years' time to know that she was often indifferent but never cruel. She had many friends but preferred to go out rather than entertain. She loved horses but merely tolerated dogs. She had very specific opinions regarding what was appropriate, or tasteful, or amusing, or ridiculous, but she never raised her voice in conversation when others disagreed. In fact, she almost never lost her composure, which was why Boxing Day stood out. Yet wouldn't any mother have got a bit teary to hear that her son had enlisted, when from the sound of it, the Western Front was cold, wet, muddy, dangerous, and in every other way awfully bad?

But as upset as Lady Rylance had been then, she was even more so now, staring daggers at her husband over the breakfast table, holding out one slim, pale hand for the letter from the War Office.

"You may read it, my dear, but it won't change a thing," Lord Rylance remarked as he placed the folded page in her hand. "I've already consented to the War Office's request to billet soldiers on the estate."

"How could you, without consulting me?" his wife said in a strangled voice, her cheeks flushed.

"There's a war on," he replied, calmly buttering his toast. "Needs must."

Lady Rylance's hands trembled as she read the letter. "The War Office gives no consideration whatsoever to the great inconvenience to ourselves. Oh! And this. We're expected to feed the soldiers as well."

"The government will pay us two shillings and sixpence per day for each man."

"A pittance that will scarcely cover the expense. Mrs. Wilson says food prices have been soaring! We might as well give the government's largesse to Monsieur Pierre to compensate him for the additional work this billeting business will undoubtedly require of him."

Lord Rylance swallowed a bite of toast and sipped his tea. "Splendid idea, Catherine. I daresay Monsieur Pierre will earn that bonus."

"Nothing about this is splendid," she retorted. "I don't like this at all, Reg, and I should like to know what you intend to do about it!"

Standing side by side against the wall as they awaited the cue to clear, April and Marjorie exchanged the tiniest of startled, sidelong glances. Never, truly never, had either of them heard the mistress raise her voice to her husband.

Lord Rylance regarded his wife for a moment, his thick black eyebrows nearly meeting at the bridge of his nose, his expression becoming ever more serious. "What do I intend to do about it." He dabbed at the corners of his mouth with his napkin. "What do I intend indeed."

His wife creased the letter, set it beside her teacup, folded her hands in her lap, and met his gaze with a silent challenge, lips pressed together, chin trembling.

"The police have allocated eight soldiers to Alderlea for billets," he said, his voice utterly calm. "We could easily accommodate twice as many."

"Yes, we have room. That's not the point."

"No, indeed, it is not. Wartime requires sacrifices of us all, my dear, and our sacrifice apparently will be to host eight soldiers while they train for battle, eight young men like our Harrison. Britain is honor-bound to go to war due to our alliances and treaties, and it is decidedly in our own self-interest to prevent Germany from controlling Western Europe. If, God forbid, Germany conquers Belgium and France, do you understand that the Kaiser's next move, if he continues to hold off Russia, will surely be to attempt an invasion of Great Britain?"

He paused for an answer. His wife gave him the barest of nods.

"You asked what I intend," Lord Rylance continued. "Well, I intend to welcome these eight soldiers as I hope another family would welcome our Harrison, and to make them as comfortable and cheerful as possible before they sail for France. I should be ashamed to do otherwise."

Silence descended.

"Well," the mistress eventually said, "you make me feel quite ashamed for complaining."

Her husband took her hand and kissed it. "My dear, let us hope this is the only sacrifice we are required to make for England in this dreadful, unsought war."

Lady Rylance murmured something April could not quite make out, but she knew the lord and lady were thinking of their young lieutenant.

In the days that followed, Lady Rylance resigned herself to her uninvited guests and worked up a scheme for how best to accommodate them, bringing to bear all the skills she had perfected over decades of planning shooting parties, dinners for twenty, and holiday balls. It was up to Mrs. Wilson and the butler to see that the mistress's instructions were carried out. For nearly a week April and her fellow servants prepared for the soldiers at a frantic pace—cleaning, preparing beds, planning meals and leisure activities—but despite the unexpected addition to her workload, April was looking forward to the soldiers' arrival. The novelty of their presence would add a bit of spice to days that were otherwise all very much the same.

Marjorie was even more delighted than April that Alderlea was preparing to welcome a more interesting bunch than the family's usual guests, and she fervently hoped their soldiers would be enlisted men, lads of their own class rather than officers. "Lads we can talk to, have a bit of fun with," she explained as she and April scrubbed the foyer's slate floor for the second time that day, since the younger Rylance children had tracked in mud.

Amused, April sat back on her heels and wiped her forehead with the back of her hand. "Better not let the mistress or Mrs. Wilson catch you flirting," she advised, dunking her scrub brush into the bucket of soapy water.

"I won't get caught."

"Don't bring any of them up to our room, either."

"I wouldn't," Marjorie protested, pushing a honey-gold curl out of her eyes with one hand while vigorously scrubbing the floor with the other. "The other girls would tell, and I'd find myself out of work without a character. *You* wouldn't tell, though, would you? Even though Mrs. Wilson is your aunt?"

"She's my mum's cousin," April said, wishing yet again that no one knew they were related, however distantly. She wouldn't want anyone to think she was getting special treatment, for she certainly wasn't. Mrs. Wilson always gave her the most difficult chores at the worst times of the day—except for the scullery girl, of course, who without a doubt had the hardest, grubbiest work of them all and was paid the least for it. If the Rylances instructed Mrs. Wilson to hire a new housemaid, then the new girl would be the lower in rank, April would earn some privileges, and—

But there was no point in hoping for it. From what April had overheard, it was very unlikely that any new servants would be hired at Alderlea, even though billeting the soldiers meant more work for everyone. Lord Rylance could certainly afford more staff, but as Lady Rylance and her friends often lamented, good servants were all but impossible to find these days. Footmen and grooms were quitting domestic service for the adventure and honor of Lord Kitchener's Army, while housemaids and scullery girls were hiring on at businesses and factories to replace men who had enlisted, doing work that women had never been allowed to try before. April and Marjorie marveled at the stories in the Rylances' discarded newspapers about girls not much older than themselves operating streetcars, delivering the Royal Mail, and operating machinery in factories. How interesting their work must be, every day something new and different, with better wages and possibilities for advancement a mere housemaid could only dream of!

"I've thought about stealing away to London and taking one of those jobs," Marjorie had admitted one afternoon as they cleaned the master's study. "Wouldn't it be grand to work at something they never let girls do before, and to be well paid for it?"

April had felt a catch in her throat. How could Marjorie even *say* such a thing, that she might run off and leave April behind? "Why go all the way to London where a German zeppelin could drop a bomb on you?" she asked, determined not to sound forlorn. "I'm sure they have those sort of jobs in Liverpool and Manchester too."

Marjorie had shaken her head. "That would be too close, and in the wrong direction. I have to go *away* from home, not toward it, or my folks might fetch me back. London would be too far for them to travel." She paused to laugh. "Oh, don't look so reproachful! I wouldn't actually do it, not without saying goodbye."

April had felt somewhat better. Her dearest friend would not abandon her anytime soon.

But that was days ago. At the moment, Marjorie was giving her a pointed stare. "Of course I wouldn't be a telltale," April belatedly replied. Marjorie was two years older than she, far more experienced in work and in life. Without her help and advice, April never would have settled in so quickly at Alderlea. How could April not adore and admire her friend, and consider herself lucky to have that friendship?

As Alderlea continued to prepare for their guests, Lord Rylance received word that two were officers, the rest enlisted men. Bedrooms were made up for the captain and the lieutenant as befitting their status, while cots and mattresses, all with thick blankets and soft pillows, were set up in the library for the others. The official instructions specifically noted that billets were required to provide their soldiers with candles, vinegar, salt, the use of the fire, and utensils for the dressing and eating of meat, so Mrs. Wilson checked and double-checked to make sure all that was sorted too.

On the day of the soldiers' arrival, Mrs. Wilson assigned the housemaids tasks downstairs so they would not be tempted to gawk and giggle as the men settled in—but any hopes she may have had to keep the maids and the enlisted men separated indefinitely were doomed from the start. Within days they had all crossed paths and exchanged names, and before a fortnight had passed, nearly everyone in one group had chosen a favorite among the other. April liked a dark-haired private from Pilsley best, but she rarely said more than a few words to him at a time, so determined was she to remain in Mrs. Wilson's good graces.

The soldiers spent most of the day drilling and training near Chesterfield, but they usually had breakfast and dinner at Alderlea, and tea as well if they returned in time for it. It soon became apparent that Lady Rylance and Monsieur Pierre had badly underestimated how much eight healthy young soldiers would eat. The weekly grocery delivery frequently ran short, obliging Mrs. Wilson to send servants into the village for supplies. A large order required footmen to go with a cart, but if only a few items were wanting, a maid or two could go on foot. Marjorie and April were always quick to volunteer, for the two-mile walk was easy and pretty, even in February, and running errands was infinitely preferable to scrubbing floors. Marjorie enjoyed flirting with the handsome young men who happened to cross their path, although she had become more discerning in recent months, and now smiled only at men in uniform.

As for April, she loved the different shops, all with their particular wares arranged so prettily to entice a customer to buy. She never had enough money of her own to buy any of the things she admired—a box of creamy ivory stationery, a pair of fine tan gloves, an oval brass picture frame etched with roses—but it amused her to make purchases for the estate and pretend she was buying them for herself. Her errands rarely sent her to the milliner, the dressmaker, the newsagent, or the confectioner, but she admired the window displays and sometimes pondered how she might rearrange the wares to make them more appealing. Marjorie admired the dressmaker's front window as much as April did, but on the whole, she preferred to keep her eyes open for handsome fellows.

"I adore having soldiers at Alderlea," Marjorie proclaimed one sunny, frosty February afternoon as they walked to the village, her breath emerging above her scarf in faint white puffs. "I think Bob is in love with me."

"Yesterday you were in love with Joe," remarked April, tightening her grip as her knitted mittens slipped on the handle of her basket.

"I never said I was in love. I still like Joe. Can't I like them both?"

"As long as they don't find out."

Marjorie laughed. "The lieutenant is the most handsome of them all, but I can't get more than a polite smile and a thank-you out of him when I lay his fire or bring him his post."

April shot her a look. "Have you *tried* to get more out of him? The mistress wouldn't like it."

"I'm just being friendly. It's not my fault he's so good-looking." Marjorie elbowed her, grinning. "Don't worry. I'm not going to make a fool of myself. But it's so hard not to admire our Tommies, don't you agree?"

"Of course." How could April not admire them? Officers and enlisted men alike, they all looked so dashing in their uniforms, so brave and strong and proud, each willing to sacrifice his life for King and Country in the defense of good old England. In December, April's elder brother had joined the Lonsdale Pals, the 11th Battalion of the Border Regiment, earning her respect and envy. She wished she too could do her bit, serve her country, make her mum burst with pride. She was less enthusiastic about risking her life, but fortunately, and alas, no one wanted her to. That bit belonged to the lads.

"I think no young man is so admirable as one who bravely answers his country's call to arms," Marjorie declared, stepping gracefully around a puddle of slush between ruts in the frozen mud. "And none is as despicable as a coward who shirks his duty."

"My," said April, surprised by the venom in her friend's voice. "Strong words."

"I mean every one of them. I consider myself a member of the Order of the White Feather."

"The what?"

"The Order of the White Feather. It's from cockfighting—you know, because a cockerel with a white feather in its tail always proves to be a timid fighter. The Order says that whenever a girl comes upon a lad in civvies who ought to be in khaki, she should give him a white feather as a symbol of his cowardice."

"That seems unkind," said April. "She wouldn't know why he isn't in uniform. Maybe he's not fit. Maybe he has bad eyes or a trick knee. Or maybe he's a soldier on leave."

"Or maybe—" Marjorie lowered her voice as they approached the village; in the past, certain curmudgeons had complained about her to Mrs. Wilson, that she laughed and talked too loudly in the streets, unbecoming behavior for any young woman, but especially a maid. "Maybe he's a coward or a shirker. Listen. I have two white goose feathers in my pocket, one for me and one for you. I'll give mine to the first slacker we see so you can learn how it's done. You can do the next."

"I don't want to," said April, a bit plaintively. "And I don't think you should either. If you ask me, the army's better off without lads who don't want to fight. Not everyone is up to it."

"Lord Kitchener doesn't agree." Marjorie smiled brightly at two strapping lads in uniform, who smiled back admiringly as they passed. "He says women can help in the national crisis by influencing our men to enlist. Lord Kitchener says every girl with a sweetheart should tell him that she won't walk out with him again until he has done his part."

"Told you that himself, did he?"

"I read the papers, now and again, and we've all seen the posters." Marjorie gestured up and down the street, where, sure enough, every wall and fence boasted a poster or two. "See there? 'Every able-bodied man is needed.' And right here, the one with the pretty girl: 'If he does not think that you and your country are worth fighting for, do you think he is worthy of you?'"

April wavered, thinking of her brother. Henry had enlisted eagerly, not only to serve his country, but also for the pay and the opportunity to better himself. Now he was training with his regiment at Blackwell Racecourse, while three of Marjorie's brothers, the ones old enough, had already sailed for France. They all had willingly volunteered. Who better than their sisters to urge other lads to do their bit too? The Tommies needed all the help they could get to defeat the Hun.

"You make a fair point," April admitted, just as they arrived at Mrs. Wilson's favorite grocer. "Are you going in or shall I?"

"You go," Marjorie said, her eyes fixed on a trio of handsome soldiers entering the pub across the street. "I'll keep watch for German spies."

April laughed and agreed.

She filled her basket with the things on Mrs. Wilson's list, charged the order to the estate account, and hurried back outside. Next they went down the block to the butcher's—who happened to be a German immigrant, though probably not a spy—and there, April waited outside with the groceries while Marjorie shopped. Uncomfortable standing around idle in the middle of the day, April kept out of the way of passersby, took in the scene such as it was, and tried to stay warm, pulling up her coat collar and scarf, tucking her mittened hands into her pockets, and marching in place a bit now and then to keep the blood flowing. The temperature had steadily dropped as a bank of thick gray clouds passed in front of the sun, and the air smelled of snow.

"So sorry, bit of a queue in there," Marjorie said when she finally bustled out of the shop, her basket full of joints and cuts wrapped in paper. "D'you want to go in and warm yourself before we set out?"

"No, I'm fine." April glanced up at the sky, where a few thick flakes drifted and swirled. "I'd rather hurry back to Alderlea before the storm."

"Right, then." Marjorie set off down the pavement, her longer stride forcing April to quicken her pace to keep up. By the last half mile, though, April knew she would pull ahead while Marjorie gradually slowed. The tortoise and the hare, Marjorie sometimes called them, which April didn't really like, since she was no slow, plodding tortoise. Marjorie might beat her in a short footrace, but in a long-distance run—

Marjorie abruptly halted, and April stopped just in time to avoid jostling her. "Look over there," she said quietly, nodding toward the footbridge just ahead, where a lanky young man with a knapsack had paused to tie his bootlace. "Here's our first slacker. He's not in the army, and he's not at work, either. Hold this."

Abruptly she handed her basket to April, who seized it just before it would have overturned. "Leave him be, won't you?" she begged as Marjorie slipped a hand inside her coat and withdrew a white goose feather from her skirt pocket. "Maybe he's on his way to work."

Marjorie laughed shortly. "Not likely." She marched ahead to confront the lad, April trailing along reluctantly after her. He straightened as they approached, his brow furrowing in curiosity. "Hello," Marjorie greeted him, smiling. "What's your name, dearie?"

"My name?" he echoed, his voice leaping an octave between the two words. "It's Peter, miss. Peter Bell."

"Well, Peter Bell—" Eyes narrowing, voice hardening, Marjorie thrust the white feather at him. "Aren't you ashamed not to be called Private Bell?"

He blinked at her for a moment, and then took the feather. "What's this about?"

"Marjorie, let's go," April murmured, aware of the other villagers who had paused to observe the scene.

Marjorie ignored her. "This is a white feather, a symbol of your cowardice," she informed Peter, as haughtily as Lady Rylance explaining the difference between silver and silver plate. "How dare you idle about here while other men fight for King and Country?"

The lad flushed. "I'm only fifteen. I'm still in school." He gestured to his knapsack, which no doubt carried his books and pencils. "I'm not allowed to enlist."

"You're fifteen?" Marjorie studied him. "How're you so tall at fifteen?"

"I'm—that is—" Voice cracking, he drew himself up even taller and glared. "Well, how'd you get so tall for a girl?"

April seized the crook of Marjorie's elbow and tugged. "Let's go. Please."

"Say, you there," a woman's voice bellowed from behind them. "Did you just give my boy a white feather?"

"Be sure to enlist when you're old enough," Marjorie ordered Peter, then clutched her basket close and turned away. "Let's go," she murmured to April, setting off briskly as the woman shouted after them. They hurried away without looking back, pretending not to hear that angry voice, moving as quickly as they could without breaking into a run.

They both agreed not to mention the incident to anyone else at Alderlea, but a few days later, Mrs. Wilson summoned them to her chamber for a scolding. Peter's father was a village councilor, his mother one of the most respected women of the community. It was she who had written, outraged and indignant, to complain about how her boy had been "unduly scolded by a pair of impertinent Alderlea maids."

"She hinted very strongly that she wanted you both sacked," said Mrs. Wilson, her gaze icy as she looked from one chastened maid to the other. "I persuaded her to settle for a formal apology from me, a fruit basket from the estate orangery, and a solemn promise that you will never again display such shameful disrespect in the village."

Eyes downcast, April and Marjorie murmured apologies. "It was me who gave the boy the feather," Marjorie added, her voice clear and steady. "April—I mean, Ann had nothing to do with it. She tried to stop me."

April felt Mrs. Wilson's eyes on her. "It's a pity she didn't try harder. This unpleasantness could have been avoided." She lectured them a bit longer about how Alderlea had always been on excellent terms with the village, and why that longstanding cordiality and mutual respect must endure. Then she turned to their punishment. April and Marjorie would no longer be sent on errands, at least not until they had earned back Mrs. Wilson's trust. If they entered the village on their afternoons off, they must be on their absolute best behavior. If a single letter of complaint attested that they had not, they would be sacked immediately and not provided with a character.

Heart thudding, tears springing into her eyes, April nodded and thanked her, knowing the punishment could have been far worse.

"Let's try to put this unfortunate business behind us," Mrs. Wilson said, sighing. "Mary, you may go. Ann, I need another word."

Marjorie threw April a worried, sympathetic glance as she curtsied to the housekeeper and hurried away. April braced herself as the door closed and Mrs. Wilson studied her over the rims of her glasses. "I realize you and Mary are friends," she said, "but you might wish to reconsider that friendship. You're a good girl on your own, but too often you disregard your better sense and follow Mary into mischief." She held up a hand to forestall any argument, not that April had any intention of interrupting. "I know you weren't the one who gave that boy the white feather, but when guilt by association is the rule, you would do well to keep better company."

April swallowed, hard. "Yes, ma'am."

At last Mrs. Wilson dismissed her. April left the housekeeper's chamber only to find Marjorie waiting for her in the passageway. "What did she say?" Marjorie whispered, glancing at the closed door. "Is she going to tell your mum?"

"I hope not," April replied quietly, ignoring the first question.

They fell in step together as they headed up the servants' staircase to resume their interrupted duties. "I only did what hundreds of girls have been doing all around Britain," Marjorie said, contrite. "Lord Kitchener himself supports the Order."

"He doesn't support enlisting fifteen-year-old schoolboys."

"I didn't know—" Marjorie inhaled deeply. "Please, don't you be angry with me too. He didn't look like a schoolboy."

"No, I suppose he didn't."

Marjorie paused on the staircase and stuck out her hand, her skin as red and chapped as April's own. "Still chums?"

"Always," April replied as she shook her hand.

As the days passed, they endured the embarrassment of knowing that everyone belowstairs was aware of their disgrace. Whispered conversations halted when they entered a room, and an occasional joke at their expense was shared at the servants' dinner table until the butler put a stop to it. By early March, as the first signs of spring brightened the estate, April had begun to hope perhaps that the whole unfortunate business, as Mrs. Wilson had called it, would indeed eventually be forgotten.

But soon thereafter, Marjorie was again summoned to the housekeeper's chamber for a scolding. April first heard about it from the scullery girl rather than from Marjorie herself. The handsome lieutenant, a married man, had become annoyed with her "chattiness and familiarity" and had requested that another maid attend to his room for as long as he remained billeted at Alderlea. He had shared his concerns with his captain, who fortunately had known to take the matter to the butler. If either of the officers had complained directly to the lord or lady, Marjorie would have endured a more serious punishment, but as it was, she would be reprimanded and assigned to other duties that would not bring her anywhere near the lieutenant. Any further violations of the household rules and expectations would result in her immediate dismissal.

April knew that if servants weren't so difficult to replace, Marjorie would have been sacked already.

"I don't know how much longer I can put up with this," Marjorie grumbled later that afternoon as they worked on the household mending. "I wasn't even flirting."

"You were just being friendly," said April wryly, threading her needle.

"That's all it ever was on my part, honestly!" Marjorie jabbed her needle into her pincushion, sank back into her chair, and folded her arms, defiant. "*He* kissed *me*."

Startled, April set her sewing down on her lap. "He *kissed* you? No one said anything about kissing."

"Well, of course he wouldn't mention *that*. He's married, and he's an officer, and I'm just a nobody housemaid from Warrington. I was making up his bed when he walked in, startling us both. He said it was a lovely surprise to find me there, and I made a joke about how it was always a surprise to find me working. He laughed, and then he just rushed over and kissed me, with my arms full of bedclothes and everything. I'm on thin ice already, as everyone knows, so I pulled away and told him I couldn't."

"What did he do? Did he threaten you?"

"No, not that. He got red in the face and begged my pardon. I told him not to mention it, no harm done, and I left." Marjorie frowned, eyes narrowing. "Obviously, he thought I might squeal on him, so he came up with his own version first, just in case."

"Maybe if you tell Mrs. Wilson what really happened—"

"That wouldn't do any good. She would still blame me for staying to talk to him instead of just bobbing a curtsey and scampering from the room the moment he walked in."

"It's not fair."

"What's ever fair for girls like us?" Marjorie asked scornfully, but April had no answer for her.

The next morning, April woke at half four o'clock, as she did every morning, and went to rouse Marjorie, only to find her bed empty. Her uniforms hung neatly on their pegs, but all of her belongings were gone.

Anxious, April quickly threw on her day uniform only to find a note in her apron pocket.

Marjorie had gone off to London, to find a new job and a fresh start. "Goodbye, chum," she signed off cheerily. "Thanks for making Alderlea more bearable, while it lasted. Think of me doing my bit of war work and be proud of your chum, and try not to miss me too much."

But a sudden pang of loss told April that it was too late. She already missed Marjorie more than she could have imagined only a day before.

Folding the letter, steadying herself with a deep breath, April finished dressing and pinned her hair up beneath her white cap, wondering how in the world she would break the news to Mrs. Wilson.

3

NOVEMBER 1910–JUNE 1915

HELEN

Helen would never forget the moment two years before when she realized with heartbreaking certainty that the only way to keep her beloved childhood home was to abandon it forever.

It was the sort of ironic tragedy that would have intrigued her father, the renowned classicist Heinrich Stahl, had he discovered it in one of the ancient Greek texts he studied and taught as a Professor of Literae Humaniores at Oxford. Helen had always adored and revered her father. His snug, paneled study, with its rich mahogany desk laden with manuscripts and floor-to-ceiling shelves packed with books and antiquities, was her very favorite room at Banbury Cottage, the only home she had ever known. The red-brick Edwardian residence on Divinity Road in the historic district was cozy but spacious enough for Helen, her parents, and her two sisters to share in comfort, with rosebushes and a pair of lovely plum trees in front and a larger, walled-in garden in back. Banbury Cottage was only a brisk twelve-minute walk to the Bodleian Library, a few minutes more to the Ashmolean. From childhood Helen had loved to stroll the route with her father, admiring the neighbors' front gardens, noting the alterations of the seasons in the foliage and flowers in the parks they passed, chatting in German, or discussing Greek myths and epic tales. Once, as Helen ran home from school, a glimpse of her home from the top of a low rise in the curving street brought her to a halt, overcome with joy and affection. She vowed right then and there never to live anywhere else.

For many years, Helen truly believed it was a vow she could keep. At seventeen she began studying at Oxford, although as a woman she could not be admitted as a full student and would not be able to claim a degree. Her happiest hours were spent working as her father's secretary and research assistant, and while her mother often hinted that she and her elder sister, Penelope, would find husbands and homes of their own someday soon, neither had any inclination to begin the hunt.

"I shall never marry," Helen confided to her sister one late summer afternoon as they walked home from a vigorous tennis match in which Helen had barely eked out a victory. Penelope's best friend's family had a court on the grounds of their estate, and they kindly allowed the Stahl sisters to play anytime they liked. "I cannot imagine I could ever love any home more than Banbury Cottage, or find any occupation I enjoy more than assisting Papa."

"Not marriage and motherhood?" teased Penelope. Her face was lightly dusted with freckles, as was Helen's own, her auburn hair a few shades darker. She was decidedly the most beautiful of the Stahl sisters, although Daphne, the youngest, was a close second. "Not even a leadership role in the Women's Social and Political Union, toiling side by side with the Pankhursts for woman suffrage?"

"The latter would be tempting," Helen admitted. "But not the former. I'd prefer to be an eccentric spinster aunt to your children and Daphne's."

"Daphne's, perhaps, but not mine," said Penelope. "I shall never marry either."

"Do you mean that?" asked Helen, surprised. "You're so lovely, you should have been the one called Helen."

Penelope rolled her eyes. "And you should have been Penelope, the faithful wife and queen?"

"No, that doesn't suit me. Atalanta, perhaps. I'm a swift runner."

"Yes, I've noticed. Without that speed, you never would have beaten me today."

"But Penny, why shouldn't you marry? You have so many admirers. When Papa invites his students home to dine with us, they barely acknowledge the rest of us, they're so busy gazing at you."

"That's not true," Penelope retorted, allowing a smile. "Well, if it is true, their attention is misplaced. None of them could ever be as dear to me as my sisters, or Margaret."

Helen nodded. Penelope and her best friend were so devoted to each other that Helen sometimes felt a bit jealous. "I often think that you would marry Margaret if you could," she teased.

"Perhaps I would," Penelope replied lightly, "but she's already engaged, and the neighbors would gossip. Nor would I be able to provide for her."

Something in her sister's tone gave Helen pause. "Is that why she's marrying Mr. Aylesworth, to be provided for? But her family is one of the wealthiest in Oxfordshire. What of her inheritance?"

"She'll receive a modest income through her mother, but her father's estate is entailed upon her elder brother. Margaret doesn't wish to be a burden to him."

"So she and Mr. Aylesworth—it's not a love match?"

"They are fond of each other," Penelope said, seeming to choose her words carefully. "They're good friends. He needs an heir, and she has always wanted children. Satisfactory matches have been based upon far less."

"I suppose," said Helen, dubious. They had reached the top of the rise, where the front gate and the roof of Banbury Cottage were visible in glimpses as the wind stirred the leafy boughs of the sheltering trees. "Poor Penny," she said sympathetically, "to be abandoned so cruelly."

"I haven't been abandoned entirely." Penelope hesitated, and again she seemed to weigh her words. "After the wedding, when the newlyweds move to Devonshire, I'll be going too, as Margaret's lady's companion."

"What? Penny—"

"Please don't say anything at home. It's all decided, but Mother and Papa don't know yet, and I should be the one to tell them."

"Certainly you should," said Helen, a bit dazed, and not from too much tennis in the bright summer sunshine. "But a lady's companion, Penny? Are you sure that's what you want?"

"It's the next best thing to what I truly want."

"And what is that? Once woman suffrage is won, which surely can't be too far off, true equality will follow, and then—"

"No, Helen, no." Penelope shook her head and laughed, a bit sadly, or so Helen thought. "Not even Mrs. Pankhurst could help me gain my heart's desire. While this arrangement may seem a bit old-fashioned, it's more likely to assure my future happiness than any other scheme." She fell silent for a long moment. "Of course, things cannot carry on as before. Mr. Aylesworth requires fidelity—"

"Don't all husbands?"

"Well, yes, but our . . . friendship will necessarily change. The . . . closeness Margaret and I have shared—" Penelope inhaled deeply, pressing a hand to her chest as if it pained her. "It cannot be the same."

"Are you unhappy because Margaret will spend less time with you? Are you worried that you'll feel like the odd one out?"

"Yes. No." Penelope closed her eyes and gave her head a quick shake. "That's only part of it."

Helen studied her. "I don't understand. You seem utterly bereft. Are you sure you'll be happy with this arrangement—which, let's be honest, seems rather archaic?"

"It's better than having no place at all in Margaret's life." Penelope managed a reassuring smile. "Mr. Aylesworth's diplomatic duties will require him to be in London or travel abroad quite often. He himself said that he's grateful I'll be there to keep Margaret company in his absence."

"Well, at least he isn't jealous of your friendship. And Margaret won't be married for another eight months. You have time to reconsider." Inspired, Helen added, "Perhaps one of Papa's students will sweep you off your feet unexpectedly."

"Believe me," said Penelope wryly, "nothing would be more unexpected."

"What about the handsome Mr. Herridge? Or the brilliant Mr. Purcell?"

Helen winced. "Not for me, thank you. They're so old, nearly thirty."

"Practically ancient."

"Also, I think Mr. Purcell prefers you."

"Nonsense. He thinks you're beautiful and I'm annoying. He's always picking apart my arguments, seeking miniscule flaws in my logic."

"That's precisely it," said Penelope, smiling, as they descended the rise and approached their own front gate. "He respects your mind enough to take your arguments seriously. Another man wouldn't want you to join the conversation at all. It's not just because you're his tutor's daughter, either."

"Certainly not. It's because he dislikes me and wants to prove me wrong at every opportunity."

Penelope laughed, a true, full-hearted laugh this time. It made Helen happy to hear it.

The days passed. Summer blazed brilliantly and faded into autumn. All too soon, as Margaret's maid of honor, Penelope became so busy with preparations for the upcoming wedding that Helen felt she hardly saw her anymore.

"Come with me to London for the suffrage demonstration tomorrow," Helen begged one November day. "We're going to march from Caxton Hall to the Houses of Parliament to protest Mr. Asquith's betrayal of his campaign promise to allow limited women's suffrage. We suffragettes supported him, raised funds for him—and how has he repaid us as Prime Minister?"

"Not well, I assume?"

"Not well at all! The suffrage bill passed its first and second readings, but he refused to let it proceed. Now he's called for a general election, Parliament will dissolve at the end of the month, and nothing more will be done on our account." Helen seized her sister's hand. "The greater our numbers, the stronger our message. Please say you'll come."

"I'm sorry." Penelope squeezed her hand before letting go. "I have plans with Margaret."

"Change your plans. Bring her along. We'll all go together."

Penelope shook her head, smiling. "I'm sorry but it's simply not possible. Margaret has a gown fitting."

"Well, bring the dressmaker too," said Helen, laughing at herself, knowing there was nothing for it. Fortunately, she had already arranged to meet some girlfriends at the train station, and together they would ride into London, march to Westminster, and return to Oxford by nightfall.

The demonstration began at noon with a rally at Caxton Hall in Westminster. Clad in a deep purple skirt, white shirtwaist, and dark green wool coat and matching hat, a satin sash bearing the phrase "Women's Will Beats Asquith's Won't" draped from her right shoulder to left hip, Helen cheered and applauded the speakers, her heart thrilling to their brave, inspiring words. Next she and her friends joined their assigned group of twelve and took their place about two-thirds of the way back from the head of the procession, which was led by Mrs. Pankhurst and the other suffrage dignitaries who formed the official delegation. The delegation would attempt to enter the Palace of Westminster at St. Stephen's entrance and demand an audience with the Prime Minister. Meanwhile, the roughly three hundred other marchers would demonstrate peacefully in Parliament Square.

A cold, brisk wind buffeted the marchers as they set off, brandishing their signs and banners with dignified resolve as they made their way to Victoria Street, which they followed to Parliament Square. The event had been widely publicized, and members of the press and curious onlookers lined every block of the route. As her group drew closer to Parliament Square, Helen observed with some trepidation that a large crowd of angry men, spectators as well as policemen, awaited them at the Westminster Abbey entrance.

"What should we do?" one of Helen's friends asked fearfully as up ahead, shouts from the men met with cries of alarm from the women.

"The women in front of us are marching on," Helen replied, more bravely than she felt. "We shall do the same." The police had protected them during previous demonstrations when angry onlookers threatened violence. She trusted they would do the same that day.

The procession had slowed, but still the women moved ever forward—until, suddenly, Helen felt as if they had slammed into a wall. All around them, men shoved the marchers back, shouting furiously, as the women struggled to stay within their groups in formation. Hats were knocked from the women's coiffures; sashes were torn off and flung into the gutters. Ears blistering from the men's ugly epithets, Helen braced herself and grimly continued on, gasping aloud when she saw one woman take an elbow to the nose, streaming blood, another a fist to the eye. To the left and to the right, her comrades were being slapped, struck, shoved to the ground, seized around the torso and hauled away by men in suits and men in laborers' clothing, their faces twisted in ugly rage.

"Be careful!" a woman cried out somewhere behind her. "They're dragging women off down the alleys!"

A frisson of alarm ran through Helen. Struggling to keep her feet as the crowd surged around her, she looked wildly about for the police, only to spot several officers shouting and jeering at the women they had been assigned to protect, striking one, shoving another, grabbing others by the arms and hauling them out of the procession. Suddenly Helen felt rough hands groping her. She cried out in pain as a man squeezed her breasts and wrenched, tearing buttons from her coat as she stumbled and fell back, shocked and disoriented.

Steadying herself, she rejoined what was left of the march. She had become separated from her friends, but she steeled herself, chest aching and anger rising. She took one step forward, and then another. Hours passed in which it seemed she progressed barely inches, but she did not turn back, even as other women, bruised and bleeding, stumbled out of the lines and headed back toward Caxton Hall. Up ahead, just beyond the gauntlet of raging men, Helen glimpsed battered suffragettes holding up torn sashes and broken signs in Parliament Square, some parading as planned, jaws set and expressions defiant. Others berated the police and begged them to intercede, only to be arrested for their trouble.

Thirsty, sore, and increasingly fatigued, Helen saw a clearing open up a few yards ahead and tried to make her way to it—only to be brought to an abrupt halt as two men seized her, one on each arm. "Steady there, lass," a policeman growled near her ear as he hauled her toward a wagon. "A few weeks in Holloway will settle you down all right."

Helen felt a surge of fear at the dreaded name and tried to wrest herself free, but she was exhausted and the men were stronger. They easily lifted her dragging feet off the pavement and shoved her inside, locking the door behind her.

For two hours or more she and her unfortunate fellow captives sat on hard benches in the back of the dim, increasingly crowded police wagon, sharing information and tending one another's wounds as best they could. Eventually, when the wagon was overfull, the captives were taken not to the dreaded, notorious Holloway prison as they had feared, but to the Bow Street Police Station in Covent Garden. Soon Helen was processed and assigned to a cell, where she and her companions refused to eat and demanded the status of political prisoners. "You'll be treated like the common hooligans you are," the scowling matron retorted.

The suffragettes resigned themselves to an indefinite wait. Night had fallen, and the stone walls of the cell radiated cold, but Helen wrapped herself in the thin, coarse blanket she had been issued and ignored the gnawing hunger pangs in her stomach. The women prayed and sang together until they grew too weary for anything but sleep. Some managed to lie down, two to a bench, while others sat and leaned back against the walls or one another, but restlessness and chagrin kept Helen awake. Other suffragettes were arrested rather regularly, but Helen had never been. She hoped her family knew what had happened and where she was. She could well imagine the worry and embarrassment she had caused them.

She was still brooding over how she could make amends to her parents without renouncing the cause when the matron approached the cell door, a ring of keys jingling in her hand. "Miss Stahl?" she barked.

Helen sat up straighter and raised her hand. "I am Miss Stahl."

"Come with me. You're free to go." The matron swung open the barred cell door. "Leave the blanket."

Helen stood, but then she hesitated. "Why am I free to go? What about my companions?"

"My orders are for you. Everyone else stays."

Helen promptly sat back down. "Then I shall stay too."

Some of the women applauded drowsily, but one of the older women hushed them. "You should leave while you can," she told Helen. "Tell everyone what happened to us out there and how we're being treated."

"Are you coming or not?" Peeved, the matron began easing the door closed but left a gap. "It's up to you. Your father paid your fine and we don't give refunds."

Helen's heart thumped. "My father is here?" She wasn't sure whether to be relieved or mortified. To think that he had come so far on her account, and on a Friday! He must have cancelled his afternoon tutorials and left Oxford the moment he learned of her arrest, and now he was missing formal hall.

"Go on, dearie," the older women urged. "Don't leave him waiting."

"Wish my father would come for me," another woman remarked, to subdued laughter. "I'd go in a heartbeat."

Helen was tempted to stay, not only to show solidarity with her sister suffragettes, but to avoid her father's weary rebukes. Still, she would have to face him eventually, so she might as well get it over with.

Offering her scratchy blanket to the older woman, she left the cell and followed the matron through a maze of grim, poorly lit corridors to an office where a bored-looking clerk shuffled papers and stamped documents with official seals and eventually indicated that she was free to go.

"Worried your father sick, you have," the matron muttered as she escorted Helen to reception, giving her a little shove through the doorway for emphasis. "You should be ashamed of yourself, ungrateful girl. He's been pacing for the better part of an hour."

Helen braced herself for an onslaught of questions and rebukes from her father—but she found herself facing a taller, slimmer, and much younger man who paced away from her, turned, and halted abruptly when he saw her.

Wavy brown hair, piercing green eyes, high, angular cheekbones—Mr. Purcell. She was so astonished, she gasped.

"Miss Stahl," he said, his glance swiftly examining her from head to toe. "Good Lord. Are you all right? Have you been mistreated?" His expression hardened as he turned to the matron, who was hastily closing the door between them. "If Miss Stahl has been in any way abused," he snapped, striding toward her, "I swear you'll sincerely regret it—"

"I'm all right," said Helen, seizing his coat sleeve to restrain him. She must look truly dreadful to have provoked such indignation. "I'm only exhausted—and hungry."

"Right. Of course you are." The matron forgotten, he patted his coat, frowning, then brightened as he took a small red apple from his front pocket. "Will this do until we can find you something more substantial?"

"Yes, thank you." Helen plucked the apple from his hand and took a larger bite than was perhaps polite. It was the sweetest, most delicious apple she had ever tasted, unless that was the influence of her hunger. She swallowed and asked, "Where's my father?"

"In Oxford, of course. He had lectures and tutorials, so I offered to come fetch you in his place."

"But the matron told me—" She broke off, but not quickly enough.

"She told you your father had come." Mr. Purcell appeared wounded. "You certainly do look very young, but I hope I don't seem so terribly old." Before she could decide whether to give reassurances or an apology, he offered her his arm. "Come. Let's get you home. If we hurry, we may catch the late train."

"And if we miss it?" she asked, taking his arm as they headed for the exit.

"Then we'll find a train going in our direction, take it as far as we can, and drive from there."

They hurried outside, where Mr. Purcell quickly hired a cab, promising the driver an ample gratuity if he got them to the train station within a quarter hour. "Done," the driver declared cheerfully, barely waiting for them to seat themselves before he chirruped to the horses and sped off.

In the first, or perhaps second, stroke of good luck Helen had found all day, they arrived with minutes to spare. Before Helen could offer to pay her own fare, Mr. Purcell purchased two tickets, and then they were running to the platform, climbing aboard as the whistle blew, and settling into their seats even as the train chugged out of the station. Immediately Mr. Purcell had a quiet word with the porter, who returned soon thereafter with a cart of tea and sandwiches. "Tuck in," Mr. Purcell instructed, and she almost forgot to thank him before doing exactly that.

By the time the train left the city behind, she was feeling much better—sated, comfortable, and relieved, but also a trifle embarrassed. "Thank you for breaking me out of prison," she said lightly.

"Happy to be of service."

"You must think I'm quite a little fool to end up in such a dreadful place."

"Not at all. I entirely support the suffrage cause—" He paused, wincing. "If not all of Mrs. Pankhurst's methods."

Helen smiled to conceal her annoyance. "Smashing windows as a form of protest is a time-honored British tradition."

"Perhaps, but when such forms of protest regularly lead to broken bones and prison sentences for the protestors, alternative methods might be considered."

"If women's bones are broken in a protest," said Helen, voice trembling, hot tears springing into her eyes as she remembered rough, groping hands, "perhaps the fault lies with the despicable men who inflicted the injuries, and the contemptible police who stood by and allowed it to happen—or worse yet, joined in, abandoning any oaths they may have sworn as peace officers!"

She turned away and stared determinedly out the window, but she felt Mr. Purcell studying her. Then she felt his hand clasp hers, resting on the seat between them. "Were you ill-treated?" he asked quietly, his voice too low for anyone else to overhear.

Still gazing out the window, she nodded. "Others endured far worse than I."

"Do you—would you prefer to go directly to a physician or nurse when we arrive in Oxford? Or perhaps this is a matter for the police?"

"Goodness, no." Shuddering, Helen turned to face him. "I've had enough to do with the police for one day, thank you very much."

"Miss Stahl—" He studied her intently for a moment, then let his gaze fall to their clasped hands. He promptly pulled his away. "I'm concerned about you, but I confess I'm at a loss for what assistance to offer. If you would like to tell me what happened today—"

"I can see that I *must* tell you," she interrupted, "or you will imagine something quite worse." And so she described, as best she could remember, everything that she had witnessed and experienced since setting out with her group of twelve from Caxton Hall.

"I'm terribly worried about my friends, the Oxford girls who attended the march with me," she added when she had finished the harrowing tale, fresh tears springing to her eyes. "None of them were arrested with me, and I don't know what has become of them."

"I can put your fears to rest on that account, if nothing else," Mr. Purcell said. "They're all home, safe and sound. When they saw you being put in the police wagon, they left the demonstration and took the next train back to Oxford. They immediately went to tell your mother, and she telephoned your father. I happened to be with him when he received the call, so I offered to come in his place."

"I'm so grateful that you did," said Helen fervently. "Were my parents very distressed?"

"I haven't seen your mother," he said. "Professor Stahl was concerned about your safety, as any parent would be, but he was not overly anxious."

"That's fortunate," she said, relieved. Her father had been experiencing some shortness of breath and chest discomfort in recent months, and his doctor had advised him to avoid emotional upset.

As the train carried them homeward, the conversation turned to other, more pleasant matters—the Stahl family's annual dinner party in early December to mark the end of Michaelmas term and herald the festive season, to which Mr. Purcell had been invited, along with her father's other students; a book her father had recently published, to significant acclaim; and Mr. Purcell's doctoral research on Philoctetes, archer and hero of Troy, which he hoped would culminate in a book as well.

"Philoctetes is best known from Sophocles's play and Homer's *Iliad*," said Mr. Purcell, quickly adding, "which, of course, you already know. I intend also to draw upon Aeschylus's and Euripides's plays about him, but only fragments remain."

"Perhaps you shall find the rest," remarked Helen, smiling.

"If I did, the discovery would likely make my academic career. At the very least, it might allow me to continue it."

"Why shouldn't you continue? You're in no danger of dismissal from the program. You're one of my father's star pupils."

"Listen at keyholes, do you?"

"Only occasionally, and it wasn't necessary in this case. My father has made no secret of his preference. Don't tell the others, though," she added hastily. "I wouldn't want any hurt feelings."

"Your secret is safe with me. And thank you for your vote of confidence, but your father isn't the obstacle."

When he didn't elaborate, Helen remarked, "As I recall, Philoctetes competed for the hand of my namesake, the princess of Sparta."

"Indeed, and he lost decisively. My focus isn't his thwarted pursuit of the most beautiful woman in the world, though, but his abandonment on the isle of Lemnos by his shipmates. They were so offended by the smell of his festering wound that they essentially left him there to die."

"With friends like those, who needs Trojan enemies?"

"Quite right. My premise is that Philoctetes's sufferings presage his—" Abruptly Mr. Purcell broke off, chagrined. "Festering wounds, agony, abandonment. What fine topics of conversation I choose to discuss with a young lady. My sister scolds me that this is precisely why I remain unmarried."

"At the ripe old age of?" Helen prompted.

"Thirty."

Mr. Purcell wasn't so ancient after all, though still ten years older than herself. "I must respectfully disagree with your sister," said Helen. "I think this is an excellent subject for conversation. There are deep human truths to discover in pain and suffering. And in the fear of abandonment, in the rejection and isolation all too intrinsic to the human experience."

He regarded her for a moment, eyebrows rising. "I'm going to tell my sister you've refuted her theory."

"Ah, but tragically, I haven't," said Helen, feigning regret. "To disagree is not to disprove. Until you marry a young lady with whom you've discussed these grim topics, your sister may yet be proven correct."

The sound of his laughter was drowned out by a deep, resonant blast of the train's whistle as they approached Oxford station. Helen found herself without anything to say as they disembarked and passed through the concourse. Mr. Purcell probably intended to escort her home, although he had not offered to, and she shouldn't presume, since he had done so much for her already. Once outside the station, she prepared to bid him goodnight and walk home alone when he gestured to a parked automobile. "Shall I give you a lift home?"

She looked at him, incredulous, then again at the automobile, a Rolls-Royce Silver Ghost, gleaming in the moonlight. "Certainly," she joked, playing along. "Do you think the owner will object if we take her for a spin, as long as we return her unharmed?"

"He won't mind at all." Mr. Purcell opened the passenger-side door and gestured for her to climb in. When she didn't, he regarded her quizzically and stepped back as if to clear her path. "Come on, then. All aboard. Never fear, I'm an excellent driver."

She didn't budge. "This is your car?"

"Yes, Miss Stahl," he replied, the first hint of exasperation in his voice that she had heard all day, which was really quite remarkable given the circumstances. "Please take a seat and I'll crank 'er up."

Feeling foolish, she did as he asked. In her defense, there was no reason why she should have known he owned such a luxurious automobile, or any automobile at all. He had always come to Banbury Cottage on foot, and he was only a student, after all.

There was more to Mr. Purcell than she had realized, Helen thought as he drove her through the darkened streets toward home. He was certainly not the irritable old curmudgeon she had imagined him to be.

When they arrived at Banbury Cottage, Helen's parents appeared in the doorway to meet them as soon as they passed through the gate. Her father began interrogating her the moment she crossed the threshold, his German accent more pronounced than usual, a sure sign of his perturbation. Her mother fussed over her disheveled appearance and asked her again and again if she was *sure* she was all right. Then Daphne darted into the room, flung her arms around her sister, and began peppering her with more questions, bursting with curiosity and excitement. Helen looked around at her family, unable to reply to one for listening to another, until Mr. Purcell's voice rose above the din. "Miss Stahl is unharmed but terribly fatigued. Perhaps the questions can wait until tomorrow?"

"Yes, yes, of course, tomorrow," her father said, seizing Mr. Purcell's hand and shaking it vigorously. "I cannot thank you enough for bringing her home to us."

Helen would have thanked him again too, but her mother and Daphne were ushering her upstairs, mandating a hot bath and a good night's rest. She glanced back over her shoulder as they led her away, but Mr. Purcell was engrossed in conversation with her father and did not notice.

The next morning, Helen soberly told her parents what she had witnessed at the demonstration and patiently accepted their rebukes, lovingly meant, that she should not have put herself at such risk. She understood their point of view better after reading the newspapers. Eyewitnesses and victims alike reported utterly brutal treatment at the hands of the mob—marchers groped and beaten, women's skirts lifted over their heads, men seizing women from behind and clutching their breasts, women hauled into alleys and assaulted. Police were observed dragging suffragettes aside, beating them, and throwing them back into the hostile crowds for further abuse, their crimes preserved forever in photographs.

In one of the most shocking incidents, the revered suffragette May Billinghurst, who had been partially paralyzed by polio as a child and was obliged to convey herself in a tricycle chair, had been overturned and dumped onto the street by the police. Two officers had picked her up, twisted her arms behind her back, causing her agonizing pain, and pushed and dragged her down a side street. Other officers followed, trundling her empty chair. They threw her to the ground in front of a gang of young street toughs, removed the valves from the chair's wheels to render them useless, and left her to her fate. The joke was on the police, though, for the rough fellows possessed gallantry the officers did not. Instead of assaulting Miss Billinghurst as the police had expected, they rescued her, repairing her chair and conveying her safely to Caxton Hall.

More than one hundred suffragettes had been arrested, but all had been released by nine o'clock, only two hours later than Helen. Two or three men had also been arrested, none of them police officers.

"Promise me you won't attend any more of these rallies," her mother begged, as Daphne chimed in her own pleas. Helen couldn't bear to do that, but she did promise to be more watchful and cautious in the future—if there ever *were* more rallies to attend. She suspected the WSPU would shift tactics in the aftermath of what the suffrage press was already calling Black Friday.

Helen hoped to see Mr. Purcell soon, to thank him again for relieving her father of the burden of fetching her home, but he did not call at Banbury Cottage, at least not when she was at home to receive him. Though disappointed, she looked forward to seeing him at her family's party at the end of the term. She had some thoughts about Philoctetes and the fragments of Aeschylus's work she hoped to discuss with him.

But when the long-awaited day came, the party had been underway for hours and still he had not appeared. Puzzled, and a bit annoyed, Helen managed to catch her father alone long enough to ask why he supposed Mr. Purcell refused to be more punctual.

"He's not late, *Liebling*," her father said. "He's not coming."

"Why not?" she protested. "He always attends our—your parties. What could have enticed him away?"

Her father shrugged, regretful. "Perhaps Mr. Purcell found himself unable to celebrate, knowing that he won't be joining us for Hilary term. I organized a farewell dinner for him a few days ago. We all said our goodbyes then."

"Mr. Purcell finished his degree?" asked Helen, bewildered. "I thought he had a year or two left."

"He does, but he has withdrawn from Oxford." Her father sighed. "What a shame. He has a very good mind and would have made an excellent professor."

"But—" Helen shook her head. "I don't understand. Why would he abandon work he loved?"

"Filial duty, *Liebling*. For years Arthur's father indulged his desire for an Oxford education, but the elder Mr. Purcell always intended for Arthur to join the family business eventually. Apparently he considered Arthur's thirtieth birthday in October to be a finish line of sorts. Time to put aside old plays and epic poems and join the working world."

"*My* father isn't the obstacle," Helen murmured, thinking aloud, remembering Mr. Purcell's cryptic remarks on the train. "*His* father is."

"Arthur convinced his father to let him finish the term, hoping his father would change his mind in the meantime, but he only became more resolute."

"Poor Mr. Purcell," said Helen, heartbroken for him. "Do you happen to know what the family business is? Please tell me they deal in antiquities, or run a museum, or plan excursions to Greece, or—"

"Manufacturing." Her father clasped her shoulder in a gesture of comfort, his face full of tender commiseration. "Sewing machines, or perhaps it was parts for industrial sewing machines. Perhaps both. Or was it looms?"

"What's the difference?" asked Helen morosely. Whatever it was, it wasn't studying ancient texts about Philoctetes. It was commerce, and as a profession it was particularly unsuitable for a scholar of the classics.

In the days that followed, Helen wanted to offer Mr. Purcell her most sincere sympathies, but he had left Oxford, and she was reluctant to ask her father for his postal address and prompt uncomfortable questions. She longed to know if working for his family's business had turned out to be unexpectedly rewarding, or at least not the miserable fate he had clearly tried to avoid. She truly hoped so.

Months passed, and then a year. Helen heard nothing of Mr. Purcell, but she thought of him now and then and wished him well, wherever he was.

Then, unexpectedly, tragedy brought him back into her life.

Her father's heart condition had worsened over time, so much so that he began to contemplate the unthinkable: retirement. He had just declared his intention to reduce his tutoring load and prepare to accept emeritus status in five years when, two days after the close of Trinity term, he suffered a heart attack at his campus office while marking exams. He died the next morning.

Helen, her sisters, and her mother were utterly bereft. Shattered by loss, they informed the university, arranged a funeral, and accepted condolences from hundreds of colleagues and friends. Several former students spoke at the memorial service, among them Mr. Purcell, grief-stricken and dignified as he honored his mentor and friend. He approached Helen at the reception afterward, and they chatted briefly, but she was barely holding herself together and she forgot to ask him anything about himself and his new career.

She trusted he would forgive her.

Since Penelope returned to Devonshire a few days after the funeral and her mother was preoccupied with poor, heartbroken young Daphne, the melancholy business of settling her father's estate fell to Helen. Working with his executor, she discovered that her father had left behind only a few miniscule debts, easily settled from the household accounts. Although this came as a tremendous relief, he had made almost no provision for future income. There would be a small pension from the university, and some earnings from his books, while they remained in print. But the latter income would be negligible, as his books were not the sort one would find on the shelves of a typical home alongside Dickens and Austen, but might, in a good year, sell a few copies to scholars and academic libraries.

Helen painstakingly went over the ledgers with her father's trusted executor, and then again with her mother, but each time she reached the same unhappy conclusion: They had scarcely enough money to live on at present, and eventually even that would run out. Their best hope—their only choice, really—was to sell Banbury Cottage, find more affordable lodgings, and live frugally from that day forward. Although it would deeply sadden them to leave their cherished home, they assured one another that it would all work out in the end.

Throughout the summer, as Helen and her mother made inquiries and arrangements, many of her father's former students and colleagues called to pay their respects, to inquire about their needs, and to offer their services. One former student who visited weekly was Arthur Purcell, who never failed to bring a welcome gift for their dinner table or the library. In fair weather, he and Helen frequently walked together through the beautiful Oxford campus with its stunning architecture and familiar gardens, every scene evoking tender memories of her father.

In early September, Mr. Purcell tactfully asked about their future plans. Knowing that he must have guessed their financial straits, Helen saw no point in embellishing the truth for the sake of politeness. She told him frankly that they could not afford to keep Banbury Cottage, and although parting from the home that held so many cherished memories would rend their hearts, they would manage.

"If I could find a position as a tutor, my earnings would help quite a bit," Helen mused aloud as they strolled along the Thames. "But what might be available in Devonshire for an aspiring young female tutor with no teaching experience is another question. I'm sure I'm capable of instructing children in Greek, German, and—"

"Hold on." Abruptly Mr. Purcell halted, placing a hand on her arm to bring her to a stop too. "Why Devonshire? Surely you don't have to go so far to find an affordable home."

"I'm sure I mentioned it," said Helen. Hadn't she? "As you know, Penelope serves as a lady's companion in Devonshire. Her employers have offered us the charming gatehouse on their estate, and for a pittance. They'd let us have it gratis except my mother refuses to accept charity."

"What about your father's collection of antiquities?" Mr. Purcell asked, brow furrowing. "I could help you arrange to sell them. I could put you in touch with an auction house, or with my contacts at various museums, private collectors—"

"Thank you, but except for a few favorite pieces he left to me and my sisters, my father bequeathed his entire collection to the Ashmolean Museum." Smiling through a pang of loss, she resumed walking along the river, and Mr. Purcell promptly fell in step beside her. "You would hardly recognize his study, the shelves and walls are so depleted."

"I'm truly very sorry that it's come to this."

"Please don't think badly of my father. He never meant to leave us in such straits, and we'll be perfectly content. Well, perhaps not *perfectly*—"

"If there is anything I might do to secure your happiness—" Mr. Purcell halted again and took her hand. "Miss Stahl, for more than a year I've longed to ask you, but you've made your love for your home and your abhorrence of marriage abundantly clear—"

"*Abhorrence* is a bit strong. My parents had a wonderful, loving marriage. You imply that I want the custom abolished, but—" Words abruptly failed her. He had longed to ask her . . . what? In context, only a few very particular options made sense.

"My dear Miss Stahl—Helen—" He took her other hand. "I love

you. I've adored you ever since you strode out of that prison cell more concerned for your father's inconvenience than your own suffering. And then you devoured that apple as if your only desire was to build up your strength for the next battle."

"I didn't devour it," Helen protested, her cheeks growing warm. "I was very hungry. I hadn't eaten since breakfast. This is a very odd proposal."

"It's my first attempt," he said wryly. "Please, dearest Helen, would you do me the honor of becoming my wife?"

Her heart was pounding, but she tried to keep her voice steady. "Are you asking me out of a sense of duty to my father, some misplaced obligation to provide for his widow and children?"

"Good God, Helen, no. I wouldn't marry a woman I didn't want to spend the rest of my life with just to keep a roof over her family's head. I would buy her a house."

He tossed that off as if it were eminently reasonable, as if one bought houses for friends every day. But that was not what riveted her attention. "You want to spend the rest of your life with me?"

He squeezed her hands, nodding. "I'll repeat myself in Greek, if it would convince you."

"I wouldn't annoy you beyond reason, always besting you in argument?"

"Even so." He raised her hands to his lips and kissed them. "Now you're deliberately tormenting me. I think you love me and want to say yes, but you want to prolong my suffering a bit longer."

"Top marks for you, Mr. Purcell." Happy tears sprang into her eyes, and she kissed him.

They married in December in Oxford, in her father's favorite chapel. Rather than choose between her sisters, Helen had two maids of honor. Margaret attended as Penelope's guest, as did Margaret's one-year-old daughter, Hestia, upon whom Penelope doted like a proud auntie. As for Helen's mother, after Arthur became her son-in-law and his generosity could not be mistaken for charity, she permitted him to pay off her mortgage. Banbury Cottage was hers, and she need never again fear losing it.

Until Helen accepted Arthur's proposal and accepted invitations to meet his family, she had not understood that his late grandfather was the founder of Purcell Products Company, which manufactured sewing machines for the home and for industry, as well as small motors and parts for other factory machines. In hindsight, it all made sense—the expensive car, the fine suits, his father's insistence that he help run the family business. Of course she had heard of the Purcell empire—she and everyone else in Britain—but it would have been impolite to inquire if he were connected to *those* Purcells.

"I assumed your family business was a small concern," Helen admitted while they honeymooned in Greece. "A wholesale warehouse or a repair shop."

Arthur burst out laughing. "You thought I would throw over my Oxford education to work in a repair shop?"

"Well, why not? The pull of filial duty is very strong, and it's really only a matter of scale."

"You're absolutely right, of course," he said, kissing her.

They kept two houses, a gracious estate in Oxford and a smaller residence in Birmingham, where the Purcell Products Company's largest factory was located, and where Arthur was required to work alongside his father and elder brother four days a week. The other three he worked from his study at their home in Oxford, which they both preferred. Often Helen did not accompany him to Birmingham, where she knew no one and had little to do but read and walk alone and wait restlessly for Arthur to come home from work at the end of the day. As the months passed, she remained in Oxford more often, contenting herself with her family, friends, books, and suffrage business in her husband's absence. She could walk to Banbury Cottage to visit her mother and Daphne anytime she wished, so although she once believed she must abandon her beloved childhood home to save it by marrying Arthur, the cottage had not been lost to her at all.

Helen and Arthur had planned to celebrate their first anniversary in Italy, but after war broke out, they decided a romantic winter holiday in Scotland would be just the thing. By that time, the British Army had added half a million soldiers to the ranks and was urgently calling for more recruits, and each one would need uniforms, weapons, and ammunition.

Industrial sewing machines were essential to the war effort, but several smaller Purcell Products factories were swiftly converted to other war manufacturing. Helen found the change disconcerting, although she well understood the necessity. Great Britain had not sought this dreadful war, but Britons would not shirk from their duty to stand by their allies and to defend their island. She saw less of Arthur as the military's insatiable need for shells and bullets obliged him to work longer, more strenuous hours. In this he was not alone. The entire munitions industry, longtime arsenals and newly converted facilities alike, could not keep up with demand. The situation collapsed into a political controversy known as the Shell Crisis, which Arthur had no time to follow in the press but Helen did, almost obsessively.

By late spring, the Shell Crisis came to a tipping point. Parliament established the Ministry of Munitions to oversee production and to ensure that the military's needs were met. "I welcome the organization of a formalized chain of command," Arthur told Helen wearily after another very late night at the factory. "However, I'm concerned about potential bureaucratic interference that could decrease efficiency."

For a moment Helen could only stare at him, astonished to hear her scholarly husband sounding so much like a businessman industrialist—but of course, that was what he had become. She often missed the clever fellow who loved to read Greek poetry aloud to her, but she told herself he would return to her after the war. In the meantime, she must support him so he could carry out his important duties. A victory for the Allies depended upon his success.

One evening in June, after a fortnight away, Arthur at last came home to Oxford. Helen hurried to meet him when she heard him enter the foyer, but as she embraced him and took his hat and briefcase, she silently noted the shadows beneath his eyes, the deep groove that worry had carved between his brows. She had held dinner back in anticipation of his homecoming, and while they ate, he told her that his father had asked him to take charge of a long-idle sewing machine and parts factory. Shuttered in the depression that had followed the Panic of 1873, it had been cleared of its antique machinery and fitted out with modern technology for the manufacture of explosive shells.

By then Helen had learned that a request from Arthur's father was actually a command, and Arthur had surely accepted the promotion on the spot. There was nothing left for her to do but praise and encourage. "Congratulations, darling," she said, reaching across the table to take his hand. "Or rather, I should congratulate your father, for having the good sense to give you your own command at last instead of assisting your brother."

"That's not quite fair. I've learned a lot from Phillip," said Arthur. "He was working in industry while I was still whiling away the days scouring Greek tragedies for elusive meaning."

Helen regarded him for a moment, taken aback. He spoke almost as if he regretted his academic pursuits, the studies that had once enriched his life and had brought the two of them together. "My apologies to my absent brother-in-law," she said, sipping her wine, carefully setting the glass down. "So. Where is this new enterprise? Birmingham?"

"Thornshire, on the Thames on the outskirts of London."

"I've never heard of it."

"Most people haven't. Let's hope the Germans never do." He sighed and clasped his other hand over hers. "Darling, I'll have to remain in London throughout the week. No more jaunts to Oxford for long weekends. I wouldn't be able to visit for months at a time."

"But Oxford is so much closer to London than to Birmingham. I would have thought you'd be able to come more often, not less."

He shook his head. "My responsibilities simply won't allow it, certainly not until this blasted Shell Crisis is resolved. Perhaps for the duration." He sat back in his chair and regarded her bleakly. "We'll keep this house, of course. We're both very fond of it, and you could never leave Oxford forever. I'll let the Birmingham place go. The question is—" He hesitated. "Will you come with me to London, or would you rather stay here?"

She was surprised he had to ask. "Of course I'll come with you, if the alternative is to spend months apart. In case you've forgotten, Mr. Purcell, I love you."

He raised her hand to his lips. "And I love you, Mrs. Purcell."

He smiled, but the relief in his eyes told her he had expected her to refuse. She was too bewildered to ask him why.

4

December 1914–June 1915

LUCY

Lucy's intuition had not misled her. At the 8 December meeting at the FA offices at Russell Square, the club representatives passed an official resolution declaring that they "heartily favoured" organizing a Pals battalion for Lord Kitchener's Army. The Footballers' Battalion, its ranks 1,350 strong, would be made up of professional and amateur players, staff, officials, and club enthusiasts. Since many footballers were small in stature, the usual army height requirements would be waived. While the battalion was undergoing military training in Britain, players would be granted leave to play for their clubs in league and cup matches. The Footballers' Battalion would officially be known as the 17th Service Battalion, Middlesex Regiment, the "Die-Hards."

On 15 December, more than four hundred footballers and enthusiasts showed up for a recruitment meeting at Fulham Town Hall, so many that at the last minute the organizers were obliged to move to a larger room. In rousing speeches greeted by thunderous applause, dignitaries from both football and politics lauded the formation of the Footballers' Battalion as a powerful and irrefutable response to critics who dared question their courage and loyalty. When the patriotic fervor in the room had reached its zenith, the chief recruiting officer for

London took the stage and appealed for men to come forward and add their names to the battalion rolls.

"Were you the first to enlist?" Lucy asked Daniel as he described the event afterward.

"No," Daniel replied. "That honor went to Spider Parker."

Lucy nodded. Fred Parker, the enormously popular captain of Clapton Orient, was married with three children. Lucy and his wife were good friends. She took a steadying breath. "And you?"

"I was tenth." He studied her, awaiting her reaction, his expression pensive. "Tenth of thirty-five."

Although she had been expecting it, his announcement struck her like a physical blow. "I see," she said tremulously. "Thirty-five out of four hundred. Perhaps the other players wanted to speak with their wives before signing up." She knew she should tell him how proud of him she was, but the proper words wouldn't come. All she could think of was the bleakness of his inevitable absence.

"Darling, don't be unhappy." Quickly Daniel enfolded her in his embrace. "I don't have to report quite yet, and even after I begin my training, you and the boys can see me at my matches."

Blinking away tears, she drew back so she could meet his gaze. "That's true," she said, forcing steadiness and courage into her voice. "Your leaving will be more bearable that way. Most soldiers' wives aren't so fortunate."

She refrained from noting that thirty-five recruits fell far short of the battalion goal. Was it unpatriotic to hope that the ranks of the 17th Middlesex filled but slowly, delaying her husband's departure? She could only hope that the war would end before he had to leave England.

There were practical matters to discuss before Daniel reported for duty. Ideally, Lucy thought, they should have discussed them before he enlisted. First and foremost were their finances. Daniel expected to receive a final payment for his work on the Henderson offices soon, but while he was in the service, he would be unable to take on any new projects and his dividends from the partnership would be reduced. He would receive his soldier's pay, and the government provided a separation allowance for families of married recruits, but Daniel's military income would not make up the difference in his lost earnings as an architect and footballer.

"I'll have to draw from our savings," Lucy fretted. "That, or impose some sort of household austerity scheme."

"All will be well," Daniel assured her, and went over the accounts with her again. Lucy acknowledged that she could manage, but only if the war didn't drag on for a year or more, and only if they avoided calamitous household expenses such as replacing the roof or the furnace. Still, every soldier's wife had to confront the same problem of household economics, and Lucy was more fortunate than most. She would manage, because she must.

The cold December days passed and the festive season approached. Lucy did her best to make the holidays merry and bright for the family, but to her, the candles seemed dimmer, the hearth cooler, the carols dissonant. On the morning of Christmas Eve, she and Daniel discovered shocking news in the early papers: On 23 December, a German aeroplane had crossed the Channel and had bombed Dover, the first time Great Britain had sustained an aerial attack. Heavy fog had obscured the view of the skies, so only a few observers had caught even a glimpse of the enemy plane. There had been no warning. A single bomb had exploded in a garden, breaking some windows in nearby homes and frightening everyone, but thankfully no one had been injured. Then, on 26 December, the newspapers reported that on Christmas Day, a German plane had flown up the Thames, presumably headed for London, but had been driven off by anti-aircraft guns near Erith.

Lucy and Daniel mentioned nothing of the thwarted attack to the children. It was unsettling to think that a malevolent German had fixed their city in his sights and might have killed them, and on the day sacred to peace and goodwill. Yet the Germans had already slaughtered countless thousands of Belgian and French civilians; Britons must be prepared to suffer the same. The English Channel afforded them some natural protection from bombardment and invasion, but centuries of history proved that the British Isles were hardly impenetrable. These two German aerial attacks showed beyond any doubt that another assault could and surely would come again, with little or no warning.

The festive season passed, and a New Year began. As recruiting efforts persisted and intensified, the ranks of the 17th Middlesex steadily grew. On 11 January, the first 250 recruits assembled outside West Africa House, Kingsway, under the watchful gaze of their commander, Colonel Charles Grantham, and other officers. Mr. Joynson-Hicks inspected the battalion, and then, as a massive crowd of onlookers roared approval, the 17th Middlesex marched through the streets of London six miles west to White City, where they would encamp at Machinery Hall.

Lucy, Jamie, and Simon were among the spectators on the pavement cheering and waving as the men marched past in a swinging stride, looking more like sportsmen than soldiers in their suits, caps, and topcoats, some with their hands in their pockets, others grinning and nodding to the crowd in appreciation for the hearty sendoff.

"There's Dad!" Jamie suddenly exclaimed, jumping up and down and waving frantically.

Craning his neck, Simon rose up on tiptoe, trying in vain to see over the taller people blocking his view. "Shall I pick you up?" Lucy offered, but Simon regarded her in horror and shook his head. She smothered a laugh. Of course he was much too old for that sort of thing, even if it meant missing one last glimpse of his father.

Not one *last* glimpse, she quickly corrected herself. They would see him in five days at White Hart Lane when Tottenham Hotspur played Bradford City. Daniel had already warned Lucy that he was ordered to return to White City immediately after the match, but he hoped to spend a few minutes with them before he was ushered away.

Lucy watched Daniel march off with his teammates until she could see him no longer. As the crowd began to disperse, she and her sons walked home, the boys talking excitedly about the parade and the war, Lucy murmuring responses when required, her thoughts elsewhere.

When they were nearly home they passed a schoolyard with a small seven-on-seven pitch. "May we play, Mum?" Jamie begged, pointing to a ball abandoned by one of the goals. "All of us, please? Just for a minute?"

"Please, Mum?" Simon quickly chimed in, tugging on her coat.

Suddenly memories flooded Lucy, impressions of the many meetings of the Dempsey Four, as Daniel called them, impromptu matches in the garden or any level field wherever they happened to be when the mood to play struck. They usually divided themselves into pairs, with Daniel and Simon versus herself and Jamie, and with Daniel playing at about ten percent and Lucy working as hard as she could, the teams were almost equally matched. The boys usually preferred to play sons versus parents, probably because they invariably won. Daniel played just well enough to challenge the boys and help them improve their skills, and Lucy couldn't bear to steal the ball away from her children or block any of their shots on goal.

Knowing her sons both wanted to play forward like their father, Lucy volunteered to play keeper. They exchanged a knowing look. "That'll be fine, Mum," said Jamie, "as long as you play properly."

"Truly try to block our shots," Simon added, for clarification. "We insist."

Lucy laughed. "Very well, I shall do my best. You may regret insisting."

The ball was slightly deflated, and the pitch was wet from melted snow, but they had good fun all the same. Lucy was quite overmatched, competing in her heeled boots and long skirt against two energetic, athletic boys in proper trousers and shoes, but she did not mind. How wonderful it felt to run until she was breathless, as the vigorous exercise and the thrill of competition drove her loneliness and worry to the back of her mind. Or so it was until Simon stole the ball from her, made an astonishing feint around Jamie, and took a shot on the goal—only to watch the ball bounce off the crossbar and soar out of bounds.

"Bad luck, Simon," Jamie called as his younger brother trudged off to collect the ball. "Brilliant run, though."

Simon did not acknowledge the praise or comically lament the missed goal, as he would have almost any other time. Lucy and Jamie exchanged a puzzled glance as, instead of kicking the ball in, Simon picked it up and carried it slowly back to them. Only after he drew nearer did Lucy realize that his eyes were red and he was breathless from the effort of holding back sobs. "It's not as much fun without Dad," he gulped, a tremor in his voice.

Lucy felt a pang of grief. "No, it isn't, is it, darling?"

Jamie jogged over and put an arm around his brother's shoulders. "It's all right, Simon. You still have me, and Dad will be back as soon as he wins the war."

Simon shrugged, eyes downcast.

"You know what we should do?" said Jamie, inspired. "We should train every day while Dad's in the army, and get really, really, really good. When he comes home and the Dempsey Four take the pitch again, he won't believe how much better we are!"

"If we play sons against parents, we'll catch him by surprise," said Simon, warming to the idea. "We could finally beat him for real! Unless—" Simon turned to Lucy. "You won't tell him about our secret plan, will you, Mum?"

"Well, I don't know," said Lucy. "After all, Dad and I would be on the same side. It seems like I ought to warn him."

"Don't warn him, please?"

Lucy pretended to consider it. "Very well. I'll keep your secret, but only because—" Suddenly she ran toward them and pretended to tickle their tummies. They darted away, yelping with laughter and surprise even though she didn't get close enough even to brush their coats with her fingertips. "Because secret plan or no, Dad and I will win because we're better!"

They headed home, pleasantly fatigued and ready for tea. Simon's spirits had improved tremendously, and he chattered happily as they walked along. Jamie caught Lucy's eye and nodded, as if to say that as the elder brother, he understood his responsibility to keep Simon's spirits up while their father was away. Her heart welled up with love, laced with gratitude and regret. A few weeks ago, Jamie's greatest concerns were football, school, and whether his parents would ever let him get a dog. At his age he should not have to bear heavier burdens, but how thankful she was that he was willing.

Lucy counted the days until Daniel's next match. There were only five, but she was as impatient to get through them as if there were five hundred. When she and the boys arrived at White Hart Lane, they passed the usual protestors with their signs and slogans, still determined to shut down the sport despite the existence of the Footballers' Battalion, not to mention the hundreds of players who had joined other regiments and were already in the fight. The protestors could not be unaware of the Footballers' Battalion either, as numerous posters within their line of sight announced its formation and encouraged spectators to join.

Simon paused to read one poster aloud. "Play the Greater Game!" his voice rang out as his eyes followed the bold black type over an illustration of three grinning Tommies kicking a German *Pickelhaube* like a football across a grassy field. "Sharpen up, 'Spurs. Come forward now to help to reach the goal of victory. Shoot! Shoot!! Shoot!!! And stop this Foul Play! Join the Football Battalion of the DIE-HARDS (17th Middlesex)!"

Lucy assumed Simon was reading aloud for her benefit and Jamie's, but then he directed a steely glare over his shoulder toward the protestors. "Maybe you should make a Grousers' Regiment," he called, but they didn't hear him over their own voices.

"That'll do, Simon," said Lucy, taking his hand and briskly leading the boys inside.

They took their seats among the other spectators. More continued to arrive up until the moment the match began, but it seemed to Lucy that attendance, though still approaching twelve thousand, was down slightly despite the clear and sunny weather, as perfect a day for football as one could hope for in January. Daniel started at center forward, but although he did not make any errors, his pace lagged. In the first forty-five minutes of play he had two shots on goal, one of which went wide, the other deftly blocked by the Bradford City keeper.

"Dad seems tired," said Simon, frowning, as the whistle shrilled to mark the end of the first half.

"No doubt he is," said Lucy, her gaze fixed on Daniel as the players cleared the pitch. "Remember he has army training every day now, too, all that marching and drilling, while most of these other players have only football."

As she spoke, two gentlemen in fine topcoats and tall black top hats approached a platform that had been erected between the pitch and the spectators' seats on the home side. The audience had been buzzing with hundreds of conversations, but the din subsided as the gentlemen took center stage. The younger introduced the elder as William Joynson-Hicks, MP, the founder of the Footballers' Battalion. Thunderous applause greeted him, perhaps less to welcome the MP than to express appreciation for the troops. Mr. Joynson-Hicks certainly did his best to add to the battalion's numbers in his rousing speech, urging all fit and loyal men to follow the example of the players they admired as well as other enthusiasts like themselves and join the 17th Middlesex.

"I tell you, whether the censor likes it or not, that we are holding our own in Flanders and no more," Mr. Joynson-Hicks warned the audience, the rapt and the skeptical alike. "Unless we are able to send enormous numbers of reinforcements by April and May, we shall do no more than hold our own. Germany has Belgium and the North of France in a vise, and she will not give up until she is forced, step by step, by the lives of Englishmen and Frenchmen."

Lucy felt a stir of unease as the audience broke out in applause punctuated by low rumbles of concern. She hoped Jamie and Simon would not follow the thread of logic and realize that one of those lives sacrificed might be their father's.

"I am inviting you to no picnic," Mr. Joynson-Hicks continued, his gaze stern and unflinching as it traveled over the audience. "It is no easy game against a second-rate team. It is a game of games against one of the finest teams in the world. It is a team worthy of Great Britain to fight!"

A louder cheer erupted on both sides of the field, and Lucy too, though silent, joined in the applause. Lastly Mr. Joynson-Hicks announced that would-be Die-Hards could enlist either there at White Hart Lane or later at West Africa House, then he bowed and left the platform.

The second half soon began, and after another hard-fought forty-five minutes, the match ended in a scoreless draw. The spectators began filing out of the stadium, and when Lucy saw Daniel breaking away from the other departing players to approach the stands, she urged the boys along and hurried to meet him at the edge of the pitch. Heedless of the amused onlookers, she flung herself into Daniel's embrace, holding him even tighter when he laughed and warned her that his sweaty, soiled jersey might offend. The boys cared even less about a bit of perspiration, if they even noticed; they flung their arms around their father, exclaiming about how happy they were to see him, how much they had enjoyed the game, how they wished he could come home with them but understood it was against the rules.

When Lucy felt Daniel pull away, she remembered that he had only minutes to spare, and she reluctantly released him. "How are your accommodations?" she asked, taking his hand and falling in step beside him as he headed toward the players' exit, glancing over her shoulder to make sure the boys were following after. "Comfortable, I hope."

"Tolerable, but only just. Machinery Hall is cavernous—vast, dim, cold, and damp. Already men are reporting sick with colds and coughs. Don't worry," he quickly added, no doubt spotting the concern in her eyes. "We're supposed to move to another building soon. How are things at home?"

"Oh, we're fine, everything's fine," she said, forcing cheer into her voice.

"I'm keeping an eye on things, Dad," Jamie interjected, jogging to catch up.

"Me too," Simon chimed in eagerly, beaming up at his father.

Lucy had to smile as Daniel laughed and tousled the boys' hair. She only wished he wasn't walking so briskly, so they could prolong their time together. "I've reunited with some old teammates," he said, turning his smile back to Lucy, warming her all over. "Vivian Woodward and Walter Tull have joined the Footballers' Battalion."

"Have they?" Vivian Woodward, an icon of British football, had represented England on the same Olympic gold medal teams as Daniel. He had played many years for Tottenham Hotspur, and had retired in 1909 only to sign with Chelsea a few months later. Walter Tull, the son of an immigrant from Barbados and an English girl from Folkestone, was one of the few Black players in the league. He had joined Tottenham the same year Woodward left it, but after two difficult seasons in which he had endured racist derision from opponents' crowds, he moved to Northampton Town, where he switched to half-back and was enjoying some of the best performances of his career, and, Lucy hoped, much kinder treatment from audiences. "It must feel good to be playing for the same side again."

"Indeed it is." Daniel squeezed her hand as they reached the players' exit. "Darling—"

"I know. You have to go." She lifted her face to his and kissed him, but the kiss lasted only a moment as the boys interposed, flinging their arms around their father again and chorusing their goodbyes. He kissed them each on the top of the head and bade them all farewell, his smile broad, his voice hearty, although Lucy sensed the effort behind it. Then he turned and hurried off to join the other players who were due back at White City within the hour, soldiers once more. The boys called out goodbyes after him, and he paused once to wave back. Then he turned a corner and was gone.

Lucy tried to ignore the hollow loneliness in her chest, the lingering ache for Daniel that their brief reunion had only intensified. "Come along, boys," she said briskly, rearranging her soft woolen scarf around her neck. "Tea and homework await us—well, only tea for me."

"Lucky," Jamie grumbled cheerfully, and she had to smile. He complained about school as boys his age were expected to do, but he was clever and earned top marks.

On the way home, Jamie and Simon reminisced about the match, so engrossed in conversation that they were oblivious to her quiet sadness, and even to the changes that the war had wrought on their city. Perhaps they had already become accustomed to the ubiquitous recruiting posters on nearly every building and shop window, on the sides of passing omnibuses and tramcars, and to the young men who had answered their summons. Soldiers clad in khaki were everywhere, while middle-aged men too old for Lord Kitchener's Army patrolled the streets in the uniforms of Special Constables. Uniformed guards were posted at railway stations, and sentries at all public buildings—excluding Parliament, where ancient law decreed the military should not be present. The bells of clock towers no longer chimed the passing of the hours, and at night their faces were not illuminated, lest the sound and light act as beacons for German aircraft. Only a few months ago, Jamie and Simon would have stopped short at any one of these unexpected sights, astonished and curious, but by now all had faded into the background noise and clutter of the city.

Not so for Lucy. As they walked along, her gaze fell upon a poster that made her breath catch in her throat. "Women of England!" the bold headline thundered. "When the War is over and your husband or your son is asked 'What did *you* do in the Great War?' Is he to hang his head because you would not let him go? Women of England, do your duty! Send your men *to-day* to join our Glorious Army."

She felt no satisfaction in knowing that the poster's admonition did not apply to her. Would Daniel have resented her someday, she wondered, if she had not willingly let him go? Not that she'd had any say in the matter. Daniel had not included her in his decision, although it profoundly affected them both. If he were killed, God forbid, would she regret not entreating him to stay? She would far more willingly bear his resentment than his death—a possibility simply too horrible to contemplate.

By mid-February, the ranks of the 17th Middlesex had grown to 850 men, still insufficient for a complete battalion. Lucy learned from another football wife that Mr. Joynson-Hicks and Lord Kinnaird, president of the Football Association, had written to every professional player in England who had not yet enlisted to encourage them to do so. "A large number of some of the finest players in the Kingdom have already joined the Battalion, but we do not see your name amongst them," the gentlemen noted. "We do urge you as a patriot and a footballer to come to the help of the country in its hour of need."

No letter had come to the Dempsey household, but Lucy's chagrined friend had shown her the copy her husband had received. "You're lucky," her friend grumbled, folding the letter and shoving it into her pocket as if she would rather burn it. "*You* can hold your head high."

"I don't feel particularly lucky," Lucy admitted, but her friend brushed that aside and declared that she wished their places were reversed.

A few days later, Lucy and the boys again traveled to White Hart Lane to see Tottenham Hotspur play Notts County, a 2–0 victory in which Daniel made a brilliant assist and several daring shots on goal which, though blocked, turned the momentum in the team's favor and electrified the crowd. When the family briefly reunited afterward, Lucy was relieved to see that Daniel seemed to be in good health and excellent spirits. He was getting used to the military training, he said, and their new quarters, while still cold and drafty, were drier and more comfortable than Machinery Hall. "We often march from White City to the West End," he said, already edging backward toward the exit, nearly late for the return journey. "We halt at Hyde Park near Marble Arch and rest for a quarter hour before marching back. It would be grand to see you all there by chance."

"Never mind chance. If you could tell us when, we would arrange it," Lucy called after him, but he had broken into a jog and was already out of earshot.

A fortnight later, Lucy was astonished by an unexpected visitor on their doorstep carrying a vegetable crate—her brother Edwin, clad in the khaki tunic, jodhpurs, service cap, and high leather boots of the Royal Flying Corps. "Edwin," Lucy exclaimed, frozen in the doorway, her hand on the latch. "What on earth—"

"Special delivery," he said, inclining his head to the crate. "Are you going to let me in?"

"Of course." She opened the door wide and moved out of the way.

He threw her a grin as he crossed the threshold, and she trailed after him to the kitchen. "Gifts from Mother's garden," he said, setting the crate on the table with a soft thud. Drawing closer, she glimpsed bundles of rhubarb, parsnips, and brussels sprouts tucked alongside jars of preserves and a paper sack from Brandt's bakery. "The bread is from me. I know how much you love their rye."

"Thank you," she replied automatically, still astounded that he was standing there, dressed as he was. "What are you doing in that uniform?"

"Showing off, obviously," he replied. "What? You didn't think I was going to let Daniel seize all the glory, did you?"

"What does Mother think?"

He shrugged. "She was a bit weepy at first, as one might expect. Alice, however, is exceedingly proud." He glanced around. "Where are my nephews?"

"At school, as you should have guessed, being a schoolmaster yourself."

"Of course." He removed his hat and tapped it lightly against his leg, peering around as if he hoped to spot the boys in a corner regardless. "Bad luck. I must be off in a moment. Looks like I'll miss them."

A trifle exasperated, Lucy gestured, taking in his uniform from boots to cap. "Are you planning to fly aeroplanes, then? Do you even know how to drive an automobile?"

"I'd make a fine pilot with the proper training, thank you very much," he replied archly. "But no, I've been assigned to an observation balloon squadron—aerial spotting, photography, and so on, or so I'm told."

"That sounds . . . ridiculously dangerous." Lucy took a steadying breath. "If you're here, who's running the village school?"

"Alice, naturally. She was a teacher before she married me."

"But she's never run a school before, surely."

"True, but many women are taking on duties they've never done before, or that no Englishwoman has ever done, for that matter."

Suddenly he strode forward, clasped her by the shoulder, and kissed her on the cheek. "I must be going. This errand cost me a few favors to be named later, and I don't want to push my luck."

She held on to his tunic and pulled him into a fierce hug. "You could have volunteered to be a mechanic and stayed safe on the ground," she admonished him, as if she were the elder sibling. "Be careful. If you upset Mother, you'll answer to me."

"I'm duly forewarned."

Then he was off, his visit so brief and his departure so sudden that she could almost believe she had imagined it, except for the crate of vegetables he had left behind. So Edwin had decided to do his bit. She should have guessed he was considering it, given the way he had lightly mocked the idea the last time they had all gathered around their mother's table back home in Brookfield. Daniel had enlisted. Her brother George was caring for wounded veterans who had returned to Surrey. Alice was running a school, of all things, truly impressive. And what had Lucy contributed to the war effort? A few pairs of knitted socks each month, a trifle.

Jamie and Simon were in school most of the day, no longer requiring her constant attention. Surely she could do more for the war effort.

She had heard that as more men quit their jobs to enlist, replacement workers were urgently needed. Earlier that month, the government had established a Register of Women for War Service, a unified list of women throughout Great Britain who were willing to work in industry and agriculture for the duration. After much careful thought, and without seeking Daniel's permission, she submitted her name to the register.

Then she waited, hoping to be called in for an interview, but weeks passed and no one from the bureau contacted her. Feeling slighted, the next time the women of the neighborhood gathered to knit for the soldiers, she mentioned the register in an offhand manner, too embarrassed to admit that apparently she had been deemed unqualified and unworthy of a response.

Her friends' responses surprised her. "The Register of Women for

War Service indeed," said Gloria scornfully. "That's useless. I don't know of a single person who's found work through that list. It's a shambles, poorly planned and even more poorly executed."

Before Lucy could ask her how she knew this, other friends chimed in assent. "For war work, all anyone needs to do is put their name down on the ordinary register at a Labour Exchange," one noted, shrugging. "My brother-in-law is a manager at Brunner Mond in Silvertown, and he says they get most of their new workers from the Labour Exchange. Some applicants just show up at the factory gate and ask around until someone directs them to the proper office."

"That's what I've heard too," another knitter replied, to a murmur of assent.

Lucy muffled a sigh, glad that she had mentioned it. She regretted that she had wasted her time with the new register, but at least she knew their silence was nothing personal.

She and the boys saw Daniel play at White Hart Lane several more times in April, but the season drew to a close with the Cup Final on 24 April when Sheffield United beat Chelsea 3–0. Daniel said it was generally understood that, having completed a full season despite heated controversy and the loss of players and spectators to the military, professional football would now be suspended for the duration. An official announcement was expected soon.

That same day, the 17th Middlesex received orders transferring them to Hombury St. Mary, near Dorking in Surrey, where they would continue their training in camp on Joynson-Hicks's country estate. The battalion now totaled fourteen hundred men, two hundred short of what the War Office now required. Although Daniel was moving farther away, it was comforting to know that he would be near Brookfield, in familiar country where she could imagine him clearly.

Again Lucy and the boys turned out to watch the 17th Middlesex march through London on the day of their departure, this time from White City to Waterloo station. Lucy searched the ranks in vain for a glimpse of her husband, but it was impossible to spot him in the lines of uniformed men marching in stoic unison, unrecognizable as the casual, smiling recruits who had paraded from West Africa House to Machinery Hall in January.

The next day while the boys were in school, Lucy went to the local Labour Exchange and hesitantly asked the clerk how she might register for war work. "You came to the right place," the white-haired man replied, smiling kindly at her shyness. "What sort of work would you like to do?"

On her way to the exchange, Lucy had passed a recruiting poster she had never seen before, an illustration of a young woman clad in a factory worker's tunic and trousers, smiling demurely as she pulled a cap over her dark hair. In the background, more sparingly sketched to suggest a great distance, a soldier loaded shells into an enormous gun. "On Her Their Lives Depend," the caption above the images declared. At the bottom left, in smaller letters, appeared the phrase "Women Munition Workers," and at bottom right, "Enroll At Once."

"Munitions work, please," she told the clerk. If soldiers' lives—Daniel's life—depended on women making shells and bombs, then that was what she needed to do.

"You want to be a munitionette, eh?" The clerk's smile broadened. "Ever done any sort of factory work before, love?"

"No, sir," she said, heart sinking. "I helped with the paperwork and books for my father's medical practice before I married, but now I keep house."

"A doctor's daughter? Why not go for a nurse?"

"No, thank you," she said emphatically. Growing up, she had seen enough to know that nursing wasn't for her.

The clerk eyed her for a moment, thoughtful. "You seem like a good lass. Clever too. If you don't mind me asking, is your need for wages urgent or can you stand to wait a bit?"

"I wouldn't say the need is urgent, but I will need to earn a wage eventually."

"Then may I suggest you enroll in a training course first? You'll make but a fraction of what you could earn if you go straight to work, but you'll learn essential skills and earn a higher wage later, when you do take a job. It's a sound investment of one's time, for those who can afford it. You'll qualify for better jobs too—cleaner, less dangerous, you know."

Lucy did *not* know, not really, but she kept that to herself. "And I'll be paid during the training?"

"Yes, love, but as I said, not as much."

A small wage was better than none, something to help make up for Daniel's lost income, and to compensate for the steadily increasing prices for everything from bread to clothing to matches.

"I'd like to enroll in the training program, please," she said. The clerk beamed and gave her the proper forms.

She felt a surge of pride as she signed her name, and yet not without a nervous fluttering in her stomach. Never in her life had she ever imagined herself working in a factory, much less as a "munitionette," surely a new addition to the English language, a word for daring girls who made bombs and shells and she could hardly imagine what else.

She would be mad to think she fully understood what she was getting herself into. She could only hope that the war would end in victory for the Allies before she finished her training course.

5

JULY 1915

APRIL

Eight weeks passed without any word from Marjorie. In the aftermath of her sudden departure, Mrs. Wilson had wrung April dry of information, asking where Marjorie had gone, how long she had been planning to quit, whether April had known, if she had tried to persuade Marjorie to stay, if not why not, if April understood what a dreadful thing it was to throw a household's smooth operation into chaos by quitting without properly giving notice—

"Don't scold *me*, ma'am," April finally exclaimed when she couldn't endure the interrogation any longer. "I've done nothing wrong. I'm still here!"

Mrs. Wilson fixed her with a hard, level stare. "Indeed you are, Ann, and until I can hire a replacement for Mary, the duties of four will be divided among three."

With that, April was dismissed. Setting her jaw so she wouldn't talk back, she bobbed a stiff curtsey and got back to work. It wasn't fair that *she* had to suffer the resentment of the entire household because she was Marjorie's best friend and Marjorie herself was out of reach. Wasn't it punishment enough that April would now have one-third of Marjorie's work added to her usual duties, without an equal increase in pay? Perhaps she too ought to be furious with her absent friend, but she couldn't be, not when everyone's surly behavior proved that Marjorie had been quite right to go. Not sneaking off in the middle of the night like a prisoner fleeing a cell—that was badly done—but taking another job if she wished. Why shouldn't she, especially if she contributed to the war effort? It was impossible to miss the newer posters in the village scattered among those urging men to enlist, the ones practically begging girls to take jobs in industry and agriculture. "On Her Their Lives Depend," one declared, calling on women to become munitions workers. "Be the Girl Behind the Man Behind the Gun," another entreated. Which was more important: ridding the floors of Alderlea of every last speck of dust, or building the weapons the Tommies needed to fight the Germans?

April hoped the answer was obvious. Even if the Rylances considered only their own son's safety, they should support any measure that kept Great Britain's military forces amply supplied.

April was aware of the urgent and growing need for war workers only because Mrs. Wilson had permitted her to run errands to the village again, not because the incident with young Peter Bell was forgiven, but out of necessity. No other servants could be spared. Soon after Marjorie quit, two footmen and a groom had resigned in order to enlist, further reducing the staff even as billeting the soldiers increased everyone's workload. At that rate, by December, neither the housemaids nor the footmen would be able to put together a side for the Boxing Day football match. Perhaps that was just as well. Marjorie had been the housemaids' goalkeeper, and they didn't stand much of a chance without her.

April thought of her fellow Alderlea servants every time she spotted a particular poster near the village hall. Titled "Five Questions to Those Employing Male Servants" and unadorned by pictures, it offered a list asking whether chauffeurs, gardeners, and other staff shouldn't be putting their skills to use in service to their country rather than a single estate or household. "A great responsibility rests on you," a caption at the bottom admonished employers. "Will you sacrifice your personal convenience for your Country's need?"

April could well imagine Lady Rylance haughtily replying, "No, indeed we shall not." Yet as far as April knew, their consent wasn't required. Likely the new recruits from Alderlea had seen the poster and realized for themselves that Britain needed them more than the Rylances did. Or maybe, like Marjorie, they had glimpsed opportunity and adventure elsewhere and had decided to pursue them.

Whatever it was that had inspired them, April marveled at their daring. In comparison, she was dutiful, safe—and as dull as dirt. Her mum and elder siblings would have reminded her that she had a guaranteed wage and a roof over her head, which was more than some people could say, so she ought not to complain. But it was hard to stay stuck when friends were setting off on adventures and she never ventured any farther than the nearest village.

Often when April was sent there on errands, she collected the post. On one such day in late May, she spotted a letter in the bundle, postmarked from London and addressed to her using her true name in unfamiliar handwriting. She longed to pause at the footbridge and tear open the letter on the spot, but Mrs. Wilson and the cook were impatiently awaiting her return, so she tucked her letter into her pocket, hastened through the errands, and hurried back to Alderlea as fast as she could with the heavy market basket weighing down her arm.

Later, after luncheon, when she was meant to be mending a soldier's torn trouser cuff and replacing a lost button on the master's favorite riding coat, she waited until the other servants had left the room. Then she draped the pile of mending on her lap, concealed the letter within the folds of fabric, and eagerly read.

The letter was, as she had hoped and expected, from Marjorie. "I hope this letter finds you well, dear friend," her friend wrote cheerfully. "I hope it finds you *full stop*, and that Mrs. W doesn't throw it on the fire on sight, the bitter old crone! Sorry, I should scratch that bit out. I forgot she's your auntie."

"My mum's cousin," April murmured out of habit, glancing up from the page to make sure no one had overheard. Marjorie wrote exactly the way she spoke, airily with a bit of cheek. April could almost hear her voice as she read on.

Marjorie was doing wonderfully well, she wrote. Leaving Alderlea had already proven to be the best decision she'd ever made. She had found work at Woolwich Arsenal, a vast factory complex on the south bank of the Thames in southeast London. She had started out in custodial, sweeping up metal filings that fell around the machines that made shell casings, but after proving herself to be prompt, tireless, and reliable, she had recently been promoted to making fuse caps. "It's a bit dull and repetitive," she admitted, "but you pick up speed once you get used to it, and it's easier on the knees than scrubbing floors. I only make caps, not a whole shell. No one makes an entire shell alone. Everyone makes their own small part and someone farther down the line puts them all together. So I never see the shell-cases being made or a finished shell with my little perfectly made cap firmly in place, but I know it's there, and that's enough for me!"

She worked six days a week, from seven in the morning until seven at night, with one hour off for lunch and a half hour for tea. Other munitionettes worked the exact opposite hours, on the night shift. The various departments staggered their shifts so that not all of the thousands of workers were arriving and departing at the same time. "Some of the girls grouse about the twelve-hour shifts," Marjorie wrote, and April imagined her rolling her eyes. "Isn't that comic? With the breaks, we work only ten and a half. It's clear *some people* have never had to rise at half four, and work fourteen hours or more, and then drag themselves up three flights of narrow stairs and collapse into a cot in the attic, too tired even to dream."

Neither did Marjorie, anymore. Now she shared a cozy room with another girl in a hostel reserved for Woolwich munitionettes. She had a bed to herself—a proper bed, not a cot—and a small bureau of her own and the entire left half of the wardrobe. "I don't have enough clothes to fill it properly yet," she noted, "but I soon may, the wages are so good."

How good? April wondered, feeling the tiniest prick of envy.

The hostel was too far from the arsenal to walk without adding hours to the trip, but it was impossible to find affordable rooms any nearer. Houses had been built for munitions workers on the Well Hall Estate in Eltham, but those had filled up long ago, and newer workers had to find lodgings farther and farther out. "Our landlady wakes us at five o'clock," explained Marjorie, "and we snatch a hasty breakfast before dashing out the door at half five to catch the tram. It's not so bad, really."

Not so bad? What a luxury it would be, April thought wistfully, to have an extra half hour of sleep every morning!

It was a long, tiring journey to and from the arsenal on the overcrowded trams, but Marjorie traveled with her new friends and they made a merry time of it, gossiping and laughing together as they rode, smiling and chatting with the Tommies if any happened to be on board, so handsome and dashing in their uniforms. Sometimes the soldiers asked the girls to write to them, and sometimes the munitionettes agreed. Marjorie already had two pen pals and her roommate had three.

"Now you understand why I've been too busy to write to you," Marjorie concluded. "Please ignore my poor example of friendship and write back to me as soon as you can."

April did as her friend asked, but the Alderlea gossip she shared was embarrassingly dull compared to Marjorie's adventures. April's letter contained mostly questions. How had Marjorie managed to get hired without a character from her previous employer? How much better *were* her new wages, exactly? Was there anything about Alderlea she missed—the lovely countryside, the cook's delicious meals, anything?

"Only you, dear chum," Marjorie responded a fortnight later. She then proceeded to describe all the larks she was having with her new coworkers, from songs around the canteen piano at tea to the impromptu football games after lunch and bicycle rides through the park on their days off. Twice she and her roommate had given white feathers to lads in civvies on the tram, embarrassing the hapless shirkers so thoroughly that they had each disembarked at the next stop, red-faced and eyes downcast. Marjorie's half of the wardrobe no longer looked quite so bare thanks to the lovely cotton summer frock she had just purchased, and she had treated her mum to a gorgeous silk shawl for her birthday. "She adored it, of course," Marjorie remarked. "She's quite forgiven me for leaving Alderlea now."

As for her wages, depending upon the number of hours she worked and whether she took on any overtime, she typically earned around three pounds a week.

April gasped. Three pounds! That was nearly thirty times what April earned as a housemaid, not including room and board and whatever came to her on Boxing Day. No wonder Marjorie could afford larks and finery and gifts for her mum.

But the wages were only part of it, Marjorie explained, turning uncharacteristically earnest. At Woolwich she felt appreciated, and she needed only to hear news from the front to understand just how essential she was to the war effort. She was "The Girl Behind the Man Behind the Gun," doing her bit for King and Country, respected and admired and independent. No one at Woolwich would dream of calling her anything but Marjorie, her own proper name, except to call her "Miss" or "Dearie" or some other fond nickname, which she rather liked. And while it was true that munitions work was by its very nature dangerous, the risks she took were far less than those her brothers and all the other Tommies faced in France and Belgium.

"There's always room for one more here at Woolwich, you know," Marjorie wrote in closing. "Your auntie will forgive you eventually, just as my mum forgave me. (The gift of a shawl may be required.) Say you'll come soon, or you may receive a white feather from me in the next post!"

"You wouldn't dare," April murmured, aghast. Quickly she folded the letter, slipped it into her apron pocket, and carried on with the mending. It wasn't fear that kept her from flitting off to London and joining Marjorie in the arsenal—well, fear of her mum and Mrs. Wilson, maybe, but not fear of the job itself. From the sound of it, munitions work was easier than domestic service, the pay was much better, and the girls were treated with respect. Why shouldn't she go? When she told her mum how much more she could earn as a munitionette, surely she would not object. Mrs. Wilson had hired a girl from the village to replace Marjorie; likely she could find someone to take April's spot, especially if April gave her two weeks' notice.

Or maybe one week would be enough. Now that April had made up her mind to go, she was eager to get on with it. She would tell Mrs. Wilson the next morning, after breakfast, and soon thereafter, her own adventure would begin.

She slept poorly, dreading an unpleasant confrontation in the housekeeper's chamber, but to her relief, Mrs. Wilson accepted the news with calm resignation. "At least you did the proper thing by giving notice rather than stealing away in the middle of the night," she said, sighing. "For that reason, I shall provide you with a character. I trust it will help you secure another position, both now and after the war, God speed the day."

"Yes, ma'am. Thank you, ma'am."

Mrs. Wilson sat back in her chair and studied April over the top of her glasses. "A word of caution, Ann. You're usually a sensible girl, but you're too easily influenced by those you admire and hope to impress. Think twice before you stumble blindly after Mary, or she may lead you into trouble."

"Yes, ma'am." April hesitated, cheeks burning. "Are you going to tell my mum I've resigned?"

Mrs. Wilson's eyebrows rose. "Certainly not. That responsibility belongs to you alone. Good luck."

April managed a weak smile as she bobbed a curtsey and hurried off to work. One week more, she reminded herself as she turned the mattresses and made up the beds. Four more days, she thought as she hauled the heavy baskets of soldiers' damp, freshly laundered clothing outside and hung it on the line to dry. She waited until the day before her departure to write to her mum, determined to be settled in London with a new job by the time her family got the news.

The same day she posted her letter, she received another from Marjorie. "If you're coming, don't look for me at Woolwich," she had hastily scrawled. "It wasn't easy to persuade the foreman to give me a leaving certificate, but now he has done, so I've hired on at Thornshire Arsenal. It's a newer factory, not the building but everything inside, and the canteen is loads better. (The food at Woolwich was dreadful. I didn't mention it before because I didn't want to put you off.) I'm in one of the Danger Buildings now, which means better pay. Grand!"

April felt her stomach drop as she read the letter over again. What was a leaving certificate—as in school?—and why did Marjorie need one? Did April need one to work? Where was Thornshire Arsenal? Was Marjorie even still in London? What was a Danger Building, and was there a place for April in it, if she could find Thornshire, and if the ominous name didn't scare her away? All of her plans seemed suddenly thrown into shambles, mere hours before she meant to depart. Maybe she should ask Mrs. Wilson if she could stay after all. That would be the safest thing, the surest thing.

Then she remembered the better wages, her brother Henry training with the Lonsdale Pals in Yorkshire, and the soldiers' desperate need for munitions. She imagined opening a letter from Marjorie only to find a white feather tucked inside, and she knew she must go through with her plans.

The next morning, she rose at the usual hour, but she put on her nicest summer dress instead of her uniform. She packed her few other belongings into the same pasteboard suitcase she had brought with her to Alderlea nearly five years before and had not used since. After one last breakfast at the servants' table, where everyone genuinely wished her well, she collected her final wages and her letter of reference, thanked Mrs. Wilson for everything, and left the estate for what was surely the last time.

The nearest train station was at the next largest town four miles away, but just outside the village, a passing farmer gave her a lift in his wagon. In another stroke of good luck, when she told the farmer her plans, he replied that he had a cousin in Thornshire. It wasn't far from Woolwich, he assured her. If she took the train into London, a ticket agent at St. Pancras could direct her to her connection.

"Ever been to London?" he asked, and when she told him she had not, he grinned and shook his grizzled head. "It's nowt like Derbyshire. A young lass like yourself could get lost. I trust you have friends there to meet you?"

"One friend," she replied, although Marjorie didn't know she was coming, and April didn't know where to meet her. If Marjorie had been obliged to leave the hostel when she quit Woolwich, the return address on her letters was no good anymore.

April had hours to consider her options on the train as it rumbled southeast through rolling countryside bright with summer greens and golds. Her carriage was packed with soldiers in khaki, a few silver-haired couples in tweeds, and several young women near her own age, perhaps aspiring war workers like herself. None of them looked as anxious as she felt, so she took a steadying breath and gazed out the window and pretended to be as nonchalant as the other girls. By the time the verdant countryside gave way to the outskirts of the city, she had decided to make her way to Thornshire Arsenal, wait outside the main gate, and search the crowd at each shift change until she spotted her friend. Marjorie would tell her what to do next.

At last the train halted at St. Pancras station, a bustling marvel of Victorian architecture with striking red brickwork and a soaring roof of iron arched trusses and glass. A lady ticket agent kindly told April which ticket to purchase to continue on to Thornshire and directed her to the correct platform, just as the farmer had promised. April had not thought to ask which direction she would be heading and how far, but, watching through the window as the train lurched forward and pulled out of the station, she figured they were traveling mostly east but also south. The train kept north of the Thames, a ribbon of brown that rippled in and out of view as spaces between the tall buildings allowed brief glimpses.

At last she reached her final stop. Suitcase in hand, she disembarked, smoothed loose strands of flaxen hair back into the roll at the nape of her neck, and appealed to the nearest ticket agent for directions to the arsenal. "Haven't a clue, love," he said. "Maybe it's near the old sewing machine factory. You might ask there." He beckoned the next customer in line, so she stepped aside before she could explain that she didn't know where the sewing machine factory was either.

Weary and stiff from sitting so long and increasingly hungry, she paused to gather her thoughts. Glancing around the small stationhouse in hope that another stroke of luck would find her, she spotted another lady ticket agent crossing the platform. April quickly hurried after her before she could disappear into the crowd, and, catching her just as she was about to enter a private office, she repeated her question.

"You're not far, dearie," the lady agent assured her, pointing her toward the correct exit and describing the route. "Now, I can't say whether you'll see signs for Thornshire Arsenal when you get there. It's newly converted, so it might have the old name still. If you see Purcell Products or anything to do with sewing machines, you've likely found it."

Sewing machines—so the first agent had been nearly correct. April thanked the lady agent and hurried on her way.

Ten minutes later, she spotted the Purcell Products sign on a smoke-stack from a block away, but as she approached and discovered high stone walls encircling a vast complex that must have covered scores of acres, her footsteps slowed. Scores of men and women streamed through an imposing entrance beneath a sign marked "Gate No. 3," while police stood guard and sentries checked identification badges. Rooftops and chimneys covered in soot were visible over the wall, but she could not tell what was making the awful metallic grinding and clattering and screeching on the other side. She hesitated on the opposite side of the street from the gate, intimidated, something acrid in the air stinging her nostrils. Clutching the handle of her suitcase in both fists, she took in the scene, swallowed hard, and suddenly remembered that she ought to be watching the departing workers for Marjorie. She was afraid to go any closer out of fear that the guards would chase her away, so she rose up on tiptoe and craned her neck, hoping desperately that her friend would appear.

Suddenly, a woman who looked to be around her mum's age noticed

April searching the crowd and paused to peer back at her, curious. April quickly looked away, but when she stole a glance back, she felt a frisson of alarm—the woman was crossing the street and striding directly toward her, with several younger women dutifully following after. April tried to look perfectly innocent, as if she had a right to be stranding there, because of course she was and she did, but her heart leapt into her throat when the woman—broad-shouldered and several inches taller than herself—halted not two feet in front of her and folded her burly arms over her chest.

"Mind yourself, dearie," she said, amused, as the younger women looked on, curious and grinning. They all had gingery hair and a faint yellow tinge to their skin, as if they spent too much time in the sun, or not enough. "Stand there gawking long enough and the police will think you're a spy. You're not one, are you?"

April gulped. "No, ma'am, I'm not."

The woman laughed, and the other girls smiled. "A lass who shows proper respect. I like that. Are you looking for a job?"

April nodded. "My friend works here already, and she said there's room for one more." Too late, she remembered Marjorie had said that about Woolwich, not Thornshire. Had April come all this way for nothing?

The younger women smiled and exchanged amused glances. "They need loads more than one," remarked the tallest, a pretty girl whose braided hair was reddish along the length of the strands but darkened to light brown closer to the root.

"Your friend might've told you that you need to apply at the Labour Exchange first," said the burly woman. She glanced over her shoulder and nodded to the tall younger one. "Peggy, you can take her."

"But I'll miss tea," Peggy protested.

"No you won't, not if you just show her the way and hurry back. That's a good girl." The woman turned back to April. "What's your name, dearie?"

"April. April Tipton." On impulse she added, "My friend is Marjorie Tate."

"Oh, yes, we know her," the eldest said. "She works in the same Danger Building as we do."

"She's probably in the canteen," another girl piped up. "She won't pass through the gates until shift change at seven o'clock."

April nodded, heart sinking. It was only four o'clock, hours to go yet.

"Mum, I only got twenty minutes left," Peggy told the eldest in an urgent undertone.

"Well, off with you, then." Her mother waved her along and threw April a reassuring grin. "Tell the clerk you want to work at Thornshire Arsenal and he'll send you right back."

April nodded. "Thank you, ma'am."

She laughed. "I can tell you were in service. It's Mabel, not ma'am. See you soon."

"Come on, then," said Peggy, not unkindly, already heading down the block. April hastened to catch up to her, wondering if she ought to apologize for making Peggy late for her tea, her own stomach growling at the thought of it. She decided to keep quiet rather than say the wrong thing.

They turned the corner, heading away from the arsenal rather than around it, and walked another two blocks before Peggy abruptly halted and pointed to an office building clearly marked with a sign. "That's the Labour Exchange. If you want to work right away, best to take whatever job they offer. Think you can find your own way back?"

"Yes, thank you."

"Good luck." Peggy offered a brief smile and hurried on her way.

Alone again, April approached the building, took a steadying breath, and opened the door, only to find that she couldn't open it all the way or it would have struck the woman standing inside at the end of a rather long queue. Slipping in through the narrow opening, she eased the door shut, took her place in the queue, and set her suitcase down at her feet. She found herself in a narrow foyer, separated from the rest of the single large room by waist-high partitions, more to keep the applicants in a corral than to create privacy. On the other side of the low walls, a dozen clerks, as many women as men, worked at identical desks sorting papers, answering phones, or chatting with applicants.

The queue led to a single, taller desk in the middle of the office, where April quickly deduced the applicants were sorted, either directed to one of the other desks or dismissed outright. As the queue inched forward, two women, one scowling, one nearly in tears, passed April on their way out, evidently rejected. An anxious shiver prickled the back of her neck. From what the posters said and Marjorie too, she assumed that the need for workers was so great that no one would be turned away. She had no factory experience, only domestic service and Mrs. Wilson's letter to recommend her. Would that be enough?

She felt lightheaded from hunger and worry by the time she reached the front of the queue. As closing time approached, the clerk had begun sending applicants to the smaller desks in groups of two or three to get through them faster. He asked April only a few cursory questions before directing her to join two other applicants at a desk near the front window, where an interview had already begun. Hurrying over, April gave her name when asked, presented Mrs. Wilson's character, and explained that she wanted to work at the Thornshire Arsenal. The other applicants' quick, appraising looks told her that they were after the same.

The agent questioned each of them briefly. The first girl, small and wiry, claimed three years' experience as a housemaid, but she looked younger than April had when she had first gone into service. The other, who looked to be perhaps ten years older, was currently employed in a hat shop, but her husband was in the navy and she wanted to do her bit too, for his sake. Neither had traveled as far as April, which gave the agent a moment's pause. Then he scanned Mrs. Wilson's letter, shrugged, and returned it, apparently dismissing whatever objection he might have momentarily entertained.

"The only positions currently open at Thornshire are in the Danger Buildings," he said, giving each of them an appraising look. "It's not hard work—no heavy lifting, I mean—but precision and a steady hand is essential."

"Is there nothing else?" the sailor's wife asked. "I have a young child. I can't take a dangerous job."

"Then perhaps munitions isn't for you, miss," said the agent sympathetically. "However, I can keep your papers on file in case something else comes in."

"I'll do it," said April quickly. "In fact, I'd prefer the Danger Building."

"I'll do anything," the wiry girl said, shrugging. "It don't bother me what, so long as the pay is good and the work is steady."

"You'll have plenty of work, no worries there." The agent studied the two of them a moment longer. "Very well. Are you willing to sign on for three years or the duration?" When they agreed, he passed them some paperwork and indicated where they should sign. Next he took two small yellow cards from a stack on his desk, filled in a few blanks, marked each in red ink with a rubber stamp, and gave one to April and one to the wiry girl along with some of the papers they had signed. "This will get you through the front gate. The rest is up to you. Good luck."

With that, all three were dismissed. The sailor's wife quickly departed, frowning, but the wiry girl lingered close to April as they left the office. "Where d'you suppose the front gate is?" the girl asked, nose wrinkling as she studied her card.

"I found Gate Three earlier," April mused. "I suppose the front gate is Gate One. We could walk around the outer wall until we come to it. Want to go together?"

The girl agreed, and they set off with April leading the way, retracing the route Peggy had shown her. The streets all around the complex were choked with traffic, automobiles and wagons and the tram, and the pavements were crowded with pedestrians, most of them workers, April guessed. Walking counterclockwise, they found Gate One, grand and imposing opposite a small park with gravel paths, flower beds, and shade trees, a bit of countryside in the city.

They showed their papers and yellow cards to one of the sentries, who waved them through the gate, which led into a covered entrance with a high ceiling. On the other side, two policemen halted them, examined their papers and passes, and directed them to the medical shed, a long, low building a few yards from the entrance. April was relieved it was so close, for the factory complex was almost a village unto itself, with industrial buildings of all sizes arranged in a grid connected by roads, where lorries and wagons vied for space. Workers in coveralls and caps bustled about, some pushing carts loaded with mysterious crates, others striding purposefully from one building to another, and more strolling and chatting with friends, evidently on break or at the end of a shift.

April and the wiry girl entered the medical shed together through double doors at the nearest end, but they quickly became separated amid the crush of other women, dozens of them, all looking as uncertain and wary as April felt. Word quickly spread from the front of the throng to those nearer the door that they must pass a medical exam before they could be hired. Sure enough, a stern woman in a nurse's dress and cap called to the new arrivals above the din, instructing them to remove all their clothing except for their shoes and coats, to leave their belongings in one of the cubicles along the walls, and to proceed to the other end of the room in groups of three.

Reluctantly, April obeyed, quickly shedding her clothing and pulling on her coat again. It was such a fine day that many of the girls had not worn an outer garment, but tried to cover their nakedness with a blouse or a sweater instead. April had worn hers only because it would not fit in her suitcase, which she stowed in a cubicle, hating to let it out of her sight.

The other end of the shed was divided into small cubicles for modesty, such as it was, and each trio was directed to wait in one, shivering and embarrassed, until two lady doctors arrived. They examined each of the girls thoroughly in turn, even checking their hair for lice. April had never been so poked and prodded before, and as she felt heat rising in her face, she knew her cheeks must be bright red and she could not bear to make eye contact with anyone. But at last it ended. One of the doctors made a few notes on April's yellow card, praised her for being so healthy and fit, and handed her a small blue book embossed with

"Thornshire Arsenal" in gold on the front. "This is your rule book," the doctor told her as she hastily pulled on her coat. "Your registration number is printed on the inside cover. Memorize your number and learn all the rules before you report for work Monday morning at seven o'clock sharp."

"Thank you," April murmured, clutching the rule book in one hand and holding her coat closed with the other. Hurrying back to her cubicle, she threw on her clothes, tucked the small blue book into her suitcase, and left the medical shed as soon as the doorway cleared enough for her to squeeze through it. She had never been more mortified, but at least she had the job. Gripping her suitcase tightly, she looked around to get her bearings, found the gate, and headed toward it.

"April!" a voice called out behind her. "April!"

She halted, turned, and sighed with relief at the sight of Marjorie running toward her, waving and beaming with delight. April set down the suitcase and held out her arms just in time for Marjorie to embrace her. "You came!" Marjorie exclaimed. "I never doubted it!"

April was so glad to see her friend that tears sprang into her eyes. "If you'd warned me about that medical exam, I might have stayed at Alderlea. I've never been so embarrassed!"

"That's why I didn't tell you! Did you get the job? Did you ask for the Danger Building?"

"Yes and yes," said April, picking up her suitcase. "I start Monday morning, seven o'clock."

"Wonderful!" Marjorie linked her arm through April's and steered her toward the gate. "We'll be working together again!"

"Just like old times," said April, smiling so broadly her cheeks ached. How swiftly the day had turned from mortification to joy, all because she had found her friend.

"You'll stay with me, of course," said Marjorie as they passed through the gate. She waved to one of the sentries, who grinned and tugged on the brim of his cap. "Our landlady can find a spare bed in another room for Edith."

"Your roommate? No, she shouldn't have to move. I'll take the other room." April would be grateful to have any bed at all. She had prepared herself to sleep on a blanket on Marjorie's floor until she could find lodgings of her own.

"Don't be silly! You're my best friend, and I just met Edith. She'll understand. I bet she'll volunteer to move before I need to ask."

"If you're sure—"

"Of course I'm sure." Marjorie squeezed April's arm. "Are you hungry?"

"Starving, rather."

"Not to worry. Our landlady always has a bite to eat ready for us when we get home. But first I'll teach you something every munitionette must know."

April felt a thrill of excitement. Yes, she was a munitionette now. How astonishing, and how wonderful! "What's that?"

"How to catch a tram home when hundreds of your fellow workers are trying to do the same. Come on." Marjorie released her arm and took her hand. "Hold on tight to that suitcase, soldier. Let's march!"

April laughed as Marjorie broke into a jog, pulling her along after her.

6

JULY–SEPTEMBER 1915

HELEN

By midsummer, Arthur had taken a lovely Georgian house in Marylebone at No. 14 Great Cumberland Place near Marble Arch, beautifully furnished and boasting every modern convenience Helen could have desired. Hyde Park was only steps away, and she enjoyed a brisk walk there every morning, admiring the flowers, savoring the fresh air and birdsong, tolerating the occasional rain or fog, smiling at the nannies and mums pushing prams or strolling hand in hand with their toddling children. She and Arthur had chosen No. 14 in part because of its charming nursery, which they both hoped would not remain unoccupied much longer. Helen saw less of her husband than she liked, and much less than she had expected, considering that the whole point of moving to London was so that they would be together. Arthur kept long hours and often collapsed into bed shortly after a late dinner, too exhausted to do anything but sleep, but when he was fully rested, he relished making love to her as much as he ever had.

"You must be very eager to be a father," she teased him breathlessly one July evening as she lay in his arms, a faint sheen of perspiration on them both. The blackout curtains were tightly drawn, a single candle illuminated the room, and they had left the door open to encourage a breeze from the hallway, but the air was warm and so still that the candle barely flickered.

"Very eager," he murmured sleepily, and then he truly was asleep. Sighing fondly, she rolled over to blow out the candle, then returned to his embrace, carefully so she did not wake him.

The truth was, though she wouldn't dream of complaining, she missed him during the long, lonely hours between his departure for Thornshire Arsenal soon after breakfast and his return, usually after twilight, often even later. Arthur frequently took a Saturday or a Sunday off, but never both, and sometimes neither. She understood that until the Shell Crisis was resolved, bringing Thornshire Arsenal up to full production capacity must be Arthur's priority. She worried to see him exhausting himself, day after day, and yet she was grateful for the essential war work that kept him in London, far from the carnage that had made widows of too many other women on both sides of the conflict.

Not that she dared express sympathy for her father's homeland. Outrage against Germany, kindled by the declaration of war and stoked by German atrocities in Belgium and the deaths of young British soldiers in the trenches, had burned searing hot ever since the devastating attack earlier that spring on the RMS *Lusitania*, a Cunard ocean liner en route from New York City to Liverpool. Nearly twelve hundred men, women, and children, including some of the world's most prominent industrialists, socialites, and entertainers, had perished in the cold waters off the southern coast of Ireland after a German U-boat fired a torpedo on the ship, sinking it in eighteen minutes. In response to international outrage, Germany had insisted that they'd had every right to treat the unarmed ship as a military vessel, since in addition to the great many civilian passengers aboard, she had also carried American-made munitions, in defiance of the German blockade.

Helen was not surprised that the British people refused to accept that justification, but she never expected the shocking riots that followed. The tumult first erupted in Liverpool, the *Lusitania*'s home port and the hometown of many of its crew, and rumor had it that friends and families of the lost sailors struck the first blows. Angry men attacked a German grocery in the North End, smashing windows and throwing food onto the pavement. Two more groceries were ransacked before police arrived on the scene, but the officers failed to contain the surging violence. Butchers' shops and shoemakers were looted, their goods stolen, their shutters and awnings torn down and broken. The homes of German immigrants were broken into, their furniture and belongings hurled from upper windows to smash on the streets below.

The furor spread swiftly through Liverpool and leapt to other cities across the country, wherever a significant German population had thrived before the war. Any business owned or believed to be owned by Germans became a potential target, although in their fury the rioters also destroyed buildings owned by Scandinavians, Russians, and their fellow Britons. When calm was finally restored days later, it was said that the windows in virtually every German-owned business in Great Britain had been smashed, nearly two thousand in London alone.

Nor was the compulsion to avenge the sinking of the *Lusitania* by punishing Germans in England confined to the working class. In London, two thousand indignant stockbrokers clad in their finest suits and top hats marched from the Stock Exchange to the Houses of Parliament to demand that the Prime Minister intern for the duration of the war thousands of Germans living or working in London. Thus far the outraged public had targeted their neighborhood butchers, bakers, and barbers, but the businessmen insisted that the real danger came from men in positions of greater influence. Naturalized Germans and enemy aliens alike must be removed from the city and confined behind high walls and barbed wire to prevent them from attacking Great Britain from within.

Suffragettes who attempted to storm Parliament had always found the doors barred to them, but the stockbrokers were permitted to march directly into the Central Hall. Prime Minister Asquith was absent, but two sympathetic MPs met with the businessmen and listened respectfully as they aired their grievances. Eventually the stockbrokers were persuaded to leave after the two MPs assured them that Cabinet was considering many of the preventative measures they demanded. "Who knows how many of these thousands of Germans, naturalized and unnaturalized, intend to assume strategic positions to assist the enemy in the event of another Zeppelin attack on London?" one of the MPs said to the press afterward. The second offered an ominous warning: "If ever Zeppelins drop incendiary bombs on London, many Germans among us would set fire to the city in twenty or thirty different places."

It was utterly preposterous, but thoroughly harrowing, and Helen waited anxiously for the government's response. Two days after the stockbrokers' protest, Prime Minister Asquith revealed new policies for the management of Germans in Britain for the duration. Naturalized subjects of enemy origin were to be left at liberty, unless there was sufficient reason for internment in individual cases, and all must register with the police and would be subject to observation. Male enemy aliens of military age, seventeen to fifty-five, were to be interned at camps on the Isle of Man and elsewhere in the United Kingdom. Men over fifty-five, women, and children were to be repatriated whenever possible.

Yet despite pressure from the opposition and even members of his own party, Asquith refused to treat naturalized citizens as enemies and spies. Most of them were surely loyal British subjects and decent, honest people, he declared. It would be disgraceful to initiate a campaign of persecution against them, not only from an ethical standpoint, but in the best interests of the country. These naturalized citizens contributed to British society, the economy, and even the national defense, for many of them, or their children, had enlisted in the military.

As heartening as it was to know that the Prime Minister refused to condemn nationalized Germans, Helen followed reports of public outrage and suspicion with increasing alarm, deeply afraid for her sisters, half German like herself, and for her mother, who bore their German surname. "Don't worry about us," her mother assured her, preternaturally calm, when Helen finally got her on the telephone. "We're Englishwomen through and through, and your father was beloved in Oxford, God rest him. No one will trouble us here."

Helen hoped her mother's trust was not misplaced. "Even so, would you warn Daphne not to speak German outside of the house?"

"She doesn't even speak it *inside* the house anymore, what with you and Penny and your father all gone." Her tone suggested that Helen's father too had only moved to another city. "You know I don't speak enough German to carry on more than a rudimentary conversation, although I do manage a few carols at Christmas."

"Would you please warn her all the same?"

"I suppose soon you'll be asking us to change our name from Stahl to Steel. Did you know the Sauers on Bankside call themselves Sawyer now? That's not what Sauer means, even I know that."

"They could hardly call themselves the Sour family, though, could they? They're confectioners." Helen closed her eyes and muffled a sigh. "Promise me you'll be careful."

"I will. You too, *Liebling*."

"Mother, honestly—"

"Got to run," she said cheerily. "Bye, now. Love you." And with that, her mother disconnected.

To Helen's relief, her family suffered no anti-German retribution, but even as her concerns for their safety subsided, she missed them all the more. In Arthur's absence, she longed for the company of her mother and sisters, for a tennis partner, for her old days full of purpose and useful work. None of the Stahl daughters had been brought up to be ladies of leisure, and Helen found the role an uncomfortable fit that constrained and chafed. Idleness made her restless, and she could walk around Hyde Park only so many hours of the day before the neighbors declared her an eccentric. Even managing their home took up little of her time. Although other ladies she had befriended in Marylebone lamented the "Servant Problem," as the alleged perpetual lack of good help was known, the Purcells' staff ran their household so expertly that Helen hardly had to do anything except approve their suggestions and make sure their wages and the bills were paid on time.

Even her service to the suffrage movement, once her greatest passion, had diminished. When Helen had first joined the Women's

Social and Political Union, the organization's call for politicians to deliver "Deeds Not Words" had resonated with her, as had its commitment to nonviolence. In the aftermath of Black Friday, however, a few militant factions had adopted more dangerous, destructive tactics, including arson and mail bombs. On one occasion a suffragette had thrown a hatchet at Prime Minister Asquith, and on another, a bomb had been discovered beneath the Coronation Chair in Westminster Abbey. Horrified, Helen had withdrawn from the WSPU entirely. She wanted the vote as badly as ever, but she could not condone violence. She was willing to suffer for her convictions, but never to kill for them.

Then, unexpectedly, Emmeline and Christabel Pankhurst abruptly halted all militant campaigning from the WSPU. The crisis required suffragettes to put their country's needs before their own, Mrs. Pankhurst declared. What good would it do British women to win the vote if Germany conquered Britain and civil rights for Englishmen and Englishwomen alike were suddenly swept aside? To avoid that terrible fate, brave Englishwomen must devote all their strength and courage to the fight to keep Britain free.

Helen welcomed the call to serve, and she had no desire to sit at home, safe in idleness, while other women responded. In recent speeches, Mrs. Pankhurst had argued that if women were unwilling to work in a time of national peril, they did not deserve the vote. Helen was inclined to agree, but what part could she play?

After considering the possibilities, she chose a Monday morning to broach the subject with Arthur, knowing he would be well-rested and in good spirits after a relaxing Sunday off. "Darling," she ventured as he sipped his coffee and scanned the newspaper headlines, "I'm determined to do my bit, and I don't mean knitting socks or distributing white feathers."

His eyebrows rose. "Is that so?" He folded the newspaper and set it aside. "What did you have in mind?"

"I thought I could put my German fluency to good use—as a translator for military intelligence, perhaps, or as a language instructor."

Arthur's brow furrowed. "No, darling," he said, shaking his head. "That would never do."

"Why ever not?" she asked, astonished. He hadn't even spared ten seconds to consider the idea. "I've grown up with the language. I speak, read, and write as a native would. I'm Oxford educated, although of course I was denied a formal degree. Is it because I'm a woman?"

"Of course not—"

"You might have noticed that all the men are preoccupied at the moment."

"It's not because you're a woman, and I'm not questioning your qualifications." Arthur sighed and rubbed his jaw, his green eyes shadowed and wary. "I believe we shouldn't call attention to your German heritage for the duration—not only to spare you abuse from the unruly and the ignorant, but so you can avoid the restrictions being imposed on German nationals. I won't have you interned on the Isle of Man or deported to Germany."

Helen was so astonished that she laughed. "For goodness sakes, I'm not an enemy alien. I'm an Englishwoman by every definition. I was born in Oxford to an Englishwoman, and I'm married to an Englishman. None of those restrictions apply to me."

"Your mother was born an Englishwoman, but according to law, if a British woman marries a foreigner, she takes on her husband's nationality." Arthur reached across the table for her hand, and she let him take it. "I doubt anyone will trouble the English-born widow of a revered Oxford professor who happened to be German. Nevertheless, perception matters. Your loyalties, which I know are above reproach, must never be doubted."

She studied him. "You mean *your* loyalties must never be doubted, because of your position. You can't have anyone thinking that Thornshire Arsenal's head man is married to a German, complete with horns and a forked tail."

"Mock me if you wish, but you can't deny that I have good reason to be concerned. The Lusitania Riots are evidence enough of that. I don't want to see my workers harassed, the factory ransacked, the equipment destroyed. Production would grind to a halt, and you can well imagine what that would mean for our soldiers."

"I thought you said Thornshire is protected by armed guards."

"Yes, to fend off German saboteurs. I don't want them firing upon misguided Englishmen in East London. Do you?"

"Of course not." After a moment, she added, "Darling, I love you, but you're being ridiculous."

He frowned and drank his coffee, his eyes never leaving hers. "One of us is, at any rate."

She felt a faint heat rise in her cheeks. "Very well. Have it your way. No translating, no instruction, no German at all. But I must contribute somehow." Then inspiration struck. "I'll come to work for you."

Her words were ill-timed; he had taken another sip and now spluttered into his napkin as he returned his cup to its saucer. "Work for me? At Thornshire Arsenal?"

"Where else? Why not?" She sat back in her chair and folded her arms. "I promise to speak only English."

"No offense, darling, but I'm not sure you've got what it takes to be a munitionette. Most of our girls have worked in factories or in service for years. You're better educated, fair enough, but as far as this work is concerned, they're qualified and you're not."

"I'll be your secretary, then, or a clerk. I worked as my father's assistant for years, handling his correspondence, attending to paperwork, minding the books, doing the occasional bit of research. I could do the same for you."

"I'm sure you'd be a marvel of efficiency and order, but I already have a secretary, and he has a family to support. How would it look if I sacked Tom to hire my wife?"

"It would be very bad form, I suppose," she replied, somewhat grudgingly.

"Yes, I should think so. I'll tell you what. If Tom resigns to enlist, I'll put your application on the top of the pile."

"I'd have to apply?"

"Of course. You and everyone else. It's only fair that the most qualified applicant should get the job."

"What if I refuse to accept a salary?" she countered. "How would that affect my chances?"

"It would improve them considerably. Yet, darling—" He sighed and reached for her hand again, and she gave it to him, along with a wry pout. "Thornshire Arsenal isn't your father's study. It isn't the Bodleian or the Ashmolean. You might not find it suitable as a workplace."

"I understand the difference between a factory and a library or a museum." Helen put her head to one side and studied him, curious. "I can tolerate mess and noise. If I didn't, I wouldn't want children. Is there some other reason you wouldn't want me about?"

He lifted her hand to his lips and kissed it, his breath warm on her skin.

"You're stalling," she said. "Stop being charming and answer the question."

He clasped her hand in both of his and brought them to rest on the table. "Helen, munitions work is inherently dangerous."

Her heart dipped. "How dangerous?"

"Very dangerous, of course. What do you want me to say? We're working with explosive materials, creating weapons to destroy and to kill. Everyone is trained to be scrupulously careful, but accidents happen."

"In that case," she said, keeping her voice even, "if it's too dangerous for me, it's too dangerous for you. Perhaps you should resign."

He fixed her with a look of fond disbelief. "You know I can't do that. Someone has to make shells, and my father accepted the responsibility on behalf of our family and his entire company. Running Thornshire Arsenal is my war service, and it shall be for the duration. Best get used to it."

Lowering her eyes, she nodded to show him she understood. "I don't like it, Arthur."

"A moment ago you wanted to be my secretary, and now you don't want either of us anywhere near the place." He leaned forward, rested his elbows on the table, and squeezed her hand. "Here's a thought. What are your plans for the day?"

"I don't have any, not really."

"Why don't I take you on a tour of the arsenal? You'll see for yourself the many precautions we take and how safe—how *relatively* safe—it is."

She barely let him finish speaking before she accepted.

Determined to make a good impression, Helen hurried upstairs to change into one of her most becoming walking suits—a sapphire blue silk moiré jacket with a pointed rear hem, rouleaux frogging fasteners, and ivory lace at the collar and cuffs, with a matching hobble skirt, discreetly pleated to allow for a more comfortable stride. She knew that while she was inspecting her husband's workplace, his workers would be inspecting her. When she descended the stairs and saw her husband gazing up at her with unmasked admiration, she knew she had chosen well.

Helen had assumed they would take the tram or the underground to East London, but Arthur wanted to drive the Silver Ghost, to give the engine a good run while they could still get the petrol. On the ride over, she queried him about the various armaments his factory produced, and he answered, patiently and amiably, with clear descriptions that avoided bewildering jargon. She couldn't help thinking that he would have made an excellent professor.

They were still a few blocks away when she glimpsed a Purcell Products sign on a tall smokestack in the distance. "Is that it?" she asked. "Why doesn't it say Thornshire Arsenal? Wait, don't tell me. Why announce to the Germans what you're doing here?"

"Might as well paint a bull's-eye on it for the zeppelins," he replied. "They might not waste a bomb on us if they think we still make only sewing machines and bobbins."

Helen fought to suppress a shudder at the thought of a bomb falling on her husband's workplace while he sat at his desk, making telephone calls or perusing contracts.

As they drew closer, she was surprised and a bit awed to discover that the factory was a complex the breadth of several city blocks, all encircled by a massive stone wall topped with barbed wire. The streets were packed with vehicles, the pavements with arriving and departing workers on foot, but the crowd parted before the gleaming Rolls-Royce as it slowly approached an imposing gate. Helen observed several of the men tugging their hats and women bobbing curtseys to her husband as they passed, and he acknowledged each gesture with a courteous nod. She was glad to see that he was apparently well-regarded by his workers. It was almost as if they knew they had the kindest and most generous of the Purcell gentlemen as their boss.

After Arthur showed his identification badge to a sentry, they were waved through the gate and drove into the complex proper, a seemingly haphazard collection of industrial buildings of all kinds and sizes, some gleaming with metallic newness, others stone and soot-covered. They drove for a few minutes down a paved road and turned onto cobblestones before halting behind a low, tan brick structure that looked more like a barracks than a manufacturer's. As she soon discovered, the building housed Arthur's offices and those of his subordinates, many of whom he introduced to her as he led her around inside. She met his much-lauded secretary, Tom, who turned out to be in his mid-fifties, a bit rotund, with spectacles and very little hair. She liked him, and given his age, she ruefully abandoned any hope that he might enlist, leaving a vacancy on the staff that she could fill. Even if he did, Tom had an assistant of his own, a young veteran who had lost a hand in the early months of the war. Surely he would be at the head of the queue, and Helen would be ashamed to snatch the promotion out from under him.

After showing her his office building—which, Arthur emphatically noted, contained nothing more explosive than tempers—he took her around to see one of the factory canteens, a clean, comfortable, spacious hall where workers enjoyed nutritious meals for a small fraction of the cost of bringing lunch from home or running out to a pub. Next they toured the infirmary, where minor injuries such as sprains, burns, and cuts were tended by a staff of dedicated nurses, who were trained to stabilize more serious cases before transport to a nearby hospital. They then viewed the loading dock, where large crates containing finished munitions were loaded onto trains with utmost care.

As they strolled from one site to the next, Helen was especially intrigued by the many munitionettes on the grounds, most clad in identical tan heavy cotton trousers and three-quarter-length jackets, their hair tucked neatly into tan broadcloth caps. She started at the sight of two women with oddly yellowish skin, laughing and chatting as they passed between buildings. "Those women seemed quite—sallow," she said in an undertone, not quite sure if that was the right word for it. "Is it a trick of the light, perhaps?"

"Your eyes aren't deceiving you," Arthur said, also quietly. "The yellow tint of their skin is a consequence of working with trinitrotoluene, a chemical component of the explosives. It seems to happen eventually to all of the girls who work in the Danger Buildings."

"That seems quite serious," said Helen, watching as the pair disappeared through the doorway of another building.

"It isn't permanent," Arthur assured her, smiling. "I hear there are creams that can prevent it—you would understand better than I—something about preventing the color from settling in the pores. But with or without such cosmetics, the color fades with time, if the workers transfer to another department or take an extended leave. The hue has earned the Danger Building munitionettes a particular nickname—canary girls."

"That's rather unkind."

He shrugged. "It's never spoken unkindly, but rather as an endearment, or as a proud boast by the canary girls themselves. They're among our most valuable employees, having taken on some of the most dangerous work, and they know it. It's said that blue-eyed canaries are especially adept at calibration work."

"I hope your canary girls are well compensated for the risks they take."

"Indeed they are, which is why positions in the Danger Buildings are so coveted." He gestured to a factory building directly in front of them, a tall stone structure next to the one the canary girls had entered. "Would you like to see some munitionettes at work?"

She most certainly did, so Arthur escorted her into a fuse factory, where dozens of women sat around long tables gauging metal rings, which other munitionettes collected and carried off to be placed into the machining part of the fuse cap. The room rang with the thudding and clanging of various machines and equipment at the far end of the vast space; though it was not deafening, Helen couldn't imagine enduring it for a twelve-hour shift, six days a week. It was somewhat quieter where the munitionettes sat. Helen was impressed by the speed and surety of their deft movements, and also by the sense of camaraderie, evident in the conversations they managed to have despite the din, occasionally bursting into friendly laughter. None of them, Helen observed as a few girls glanced up to smile or nod as she and Arthur passed, presented with the yellow skin of the so-called canary girls.

"May I see one of the Danger Buildings next?" she asked as Arthur led her to the exit.

"I'm sorry, darling. That's forbidden."

"I thought you were in charge here," she protested, smiling. "Can't you make an exception?"

"Not even once, not even for you." He regarded her with mock severity. "We operate under strict military regulations. Even if I were willing to allow a civilian such as yourself into the Danger Buildings, it would greatly inconvenience our overseers. Are you wearing any silk or nylon?"

"Silk," she replied, waving a hand gracefully to indicate her pretty ensemble.

"Well, that would have to come off."

"What? Here? Goodness, darling. You astonish me."

"In one of the shifting houses," he said emphatically, inclining his head toward a low white wooden structure centrally located amid the factories. "Friction from silk and nylon could cause a spark that would set off a conflagration. Same with metal." He looked her up and down, examining her attire. "I spy metal clasps, hooks, and eyes on those boots you're wearing. The workers swap their shoes for wooden clogs. You'd also need a cap instead of that charming hat, which, if I'm not mistaken, boasts a silk ribbon and a metal buckle. But your jewelry and every hairpin in that lovely coiffure would have to go, as would your brassiere, because I happen to know yours has metal fasteners."

"Well, safety first," she replied. "I wouldn't want to sacrifice the entire arsenal to fashion. Do you have a spare uniform, clogs, and cap I could borrow?"

"I'm afraid not. You'd have to take them from another girl's cubby, which means she'll find herself without a uniform when she arrives for her shift, which means I'm down one worker and she won't earn any wages."

She studied him through narrowed eyes. "I think you're just trying to discourage me, Mr. Purcell."

He smiled. "I might be. I'm also telling you the absolute truth."

She let out a loud, exasperated sigh and took his arm.

"Has it eased your worries at all, seeing the place?" Arthur asked as he led her back to the Silver Ghost.

"A bit," she admitted. "I'm glad to know that your office isn't in the same building as the explosives."

"Did you imagine my desk was a plank balanced on two kegs of gunpowder?"

"Something like that." Arthur opened the passenger door for her, but she hesitated before climbing in. "I do wonder about your canary girls, though. Anything that turns the skin that lurid shade of yellow cannot possibly be good for them."

"As I said, it's not permanent. It fades with time, after they're no longer exposed to it daily." He took her hands. "Weren't your fingertips always stained with ink when you worked for your father?"

"Not constantly, no, but ink isn't tri-ni-whatever-you-call-it—"

"Trinitrotoluene. Just call it TNT. Everyone does."

"Ink isn't TNT, and ink doesn't explode. Do your canary girls carry that yellow powder on their skin and clothing home to their families, to their children?"

"No, darling, they don't." His voice was gentle, patient. "When workers arrive, they change into their uniforms in the shifting houses and stow their own clothing and other belongings in their cubbies. At the end of the shift, the Danger Building workers leave their soiled uniforms in the shifting houses, have a good wash-up, and wear their own clothes home, or to their hostels, as the case may be. The soiled uniforms are laundered on-site, and when the workers arrive for their next shift, they find a clean kit waiting for them in their cubby."

Helen mulled it over. "I suppose you can't be any more careful than that."

"Remember, too, that no one is required to work in a Danger Building. Any worker who feels unsafe can request a transfer to another department." He held her gaze, his green eyes intent. "Munitions work is hazardous, Helen. Our girls know that. I do what I can to minimize the danger."

"I'm sure you do."

"You don't sound convinced." He gestured to the passenger seat. "Climb in. I'll show you something that may make you feel better."

She raised her eyebrows at him, skeptical, but did as he asked. Soon they were driving slowly through the complex, until they halted at a small courtyard that opened onto a grassy field, still within the stone walls. There, a dozen or so women in munitionettes' attire kicked a football back and forth, running the length of the makeshift pitch, laughing and calling out to one another. Even from a distance, Helen could see that nearly half of them were canary girls.

"Very well, darling. I see the point you're trying to make," she said. "These girls are clearly vigorous and healthy."

"They're thriving," he said emphatically. "All of our girls are. Do you know, most of our workers are eating better now in our canteens than they ever have in their lives?"

"I'm not surprised," she replied. "I'm glad you provide for them." The lack of plentiful, nutritious food in working-class households was a terrible societal ill. Suffragettes wanted the vote in large part to gain the influence necessary to remedy such problems. She regarded Arthur fondly. "I know you would never intentionally let any harm come to your workers. I'm sure you're doing everything you can to keep them safe."

The relief in his eyes surprised her. "Thank you, darling," he said, a catch in his voice. He cleared his throat and smiled. "Shall I take you home? I'll have to come right back, of course."

"I could stay and help you around the office, save you the trip."

He threw her a wry look, adjusted the throttle, and drove on.

A week passed. Helen troubled Arthur no more about coming to work for him at Thornshire Arsenal, but the tour had another consequence that he had perhaps not expected. After seeing the munitionettes hard at work, even risking their lives, she had become even more determined to do her bit.

Then, sometime around eleven o'clock on the evening of 9 September, as she sat up in bed reading one more chapter while Arthur slept beside her, the stillness of the night was suddenly shattered by the thunderous roar of bombs and gunfire.

Arthur immediately woke and bolted out of bed to the window; heart pounding, Helen quickly doused the light just as he pulled back the blackout curtain. Hastening to his side, she stood beside him and watched, horrified, as distant searchlights illuminated a long, silvery object drifting over the city.

She flinched as another explosion boomed; Arthur put his arm around her. "It appears to be over the City, near St. Paul's Cathedral," he said, his gaze following the gleaming object until it disappeared behind a cloudbank or a building—in the darkness, at that distance, she couldn't be sure. Police and ambulance sirens wailed; the searchlights swept the skies. "It's not coming this way. Go back to bed. I'll wake you if there's any danger."

"But darling—"

"Please, Helen."

There was a strange undercurrent to his voice she had never heard before, one that would allow no argument. She nodded and returned to bed, but she could not sleep until the sirens faded and silence returned, and Arthur lay beside her once more.

The next morning, the papers, bound to obey the censors, reported only that "a London district" had been "visited" by a zeppelin, which had dropped incendiary and explosive bombs. Arthur left for work early in case the streets were impassible, so she had to wait impatiently for him to return that evening for any real news.

The reports he had managed to gather were grim. Wood Street and Silver Street behind Cheapside and near Guildhall had been struck in the raid; warehouses and other structures were gutted and smoldering ruins, and large craters marred the roads nearby. Another bomb had destroyed an omnibus on the way to Liverpool Street station, killing all twenty people on board and a housekeeper standing outside on the steps of her workplace. The zeppelin's bombs also struck Bartholomew Close, smashing the windows of Bart's Hospital, but sparing the ancient church of St. Bartholomew the Great. Later Arthur learned from his father's well-placed sources that thirty-eight people had been killed in the raid, including two policemen, and 124 had been wounded. As ever, the papers were required to be vague lest reports of deaths hearten the enemy.

Afterward, Helen and Arthur said little about that harrowing night, except to note that they had never been in any real danger. Nor were they likely to be, Arthur said, as there was nothing of military interest in Marylebone.

"What of Buckingham Palace?" Helen ventured. "It's only a mile and a half away."

"Let's trust that our ground defenses will prevent an attack there," Arthur said. "We might expect more interruptions of our sleep in the weeks to come, but likely nothing worse than that."

So he said, and so she pretended to agree, but Helen didn't think Arthur believed it any more than she did.

7

September–November 1915

LUCY

Jamie and Simon thought it terribly exciting that a zeppelin had struck at the very heart of London, only about five miles from their home. Sorely disappointed to have slept through the attack, they begged Lucy to take them to see the ruins, but she firmly declined. "You have school, and I have my training course," she reminded them, and received a chorus of groans in reply. Without thinking, she snapped, "Show some respect. Thirty-eight people died."

"Sorry, Mum," Jamie said contritely, and Simon nodded. Ashamed, Lucy swept them into a hug and apologized for losing her temper. She caught herself before pointing out that they would likely have far too many opportunities to observe the wreckage of German bombs before it was all over.

"Be thankful your boys weren't terrified," Gloria advised as they stood outside on the pavement watching their children head off to school together. "My Katie was inconsolable. I couldn't get her back to sleep for hours."

"Which means you hardly slept either," said Lucy sympathetically.

"No, but I wouldn't have regardless. I was as frightened as Katie. I just couldn't show it."

Lucy put her arm around her friend's shoulders as they watched their children disappear around the corner. She couldn't bring herself to admit that she had been desperately afraid during the raid too, that her heart had pounded and her hands trembled and tears filled her eyes. She had wanted Daniel there with an intensity that felt very close to anger. Ever since they were children together, Daniel had looked out for her, helping her with her schoolwork, fending off her teasing elder brother, encouraging her to conquer her fears, whether it was swimming or climbing trees or riding a horse. But last night, when she had felt so helpless and imperiled, he had been miles away, and she had been denied even the comfort of his voice on the telephone.

She longed for those afternoons at White Hart Lane when she and the boys had briefly reunited with Daniel on the sidelines after a match. They had not seen him since late May, when most of the 17th Middlesex had been granted a few days' leave to recuperate from a series of essential vaccines. They stayed at the Evans family home in Brookfield with her mother, George, and Eleanor as always, and other friends and relations turned up for tea and dinners. Everyone had news to share. Edwin was training in aerial reconnaissance with the Royal Flying Corps, and he expected to be sent to France any day. Alice was running the village school marvelously, although she admitted concern for the older boys' classes, which were steadily diminishing as one student after another ran off to enlist as soon as they came of age. Dieter Brandt had been interned at the Lofthouse Camp near Wakefield, and his wife and eldest daughters were struggling to keep the bakery open without him. Nearly everyone in the village had signed a formal petition attesting to Mr. Brandt's loyalty and excellent character and requesting his immediate release, but so far they had received no reply except for a perfunctory letter acknowledging that the petition had been duly filed.

As for Lucy, she wouldn't have mentioned her own small contribution to the war effort, but Daniel proudly boasted about her munitions training course at dinner their first night back in Brookfield. Her sisters-in-law were so delighted by the news that they actually applauded.

George, ever the physician, furrowed his brow and asked what tasks Lucy was preparing for, because reports were circulating in the medical community about peculiar symptoms and troubling illnesses possibly linked to munitions work.

"I've been trained to run a lathe and a hydraulic press," Lucy said, "and to use various sorts of gauges and calipers. Next week, we're going to begin on the milling machine."

"But no chemical work?" her brother queried. "Just machinery?"

"Yes," said Lucy, "at least, so far. Why? Is something wrong?"

"No, sorry, don't mean to alarm you." Some of the tension left George's expression. "If you're working with tools and equipment, you should be fine, as long as you're properly trained to use them."

"I should think chemicals would be safer than those enormous, clanging metal machines," Alice remarked. "The potential for injury—well, it just seems terribly dangerous."

"I'm sure our Lucy is up to it," said her mother. "What I wonder is who is minding my grandsons while you're at your course?"

"We can mind ourselves," Simon piped up from the other end of the table.

"Oh, I'm certain you can," said Daniel wryly as the other adults smiled indulgently.

"My neighbor Gloria looks after them," Lucy explained. "It's only a half day."

"Yes, but what about when you take an actual job?" A slight frown appeared on her mother's soft, dear face. "I've heard that munitionettes work twelve-hour shifts. That's a lot of childminding to ask of even the most generous neighbor."

"I haven't quite worked that out yet, but I will."

"It sounds to me like you may need a granny on the premises."

"What's this?" asked Eleanor, suddenly wary. "*We* need a granny on *these* premises."

"Your children are older than Lucy's, dear," Lucy's mother replied kindly, "and you have George around to help you."

"Let's not fight over Mother," Lucy pleaded, to head off any hurt feelings. "It may not be necessary. I might not even be able to find work."

"Well, keep me apprised, dear. I'll be ready to answer the call, should you need me." Her mother turned to Daniel. "None of this would be necessary if you hadn't run off to enlist with all your footballer friends. And at your age! I've known you since you were a lad, and I thought you had better sense."

"Mother," Lucy chided, but not too severely. She couldn't fault her mother for saying aloud what she herself had often thought.

The visit, even with its awkward moments, had passed all too swiftly. In the first week of July, the 17th Middlesex transferred to Clipstone Camp near Mansfield, and moved again a few weeks later to a large encampment at Perham Down on Salisbury Plain. Though he was now only eighty miles west of home, Lucy knew better than to hope to see him, for the 17th Middlesex had begun the final stages of their military training. For the first time, Daniel wrote, the men of the battalion fired rifles on long ranges, and all three of the division's brigades engaged in joint live fire exercises. Yet their commanders still made time for football, arranging matches between a side from the 17th Middlesex and several teams from the league and other army battalions—no doubt with the expectation that the spectacle would increase recruits. Lucy hoped that a match would bring Daniel to London, but by early September the battalion team had played no games any closer than Reading.

Daniel was still in Perham Down in late September when Lucy finished her training program. On the last day, she and the other graduates were issued certificates and a yellow card with the name and address of a munitions factory that had requested trained workers, and where they would almost certainly be hired, pending an interview. Lucy's assignment was at Thornshire Arsenal. When she searched for its pin on the large map posted on the common room wall, she was dismayed to learn that it was all the way in Barking in East London, northeast and across the Thames from the better-known Royal Arsenal in Woolwich. Taking a steadying breath, she returned to the instructor who had given her the card. "Excuse me, but isn't there anything closer?" she asked. "This must be fifteen miles from my home. It would take me ninety minutes each way."

"More like two hours, I should think," the instructor replied. "This is all we have at the moment for your training and experience. You could decline this job and try your luck applying elsewhere, but there's no guarantee they'd have anything for you. If they did, you'd likely have to start at the bottom, with a lower wage."

"I see," said Lucy, heart sinking.

"If you can't bear the commute, many girls stay in hostels closer to work. The superintendent at Thornshire could help you find a place."

"I'm afraid that's not possible. I have two children at home."

"Oh, that *is* a problem." The instructor frowned, sympathetic. "Do you want me to ring the arsenal and cancel the interview? Shame to waste your training."

"No, thank you. I'd better keep it." Lucy needed the wages, and if she could do anything to hasten the end of the war by making munitions, she had to try. "I'm sure it'll work out somehow."

She repeated that to herself the following morning as she dressed with care, saw the boys off to school, and raced to catch the train, which was overcrowded and slow. Despite her best efforts, she arrived fifteen minutes late, breathless after the sprint from the station. Guards at the main gate inspected her papers and directed her through to a small, square building resembling an oversized shed not far from the entrance. A queue stretched out from the front door, and she hurried to join it, breathing a sigh of relief when she overheard that the girls in front of her had appointments at eleven o'clock too. The superintendent was evidently running behind, and would be none the wiser that Lucy had arrived late.

The queue moved steadily forward, and before long Lucy crossed the threshold and entered a hallway with one large rectangular room to the right and several small offices along the left. In the room to the right, which was separated from the entry by a waist-high partition, a large central table faced a bench along the wall. Shelves lined the walls between the windows, packed with file boxes, stacks of forms, and boxes of stationery and other supplies. A middle-aged woman in a dark brown suit sat at the table, and the younger woman who had opened the door quickly ushered Lucy and seven other applicants to the bench and invited them to sit. They just managed to squeeze onto it without anyone tumbling off the ends. At the far end of the room, four clerks worked at two smaller desks, one seated on either side, typing reports and sorting blue forms and yellow cards, apparently too engrossed in their tasks to notice the interviews happening nearby.

The woman at the center desk introduced herself as Superintendent Carmichael and beckoned the two girls on the right end of the bench forward. She examined the papers they had brought along and looked over both applicants appraisingly. "How old are you?" she queried the first.

"Twenty-three, miss."

The superintendent turned to the other. "And you?"

"Twenty-one, miss."

"Given your qualifications, I have no vacancies for you except in the Danger Buildings. Are you willing to work with mercury?"

"Yes, miss," the two applicants said in unison.

"Very well." From the neat piles of different-colored forms on the table, she plucked two blue pages and handed one to each woman. "Take these to the office across the hall, and you'll be shown how to fill it in." She beckoned to the next two girls on line on the bench. "Come forward, if you would."

The first two girls left, and the next two quickly took their places before the table. Again the superintendent scanned their papers and looked them over. "How old are you?" she asked the first, tall and sturdy.

"Twenty-eight, miss."

"And you?"

"Nineteen, miss," the younger replied meekly.

To the elder, the superintendent said, "Are you willing to work in yellow powder?"

"What's that?"

"Trotyl. TNT. You'd be filling shells."

"Well, miss . . ." The woman hesitated. "My husband's at the front, and I have my children to look after. I don't feel I ought to run the risk."

"That's all right," the superintendent said briskly. "You look strong. Would you have no objection to heavy work? You might undertake trucking, moving trolleys of shells or parts from one site to another."

"Yes, miss, I could do that." She indicated the younger woman with a tilt of her head. "May my friend work with me?"

The superintendent took in the younger woman's slight frame and narrow shoulders. "No, I shan't put her to that task, but there's a place for her in the fuse factory, and you'll be able to take your meals together in the same canteen."

The pair agreed, and they carried off their blue papers, smiling.

The next two applicants to approach the desk were a gray-haired woman with stooped shoulders and a perky blonde about half her age. The superintendent examined their papers, studied the elder for a moment, and asked, "How old are you?"

The woman shrugged and said nothing.

The superintendent's eyebrows rose. "How old are you?" she asked, slightly louder.

"A lady don't like to give her age," the older woman protested. The young blonde snickered, and the older woman shot her a withering look.

The superintendent sighed. "What is your age?" she asked again, enunciating each word precisely. "We cannot proceed without that information. There are rules and regulations, you see."

"Forty-nine," the woman replied grudgingly.

"There, now, was that so difficult? What were you doing for employment before you came to us?"

"I take in a bit o' washin' now and then."

"Indeed? It just so happens we have a laundry on-site, for washing the workers' uniforms and such. Would you have any interest in that?"

"The wages any good?"

"Better than you'd earn taking in washing at home. They can tell you more across the hall."

"All right," the woman said, nodding thoughtfully. "Won't hurt to hear more about it."

"I'm eighteen," the blonde spoke up, without waiting to be asked. "I was a girl-of-all-work at a hotel in Mayfair. I want a job in a Danger Building, please. I hear that's got the best wages for a girl just starting out."

"You've heard correctly." The superintendent studied the girl's documents. "I see your current employer has provided you with a character. I wonder why they should be so eager to see you go?"

"They weren't eager, miss," the blonde quickly replied. "But hotels don't require leaving certificates, so they couldn't keep me if I had a mind to go."

"I'm pleased you understand the rules," the superintendent remarked, handing her and the laundress their blue cards. "I'll put you to work filling shells, then."

The pair stepped away, but the blonde suddenly turned back. "One more question, miss. Must I wear a cap?"

"A cap," the superintendent echoed, eyebrows rising. "Are you referring to the regulation cap which is a required part of every worker's uniform?"

"Yes, miss. I don't much like caps."

"The wearing of a cap assures your safety and that of your fellow workers," the superintendent said, a touch of frost in her voice. "If you'd ever worked in a mill, and had seen a woman after her hair got caught in the machinery, you wouldn't question the necessity."

"But I won't be working near machinery, will I? Caps aren't comfortable. They make my head itch."

"It's just a bit of fabric, dearie," the laundress said, incredulous. "Hardly worth such a fuss. Wear a cap or go work somewhere else."

The superintendent held up a hand. "That will be all, thank you." The laundress shrugged and muttered under her breath as the superintendent turned back to the blonde. "Caps are required. This is where duty comes into war work—not only performing your assigned tasks, but willingly submitting to discipline. We are His Majesty's servants, and like the men at the front, we must be obedient to regulations."

"Yes, miss," the blonde replied, resigned. Eyes lowered, she followed the laundress into the office across the hall.

Sighing wearily, the superintendent beckoned Lucy and the last applicant forward. When she examined Lucy's credentials, she nodded approval. "I see you've been through the training course, and you've finished school." Her eyebrows rose as her gaze lit upon another detail. "Your previous work experience was helping at your father's medical practice?"

"Yes, miss."

"A doctor's daughter, and yet you're not going into nursing?"

Lucy's heart dipped. Must everyone ask her that? "No, miss. It's not for me."

"How unfortunate. Nurses are desperately needed." The superintendent glanced up and smiled faintly. "But so are munition workers. What is your age, Mrs. Dempsey?"

"I'm thirty, miss."

"Mature, educated, with practical office experience. I'd quite like to hire you to work here, in administration, but we don't have anything open at the moment." The superintendent frowned thoughtfully. "With your machinery training, you could work in the Finishing Shop. How would you feel about a job in the Danger Building? It does mean better pay, as befitting more skilled work."

Lucy remembered George's warning. "I'd be operating machines, not working with chemicals?"

"You'd be using machines to finish the shells, but the shells do contain explosive chemicals. It's the girls in the Filling Shop who pack the explosive powder into the shells before they come to you." She waved a hand dismissively, ready to move on. "It will make sense when the foreman demonstrates. Never fear; you won't be left to finish shells alone until he's certain you're properly trained."

Lucy wanted to think it over, but the next eight girls were squeezing onto the bench behind her, and she knew that if she didn't take the better-paying, more skilled job, someone else would snatch it up and she might be dispatched to the laundry. "I'll take it," she said. "Thank you, miss."

The superintendent nodded, handed her a blue form, and turned to the last girl in the group. "What is your age?" she asked, taking her papers.

"I'm twenty-two," the girl said, flashing a cheerful grin. "I want to work in the Danger Building too if that's where the higher wages are. I'll work hard at any task given me, I don't complain, and I rather like caps."

"That's the spirit," the superintendent said, smiling as she handed her a blue form. "Very well, you two. Off you go. Welcome to Thornshire Arsenal."

They stopped at the office across the hall, where they were officially registered and directed to the medical shed next door for a health exam. The examination wasn't painful, merely unexpected, and all the more embarrassing for that. Afterward Lucy was declared to be in excellent health and was provided with a small blue rule book she intended to learn by heart before she reported for work in two days' time.

The first thing she did upon returning home was to ring her mother. "I got the job," she told her, feeling a surge of pride at her accomplishment. "I start Thursday."

"Congratulations, dear," her mother said warmly. "I'll pack straightaway, and I'll see you and my grandsons tomorrow."

Lucy had a day and a half to prepare herself, and the boys, for her first day as a munitionette. She went over the boys' schedule with her mother and studied the slim rule book, which contained lists of administrative policies, safety guidelines, and military regulations. She slept restlessly Wednesday night, and woke, more apprehensive than excited, to find her mother already in the kitchen preparing a strong pot of tea and raisin buns from Brandt's bakery, which she must have smuggled into the house in her suitcase or knitting basket, a lovely surprise.

"I'm so grateful you're here," Lucy said in the foyer as she threw on her coat and carefully tugged a hat over her long dark hair, which she had neatly braided and coiled with nary a hairpin. "I couldn't do this without you."

"The pleasure is all mine," her mother said, smiling as Lucy kissed her cheek. "Now I'll get to see more of Jamie and Simon as well as my only daughter. I should be thanking you."

"We'll see how you feel in a week," Lucy teased, giving her mother a little wave as she headed out the door.

It was not yet dawn. The streetlamps were doused for the blackout, but there was enough rosy morning light to see by, with the help of luminescent paint marking curbs and other obstacles. Small blue lightbulbs faintly illuminated the building and platforms of Clapham Junction station, already quite busy despite the early hour. Lucy's train reached Thornshire station at half six o'clock, and it was a ten-minute walk to the arsenal, so she passed through Gate Four twenty minutes before her shift began, just as the rule book recommended. It was not enough to be on the arsenal grounds at seven o'clock when her shift began; workers were required to change into uniform, pass through inspection, and be at their stations at the appointed hour or they were considered late.

Unfortunately, Lucy lost a bit of her head start by going to the wrong shifting house, and by the time a helpful passerby redirected her, she had lost a good measure of her nerve too. The warning bell by the main gate tolled just as she hurried into the shifting house. There she found dozens of other women changing from their street clothes into their trousers and jackets, chatting and joking. Some yawned and looked as if they had risen only moments before, while others were bright-eyed and eager, laughing with friends. Heart thudding as the minutes ticked away, Lucy searched the rows of cubbies until at last she spotted one marked with her own name. As she hurried toward it, shrugging out of her coat on the way, she accidentally struck another worker with her coat sleeve, knocking her cap off her honey-gold curls.

"Oh, I do beg your pardon," Lucy exclaimed, stooping to pick up the cap.

The younger woman snatched it from her. "Watch what you're doing," she snapped, then paused to look Lucy up and down. "Oh, look, it's Lady Prim, all dressed up for her first day of work. Watch out, girls. This one's clumsy."

A smattering of laughter broke out, and Lucy felt heat rise in her cheeks. "I *am* sorry," she murmured, continuing on to her cubby. In what way had she overdressed? Her long gray wool skirt, white blouse, and black-and-gray cardigan appeared to be finer quality and better tailored than the light blue chambray dress the other girl had removed, but it was hardly Savile Row. As Lucy changed into her uniform and tucked her coiled braids beneath her tan broadcloth cap, she managed a friendly nod for the girls to her left and right, but only one returned a cheerful grin. The others ignored her.

Her heart sank. She had signed on for the duration. Had she ruined her chances to make friends among the other girls all because of one careless misstep?

"Don't mind Marjorie," a voice spoke quietly behind her. Lucy turned around to find a young flaxen-haired girl regarding her with sympathy. "She doesn't mean any harm, but she always says exactly what she thinks."

"Oh, it's fine," Lucy said, smiling and forcing good cheer into her voice. "It's all in good fun."

"But maybe it's not the nicest way to welcome you on your first day." The girl stuck out her hand. "I'm April Tipton. I've been here only a few weeks myself. I can help you along until you find your feet."

"Thank you. That's very kind." She shook April's hand. "I'm Lucy Dempsey." She glanced toward the opposite end of the room, where the other workers were lining up at an exit. "I assume that's the way to inspection, and then on through to the Danger Building?"

"It is, but you might want to put that away first." April pointed to Lucy's left hand. "Rules are rules."

Lucy gasped as her gaze went to her wedding ring. "Oh, goodness. Metal is forbidden, of course." Tugging it off her finger, she thought for a moment, reached into her cubby, and tucked the precious ring into the toe of her boot. "Tomorrow I'll leave it at home. I'm so used to wearing it, I didn't think."

April nodded understandingly as they headed to the inspection line. "Been married long, have you?"

"Ten years." Lucy smiled, thinking of Daniel. "We have two sons. My mother's looking after them while I work," she added, lest April think she was neglecting her children.

"What does your husband think about you taking on munitions work?"

"Oh, I think he's proud of me for doing my bit," she said, stepping carefully as they joined the end of the queue. Those wooden clogs would take some getting used to. "He's in the army."

"So's my older brother, Henry. He's in the Lonsdale Pals, the Eleventh Battalion of the Border Regiment."

"My husband is in a Pals regiment too. The Seventeenth Middlesex. They're still training in Surrey."

"Henry's still training too, though he expects to go to France soon." April paused, thinking. "The Seventeenth Middlesex? Isn't that the Footballers' Battalion?"

"Yes, that's right." They had almost reached inspection, and Lucy watched carefully as the girls ahead of her split into two queues on either side of a table, each with an overseer. As each girl reached the front, the overseer asked her if she had anything to declare, giving them one last chance to rid themselves of anything they were forbidden to carry into the Danger Building. According to the rule book, men were not given the option to declare, but were patted down by their overseers, their pockets scrupulously searched for matches and tobacco.

"Dempsey." April threw her an appraising glance over her shoulder as she moved up the line. "Is your husband Daniel Dempsey, of Tottenham Hotspur?"

"Yes," Lucy said, lowering her voice, "but perhaps we can keep that between us? After what happened before, I don't want anyone to think I'm putting on airs."

"Of course. Your secret's safe with me." April puffed out a breath and shook her head. "Daniel Dempsey. He's brilliant."

"Well, naturally I think so, but I'm hardly impartial."

April laughed, and Lucy suddenly felt very much better.

The good feeling lasted as she passed through inspection and into the Danger Building, where the munitionettes timed in with their punch cards and dispersed to their stations. April and most of the other girls entered an open doorway beneath a sign that read "Filling Shop," but Lucy heard mechanical noises coming through a doorway at the far end of the room, so she decided to head that way. As she passed the Filling Shop, she glimpsed April taking a seat beside the unfriendly girl—Marjorie, April had called her—chatting and smiling as if they were old friends. Maybe April would convince Marjorie to give her another chance. It would certainly make the long shifts more pleasant if they all got along.

The other doorway, above which was painted the words "Finishing Shop," turned out to be a wide, short brick-and-concrete passageway leading into a different building altogether, one noticeably warmer. Entering a vast room about the size of a warehouse, Lucy observed rows of industrial machines, still with the shine of newness upon them, and a dozen or so munitionettes and at least as many men and boys turning out copper caps on huge pressing machines. As she approached the first row, a burly, dark-haired man in his mid-forties broke off an intense discussion with a slighter fellow who was wielding a wrench in the guts of the only silent machine in the room. The burly man scowled at the sight of her, which told her with immediate, dismaying certainty that he was the foreman. "I s'pose you're the new girl who fancies herself an operator?" he fairly growled at her.

"I'm Lucy Dempsey, sir," she replied, offering a courteous nod. "I've been assigned to finish shells. Superintendent Carmichael ordered me to report to the foreman."

"Aye, that's me." To the man with the wrench, he added, "Keep at it. I'll ask Donovan about those parts." Turning back to Lucy with a glower, he said, "Let me make something clear: This is no place for a girl. Filling shells is one thing, but the Finishing Shop is for skilled work, union labor, and if not for the war I'd have nowt to do with training you."

"Well, naturally." She was too surprised to be intimidated. "If not for the war, I wouldn't be here."

"That's true enough. Don't think you're keeping this job after we thrash the Germans, either. You'll be out, and a soldier returned from the front will get his job back."

"He'll be welcome to it," she said, bewildered. When the war was over, the government wouldn't need munitionettes because it wouldn't need munitions. In the meantime, she needed the wages, and surely the foreman didn't think the jobs should go unfilled because there weren't men enough to fill them. "I understand this job is only temporary. I'm just trying to do my bit, and to earn a wage while my husband is in the service."

The foreman eyed her suspiciously, but apparently she had said the right thing, because his scowl became slightly less fierce. "Don't think you can pass off second-rate work just because you're a novice," he warned. "A badly made English shell can kill our Tommies just as dead as a German bomb. I'll be inspecting your work very carefully, and you'll do over any mistakes until you get it right."

"Thank you, sir." Why did he keep thundering the obvious at her as if he expected an argument? This was her first factory job ever. She had hoped for close supervision, at least in the beginning, but she had been reluctant to ask for it, since the foreman was evidently very busy.

"No need for 'sir.' Save that for the boss. Call me Mr. Vernon." He turned away, gesturing sharply to indicate that she was to follow. "Come on, then. We'll get you started."

He led her past the pressing machines to a more open space at the far end of the building, where munitionettes worked with a variety of hand tools before rows of large shells set up on their back ends. She expected to be taken through to another chamber containing the sort of machines she had trained on in her course, but instead he halted before the row of eight-inch shells in the far corner of the room, as if he meant to tuck her away out of sight.

"Other girls will bring you the filled shells on a trolley, and you'll help unload them to your station," he said brusquely. "Before you finish the shell, you need to fit the exploder." He selected a hand drill from a rack of tools and demonstrated how to attach it to the tip of the shell and remove the plug with several vigorous turns of the handle. Setting the plug and drill aside, he inserted the nozzle of a long black rubber hose into the shell and tamped down the powder, scraping around the interior sides, presumably to catch every grain. "Where are your gloves?" he asked, eyeing her slender hands.

"I haven't got any."

He grumbled deep in his throat and jerked his thumb, showing off his own gloves, toward a wall of shelves where various caps, aprons, gloves, masks, and other gear were sorted. She hurried over and quickly searched through the gloves for the smallest pair, grabbed a heavy apron for good measure, and hurried back. Mr. Vernon had picked up a metal pitcher with a spout and was stirring the contents with a metal bar, scraping down the sides. "Don't get any on yourself or you'll be badly burned," he advised curtly as he carefully poured a thick liquid into the shell. "You've got to top off the case with molten explosive."

She nodded, feeling the heat of it on her face.

Setting the pitcher aside, he selected a tool that resembled a metal bar with an open, circular cutting tool in the middle. "This is a die stock," he said, fitting the circular part on the head of the shell and grasping the two metal bars like handles. "When you turn it like so"—he paused to demonstrate several firm, clockwise turns—"it cuts uniform threads. Do you know what that means?"

As it happened, since her father-in-law was a carpenter, she did. "Yes, sir—Mr. Vernon."

"Next, use the wheel to clear the screw threads." Taking another tool from the rack, he inserted the shaft into the tip of the shell and turned the wheel clockwise several times with an effort she wasn't sure she could match. Removing the wheel, he inserted a sort of spike into the tip and pounded the end again and again, then withdrew the spike, inserted a rod to measure the depth, and repeated the process until its contents reached the proper level, a number he ordered her to memorize. He then demonstrated how to insert the detonators, narrating the process with warnings and admonitions. Lastly, he stenciled the destructive mark on the shell. "Trolley workers will carry the finished batch away," he added. "You'll help them load. Then you catch your breath, if you can, before the next trolley of shells to finish arrives." He eyed her, frowning. "Are you ready to give it a go?"

There was no other acceptable answer but yes, so she nodded.

For the first two hours Mr. Vernon hovered over her shoulder, observing her every move, correcting and reproving her whenever she fell short of perfection. From the corner of her eye she observed other munitionettes working deftly and assuredly at the same tasks, seeming to complete three shells to her every one. The work was physically demanding, especially turning the wheel and pounding the mallet, and by the time Mr. Vernon trusted her enough to proceed on her own, checking in only every ten minutes, and then every fifteen, her arm and shoulder ached.

Once, when the foreman stepped away, Lucy paused to wipe her brow with her jacket sleeve. "Don't mind Mr. Vernon," another munitionette called to her from the next row. "He doesn't like any of us. You're doing well."

"Thank you," Lucy called back, grateful for the first kind word she'd heard in hours. "Is it normal to feel like your arm is about to fall off?"

"Only on the first day," the other woman replied, grinning. "Maybe the second too."

Lucy was glad for the gloves and the apron, but she would have liked a veil too, or nose plugs. Traces of yellow powder lingered on the tools and her station, and the acrid odor of the molten explosive stung her nostrils and eyes. It was a relief when the bell rang to signal the dinner hour. She quickly finished stenciling a shell, left her gloves and apron on the table, and followed the other munitionettes back into the other building, where the workers were passing through a stone archway into a washroom with several rows of sinks. Chatting and teasing, they lined up to scrub their hands and faces, some vigorously, others indifferently, then passed through a doorway on the opposite wall into the canteen.

Lucy was one of the last to get a turn at a sink, and she took extra care to wash every yellow speck off her hands; somehow, despite the gloves, powder had gotten on her skin. By the time she entered the canteen, most of the other workers had already collected their food and had claimed seats at one of the many tables for four or six.

"Oh, look, it's Lady Prim," a familiar voice sang out from a table somewhere to her left. Heart sinking, Lucy did not spare Marjorie a glance but walked straight on to the dinner queue. "Why does *she* get the Finishing Shop, I wonder? Why shouldn't she start in the Filling Shop and work her way up like everyone else? Mabel's been here longest."

Cheeks flushed, Lucy ignored Marjorie and the murmurs of agreement her taunting evoked. Tray in hand, she looked around the room for the friendly girl from the Finishing Shop, or the cheerful girl from her interview, but she spotted neither. She found a table where three other workers were nearly finishing eating; they smiled kindly as she seated herself, but they forgot to introduce themselves or ask her name, too busy conversing in hushed, scandalized voices about a lathe operator named Mary who had been sacked that morning after a matchstick had tumbled out of her pocket.

"A matchstick," one of the girls exclaimed. "Why on earth would she bring a matchstick into the Danger Building?"

"Mary said she forgot it was in her pocket from lighting the fires at home before she left this morning," another replied.

"That doesn't make any sense," the third scoffed. "It would have been in the pocket of her regular clothes. She changed into her uniform in the shifting house like everyone else."

"I'm just telling you what she told the superintendent," the second girl said as they all rose and gathered their trays. "What a sight that was, Mrs. Carmichael swooping in and snatching her up and hauling her off."

"Poor dear," the first girl said as they walked away. "So humiliating, to be sacked in front of everyone! She won't get a leaving certificate either, so she won't be able to find a job somewhere else."

Lucy felt a chill. That disgrace could have been hers, if April hadn't reminded her about her wedding ring.

The afternoon went better than the morning, and at teatime Lucy made sure to stay close to the friendly girl from the Finishing Shop as the crowd passed from the washroom to the canteen. To her relief, when she asked if she might join her and her two companions at their table, they cheerfully agreed. As soon as Lucy introduced herself and learned all their names, the conversation quickly turned to the unfortunate Mary, sacked for a matchstick. "She could have blown up the whole building," one of the girls said, her eyes wide with alarm.

"Not unless she dropped a lit match onto a pile of TNT," said the friendly girl, whose name was Daisy.

"I'm not so sure. That yellow powder drifts everywhere." The third girl held out her hands, the yellow tint of her skin proving her claim. "If someone had stepped on that match with her wooden clog, and scraped it on the floor, and it had sparked—" She pressed her lips together and shook her head, and they all exchanged uneasy looks.

"It could have happened, but it didn't," Daisy said reassuringly.

"Can I look forward to this much excitement every day?" asked Lucy, smiling, hoping to chase away the gloom. They smiled back and assured her that grave mistakes like Mary's almost never happened. Everyone was scrupulously careful. The consequences would be too dire if they were not.

Although she was tired, Lucy finished out her shift well, spirits lifted by the promising new friendships she had made. The next day her muscles were sore, but she had become accustomed to the tasks and worked through the aches, earning a grudging grunt that might have been approval from Mr. Vernon. But at dinner, Marjorie trod on her heel as Lucy carried her tray to her table, causing her to stumble in her wooden clogs and spill her soup. Daisy fetched her another bowl as Lucy cleaned up the mess, so she still had time to eat before she was due back at her station.

The days turned into weeks. Lucy's muscles lost their soreness and she felt her arms growing stronger and leaner. She did her best to avoid Marjorie, but she still endured the occasional ill-timed jostling. More frequent were disparaging remarks about how she had stolen a prime job from Mabel, who, Lucy learned, turned out to be the most popular worker in the Danger Building, a motherly figure to many of the girls who did not get along with their own mothers, or who had moved to London from distant villages and struggled with bouts of homesickness. Mabel and the other longtime workers bore their yellowed skin proudly, evidence of their seniority.

Lucy told herself that invented workplace tiffs were beneath her notice considering the very real dangers all around. In the middle of October, a zeppelin again bombed "a London district," as the papers put it, the censors having forbidden anything more specific; more than three dozen people had been killed and eighty-seven injured. Exhausted from long shifts at the arsenal, Lucy had slept through the entire raid. She was immeasurably grateful to her mother, who kept the boys fed and all of them in clean clothes. On the last Saturday of October, her mother even took Jamie and Simon out to the Dell to see Daniel play for the 17th Middlesex team against Southampton. Lucy could not go because of work, but she hung on every detail of the boys' description of the match the next morning.

"Dad said they expect to get their orders soon," Jamie said, watching her closely for her reaction.

"Just as we've all expected," she said evenly, concealing her dismay. "It's nothing to worry about. Haven't we been fortunate to have seen him so often, even after he enlisted? Most families aren't that lucky."

The boys agreed, but Lucy's mother gave her a sympathetic look, knowing how she really felt.

A week later, Lucy was in the inspection queue, lost in thought, brooding over a recent letter from Daniel in which he said he believed their deployment to France was imminent. Just then her gaze was drawn to a glint at the nape of the neck of a tall, pretty young woman who wore her reddish-brown hair in a braid beneath her cap. She was four people ahead of Lucy, and was next up to approach the overseer. Without thinking, Lucy quickly stepped around the girls in front of her, snatched the hairpin from the taller girl's locks, evoking a cry of surprise. Concealing the hairpin in her fist, Lucy strode back to the shifting house and dropped it into the nearest waste basket. Pausing a moment to settle her nerves, she joined the end of the queue and tried to keep her expression serene despite the curious, astonished glances of the munitionettes who had clearly witnessed everything. The tall girl, hand clasped to the nape of her neck, stood frozen at the head of the queue, her face pale wherever the yellow tint had not touched it. The overseer had to prompt her twice before she responded that she had nothing to declare and moved on into the Danger Building. Right behind her, Mabel held Lucy's gaze for a long moment, offered her a slow nod, and took her turn with the overseer.

Later, Lucy was enjoying dinner with Daisy and her friends when Mabel approached the table, the tall girl trailing after. Everyone respected Mabel, so their conversation abruptly ceased when she halted by Lucy's chair and planted her hands on her hips. "You did a good turn for my Peggy today," she remarked, inclining her head toward the younger woman. "If the overseer had seen that hairpin, my girl would have been sacked, and there would've been no leaving certificate coming to her."

"I'm glad I saw it first," said Lucy, offering Peggy a nod and a smile.

Mabel regarded her knowingly. Behind her, other munitionettes had drawn closer to observe the scene—Marjorie near the front, scowling, April just behind her, absently fingering the cuff of her jacket, her expression hopeful. "If anyone had caught you with it before you tossed it, you would've been the one sacked."

"I'm glad I was quick," Lucy replied.

"I hear you're Daniel Dempsey's wife."

"Did you, now?" Lucy raised her eyebrows at April, who had the decency to look embarrassed.

Mabel paused to cough and clear her throat. "Do you play football as well?"

"Not as well as my husband."

A ripple of laughter rose from the group. "I wouldn't expect that," said Mabel, grinning. "Some of us play now and then after dinner and after tea, in that field near the motor pool. That doesn't leave us much time, of course, not enough for a real match or even a half, but it's a bit of fun in the fresh air. You're welcome to join us anytime, if you like."

A wide-eyed look from Daisy told Lucy this was a great honor indeed. "I'd enjoy that," said Lucy. "Thank you."

"Thank *you*," said Peggy, stepping forward and reaching for Lucy's hand. "I would've been out on my ear if not for you."

Lucy gave her hand a quick squeeze. "Maybe the overseers wouldn't have noticed."

"If not them, *someone* would have noticed," Peggy said, frowning as she glanced over one shoulder and then the other, as if she expected to find a spiteful coworker or a German spy eavesdropping. "Someone would have reported me."

On her way home that evening, Lucy was fatigued but elated. She had helped a fellow munitionette and had earned Mabel's approval, which surely meant that Marjorie would quit bullying her. She couldn't wait to tell her sons, and Daniel, that she had been invited to join the Thornshire Arsenal football club, even if it was just a few friends having a bit of fun on a makeshift pitch.

But when she arrived home at nine o'clock, her mother met her in the foyer, her expression grave, an envelope in her hand. "Daniel addressed it to the family," she said, holding out the letter, "so Jamie convinced me to let them read it. I apologize if I should have waited, but they were so eager to hear from their father."

"Of course they were," Lucy said, setting down her bag and taking the envelope. "As you say, Daniel sent it to all of us. It would have been cruel to make them wait."

Daniel had written the letter himself, she reminded herself as she removed it from the envelope. He must be well, safe, uninjured. She mustn't always imagine the worst.

But as she read, her heart plummeted. Though his news wasn't the worst she could have imagined, it was still quite bad indeed.

The 17th Middlesex had received its orders. In less than a fortnight, Daniel would be sailing for France.

8

November–December 1915

APRIL

"I don't understand why Mabel had to invite Lady Prim to play football with us," Marjorie grumbled early one mid-November morning as she and April rode the overcrowded tram to Thornshire.

"Mabel owed her a favor," April reminded her. "Peggy would've been sacked if not for her."

Marjorie rolled her eyes. "I might've seen that hairpin too, if I had been standing as close as she were."

"Maybe, but you weren't and you didn't." Marjorie seemed determined to dislike Lucy, but everyone else thought well of her. Even Mabel didn't begrudge her the job in the Finishing Shop, once word got out that Lucy had undergone weeks of training and was more qualified despite being new. "Lucy risked her own job to save Peggy's. Inviting her to join the club was the least Mabel could do."

"It's all very well that Lady Prim's husband is a footballer," Marjorie went on, her voice shaking as the tram hit a bump that jostled April and Marjorie against each other. They were lucky to have seats at all; most mornings the cars were so packed that they had to stand, and on particularly bad days, they couldn't squeeze aboard at all, but had to wait for a later tram. "But it's not like *she* plays for Tottenham."

"Would you stop calling her that name? What has she ever done to you?"

"Why are you defending her? She's stuck-up."

"No she isn't. She's just shy. If you'd give her a chance, you'd see that." April heaved a sigh, exasperated. "Anyway, even you have to admit she's a great player. It has nothing to do with her husband. Maybe he taught her a few things, but she came by her natural ability herself."

"She's not bad," Marjorie admitted grudgingly. "All I can say is she had better not want to play keeper. That's Mabel's job, and I'm next up."

"You have nothing to worry about. Lucy's a striker." April shook her head. "Honestly, Marjorie. Lucy didn't steal Mabel's job at the arsenal and she's not after anyone's spot on the pitch. You should be glad Mabel's adding more players. If we get enough for two full sides, we can play a proper match, and we'll need two keepers, you and Mabel both."

"That would be grand, if Mabel keeps playing."

"Why wouldn't she? She founded the club." Marjorie only took over in the goal when Mabel needed a rest. Otherwise, which was most of the time, Marjorie played defense.

"Yes, but her cough is getting much worse. Haven't you noticed?" Marjorie shook her head, frowning. "Seems to me she rests a lot more than she used to."

"Maybe, but she's older than we are."

"Yes, but she's hardly *geriatric*." Marjorie rose as the tram slowed to a stop. "Come on, then. Mustn't be late again or we'll both be out of the club, since we'll be sacked."

Nodding, April too rose and grasped the handrail for balance. Her mother's letters made it clear how much April's increased earnings mattered to the family, providing more and better food and new clothes and shoes for her siblings. She couldn't let them down by losing her job. Even transferring out of the Danger Building to an easier position elsewhere at Thornshire would be giving up too much.

As they disembarked the tram and headed toward the arsenal, Marjorie suddenly turned her head, covered her mouth with her mitten, and coughed hoarsely. "Wouldn't you know it? I think I've caught Mabel's cold."

"Something's definitely going around," April agreed as they were swept up in a stream of other workers all heading to Thornshire. Their entrance, Gate Four, was on the far side of the complex, so they took their usual shortcut on the gravel paths that wound through the park in front of Gate One. The park was usually quite pretty, but it was less so now with the flower beds covered in mulch and the shade trees barely holding on to their last dry, brown leaves. April enjoyed walking through the park in any season, and she inhaled deeply, imagining for a moment that she was back in the countryside. London had its delights, and now that she was earning enough to keep back a bit of spending money from the wages she sent home, she occasionally indulged in a visit to a teashop or a confectioner's, or enjoyed a night out at the theatre. Even so, she preferred the green moors, murmuring brooks, deep forests, and rich farmlands of Derbyshire to the odors and noise of the city.

Until she could manage a visit home, which looked unlikely for the duration, the park would have to do.

As they hurried along the gravel path, Marjorie nudged April and nodded to a young man seated on a bench near the center of the park, hands in his pockets, legs stretched out before him and ankles crossed. "Look, there he is again," Marjorie said, scornful. "Lazing about on a park bench while our brothers are fighting in trenches somewhere in France. If this one can't put on a uniform, the least he could do is find a job."

April studied the young man for a moment, not long enough to draw his attention. "Maybe it's his day off."

"Then he has a great many days off," Marjorie retorted. "He's here nearly every morning we are. There's a war on. Everyone who can contribute something in the way of war work ought to do so. In his case, he should enlist. He's a bit scrawny, but he's strong enough to carry a rifle."

"He's not scrawny. He's slender, but he's at least as tall as you." April actually thought he was quite good-looking, although she wouldn't admit that aloud, not after Marjorie disparaged him. He had thick auburn hair so dark it appeared brown from a distance, an oblong face with a straight, narrow nose, dark brown eyes, straight eyebrows often drawn together as if he were concentrating, and a mouth that was neither too narrow nor too full. It tended to quirk upward on the right, which April supposed sufficed for a smile for a serious young man. His suits and coat were not flashy, but they were clearly a better cut than what the workmen who passed him on the way to the arsenal could afford. Curiously, several of those workmen exchanged nods or a few pleasant words with him in passing, something she had noticed on other mornings. Evidently they didn't share Marjorie's disgust for shirkers.

"Maybe he has bad eyes," April suggested as they passed him, lowering her voice so he would not overhear.

"Then where are his glasses?"

"In his pocket? Or maybe he has a heart condition."

"Or maybe he's a coward." As they left the park, Marjorie caught the sleeve of April's coat and brought her to a halt. "Why must you always defend every shirker, or is it only the good-looking ones? Are you a conchie?"

"Of course not," said April, tugging her sleeve free. So Marjorie thought he was good-looking too. "Would a conscientious objector work in munitions? I support our Tommies every bit as much as you do. I'm just not as obvious about it."

"Maybe you should be." Marjorie started walking again, so briskly that April had to run a few steps to catch up. "What good is your silent disapproval? It doesn't get a single shirker to enlist."

"Well, no, but if the government passes conscription—"

"You don't know that they will, and that doesn't let you off the hook in the meantime. You've seen the posters. Getting able-bodied young men to enlist is one of the most important bits of war work we women can do."

"I'd rather fill shells."

"You can do both. I do."

April muffled a sigh and let it go. She didn't want to spend the bright, clear morning arguing, not when they were about to spend the next six hours indoors.

They didn't exchange another word as they approached Gate Four, showed their identification badges to the sentries, entered the shifting house, and changed into their uniforms. Only when they stood in line for inspection did Marjorie turn around, peer at her for a moment, and remark, "You know, in this light, your hair looks a bit ginger."

"Yours too." April wondered why she hadn't noticed before.

"Yellow skin and ginger hair. I s'pose we're fully fledged munitionettes now."

"Canary girls through and through," April agreed. The more experienced workers had told them that changes to their skin and hair were only a matter of time, and not to worry about them. They weren't harmful, and the color would fade soon enough if they quit work or transferred out of the Danger Building. April's skin had taken on a faint yellow tinge, as had Marjorie's, not yet as vivid as Mabel or Peggy or the others who had been munitionettes longer. April had become so accustomed to the canary girls' peculiar coloring that she scarcely noticed it in her friends. She nearly always forgot about the changes to her own appearance unless a passerby on the street gawked, which made her feel both embarrassed and angry, or if a fellow passenger on the tram thanked her for her service, which filled her with pride. Some of the girls lamented their diminished beauty, but all agreed that any fellow who spurned them for a temporary imperfection was beneath their notice.

In addition to the changes to their appearance, they all seemed to acquire the same hoarse cough eventually. Most of the canary girls supposed that it was just a bad cold they kept passing around, working so closely together and often sharing the same hostels. The women who had come to munitions works from the mills disagreed, though, recalling the similar, persistent coughs that girls developed after months at the looms, breathing in all those stray fibers. Yet all of the munitionettes believed that they had more to fear from accidents and bombings than anything floating in the air. They didn't often talk about it because it was simply too grim, but the possibility of a deadly explosion, whether set off by a careless worker or a German bomb, was something they had to accept every time they passed through the arsenal gates.

For her part, once she entered the Danger Building, April pushed her worries to the back of her mind and focused on the task at hand. Although the work was monotonous and repetitive, it was less exhausting than being a housemaid and the hours were much better, just as Marjorie had promised. Sometimes April thought wistfully of the walks into the village from Alderlea, and the happy stolen moments browsing the shops or admiring the storefront windows, but mostly she took pride in her contribution to the war effort and delighted in her substantially higher wages.

Filling shells wasn't difficult, and with experience April had improved her quality and speed, earning praise from the overlookers. The munitionettes in their section worked mostly on eight-inch shells, although sometimes they were shifted over to other armaments depending upon need. Each girl had a tin of TNT, a funnel, a mallet, and a smooth wooden rod that looked like it might have been cut from a broom handle. Inserting the funnel into the tip of the empty shell to prevent spillage, a worker would scoop a measure of the yellow powder inside, then tamp it down firmly with the wooden stick—but not too firmly, or it might detonate. They would repeat the process, again and again, until the shell was sufficiently filled, which they measured by marks on the stick. Then they would fix the cap firmly in place and load the shell carefully onto a trolley, to be wheeled off to the Finishing Shop.

They were as tidy and efficient as they could be, but the yellow powder drifted over everything—not massive piles of it, they were not so wasteful as that, but traces here and there and everywhere. It was little wonder their faces and hands turned yellow, but April didn't understand why blondes should turn into gingers, and in such a patchy fashion. Typically the length of the strands changed, with only the roots showing the natural color. The dark-haired girls looked even more peculiar. Brunettes' hair sometimes took on a greenish hue, while girls with very dark hair saw it leached of color until it became almost white, but usually only in the front, where they faced the shells. The workers' caps, which were meant only to keep their hair out of the way, did nothing to prevent the discoloration.

A week before, Lucy, who had the most enviably beautiful, silky black hair, ruefully lamented the light patch appearing above her brow. She had assumed she was going gray due to stress and lack of sleep until Mabel set her straight. "The color will come back, though, won't it?" Lucy asked worriedly as they kicked the football around after dinner. "I don't want Daniel to see me like this."

"I don't blame you," said Marjorie, voice dripping with false sympathy. "You look a bit like a badger."

"Shut up, you," Peggy said fondly, giving her a shove. Most of the other girls laughed as if it was all in good fun, but not Lucy, and not April.

Marjorie's casual cruelty did not escape Mabel's notice. While she didn't call Marjorie out in front of the club, she did move her from defense to midfield, which Marjorie hated. She liked to see the whole field spread out before her without having to turn this way and that or look behind her, and nobody ran more than the midfielders, as April could attest.

"The color won't come back, but the bleached parts will grow out," April reassured Lucy later as they headed back to the Danger Building. "The hair will grow back the proper color."

"Only to be bleached anew," Lucy said mournfully, but she managed a smile. "It's only for the duration, though, right?"

"Right," April said, putting an arm around her shoulders.

As the weeks passed, the white patch in Lucy's hair spread and her skin turned as yellow as April's. Almost all of the girls in the Filling Shop caught Mabel's cough, and several, including April, were also bothered by persistent sore throats and sneezing. Over a fortnight, three players reluctantly quit the football club, complaining of shortness of breath, fatigue, and chest pains. Mabel told them they were welcome back anytime and urged them to visit the infirmary. Later April learned that two of the girls had transferred out of the Danger Building, while the third had left munitions work altogether.

"If they're tired and short of breath, it just means they aren't fit," Marjorie declared one evening as they rode back to their hostel. Her voice was barely audible over the rattling of the tram, and she rested her head on April's shoulder, eyes closed. "They need more exercise, not less."

"Maybe they do need rest," April said. "Sleep helps you with your migraines. Isn't this your second this week?"

"Third, if you count Monday." Marjorie sighed and clasped a hand to her forehead. "It's the weather. Ever since it turned colder, these stupid migraines—" She winced as the tram jolted. "I'll be fine in a couple of hours."

"It got colder than this in Derbyshire, and you never had migraines there."

"Only because Mrs. Wilson wouldn't allow it." A wan smile appeared on Marjorie's sallow face, which had always been porcelain and roses before. "D'you mind if we don't talk for a while? My head is throbbing, and I feel like I have to think very hard to keep my words in the right order."

"Of course," April replied in a whisper. She stroked her friend's head gently, smoothing her gingery curls away from her forehead until the tram reached their stop.

The next morning, Marjorie felt perfectly fine again. She had slept a solid eight hours and the weather was a good deal warmer and sunnier than it had been the day before, and so, as they raced off to the arsenal, she and April argued good-naturedly about who was right about the cause and the cure for her migraines. The auburn-haired lad was in the park by the front gate again, but on a different bench this time, chatting with an older fellow April recognized as the foreman of the shell casting shop.

"Maybe Mr. Townes will convince that shirker to get a job, if he won't enlist," Marjorie said, raising her voice as they approached.

"Hush," April murmured. "They'll hear you."

"So what if he does?" Marjorie replied in a somewhat quieter voice. "If he understood how girls like us despise shirkers like him, he might do something about it."

"I'm more concerned about offending Mr. Townes." April linked her arm through Marjorie's and propelled her forward, watching the men from the corner of her eye as they passed. Neither the foreman nor the auburn-haired lad took any notice of them.

When they had left the hostel earlier that morning, April and Marjorie had been obliged to wait nearly a quarter hour until a tram with room for them to squeeze aboard had come along, so they were cutting it close as they raced into the shifting house, threw on their uniforms, timed in, and took their stations with only seconds to spare. The foreman worked them at top speed all morning, which told them a big push was in the works over in France. Whenever demand for armaments surged, within a fortnight the newspapers would report some new offensive, an assault on German forces entrenched near a battered French village or a bold attempt to break through a German salient.

At dinner, April, Marjorie, Peggy, and a trolley girl named Louise discussed what might be happening abroad. They were piecing together details from their brothers' and sweethearts' letters from the front when Peggy suddenly set down her fork and grimaced. "Does this taste strange to anyone else?" she asked, eyeing her meal suspiciously. "It tastes like someone sprinkled iron filings in it."

"Maybe it's copper shavings from the fuse cap shop," Louise said with a grin. "A little accidental extra seasoning."

"I taste it too," said April, sampling her mashed turnips thoughtfully. "Something metallic."

Marjorie studied her plate. "I don't think it's the food," she said, pulling a face. "I've been tasting metal for days, whether I'm eating in the canteen or the hostel or that posh teashop April's always dragging me to. It comes and goes. I figured it had something to do with my migraines, but if you have that metal taste in your mouths too—"

They exchanged uneasy looks.

"Do you think that's why they're always telling us to drink more milk?" asked Louise, nodding to the dispensers and pitchers recently installed near the food queue. "I have friends in other buildings, and they say no one constantly nags them to drink milk the way they do in our canteens."

Peggy folded her arms over her chest as if warding off a chill. "That's another thing. Don't you think it's odd that we have to eat separately now?"

As Louise nodded, April and Marjorie exchanged puzzled glances. "Oh, this was before your time," Louise explained. "We used to be allowed to eat with our friends at other canteens, or they could eat here with us, but early this summer, Superintendent Carmichael put a stop to that."

"She said it was to improve efficiency, so we didn't waste time walking from building to building," added Peggy. "That made sense to us, since our breaks are so short as it is, but I heard that some workers complained about canary girls leaving yellow powder everywhere, on the tables and chairs, the cutlery, the doorknobs."

"I heard it was because we're always coughing," said Louise, frowning. "That, and our yellow skin made them lose their appetites."

The other girls gasped at the rudeness of it all. "I hope that's not true," April said. "Some of our footballers work in other buildings, and they treat us the same as anyone else. Although—" She hesitated, thinking.

"Although what?" Marjorie prompted.

"It's probably nothing, but on the trams, other passengers gawk at our skin and hair—you know those looks." April glanced around the table for confirmation, and they all nodded. "Some people thank us for our service, but they all keep their distance."

"That's true," said Marjorie. "They edge away from us as much as anyone can on a crowded tram."

"I thought it was out of consideration or respect for us as war workers," said April, "but maybe they're afraid it's catching?"

"Something's not right," said Peggy, shaking her head. "You know what I think? That yellow powder is making us sick. Everyone calls my mum's cough a cold, but no one has a cold for weeks and weeks."

"Maybe we should talk to Superintendent Carmichael," said Louise. "She might tell us what's what."

"Doubt it," Marjorie scoffed.

"I'd be afraid to be labeled a troublemaker," said Peggy. "I can't lose this job. Is there anyone we can ask who doesn't work at Thornshire Arsenal?"

"Lucy's brother is a doctor," said April. "Maybe he knows something."

"Wouldn't hurt to ask." When Louise stuck her hand in the air, April followed her line of sight and spotted Lucy seated at a table a few yards away with Daisy and two other Finishing Shop girls. "Lucy! Lucy!" Lucy glanced up, her eyebrows rising as Louise beckoned. "Do you have a minute?"

Lucy nodded, exchanged a few words with her friends, and then she and Daisy came over. "Hello, girls," she said, a bit warily. "What's going on?"

"Lo, Miss Prim grants the peasants an audience," said Marjorie, smirking.

"That's *Lady* Prim to you," said Lucy, without sparing her a glance. "Is this about that header? I promise you it was an accident. I don't think I could repeat it."

"Never explain away a header that leads to a goal," said Louise. "No, this is about something else. Your brother's a doctor?"

"Yes, but not in London. He's taken over our late father's practice in Brookfield. In Surrey," Lucy added, which was useful to April if no one else, for she had never heard of it.

"Could you ask him if we should be worried about working with the yellow powder?" Peggy grimaced and tugged on the end of her braid. "We all can see what it does to our skin and hair, but is there more to it? Things on the inside, things we can't see?"

"I don't like to break a confidence," said Lucy, lowering her voice, "but some of the girls in the Finishing Shop have been complaining about upset stomachs and sore breasts. We all assumed it was just the monthlies."

"Speaking of women's troubles—" Daisy inhaled deeply, grimacing. "I heard from a friend who works at Woolwich that one of their canary girls gave birth to a bright yellow baby."

They all gasped.

"That can't be true," said Marjorie.

"Can't it?" said Daisy, planting a hand on her hip. "How would you know?"

"Something so odd would've been in the papers."

"The same newspapers that tell us, 'A zeppelin visited a London neighborhood last evening,' instead of just coming out with it, that the Germans bombed Covent Garden and left a massive crater in Wellington Street?"

Lucy raised her hands for peace. "I'll call my brother tonight," she promised. "He did warn me not to work with chemicals. He won't be pleased that I didn't heed him, but he'll help us if he can."

They were due back at their stations, so the conversation ended there with no one's curiosity satisfied. Even Marjorie looked uneasy, despite her flippant dismissals. Throughout the afternoon, April glimpsed a crease of worry between her friend's eyebrows whenever she happened to glance her way.

The next morning, Marjorie seemed unusually prickly as they hurried from the hostel to catch their tram. When April timidly asked her if she were suffering another migraine, Marjorie snapped, "I wasn't until you started nagging me, but now I feel one coming on."

Stung, April nevertheless decided to excuse her friend's temper on account of nerves. Weren't they all feeling a bit unsettled after the revelations at dinner the previous day? "Maybe Lucy will have some word from her brother to put our minds at ease," she said. "It could be just colds and migraines and the monthlies, as you said. We could be making something out of nothing."

"I never said it was *nothing*," said Marjorie, somewhat contritely.

"Canary girls are definitely feeling ill. The question is why." Then she tossed her head. "Honestly. A canary baby? Sounds like a creature out of an old-fashioned penny dreadful. Did it have a beak and feathers, I wonder?"

"Now that *surely* would have made the papers."

Marjorie rewarded her with a grin.

When they reached the park in front of Gate One, Marjorie suddenly clutched April's arm and halted. "There he is *again*," she said, an edge to her voice, lifting her chin to indicate the bench up ahead where the auburn-haired lad sat alone, hands in his pockets, collar turned up against the December cold. "I have absolutely no patience for shirkers today. Fortunately, I came prepared." Digging into her coat pocket, she withdrew a white feather and held it up, the shaft pinched between her thumb and forefinger. "I've done my bit. Now it's your turn. You're going to give this to him."

April yanked her arm free. "I most certainly am *not*."

"Think of your brother, risking his life in the trenches. What would Henry say if he knew you were willing to let this shirker off easy because you were afraid to hurt his feelings?"

"Leave my brother out of this. He'd say let the poor fellow alone if he isn't brave enough to enlist on his own."

"Sometimes the lads need a little incentive to be brave."

"Put that silly feather away. I told you, I'm not doing it."

"Fine." Marjorie linked her arm firmly through April's. "Just come with me, then, and I'll take care of it."

"Marjorie, no," April protested as her friend steered her toward the unwitting target. "Let's go find Lucy."

"Lady Prim can wait. This can't."

April tried to slow her down, but it was hopeless. Marjorie was purposeful, unrelenting, and the best April could hope for was to get the embarrassment over quickly.

The lad looked their way as they approached, a corner of his mouth turned up in a faint smile. "Good morning," Marjorie said crisply as she brought April to a halt in front of him. “My friend has something to say to you.”

“I most certainly do not,” April retorted hotly. The young man’s eyebrows rose inquisitively as he regarded them both. “You’ll have to forgive my friend. She’s barking mad.”

“If you have something you need to say, you can tell me.” He didn’t rise to speak with them, but leaned back against the bench, hands still in his pockets. “You both work in the Danger Building, isn’t that so?”

“Yes,” said April quickly, “and we’re going to be late, so we’d better go. Sorry to trouble you. Come on, Marjorie—”

She tried to walk on, but Marjorie planted her feet and held her back. “I have something for you,” Marjorie told him sweetly. “A gift, and some advice.” She held out the feather. “You should be in uniform.”

He studied her for a moment, the smile in the corner of his mouth turning contemptuous. “Let it never be said that I refused a gift from a lady,” he said, removing his hands from his pockets as he stood. He held out his right hand to accept the feather, and Marjorie gave it to him, visibly surprised, for no one had ever submitted to her torment so willingly before. Mortified, April looked away, her gaze falling on the wooden prosthetic where his left hand should have been.

She gasped aloud.

The sound drew his attention. “Put it in my buttonhole for me, would you, love?” he asked coolly, holding out the feather.

“I—I’m so sorry,” April blurted. “We didn’t know. We’re sorry.”

Marjorie’s laughter trilled. “Oh dear. What a dreadful mistake, and what an excellent joke on us! Is that why you keep your hands in your pockets all the time?” She reached for the feather. “Here, I’ll take that back. You obviously don’t deserve it.”

“Does any man?” He held up the feather in front of his face and examined it, front and back. “I think I’ll keep this. You clearly don’t understand how dangerous this little white feather can be.”

“Don’t be angry.” Marjorie offered him her most disarming smile, but it failed to work its usual magic. “It was an honest mistake.”

"Marjorie, let's go," April nearly shouted. "We're going to be late."

"Hey there, you two!" a deep voice bellowed behind them. Instinctively the girls turned and discovered a very large, very angry man storming toward them. "What's that you're on about, handing Mr. Corbyn a bleedin' white feather? He's a veteran, ye daft bints! He works for the boss!"

"I think it's time to go," Marjorie murmured. She seized April's hand and took off running. April quickly outpaced her, and by the time they reached the edge of the park, they had lost themselves in the throng of workers. Only after they passed Gate One and rounded the corner of the complex did they slow to a walk.

"Do you think he'll recognize us if he sees us again?" Marjorie asked, breathless.

"Of course he will! How could he ever forget us?"

"Not the small one. The large, terrifying one."

"I don't know!" Glaring, April lifted her hands and let them fall to her sides, exasperated. "One glance at us and anyone would know we work in the Danger Building."

"But did either of them get a *good* look, d'you think?"

"I don't know," April snapped again. Shaking her head, she strode off toward Gate Four. "You're going to get us sacked, if you haven't already. You heard what the bigger fellow said. Mr. Corbyn works for Mr. Purcell. All he has to do is say one bad word about us, and we'll be out on our ear."

"Why should he do that?"

"Why shouldn't he?"

"It was an honest mistake!"

"It was a mistake, sure enough. I told you that before you did it, but you didn't listen. You never do."

They fell silent as they joined the queue at Gate Four, showed the sentries their identification badges, and moved on to the shifting house. April's jaw ached from clenching it, and she resolutely ignored the contrite sidelong glances Marjorie threw her as they changed into their uniforms.

"How was I supposed to know he works for the boss?" Marjorie murmured in April's ear as they queued up for inspection. "How was I to know he works at all, when we only ever see him sitting around the park?"

"Sitting in a park in a nice suit and coat in front of a munitions factory that employs thousands of people," said April flatly. "You're right. Who could have imagined that he works here, same as us?"

"You don't have to be sarcastic," said Marjorie, subdued. "I said I was sorry, to him and to you."

"You did *not* say it to me," April retorted over her shoulder as she reached the top of the queue.

"Well, now I will. I'm sorry."

"Have anything to declare?" the overlooker inquired.

April shook her head. "No, miss."

"Be sure to drink a full glass of milk with your dinner today." The overlooker waved her through. "Next."

"I truly am sorry," Marjorie said earnestly, catching up to April at the doorway to the Filling Shop. "Will you ever forgive me?"

"If we don't get sacked, I'll consider it."

As they took their stations, side by side as always, and the trolley girls arrived with the first shell casings of the shift, Marjorie smiled, relieved, as if the whole ugly business was not only already forgiven but entirely forgotten.

Perhaps in Marjorie's case it nearly was, but whenever April remembered the look on Mr. Corbyn's face when Marjorie had handed him the feather, she felt sick with remorse. April might forgive Marjorie, but she could not forgive herself.

9

JANUARY–MARCH 1916

HELEN

"Listen to this, Arthur," said Helen, reading the *Times* over breakfast on the last day of January. "'Elsie Mary Davey, aged seventeen years, who has been missing from her home at Fleet-road, Hampstead, since January tenth, has been found engaged on munition work in a factory at Woolwich. In trying to obtain assistance from the Marylebone magistrate on Monday, the mother—a widow—said the girl was "mad on munitions."' Can you imagine?"

"I certainly can," replied Arthur, his eyes fixed on the financial papers as he raised his teacup to his lips. "Munitions factories offer young women steady work, lucrative wages, and independence—financial and otherwise—that most of them have never known before."

"You sound like a recruiting poster, darling."

"Add to that the canteens providing good square meals twice a shift, the camaraderie, the social and recreational activities, the pride that comes from patriotic duty, and the only real curiosity is why *more* young women don't run off to become munitionettes."

"Miss Davey left home without informing her family and was missing for three weeks," Helen pointed out. "From the sound of it, her poor mother was frantic. Surely you don't condone that."

"Well, no, not that," he conceded. "That was inconsiderate. But one has to admire the girl's pluck."

"I suppose." Helen silently read the rest of the article, which described Mrs. Davey's great relief upon reuniting at home with her willful daughter. "Miss Davey is seventeen. I thought the minimum age for munitions workers is eighteen."

"It is, officially, but factories have long employed lads as young as twelve and no one bats an eye. Not Thornshire," he quickly added. "Turn that accusing glare upon someone who actually deserves it."

"I'm not glaring at you. I don't glare."

"Perhaps it doesn't feel that way to you, but it certainly resembles a glare on the receiving end."

"Oh, stop. It's too early to endure your teasing." But she smiled, amused. "I suppose young women go into domestic service much younger than seventeen."

"And let's not forget the lads who lie about their age in order to enlist in the military." Sighing, Arthur set down the newspaper, closed his eyes, and rubbed the bridge of his nose as if to pinch off a headache in the bud. "If more young men in their twenties shared the younger boys' zeal, the government wouldn't need to enact conscription."

Helen paused, teacup cradled in her hands, thinking of Mrs. Pankhurst's recent calls for a national scheme to conscript young men into the military and young women into national service. "I don't approve of conscription," she said. "I think it goes against our great English principle of free will. Nor do I like this ugly business of shaming men into enlisting. One can hardly blame a fellow for not wanting to risk his life on a battlefield if he doesn't have an aggressive nature."

"Enlistments are down," Arthur reminded her. "How do you propose the government make up the numbers?"

"I surely don't know." Yet after a moment to consider, she said, "Why not improve the incentives? Better pay. Increased separation allowances for soldiers' families. Substantial life insurance policies to be paid out to widows and children. The option to volunteer strictly for the home defense. A man reluctant to go abroad may nonetheless be very willing to defend English soil against an invasion."

"Excellent proposals, darling." Arthur wiped his mouth with his serviette and set it on the table. "We should send you to Parliament to talk some sense into those squabbling politicians."

"I tried that as a suffragette and I ended up in prison."

"Right. Let's avoid repeating that, shall we?" Rising, Arthur passed behind her chair, rested his hand on her shoulder, and kissed her on the top of the head. "I'm off to work. What are your plans for the day, aside from solving all the nation's problems?"

"I'm planning to meet some Marylebone friends for luncheon, and then I thought I might visit the British Museum."

"Splendid idea." He paused, his hand warm on her shoulder. "Darling, if there are any other museums you've been meaning to visit, don't put it off much longer. The war costs the government around five million pounds a day, and there's been talk about closing public museums and art galleries for the duration to cut expenses."

"Oh, but they can't," Helen protested. "What a blow to morale that would be!"

"I agree, darling, but what would you suggest shutting down instead? The hospitals, the railways, the post?"

"Of course not." Suddenly inspiration struck. "They could close Alexandra Palace and send everyone home."

"I'm sure there are many who would argue that internment camps are essential to the war effort," he said wryly.

"For prisoners of war, perhaps. Not for civilians who have committed no crimes other than being German in Great Britain." Once Alexandra Palace in North London had been a vast, seven-and-a-half-acre entertainment and recreation venue. In the early weeks of the war, it had been used as temporary housing for refugees from Belgium and the Netherlands, but in recent months it had been converted into an internment camp for around three thousand German, Austrian, and Hungarian men deemed enemy aliens. "If my father were alive today, he might have been interned there, if he had not been deported outright."

"I never would have let that happen."

"Yes, I know, darling, but what of all those poor people who don't have your connections?" Helen shook her head. "Never mind. I don't expect you to fix it. I complain to you because no one else will listen."

"I'm on your side, Helen," he said. "Fear and suspicion have even sensible people imagining spies lurking everywhere. As I see it, our best hope is to end this war as swiftly as possible so that things can go back to the way they were."

She could tell from the way he shifted his feet that he was running late and needed to set out, but he hated to leave her there alone, disconsolate and frustrated. "Off with you, then," she said briskly, squeezing his hand and forcing a smile. "Go do your bit, and your proud little wife will greet you at the door with a kiss when you return."

"Are you sure you'll be all right?"

She assured him she would, and kissed him goodbye.

After breakfast, she took a brisk walk through Hyde Park hoping to improve her mood. She would have cancelled her luncheon date rather than inflict her ill humor on others, but she had met few women in her new neighborhood with whom she felt she had much in common, and she didn't want their invitations to stop coming.

At one o'clock, she met her friends at the luxurious Great Central Hotel on Marylebone Road, which had been requisitioned by the military for use as a convalescent hospital for wounded officers, but still kept its smaller dining room open to the public. Like herself, her three new friends were educated, married to wealthy, accomplished gentlemen, and determined not to idle away their lives as ladies of leisure. They all had been married longer than Helen, and all were mothers. When Helen confided her worries about her own empty arms, they assured her it was only a matter of time until she and Arthur were similarly blessed. Until then, it was simply a matter of try, try again.

At the moment, her friends' children were either in school, with nannies, or old enough to fend for themselves, so all four friends could relax and savor their consommé fermier, eggs with asparagus tips, fillets of salmon, and several excellent cheeses. It was impossible to avoid some talk about the war, as it influenced nearly everything about daily life in London, but since Beatrice had a son in the British Expeditionary Force, they refrained from discussing its more disturbing aspects.

As their soup bowls were cleared, Helen's next-door neighbor, Evelyn, announced in a whisper that she had heard an astonishing tale about the East End charity hospital where her husband was a trustee.

"Do divulge all," said Beatrice, an amused grin playing on her lips. "I love an astonishing tale."

After a sip of water and a dramatic pause, Evelyn said, "The hospital sees a good many patients from the Royal Arsenal at Woolwich and the Brunner Mond munitions factories at Silvertown—accidents, injuries, burns, and such."

Helen's heart dipped at the mention of munitions, but she said, "Go on."

"Occasionally young women who work with the explosive yellow powder show up in the maternity ward."

"Yes, with yellow skin and ginger hair," said Violet, the eldest of the four, a baroness. "I understand they are often called canary girls, but not in any derogatory sense."

"All true, but that's not the astonishing part." Evelyn inhaled deeply, letting the suspense build. Helen braced herself. "Several of these canary girls have given birth to yellow babies."

Her three companions gasped.

"You're teasing us, surely," said Beatrice.

Evelyn shook her head. "I have it straight from my husband, so unless he was misinformed, it's absolutely true."

"Could it be jaundice?" asked Violet, eyebrows drawing together over the bridge of her nose. "That's not uncommon in newborns. My own Cecily was born with jaundice, but the yellowish tint to her skin faded in a matter of days. Exposure to sunlight was the cure."

"It's not jaundice," said Evelyn. "This is a bright yellow, the same hue as the mother's skin. The babies appear to be in perfect health in every other regard, and the color fades within a few weeks, but I imagine it must be quite disturbing while it lasts."

Helen sipped her water to conceal her dismay. Was this condition afflicting Thornshire's canary girls too? Did Arthur know?

"Even if it is only simple jaundice," said Violet, "an increase in cases among so specific a group of mothers should be cause for concern." She turned to Helen. "Has your husband ever mentioned anything about this?"

"No, he hasn't," Helen replied, carefully setting down her glass. "He wouldn't permit any expectant mothers to work in the Danger Buildings, if he knew. He would have them transferred to a less dangerous site—the laundry or the canteen, perhaps—as long as it was safe for them to continue working."

"I'm sure he would," said Violet sincerely. "One would hope all gentlemen in his position would be as scrupulous as he."

After that, the conversation turned to other matters, but Helen was so troubled she scarcely heard a word of it.

Arthur did not return home until nine o'clock, but Helen had held dinner for him, although she herself had no appetite. She greeted him at the door with a kiss, as she had promised earlier that day, but she could not disguise how heavily her thoughts weighed upon her.

"What's the matter, darling?" he asked, cupping her cheek in his hand. "You look pensive."

She saw the strain and weariness in his eyes, and she could not bear to worsen it. "It's nothing," she said, helping him out of his coat and taking his hat.

"Whenever you say something is nothing, I know it's anything but."

Helen sighed, hung up his coat and hat, and took his hand. "It's just something one of my friends said at lunch today." She led him into the dining room and rang the bell. They had barely seated themselves when the cook bustled in with their first course.

"I see." He studied her. "Out with it."

Reluctantly, she told him what Evelyn had said about the canary girls' babies. Arthur's dinner cooled on his plate as he listened, his expression becoming increasingly solemn. When she finished, he sighed, took his fork in hand, but did not taste his food. "I can't tell you whether this story is true or false," he said. "But I promise you, I haven't heard one word about canary babies, and I trust someone would have informed me."

"I would certainly hope so."

"I can't speak for any other arsenal, but at Thornshire, if a worker is known to be in a delicate condition, she is immediately transferred out of the Danger Building." He paused for a moment. "Of course, as you know, no place on the arsenal grounds is entirely safe. That is an irrefutable fact of the manufacture of explosives."

"I understand that."

"It's clearly stated in the rule book that workers are required to inform their supervisor if they're expecting, but often they say nothing until their condition is too obvious to conceal."

"Why? Because they don't want to be transferred to a job that pays a lower wage, despite the risks?"

"Exactly so." Arthur took a bite of his dinner, then apparently realized that he was ravenous, for he tucked in with earnest. "We rely upon our overlookers and foremen to keep an eye out. Sometimes a concerned friend will quietly inform the superintendent."

Helen mulled that over as he finished his dinner. "Arthur," she asked carefully, "why *is* this explained in the rule book? If you don't believe that TNT is harmful to canary girls and their babies, why are expectant mothers prohibited from working in the Danger Building? You must have made that decision months ago, before the books went to print."

He set down his fork and took a long drink of wine. "My father and brother wrote the rules for all of the Purcell factories before Thornshire opened. Munitions work includes many tasks and conditions that expectant mothers should avoid, not only TNT exposure. Heavy lifting, long shifts, the potential for accidents—"

"Of course."

"If I had my way, expectant mothers wouldn't be permitted to do any munitions work, but it's not my decision. I have to answer to my father, and he has to answer to the Ministry of Munitions."

Helen managed a wan smile. "I wasn't aware that your father answered to anyone, except, perhaps, the King."

Arthur laughed shortly. After a moment, he reached across the corner of the table for her hand. "I know you're worried about the munitionettes. Would it reassure you to know that in late December, the Ministry of Munitions established a Welfare Department expressly to look out for munitionettes' interests?"

"I'm pleased to hear that," said Helen. "Dare I hope it won't be all talk and no action? What precisely will this department do?"

"Among other things, the committee has advised factory managers to hire women welfare supervisors, responsible matrons whose sole duty is to look out for the well-being of their munitionettes." He shrugged. "Granted, my father isn't too keen on the idea, but he might come around. In the meantime, my assistant secretary—you met him, Oliver Corbyn—"

"Oh, yes. The veteran. The good-looking chap."

"Yes, him," said Arthur, amused. "For the past few months, I've asked him to spend the first hour of his workday at the park in front of the main gate. That's our busiest shift change."

"So he's your spy? He eavesdrops and takes notes?"

"Nothing as sinister as that. His task is to be available to listen to any complaints or concerns the workers might have. He's from a village northeast of the city, he served in the army, and he's an amiable fellow, so naturally the men feel more comfortable talking to him in a casual setting than coming to my office and speaking with me, or even with Tom."

"Oh, yes," teased Helen. "You and your secretary are both so stuffy and intimidating."

"Perhaps not to you, my dear, but consider that no man wants his fellow workers to think he's a telltale. And if the workers' comments and complaints come through Oliver, they remain anonymous. I can learn what the workers are thinking, and they need not fear retaliation if I hear uncomfortable truths."

"And everyone at Thornshire is aware of this arrangement?"

"Through word of mouth, yes. It's strictly unofficial, based entirely upon trust. Over time, the men have learned that if they share their thoughts with Oliver, before long, problems get sorted."

Helen nodded, but a doubt still nagged at her. "You keep saying 'men' interchangeably with 'workers.'"

"I don't mean to. Force of habit. We employ at least as many women as men these days."

"Yet you might have chosen the correct word at that. How certain are you that the young women you employ feel as comfortable as the men do, approaching Oliver in the park and confiding in him?"

Arthur thought for a moment, frowning, and then shrugged. "As you said, he is rather good-looking."

"All the more reason the girls might be shy. They might not feel comfortable approaching a young man in a park if they haven't been properly introduced."

"Maybe not in Oxford or Marylebone, but this is at an arsenal in wartime. The customs are different than ordinary times."

"I doubt they're as different as you think. These are respectable girls, are they not? And this is a public park in front of the arsenal, not on the grounds of their workplace proper." Helen shook her head. "I suspect the munitionettes are completely unaware of this arrangement, and they might not take advantage of it even if they did know."

Arthur ran a hand through his hair, grimacing. "Fair point," he admitted. "All the more reason to convince my father to permit me to hire a welfare supervisor. In the meantime, if the munitionettes would prefer to confide in a woman, they can speak to Superintendent Carmichael."

Helen regarded him, skeptical. "The woman responsible for hiring, placing, and firing them?"

"It's not a perfect system."

"No, I can see that." Helen could also see that Arthur meant well and was doing his best, but she suspected the munitionettes might have many legitimate grievances that he simply never heard about.

February passed and winter faded as March brought steady winds, spring rains, and grim reports of casualties and terrible losses in France and Belgium. In Britain, zeppelin attacks persisted, especially on moonless nights, when the airships were more difficult for ground defenses to spot. Meanwhile, Arthur seemed to become more exhausted by the day, and as far as Helen could tell, nothing new had been done at Thornshire to ensure that the canary girls were being looked after.

Eventually she decided that she would have to investigate where Arthur could not, or would not, himself.

One morning, she waited until he left for work, then changed into a brown wool walking suit, taking care to leave all hairpins and undergarments with metal fastenings and wires behind. Carrying an empty satchel, she made her way by train and tram to Thornshire, arriving at the park in front of the arsenal just as the last dozen workers were hastening through the gates. For one alarming moment she thought she had missed her chance, but then she spotted Oliver Corbyn, hands in his pockets, striding away from her toward Gate One.

"Mr. Corbyn," she called out, hurrying after him. "Mr. Corbyn!"

He halted and turned, but he did not recognize her until she drew closer. "Mrs. Purcell," he said, surprised. "Hello."

"Good morning," she said, smiling, hands on her hips as she caught her breath. "I'm so glad I found you. I'm on an errand, but I don't have an identification badge, and as you can see, my husband isn't here to escort me. I was wondering if you could vouch for me to the sentries?"

His brow furrowed. "Where is Mr. Purcell?"

She waved a hand. "Oh, he went on ahead. You know him. The early bird seizes the day and all that." Mindful that her nervous babbling might raise his suspicions, she paused and took a quick, steadying breath. "So, what do you say? Can you get me in?"

He studied her warily. "I wouldn't be aiding and abetting any sort of crime, would I?"

"Certainly not. Why would you think so?"

"I understand you have a . . . rather storied past."

Helen trilled a laugh. "Oh, my husband has been telling tales, has he? I assure you, I have no intention of shattering any windows or staging any protests today. I'm confident my husband would fully support my visit."

"If he knew you were here."

"Well, yes." Beseechingly, she added, "Please, Mr. Corbyn. I only want to look around a bit, anonymously, so people won't conceal problems from me. You understand how important that is, don't you?"

"I suppose I do." He hesitated. "I can get you through the front gate, but after that, you're on your own."

"Perfect," she exclaimed, relieved. "I'm in your debt."

"Let's hope I don't get sacked for this and need to call in that debt." He offered her his right arm. "Shall we?"

She tucked her hand in the crook of his elbow and thanked him with a smile. He escorted her to the queue at Gate One, where he showed the sentry his identification badge and introduced her as Mrs. Purcell. The sentry offered her a cordial good morning and waved them through without hesitation.

"That was easier than I thought," she remarked as they cleared the passageway and emerged on the other side of the stone wall. "Thank you so much."

"You're welcome," he said, inclining his head politely, his expression still wary. "Good luck."

They parted company, Oliver heading quickly toward headquarters, Helen walking slowly down the pavement, looking around to get her bearings. Remembering the way to the laundry, she set off with a confident stride, as if she had every right to be there.

When she reached the laundry, she held the door open for a worker pushing a cart loaded with soiled clothing and followed her inside to a vast room where women toiled near vats of hot water billowing steam, the scents of soap and bleach in the air. One stout woman, older than the others and red-faced from the heat, walked about looking over shoulders and issuing commands, which clearly marked her as the supervisor.

Aware that her time on the arsenal grounds might be cut short the moment someone recognized her, Helen quickly approached the laundress in charge. "Good morning," she said pleasantly. "Could you assist me, please? I need a uniform for the Danger Building."

The older woman's eyebrows rose. "*You* need a uniform for the Danger Building?"

"If you would you be so kind."

"Right." The laundress studied her, dubious. "Weren't it in your cubby in the shifting house?"

"I didn't see it there," said Helen, which was not a lie.

"Likely a girl on the night shift misplaced her own and pinched yours. Wouldn't be the first time." She headed deeper into the stifling, steam-clouded room, gesturing over her shoulder for Helen to follow. "Come along, miss. We'll get you sorted."

Soon thereafter, Helen departed, a uniform, cap, and wooden clogs stowed neatly in her satchel. Arthur had omitted the Danger Building from his tour, but she had seen yellow-hued munitionettes entering a particular factory, and after wandering around a bit, she found it. A few workers with the characteristic yellow skin and ginger hair of canary girls were hurrying through the doorway, which told Helen that she had arrived fortuitously in the last minutes of a shift change. With some effort, she might be able to slip in with the stragglers.

Inside the shifting house, Helen changed into the uniform, stowed her own clothing in her satchel in an empty cubby, and joined the queue for inspection. "New girl?" one of the overlookers inquired, frowning, after asking whether Helen had anything to declare.

"How did you know?" Helen asked, wary.

"No yellow, of course." She gestured to Helen's face and hands, her only visible skin. "It's bad form to be late on the first day."

"I'm very sorry," Helen replied humbly. "It won't happen again." The overlooker harrumphed and waved her on through. Helen paused by the racks of punch cards but quickly moved on again, hoping no one noticed that she had not timed in.

For the next hour, she wandered from the Filling Shop to the Finishing Shop, from the loud, clanging machinery section to the canteen. She lingered on the periphery, listening, occasionally breaking into a conversation to ask questions, moving on with a respectful nod whenever a supervisor peered at her with curiosity or suspicion. Once, the burly, dark-haired foreman of the Finishing Shop, whom she had briefly met on the arsenal grounds during her tour, paused to give her a second look when they crossed paths in the corridor connecting the two buildings, but she quickened her pace to appear as if she were walking with a pair of workers just ahead of her. She even laughed at one of their jokes. When the foreman did not confront her, she took that as a sign that her ruse had worked, but she sensed that her time was running short.

Even if it was, she had learned so much already. None of the munitionettes she spoke to had ever seen a canary baby, but all had heard rumors from workers at other arsenals. They accepted the changes to their skin and hair with resignation and, occasionally, humor, but the list of other symptoms they mentioned left Helen nearly breathless from alarm: chest pains, fatigue, wracking coughs, painful sore throats, migraines, diarrhea, vomiting, and a persistent metallic taste in their mouths.

"How can you bear it?" Helen asked a group of workers in the Filling Shop, forgetting for a moment that she was supposed to be one of them.

"I need the wages," said one canary girl, tall and strikingly pretty despite her yellow skin, with curls that had probably once been lustrous honey-gold peeking out from beneath her broadcloth cap.

"It's not only the wages," said the woman called Mabel, to whom the others deferred. "Britain can't win the war without the shells we build. The lads risk their lives in the trenches. We risk ours in the arsenal."

Helen found herself too overcome to speak as the other brave girls nodded and chimed in agreement.

"I beg your pardon, Mrs. Purcell," a woman spoke briskly behind her. "Might I have a word?"

With a start, Helen glanced over her shoulder and discovered Superintendent Carmichael regarding her sternly, mouth pinched. When

Helen turned back around, the canary girls were staring at her, dumbfounded. She offered them a quick, apologetic smile as she rose. "Certainly," she told the superintendent. "Lead the way."

She was not surprised to be led promptly back through inspection to the shifting house. "With respect, Mrs. Purcell, you are not authorized to enter the Danger Building," Superintendent Carmichael said crisply. "Mr. Purcell requests your presence in his office immediately."

Helen smiled brightly as she retrieved her satchel from its cubby. "Oh, I don't wish to trouble him," she said. "Please let him know that I'll see him at home this evening."

"I'm afraid he insists, ma'am."

Helen muffled a sigh. "Give me a moment to wash and change."

The superintendent nodded assent and moved off to stand by the exit. There was no getting past her, unless Helen wanted to knock her down and scramble over her, but after that she would still have to get past the sentries at the gate. No, there was nothing for it but to comply.

She scrubbed herself clean of yellow powder and changed back into her own clothes, leaving the soiled uniform in the laundry bin. She felt like a naughty schoolgirl being marched to the headmaster as Superintendent Carmichael escorted her to Arthur's office. Helen tried to chat along the way, querying the superintendent about working conditions in the Danger Building, but she gave only curt replies, apparently preferring smoldering silence.

The superintendent left Helen in the antechamber to Arthur's office, offering a haughty sniff in farewell. Alone, Helen paced the length of the small room, alternating between chagrin and indignation. When the door to her husband's office opened and Oliver walked out, they both started at the sight of the other. "Well, hello, Mr. Corbyn," said Helen, too loudly and with false cheer. "I haven't seen you in ages! How are you?"

"Fine, thank you," said Oliver, also loudly, for Arthur's benefit.

Then Arthur himself appeared in the doorway. "Helen, darling, would you come in please?" he asked, his voice quiet and oddly formal.

She nodded, threw a quick, commiserating glance to Oliver, and strode into Arthur's office. "Before you scold me—"

He seized her by the shoulders and swiftly looked her over from head to toe. "Are you feeling unwell?" he asked urgently, his voice strained. "Did you handle the yellow powder?"

"No, I'm not a complete fool. I just had a look around. I wouldn't attempt to do munitions work I wasn't properly trained for."

He released her and ran a hand over his jaw. "What were you thinking?" he asked. "We're trying for a baby, and you just—" He gestured, frustrated, grasping for the words. "You just strolled into the Danger Building like a tourist at the British Museum!"

"Yes, I suppose I did. Imagine if I had been a saboteur. You might want to improve your security measures."

"Don't deflect from the subject at hand. Why would you knowingly and recklessly expose yourself to such danger, when you might be with child?"

"Well, I don't believe I am with child," she said, taken aback by his fervor, by the fear and worry in his eyes. "You realize, Arthur, that you're tacitly admitting that the Danger Building is hazardous for expectant mothers."

He gestured impatiently, brushing that aside. "I've always said that."

"I'm not talking about explosions or lifting heavy loads." She studied his face. "You believe that the yellow powder is poisonous, don't you?"

He began to speak, hesitated, and sat down wearily on the edge of his desk. "I suspect it is," he admitted. "After you shared what your friend told you about the newborns with yellow skin, I made inquiries. It's true. At least six babies with bright yellow skin have been born to canary girls in London alone."

Helen shuddered and clasped her arms to her chest. "And you didn't tell me?"

"I wasn't aware that you were planning an unauthorized inspection, so I didn't think it would affect you. I didn't want you to worry."

"I was worried then and I'm more worried now." Leaning against the edge of the desk beside him, she told him what she had learned from the women in the Danger Building—their symptoms, their dedication to the war work, their brave fatalism, their trust in their supervisors, who they believed were looking out for them.

By the end of her report, Arthur's shoulders were slumped, his expression haggard. "Until we know more about trinitrotoluene, we must assume it's a hazardous substance," he said. "I don't believe its effects are merely cosmetic."

"Nor do I."

"Nor, I think, do our workers. Surely this is no secret. Our munitionettes are clever and observant. They understand the dangers and accept them."

Helen remembered what Mabel had said in the Filling Shop. "Yes, I believe that's so. But even if they're entirely willing to risk their lives, more must be done to protect them."

"Agreed. Absolutely."

She laid her hand on top of his. "What do you intend to do?"

"We can't stop using TNT. We must produce shells, or the war is lost. France, Belgium, England—all will be lost." He brooded for a moment in silence. "The girls told you more in an hour than they've told me, or Oliver, in months. It's obvious what to do next. You should become Thornshire Arsenal's welfare supervisor so you may personally look out for the girls' interests."

"*I* should?" said Helen, incredulous. "Someone should, certainly, but I don't have any qualifications."

He smiled wryly. "On the contrary, I think today you proved you do. You're caring, you're clever, you attack problems with a vengeance, and you've already said you wanted to work here."

"That was when I thought I could be your secretary—bringing you tea, organizing your papers, tormenting you—"

"As delightful as that would be"—he interlaced his fingers through hers—"don't you think this other job would be eminently more rewarding?"

She took a moment to consider. "May I have my own office?"

"Of course. You'll need a place to work, somewhere you can confer privately with the munitionettes."

"Can my office be larger than yours, with a better view?"

"No, it cannot. There's nothing larger, and this is an arsenal. There aren't any good views."

She smiled at him, her gaze holding his. "This one, right here, is rather excellent. But what about your father? You said he wasn't keen on hiring welfare supervisors."

"You let me worry about my father." He leaned closer, a faint, tired smile appearing. "So what'll it be? Do you accept?"

"I'll need to see my office before I decide," she teased, and leaned in to kiss him.

10

APRIL–JUNE 1916

LUCY

Lucy cherished Daniel's letters from France, even though he provided frustratingly few details about his specific location and experiences. The first were omitted to satisfy the censors, the second, she suspected, to avoid upsetting her and their sons. He wrote of football matches, the endless rain and mud, and his chums, but almost nothing of the danger and hardships he faced every day. No doubt he intended to spare her worry, but sometimes he only made her imagine the worst.

If his letters were addressed to the family, Lucy's mother let her grandsons read them when they arrived home from school. If the envelopes bore Lucy's name alone, her mother hid them in a kitchen drawer so Jamie and Simon wouldn't be tempted to open them. Lucy would return home around nine o'clock, exhausted from a long day finishing shells, but after she went upstairs to tuck the boys into bed, and returned to the kitchen to find a cup of tea, a scone or a biscuit, and an overseas letter arranged at her place at the table, her fatigue would suddenly disappear. With a grateful look for her mother, she would snatch up the letter and lose herself in her beloved husband's words from far away.

In early January, he had written of the 17th Middlesex's football team's 6–0 victory over the 2nd South Staffords to win the 6th Brigade Final. Five days later, the Footballers' Battalion team played a "Best of Brigade" side and won 3–1. Then followed a fortnight of no letters at all, and since the extraordinary Royal Mail could deliver letters back and forth between soldiers and their families in Britain in about three days, Daniel's silence told Lucy the 17th Middlesex had returned to the front.

When at last another letter arrived in the beginning of February, Daniel sent dreadful news about his friend and fellow Olympian Captain Vivian Woodward, who had been wounded in the right thigh from grenade splinters. His injuries were so serious that he had been sent back to England for treatment, and it was expected that his football days were over. A few days later Lucy read in the papers that Woodward's wounds were not as serious as the battlefield doctors had believed, and that he might indeed take the pitch for Chelsea again after the war. Lucy was greatly relieved for her husband's friend and former teammate, and her heart went out to his wife, who must have suffered terribly when the dreaded telegram arrived.

Yet despite the fortunate outcome, Lucy brooded over the terrible reminder of the risks the footballers confronted every day. What if the shrapnel had struck Daniel instead? What if the injury had been so severe that he could never play football again? He must have resigned himself to that possibility when he enlisted, but Lucy could not imagine how he would endure the rest of his life without football bringing joy, excitement, and the thrill of competition to it.

In the middle of February, Daniel had sent the family a humorous account of two young German soldiers who had crossed no-man's-land unarmed, hoping to be taken prisoner. They had approached the British lines with their hands raised, shouting, "*Kameraden! Kameraden!*" and had surrendered to Private Tim Coleman, a forward with Nottingham Forest. "The Germans were strapping, square-jawed chaps," Daniel had written, "disgruntled officers' servants who had become so frustrated with their maltreatment that they had deserted.

As a parting rebuke, they had carried off their former masters' ample supply of cigars, cigarettes, schnapps, and wine, which they offered to us gratefully in exchange for taking them prisoner. We turned the men over to the authorities and had a jolly celebration in our trenches."

Jamie and Simon loved that story and often asked her to read it again, but although Lucy found the anecdote amusing too, she had to wonder about everything Daniel was leaving out of his letters. She understood why he would not want to upset the boys with too much harrowing detail, but why would he not confide in her? When they were growing up together in Brookfield, Daniel had never played at soldiering or expressed any desire to join the military. For all his courage, intelligence, and inherent steadiness, the experience of war must be jarring. It could not be all football and comic enemy deserters and liberated cigars. Why would Daniel not allow her a glimpse inside his real war, so she could offer him all the comfort and strength she could?

She did not want to upset him with complaints, so she lovingly told him that she knew he was protecting her from the truth, but she neither wanted nor needed him to. "Let me share whatever part of your burden I can," she implored. In his next letter, which he sent at the end of February and addressed to the family, he described a surprise visit from MP Joynson-Hicks, the founder of the 17th Middlesex, who had brought with him a letter from King George, which expressed His Majesty's best wishes to the battalion. Lucy, the boys, and her mother delighted in the story, and Lucy was very proud that the King had honored Daniel and his comrades so splendidly. Yet Daniel's letter still left her feeling disappointed, despite the warm assurances of his love and his longing to be near her again, with which he closed every letter.

"He's an Englishman, dear," her mother said when Lucy confessed her ongoing frustration and bewilderment. "He's not going to whinge and complain. He's going to quash his fear, do his duty, and soldier on."

"But I'm his wife, not one of his men. He can confide in me."

"Perhaps it's simply too hard for Daniel to put into words the fear and horror that have become a part of his daily life."

"I could bear the horror," Lucy said. "What I can't bear is this distance. Not the miles, but his—remoteness."

Her mother took her hand and held it, her face full of compassion. "Of course you could bear it. You would bear anything for love. But imagine what it must be like for Daniel, to witness and experience these dreadful things, only to relive them as he describes them to you. Perhaps the kindest, most loving thing you could do is to respect his silence, and trust that he will unburden himself when he comes home."

"But what if he—" She could not finish the sentence. Instead she choked out a laugh. "See how foolish I am. I demand honesty from Daniel, even if it pains him, but I want you to lie to me and tell me you're absolutely certain he'll be coming home."

Her mother smiled sadly. "Could you tell me that you're absolutely certain your brother will come home?"

Lucy inhaled shakily, thinking of Edwin, somewhere in France working in aerial reconnaissance—the interpretation of photographs, not the actual photography, a somewhat safer occupation in that his feet remained firmly on the ground. "No," she admitted. "Nor would you wish me to."

Her mother sighed. "Uncertainty is the hallmark of these perilous times, my dear. We must somehow scrape together enough courage to endure it, every day, because raging against it changes nothing."

Lucy knew her mother was right. Some things, the most important things, *were* certain: She loved Daniel, and he loved her, and their children were safe, happy, and beloved. She also knew, and felt pangs of guilt for it, that she was not offering Daniel the same honesty she sought from him. He knew she had become a munitionette, of course, and he was tremendously proud of her, but she had not told him of her yellowing skin, her whitening hair, her sore throats and headaches, nor the accompanying fears that her symptoms would only worsen with time. Nor had she shared what George had told her about rising concern among physicians that TNT and other chemicals munitionettes worked with were hazardous to their health. Often a canary girl's symptoms would fade over time if she left munitions work, but sometimes they persisted indefinitely, or even worsened, a troubling and unexpected development. A government study was underway, but George urged Lucy to limit her exposure in the meantime, just in case. She passed this information along to the other Thornshire canary girls, but not to Daniel. Why worry him, she reasoned, when he could do nothing about it? Why unburden herself at his expense?

No doubt Daniel asked himself those same rhetorical questions when writing home about life in the trenches.

And so when Daniel wrote in mid-April to announce that the 17th Middlesex had defeated the 34th Brigade, Royal Field Artillery 11–0 to win the Divisional Cup, Lucy and the boys cheered and sent him a marvelous congratulatory letter in reply, complete with illustrations. They had never doubted the outcome, they declared, and their only regret was that they had not been able to see it for themselves. Privately Lucy and her mother agreed that Daniel's stories of the tournament were as reassuring as they were entertaining, for he could not be enduring too much hardship in too dangerous a setting if the regimental team could play football. For that brief interval, they knew he was safe.

Lucy sent Daniel another letter of her own the next day, a breezy, comical account of her adventures with the Thornshire girls' football club—the whimsically uneven surface of their makeshift pitch, which was barely large enough for their small-sided matches, with a single goal built of discarded wood and no net; the challenges of playing in their Danger Building uniforms instead of a proper kit; the admiring fans and jeering critics they had acquired among the other arsenal workers, men and women alike who watched their practices and shouted advice from the sidelines; how she had emerged as the club's top striker, always the first to be chosen when they made up sides; and how jealous this made Marjorie, the younger woman who had decided, much to Lucy's regret, that they were fierce rivals. The conflict was all one-sided and had been since that first day when Lucy had accidentally knocked off Marjorie's cap in the shifting house, but Lucy had grown steadily more annoyed by Marjorie's derisive comments and nicknames, and the elbows in the ribs in the dinner queue, and the treading on her bare toes in the shifting house. Now, she admitted sheepishly, she got a bit of her own back by placing a beautiful shot in the top right corner, just beyond the lanky girl's long reach, or low and swift on the left, just skimming the grass, forcing Marjorie to dive and miss. And that header into the goal—Marjorie had glowered, red-faced, for an hour after that. Lucy emulated her husband's cool, dignified manner on the pitch and never grinned or gloated after scoring. Although she hadn't intended it, her humility encouraged the other girls to praise her effusively to make up for her diffidence, which infuriated Marjorie all the more.

As Lucy had hoped, Daniel delighted in her football stories. "Just play your game, and eventually your hot-headed keeper will come around," he assured her. "I'm not surprised you've turned out to be an excellent striker. I still recall the first time I saw you kick a football, in the schoolyard. Your talent and mettle were evident even then. Do you remember?"

Of course she did. That might have been the day she had fallen in love with him, although at eight years old she would not have known it for love.

It was a crisp, sunny autumn day, and it must have been after lunch because all the children were outside playing, the boys on their side of the schoolyard, the girls on the other, separated by a wooden slat fence, low enough for all but the youngest children to see over.

Lucy was playing hopscotch with her friends when suddenly a football sailed in front of her face. Startled, she landed two-footed and glanced over her left shoulder to find a group of boys, eleven-year-old Daniel and her brother Edwin among them, approaching the fence on the boys' side.

"Lucy, get the ball," Edwin shouted, gesturing. Lucy looked to her right and saw the football first bounding and then rolling away. "Go on! Hurry!"

"She's busy," her friend Nettie shouted back, fists planted on her hips, "and you didn't ask nicely!"

Edwin scowled. "Get the ball, Lucy."

She glanced down at her feet, her scuffed brown shoes both planted in the same square. "I'm out anyway," she told her friends. She picked up her marker and trudged after the ball.

"We don't have all day," her brother shouted irritably.

Lucy picked up her pace, only to hear Daniel shout, "It's fine, Lucy. Take your time." A small smile came to her face, and knowing Daniel was watching, she broke into a run. She snatched up the football and ran back, determined to show off her speed, but for some reason—later, she never knew why—she halted abruptly when she reached her friends at the hopscotch court.

"Come *on*," Edwin shouted when she didn't move except to pass the ball from one hand to the other. "You can't throw it that far."

"Maybe she's going to keep it," said Nettie gleefully, and their friends giggled.

"That's stupid!" Edwin scoffed, but there was worry in his eyes. "Girls can't play football."

Daniel looked at him, bemused. "Of course girls can play football."

"Well, my sister can't." Edwin gestured, scornful. "Look at her, all skinny arms and legs."

Daniel *was* looking, and his smile made Lucy feel warm all over. "Thanks for getting the ball, Lucy," he called. "Could we please have it back now?"

"Well," she said, pretending to consider, "since you said please." She meant to run it back and hand it to him, but the same sudden impulse that had brought her to a halt before now compelled her to hold the ball with her hands outstretched, release it, and punt it over the fence, over the boys' heads, so they all instinctively turned to watch it soar. Her friends clapped and cheered. Several of the boys chased down the ball and immediately resumed play, but Edwin turned around to glare at her, and Daniel caught her eye and grinned. That look, which sent a warm glow from her chest that spread to the tips of her fingers and toes, was worth every bit of Edwin's anger.

And Daniel wondered if she remembered that day.

In that same letter, he told her he wanted to hear every detail about her football club, and after dinner during her shift the next day, she learned that she would soon have exciting news to share.

No sooner had the players arrived for practice than Mabel called them together. "The girls at Brunner Mond in Silvertown heard that we have a football club," she said as they gathered around her in a half circle. "They've challenged us to a match."

Gasps and exclamations of surprise went up from the group. "But we've never played a proper match," said Louise. "We don't even have a proper team."

"We have fourteen girls, don't we?" said Mabel. "That's eleven on the pitch and three reserves."

"We've never played as a team of eleven before," said Daisy, though her eyes were bright with excitement at the possibility.

"For all we know, they haven't either," said April, hands on her hips. "I think we can trust Mabel to get us sorted." Her words met with a murmur of agreement.

Mabel turned her head aside to cough hoarsely, and when she turned back, her gaze was sharp. "I certainly hope we're all feeling up to it. I wouldn't have it said that Thornshire Arsenal had shown the white feather, so I accepted the challenge on the spot."

"Hear, hear," said Marjorie stoutly, clapping her hands. Several others joined in, and then a few more, until all were applauding with steadily increasing enthusiasm as the girls exchanged glances around the half circle, excited, eager, wary.

"How much time do we have to prepare?" Lucy asked, already mulling over who would be best suited for each position.

"Plenty of time." A smile played in the corners of Mabel's mouth. "The match is a week Sunday."

The murmurs turned apprehensive, the glances alarmed and furtive.

"We can do it," April insisted, looking around the half circle. "Today's only Monday. We have nearly a fortnight."

"We *have* to do it," Marjorie pointed out. "We can't back out now or we'll look like cowards. I for one am looking forward to a proper match, on a proper pitch." She turned to Mabel. "They have one, don't they? Because we certainly don't."

"Yes, they've organized a pitch," said Mabel, her smile broadening. "Stratford United agreed to let us have the Globe for the day." When the girls exclaimed, astonished and delighted, she shrugged again. "And why shouldn't they? Nearly all the Upstart Crows have enlisted, so the club isn't using it much, and the gate will go to the Prince of Wales's National Relief Fund."

Lucy clapped her hands twice for attention and to settle the girls down. "We need to practice. If we're going to play on a professional pitch, we need to look the part."

"How can we look the part if we don't have proper kit?" Peggy asked as they ran out onto the grassy field. "We can't play in our munitionette uniforms."

Everyone but Lucy laughed as they took their favorite positions, adjusting for a full side and deferring to Mabel's judgment. They practiced harder than Lucy had even seen them do, but all the while, her mind churned over the question of what they were going to wear at their first real match.

By the time they hurried back inside to their stations, she had a plan.

She caught Mabel at the entrance to the Filling Shop, hoping to speak to her alone, but Peggy, Marjorie, and a few others lingered. "I think we should ask Mrs. Purcell for help," Lucy said. "She's often said that if there's anything we need—"

"The boss's wife?" Mabel pulled a face, disgusted. "The great lady who pretended to be one of us, and why? To spy for her husband, no doubt. Now here she comes, every week like clockwork, skulking about, asking us how we're feeling, inviting us to chat in her posh little office, always reminding us to wash our hands, as if we don't already, or urging us to get some fresh air whenever we want to, as if we could and still meet our quotas." Mabel shook her head and continued into the shop. "She's insufferable. I don't have anything to say to her."

"You don't have to," said Lucy, trailing after her. "I'll speak to her on the team's behalf. Remember, she's not only the boss's wife; she's also our welfare supervisor."

"Welfare supervisor," Mabel echoed, as if the words left a bad taste in her mouth. "I know that sort. Wealthy do-gooders meddling where they aren't wanted."

"We're the Thornshire Arsenal football club," Lucy persisted, knowing that whatever Mabel decided, the other girls would go along. "Thornshire Arsenal should sponsor us. They've got government money for programs to improve worker morale. This match certainly qualifies, don't you think?"

"It would help *my* morale," Peggy ventured. "Come on, Mum. You know those Brunner Mond girls will take the pitch perfectly kitted out. We can't show up all mismatched. It would be a disgrace."

"Exactly," said Lucy, throwing Peggy a grateful look. "At the very least, we all need decent shoes and shin-guards."

"In the end no one cares how we look, only how we play." Then Mabel hesitated. "Though you make a fair point about the shoes and shin-guards. The other side will have the advantage if we play in our regular shoes. And no one wants to break a shinbone."

"Might as well go for the whole kit, then," said Marjorie, the last person Lucy ever expected to support any idea of hers. "In for a penny, in for a pound."

"And it's not even our pounds we'll be spending," Lucy reminded Mabel, glancing around to include all the onlookers. "Imagine that: a lovely new kit, entirely free of charge, all for the honor of representing Thornshire Arsenal."

"Please, Mum?" Peggy implored.

Mabel sighed. "Very well, then. If they're going to spend the money, they might as well spend it on us."

Lucy thanked her and hurried to the Finishing Shop seconds short of being late, earning a curmudgeonly scowl from Mr. Vernon. When teatime came around, she raced off to the administration building instead, where she found Mrs. Purcell's office amid a warren of small rooms on the ground floor. The welfare supervisor was disarmingly pleased to have a visitor, and she welcomed Lucy into a small, windowless office, really more of an oversized closet, with barely room for a small desk with a chair on either side, a filing cabinet, and a bookcase loaded with files and forms. On the wall hung a landscape painting of the Surrey countryside, so reminiscent of home that Lucy paused to admire it.

"Lovely scene, isn't it?" said Mrs. Purcell, gesturing to the chair in front of the desk while she seated herself in the one behind it. "I've considered hanging curtains around it and pretending it's a window. What can I do for you today, Mrs. Dempsey?"

Lucy took her seat and told Mrs. Purcell about the challenge from Brunner Mond and the Thornshire club's lack of proper uniforms and gear. To her relief, the welfare supervisor was quite keen on the idea. "I believe we have funds for that," she remarked, leaning forward to rest her arms on her desk. "Let me see what I can do. Any thought to team colors?"

"We'll take whatever we can get, honestly."

"That's not the way to ask for something," Mrs. Purcell protested, smiling. "Tell me what you want, firmly and confidently."

Lucy hesitated. "We didn't discuss team colors." Then it came to her. "Yellow and black. We're the Thornshire Canaries, after all."

"Player names and numbers on the back of the jerseys?"

"Numbers only. And a variety of sizes. I'm not sure who will be wearing what."

"I'm on it," said Mrs. Purcell. "Is there anything else I can do for you?"

When Lucy declined, they rose and shook hands, and Lucy raced back to work.

A week later, the Thornshire Canaries were practicing after dinner when Mrs. Purcell and a handsome young man with a prosthetic arm arrived, pushing a cart full of large boxes. Abandoning their scrimmage, the players hurried over and discovered a marvelous assortment of yellow jerseys numbered on the back in distinct black printing; short trousers, also black; long, black wool socks; proper football boots; and shin-guards.

"Take what looks to be your size and we'll get sorted in the shifting house after work," Mabel called out above the clamor as the girls dug through the boxes, exclaiming with delight or nodding in satisfaction as they chose items for their kit. "Remember to thank our benefactor."

"It was my pleasure," said Mrs. Purcell, looking embarrassed but pleased as a chorus of gratitude rose from the players.

"Thank you, Mr. Corbyn," April added, and Mrs. Purcell's handsome assistant, expressionless, nodded once in reply. Inexplicably, Marjorie trilled a laugh, but that was Marjorie for you.

The rest of the shift dragged on endlessly as the Thornshire Canaries—fortunately everyone approved of Lucy's impromptu choice—awaited the moment they could race off to the shifting house and try on their uniforms. It took nearly a half hour, but eventually everyone had something that fit well enough, although the men's cuts were rather boxy on them, and a few of the girls expressed some dismay at exposing their knees.

"You'd expose more in a bathing costume," Marjorie remarked, admiring her own uniform and pretty knees.

"That's at the seaside," Daisy countered, "not in the city, with thousands of gawking spectators."

"No one will be looking at our knees," Lucy interjected quickly, before anyone got nervous and decided to quit. "They'll be too busy admiring our brilliant play."

As the week passed, she dared hope that her prediction would come true. Their passes were quick and accurate. April sprinted up and down the pitch as if she had boundless reserves of energy, racing for each loose ball and feeding it up to the forwards. Daisy could steal a ball so deftly her opponents sometimes kept running before they realized it was gone. Marjorie prowled the defending third with a ferocity that would strike fear into the heart of any attacker who dared venture too close. Lucy and Peggy alternated between taking shots on goal and passing in front of the box so the other could put it in. Mabel was a bit slow in the goal, perhaps, but although Marjorie clearly ached to be asked to replace her, no one dared suggest Mabel step aside.

At last, and all too soon, the day of the match arrived. Jamie and Simon were beside themselves with excitement as Lucy's mother ushered them aboard the train to Stratford, while Lucy, her stomach in knots, followed a pace behind carrying her kit in Jamie's knapsack. She was familiar with the Globe, having attended several of Daniel's games there, and she remembered it as being older and smaller than White Hart Lane, but still able to seat eight thousand spectators. The rounded, half-timber façade over brickwork was meant to evoke the old Globe Theatre of Shakespeare's day in London, which of course no one had ever seen except in sketches and paintings, but the resemblance ended there.

Lucy parted with her family at the main gate, then joined the rest of her team in the players' changing room. They were all nervous and excited, with the possible exception of Mabel, who seemed perfectly sanguine, even after they took the pitch to warm up and observed how swift, strong, and skilled their opponents were as they ran through warm-up drills with the grace and power of a choreographed dance.

"They're going to crush us," a Canary muttered as the official blew the whistle and raised a flag to signal that the match was about to begin. Lucy did not know who spoke, but no one contradicted her.

After ninety minutes of play, it did not turn out quite as badly as all that. Lucy estimated the size of the crowd to be about five thousand, far short of capacity but really quite remarkable for a women's match. Lucy supposed most of the spectators were friends, family, and coworkers of the players, with the rest turning out to support the charity. A good portion of the audience had been drawn by curiosity, she suspected, since women's matches were quite a novelty. Perhaps a few—a very few, she hoped—had come to steal a glimpse of their knees.

The Canaries had started out strong, moving the ball well and attempting several shots on goal, but eventually their more experienced opponents wore them down. The Brunner Mond Belles scored after the first seventeen minutes, but Lucy put an equalizing goal in with thirty seconds left in the half. The Canaries were ecstatic in the changing room during the interval, reminding one another that it was an even game now and they could come out ahead in the end, but their hopes were short-lived. The Belles' incredibly swift center forward scored on the first play of the second half, which took the wind out of the Canaries a bit. Twenty minutes later, the Brunner Mond left forward scored following a scramble in the box after a corner kick, admittedly a difficult shot to block. After that, nearly all the action was on the Canaries' side of the pitch, except for one breakaway run late in the game which made their fans leap to their feet, cheering and shouting. Lucy even thought she heard Jamie's and Simon's voices carrying above the din. But the Belles' brilliant keeper leapt high in the air to block Lucy's shot just when she was sure it would go in, and neither she nor any of the Canaries had as good a chance to score after that.

In the end, the Belles held on to win 3–1, but the Canaries had made a respectable showing for their first time out, and the Belles were so friendly and encouraging as they shook hands afterward that none of them, except possibly Marjorie, felt dispirited over the loss. "You should join the Munitionettes' League," their captain said, addressing her remarks to Mabel, but smiling around the circle to include them all. "It's jolly good fun, and if we can raise a few shillings for the Prince's Fund, so much the better."

"Say yes, Mum," Peggy exclaimed, while the other Canaries either chimed in assent or nodded, beaming, bouncing up on their toes with excitement.

Mabel shrugged. "You hear them," she said, extending a hand to the Belles' captain. "Count us in."

The Belles' captain must have done, for two days later when Lucy reported for her shift, Mabel announced that they would have another match on the following Sunday, this time in East London versus the Hackney Marshes National Projectile Factory Ladies. Determined to improve upon their first appearance, they resolved to train harder, but their dinner and tea breaks did not allow much time for it. Marjorie suggested that they stay after work for an hour or so three times a week to run plays, but very few of the Canaries were willing. Some pleaded exhaustion after their twelve-hour shifts, while others had long commutes or families who needed them at home, and some, like Lucy, had both.

Their brief training sessions already taxed them enough, and not only because all the players who worked in the Danger Building had developed coughs by then, and those who had had coughs for months found theirs worsening. They were worn out even before they took the pitch because their foremen had increased production to a nearly breathless pace, increasing their quotas and urging them to pick up speed, but without sacrificing accuracy or safety, a rather fine needle to thread.

"If the Kaiser's spies want to know when a big push is on the way," Daisy remarked to Lucy, huffing with effort as they loaded a trolley, "they need only look to the frantic pace of the Thornshire munitionettes to know that something's in the works."

By late June, Lucy was often so fatigued at the end of a shift that she plodded, yawning, through washing up, changing out of her uniform, and walking to the station, which meant she had to catch a later train and would not arrive home until after her mother had put Jamie and Simon to bed. Lucy regretted every goodnight she missed. She always tiptoed into her sons' bedroom and kissed them softly while they slept, but she missed having them greet her at the door with hugs, and asking about their day. Their stories always brought a smile to her face no matter how exhausted she was.

One night she arrived home to find her mother pacing around the kitchen, compulsively scrubbing the already sparkling counters, her face pale and lips tightly pressed together.

"What's wrong?" asked Lucy, sick at heart, thinking of Daniel and Edwin and everyone else their family knew who was in peril.

Her mother gestured to the table, where a newspaper had been folded in an awkward fashion to emphasize a particular column.

"What is it?" Lucy asked, picking up the paper, scanning the page. It was all war news, all dreadful, but nothing leapt out at her as a particular concern of her mother's.

"What is it?" her mother echoed, striding over and snatching the paper, then remembering she could not read the small print without her glasses, retrieving them from her pocket, trying again. "'Woolwich Worker's Death,'" she read aloud, shooting Lucy a pointed look over the rims of her glasses. "'At an inquest at Woolwich on the body of Gwendoline Darrell, twenty-four, employed in the Royal Arsenal, it was shown that her death was due to acute jaundice caused by poisoning by tri—'" Her voice faltered, but she inhaled deeply and carried on. "'—trinitrotoluene, TNT, with which the girl had been working.' That's the same yellow powder you work with, isn't it?"

Lucy nodded and sank into a chair, resting her elbows on the table and cradling her head in her hands.

"'The coroner said that the workers in explosives wore masks, and he hoped it might be possible in the future to treat the masks with chemicals to counteract the effect of the fumes.' Do you wear masks? Tell me you do."

"We do," Lucy replied quietly, hoping her mother would also lower her voice, then amended, "We haven't always done. Our new welfare supervisor made it policy for all Danger Building girls a few weeks ago."

Her mother's chin trembled and her eyes glistened as she returned her gaze to the newspaper. "'A verdict of Accidental Death was returned.' What was accidental about it? She wasn't struck by a lorry on a blind curve. This was steady poisoning over God knows how many weeks or months!"

"I think they mean accidental in the sense that it wasn't intentional. It wasn't deliberate, it wasn't planned—"

Her mother slapped the newspaper down on the table. "But surely the men in charge knew what might happen. Look at your skin, your hair! You cough constantly. I'm going to have George come to London to examine you."

"You know he can't leave his patients. He's the only doctor for miles anymore."

"Then you shall go to him."

"I can't leave my work either."

"You can, and you must." Her mother sat down adjacent to her and put a hand on her shoulder. "Think of your children. Think of your husband. If you won't quit for your own sake, do it for them."

"It's because of them that I can't quit," Lucy said, letting her hands fall, sitting upright and regarding her mother bleakly. "If every canary girl quit, who would make the shells? Without munitions, how could England hope to win the war? Don't you see? I have to do this to bring Daniel home. If my work can shorten the war by even one day, that's one day Daniel isn't in danger."

She felt tears gathering in her eyes, and before she could blink them away, exhaustion and fear and worry made them well up too fast and they spilled down her cheeks. She folded her arms on top of the table, rested her head upon them, and closed her eyes, pressing her lips together to hold back sobs that might wake the boys.

She felt her mother's gentle hand on her back. "Oh, my dear girl," she murmured soothingly. "I'm so sorry. I shouldn't have railed at you like that."

"It's fine." Lucy tried to laugh, but it sounded rather desperate. "I'm sorry for the wretched tears."

"You have every right and reason." Her mother patted her twice on the shoulder, then pushed back her chair. "You go wash your face and put on your nightgown. I'll make you a cup of tea, and I have some of those shortbread biscuits you like."

Lucy managed a wan smile. "Thank you, Mum." She inhaled shakily. "You know I wouldn't be able to do any of this without you."

Her mother waved that off. "You'd manage well enough," she said over her shoulder, already busy with the kettle.

Lucy doubted that very much. She was just pushing herself through the days, longing for Daniel, finding joy in her children, taking heart from the camaraderie of her fellow munitionettes and her teammates.

For months she had tried to ignore her own symptoms and to conceal them from her family, but she could almost hear a clock ticking down the minutes until she too succumbed to the mysterious, terrifying illness that had claimed the poor young woman from Woolwich.

"On Her Their Lives Depend," the posters said of the soldiers and the munitionettes, but it was incomplete. Not only the soldiers' lives, but the outcome of the war, and therefore the future of Great Britain, depended upon the munitionettes' work.

How could Lucy walk away?

11

JULY–AUGUST 1916

APRIL

On the first Sunday of July, the Thornshire Canaries beat the Hackney Marshes National Projectile Factory Ladies on their home ground in East London, 3–2, with Lucy scoring two goals and Peggy the third. A week later, buoyed by high expectations of themselves, they were soundly thrashed, 4–1, by the Associated Equipment Company Ladies from nearby Beckton. Mabel allowed three easy goals in the first half, while Lucy scored the Canaries' only goal of the match. At the interval, a wheezing, coughing Mabel told Marjorie to put on the keeper's jersey, while Louise took over for Marjorie at center back. Marjorie allowed only one goal in the final minutes of the game, when she was worn out from fending off the AEC Ladies' aggressive offense, but Lucy and Peggy were so pinned down that they could barely get off a shot. For April's part, she had never run so hard so long in her life.

As they trudged off the pitch after congratulating their opponents—who, it had to be said, were gracious in victory—Mabel tried to assume sole responsibility for the loss. "I was winded after the first five minutes," she said, as if she couldn't quite believe it herself. "Aging is demoralizing, but I s'pose it's better than the alternative." She nudged Marjorie with her elbow, inadvertently making the lankier girl stumble sidewise. "You hold on to that keeper's jersey. I'm relegating myself to team manager."

"If you think that's best, Mabel," said Marjorie deferentially, but her eyes shone with excitement.

"I do." Mabel sighed wearily as they filed into the changing room. "For me, the role of goalkeeper has been officially declared a dangerous trade."

The response was split between protests that Mabel was the very heart and soul of the team and must never forget it, and derisive snorts and wry chuckles from those who understood the grim humor of her reference. A few weeks before, Mrs. Purcell had informed the Danger Building girls that the government had classified TNT work as a "dangerous trade," which would allow the Home Secretary to impose regulations to protect workers. Mrs. Purcell had already distributed two masks apiece to every Danger Building girl, each a piece of cloth with a pair of ties, one that fastened on the back of the head and the other at the nape of the neck, which covered the nose and mouth and were meant to keep the wearer from inhaling the yellow powder. The masks were deemed as essential as every other part of their uniform and were required to be worn whenever they were in the Danger Building, but many of the girls considered them a nuisance. Out of the welfare supervisor's hearing, they declared that they would take their chances, and they kept their ties slack so the masks barely stayed on, tightening them only when ordered to by an overlooker. Even the girls like April who wore theirs properly learned that the yellow powder could slip in through the narrow spaces between the mask and one's face. Still, April figured that every bit of TNT she kept out of her nose and lungs had to be a good thing.

It seemed likely that more precautionary measures would be coming. Mrs. Purcell visited the Danger Building at least once a week, walking the floors, inquiring after each girl's health and well-being, urging them to come to her with any concerns or complaints. Their teammates who worked in other buildings said that Mrs. Purcell made the rounds of their workplaces too. April appreciated the welfare supervisor's concern, and she, Lucy, Daisy, and a few others believed she meant well, but many of the girls were skeptical of the boss's wife and her intentions. Surely she would report any complaints to her husband, who might retaliate or even sack them—or so they believed, even after time passed and only good came of Mrs. Purcell's interventions.

"Have you forgotten that she's the one who organized our uniforms and gear?" Lucy challenged some of her teammates after a match in mid-July. To everyone's surprise, Mrs. Purcell had attended, and she had come down to the pitch afterward to congratulate them on a hard-fought victory clinched by Peggy's fantastic penalty kick. After Mrs. Purcell walked away, some of the girls had joked that it had likely been her first football match, and she probably hadn't understood a thing. They could only hope she hadn't accidentally cheered for the other side. But although the reminder of Mrs. Purcell's support quieted the mockery, those who had mistrusted her before remained just as skeptical after.

Yet even as Mrs. Purcell warned of hazards and took precautions, Superintendent Carmichael cheerfully urged the canary girls to invite all their friends who were not yet doing their bit in war work to register at their local Labour Exchange and request assignment to Thornshire. It wasn't hard to figure out why. Demand for munitions soared as the war churned on, and production had to increase to keep pace, but canary girls frequently resigned due to poor health—not in droves, but enough that replacements were needed continuously. As for the men, Thornshire held on to their skilled, experienced workers as tightly as they could, but that had become more difficult now that conscription was the law of the land. At first the Military Service Act had decreed that all unmarried men aged eighteen to forty-one could be called up for active service, unless they were widowed with children, physically unfit, or members of a "reserved profession," which apparently included every man working at Thornshire. But a few months after the law went into effect, it was changed to include married men as well, and the government also reserved the right to reexamine men who had been previously declared physically unfit. New posters sprung up among the familiar recruitment adverts to announce the changes in the law, noting that men who sought an exemption should apply to their local tribunals.

For male munitions workers, it was a bit of an inconvenience to spend their day off down at an office applying for a Scheduled Occupation Certificate, but at least they were guaranteed to get one. For men whose status was more questionable, they might bring their employers along to vouch for how indispensable they were to a business that would likely collapse without them. If they argued well enough, a fellow might walk out smiling, certificate in hand, but success was not certain. In her last letter, April's mother had written that the tribunal in Carlisle was quite lenient toward the lads who were needed on their family farms as long as they had the proper forms, but their neighbors, the Collins family, worked a rented farm, and the landowners refused to confirm that the two eldest sons were required laborers. "Both lads were taken into the army, and both were sent to France," April's mother wrote. "They were killed within a fortnight of each other. All for the want of a signature on a paper. Why the landowners refused to sign, no one here knows." If the war dragged on much longer, she concluded pensively, she would fear for April's younger brothers, who would soon be old enough to join Henry in the trenches, God forbid.

April imagined the war to be a ravenous beast whose hunger for young men was never sated. As many soldiers as were sent, more were demanded. Since munitions work spared men from conscription and paid well besides, such jobs were eagerly sought after, and April would have expected each one to be filled. But there were some highly skilled positions that women were not allowed to do, even in wartime, and as men continued to leave Thornshire to answer the call to arms, it became increasingly difficult to replace them.

"Eventually they'll have to let us girls take those jobs," said Marjorie one morning as they rode the tram to work in a rainstorm. "If conscription keeps expanding, and if they stop exempting male munitions workers."

"I hope the war will be well over before that happens," said April, thinking of her younger brothers. When the war started, she had never imagined they'd ever have to go, but now it seemed increasingly likely that they might. "You do understand that if conscription expands, it's because they need to replace the soldiers they started with, which means that a great many lads have been killed or wounded and the war is going very, very badly for us, right?"

"Don't get excited," Marjorie protested as the tram reached their stop. "Of course I understand that. I want victory and peace as much as you do. I want my brothers and all the lads to come home."

"I know you do," said April as they disembarked and raised their umbrellas. "Don't mind me."

"I never do," Marjorie replied, flashing a grin.

"I suppose the good thing about conscription is that they're taking every eligible man." April gave her friend a sidelong look as they hurried along the pavement. "No young men are left in London except those who are officially exempt from military service. That means the end of the Order of the White Feather. You've been made redundant."

"You're forgetting the conchies," said Marjorie as they entered the park. "So-called conscientious objectors are liars and cowards and belong in a uniform or behind bars. If I have to pluck every last goose in the city to shame them, I shall."

"You're incorrigible," said April. "I'm not having this argument again."

"You brought it up."

So she had. Just then, April spotted Oliver Corbyn on the edge of the park closest to Gate One. He was alone, not surprisingly since no one wanted to linger for a chat in that weather. His prosthetic hand was concealed in his coat pocket while the other held up a black umbrella that provided a very inadequate shelter, as wind-driven rain had soaked his trousers from the knees down.

"You go on ahead," said April, pausing at the gravel path that wound in his direction. "I'll see you in the shifting house."

Marjorie followed her line of sight, saw Oliver, and groaned, exasperated. "Don't bother. He never talks to you."

"One day he might, and that day could be today."

"Doubt it," Marjorie called over her shoulder as she hurried on her way. "Don't be late."

April steeled herself, gripped her umbrella tightly, and approached Oliver, quickening her pace when she realized he had spotted her, in case he made a run for it. "Hello, Mr. Corbyn," she said, halting before him.

He nodded, once, then glanced at the clock above the main gate. Never before had she met a young man who so obviously wanted her to go away—except for Peter Bell, the boy from the village in Derbyshire, and that too had been on account of Marjorie and her stupid white feathers.

"We should get out of this rain," she said, trying to smile, inclining her head toward the gate. "Shall we go?"

"*You* should go," he said flatly. "I have another thirty minutes."

She was so astonished to hear his voice that she forgot what she had planned to say next. "Right," she said. "Mustn't be late."

He nodded again, looking beyond her to reply to a man who had greeted him in passing.

"Very well, then." She turned away, then halted. "Although—" She could have sworn he stifled a groan as she turned back to face him. "I know you don't want to talk to me, but I just want to say, again, how truly very sorry I am about that whole awful business with the feather."

He regarded her levelly. "You're right."

Her heart leapt. "I am?"

"You're right, I don't want to talk to you."

And plummeted again. "Oh, I see."

"Look, Miss Tipton, just leave it alone, would you? I understand that you're embarrassed, but don't look for forgiveness from me."

"Where should I look for it, then?"

"Try the good men who've been blown to bits because a girl like you gave them a white feather and shamed them into enlisting." He laughed shortly. "Hold on, no, you can't. They're gone. Ask their grieving mothers instead."

She felt as if she had been struck. "Right." She cleared her throat and drew in a shaky breath. "Well said." Stiffly, she turned and walked away, shoulders drawn up nearly to her ears against the unseasonable chill, umbrella trembling.

Marjorie was already pulling on her wooden clogs by the time April joined her in the shifting house, dripping and miserable. Marjorie took one look at her face and shook her head. "I told you he wouldn't talk to you."

"Right again," said April, more curtly than she had intended. As she slipped out of her wet raincoat, she smiled and shrugged to take the sting out of her words, and Marjorie smiled ruefully back.

The canary girls' shift seemed especially grueling that day, but as hard as the foremen drove them, they pushed themselves harder. Since the beginning of July, stories had appeared in the press about fierce fighting along both sides of the river Somme, and the usually vague reports from the front spoke of tens of thousands of casualties. Some of the girls in the Danger Building had received telegrams during their shifts bearing the terrible news that their husbands had been killed or wounded, or were missing in action.

"Why would the War Office deliver the telegrams here, instead of their homes?" April wondered aloud as a young woman from the machinery section was escorted from the canteen, weeping, by Superintendent Carmichael and a cluster of sympathetic friends.

"Better to hear it here surrounded by friends than alone," said Marjorie, watching the women pass. "If anything happens to our brothers, our mums will get the bad news, not us. We won't know until days after."

April imagined her mother standing in the doorway of their cottage, the little ones clinging to her skirts, clutching a telegram to her chest and sinking with despair. April felt sick at heart at the thought, but what could she do except keep filling shells and hope it made a difference?

On a warm, sunny Wednesday morning at the end of July, April and Marjorie again came upon Oliver in the park on their way to work.

He was seated on a bench beside the main pathway, two men sitting on either side of him and several others gathered in front, all engrossed in a serious discussion, from the looks on their faces. Marjorie was chatting away about strategy for their upcoming match on Sunday, but April caught enough of the men's conversation in passing to figure out that they were discussing the Footballers' Battalion, which had suffered heavy losses on the Somme. Clapton Orient forward William Jonas had been killed, April overheard, as had Norman Arthur Wood of Chelsea and Stockport County, and more players had been seriously wounded. She braced herself, but to her relief, no one mentioned Daniel Dempsey.

They had already passed the men when a change in their voices told April the group was parting ways. "Marjorie, you go on ahead. I'll—"

"I know, I know." Marjorie gazed heavenward and shook her head. "You'll meet me in the shifting house."

April squeezed her arm in thanks and hurried back to Oliver, who was rising from the bench and regarded her with wary resignation. "Mr. Corbyn," she began, falling in step beside him as he strode toward Gate One. "I wanted to tell you that I truly admire your determination to keep doing your bit despite your—" Voice faltering, she gestured quickly to his prosthetic limb, mostly concealed in his pocket.

"Despite leaving a hand behind in France?" he finished for her.

"That's not how I would have put it," she said, stung. "Be rude if you want. Maybe I deserve it. But I really do admire you."

He halted. "Miss Tipton, I don't want or need your pity. I lost my hand, not my life. I'm much better off than most of my pals, as it turns out." He looked at her keenly for a moment, and something like sympathy softened his expression for a moment. "Better off than you are, I fear."

"It isn't pity." Then his words sank in. "What do you mean, you're better off than me?"

"I think you know." He inhaled deeply, frowning, as if he were weighing his words. "Listen. Whatever Mrs. Purcell tells you to do for safety's sake, do it. She knows you girls don't trust her, and that you mock her when you think she can't hear, but she's very clever, and she genuinely has your best interests in mind." He nodded rather than saying goodbye and continued on his way.

"I trust her," April called after him. "I don't mock her."

Without turning around, Oliver lifted his good hand in a dismissive wave. Of course he didn't believe her. Why should he? He only knew her as Marjorie's meek shadow.

It was time to change that. He might never think well of her, but at the very least, she could show him that she wasn't always marching in lockstep with her bolder, occasionally reckless, and often impetuous friend. Oliver wasn't the only one who could contribute more than people expected.

She needed two weeks to speak with all the Danger Building girls on the day shift, and two weeks more to speak with nearly everyone on the night shift, which was a more difficult task since she had to arrive an hour early or stay an hour late to catch them in the shifting house. By the end of August, she had put together a chart several pages long dividing the canary girls according to work assignment. She listed each worker's age, every symptom they suffered, how long the symptoms lasted, if they seemed to be improving or worsening over time, and what remedies, if anything, made them feel better. Everything was strictly anonymous, especially the section at the end where April included lengthier complaints, comments, and questions that could not fit on the chart. She used a ruler and good paper and her very best handwriting, until she was as satisfied with its appearance as with the valuable information it contained.

One day in late August, she left the hostel early, alone, and arrived at the park before Oliver did. She seated herself in the center of his favorite bench and waited. It occurred to her that he might not be scheduled to work that morning, or that he might see her and veer off to another part of the park, but he rarely missed a day, and she figured curiosity more than anything else would compel him to find out why she was there.

Sure enough, about an hour before her shift would begin, Oliver strolled up, halted right in front of her, and peered at her quizzically. "Something on your mind, Miss Tipton?"

For the first time, she realized that he knew her name. She had never given it to him, and yet he knew it, and had used it at least twice before that she recalled. "You're the one we're supposed to go to with concerns, isn't that right?"

"Yes, unofficially," he said warily. "Do you have a concern, and am I going to regret asking?"

Ignoring the question, she scooted over and gestured to the bench beside her. Muffling a sigh, he sat down and regarded her speculatively. Opening her bag, she withdrew the document, which she had carefully wrapped in brown paper, and held it out to him. When he did not take it, she brandished it, frowning, until he did. "You and Mrs. Purcell both say you want to know what's going on with us canary girls," she said. "Well, the girls aren't going to talk to you here in the park like the men do, they don't know you well enough, and they aren't going to speak with Mrs. Purcell when she comes to the Danger Building, not in front of the other girls, and there's hardly any time to run off to her office during our breaks, unless we skip dinner or tea, and no one wants to do that, or to be seen doing it and be called a telltale."

It had all come out in a breathless rush. Oliver was staring at her, eyebrows raised. "Go on," he said. "I'm listening."

She took a deep breath. "Well, since there are things Mrs. Purcell ought to know, and the girls aren't likely to come to *you*, I went to them. I asked them all the things Mrs. Purcell usually asks, but they don't have to be careful with me, so they said what they *really* think. I wrote everything down—except for their names, because I promised I wouldn't—and it's all there, all organized, their answers."

He looked at the document, then back at her, amazed and disbelieving. "You surveyed all the canary girls?"

"Maybe," she said, uncertain. "I don't know. I just asked questions."

He weighed the document in his hand. "This could be very useful."

"Of course. That was entirely the point." Then she remembered something else. "I wrote things down as they were speaking, and I got their words as close as I could. Some of them didn't hold back. Maybe Mrs. Purcell isn't used to indelicate words or too much criticism. You might want to tell her to brace herself first."

A corner of his mouth quirked upward. "I'll do that. Thanks for the warning."

"And you didn't get this from me," she added, closing her satchel with a snap.

"As you wish." He glanced pointedly around the park, where dozens of workers were passing them on their way to the arsenal, and then returned his gaze to her. "Fortunately, there are no witnesses."

She felt heat rise in her cheeks. "Just don't give the boss's wife my name, all right?"

"She may want to thank you."

"She can thank me by helping us not get sick." Abruptly she stood, clasping the handle of her bag in both hands and holding it in front of her. "One more thing, and this is important. Don't let her use this to sack anyone. None of the canary girls wants to quit. We all need the wages and we're willing to do the work. We just don't want to die like that poor girl at Woolwich."

He nodded and rose, all traces of doubt gone from his expression. "Understood."

"I'm trusting you."

"You *can* trust me."

"Well, you can trust me too," she retorted hotly, then bit her lips together, immediately regretting it. "Good day, Mr. Corbyn."

"Good day, Miss Tipton." He lifted the paper-wrapped document. "And thank you for this."

She nodded in reply, then quickly turned and hurried off as if she were expected somewhere else, although she had plenty of time before her shift and he probably knew it.

Two days passed in which April either did not see Oliver in the park, or she saw him from a distance but did not approach. Why she was shy now when she had not been before, she couldn't say, but a part of her feared that all her hard work had been a foolish waste of time, that her survey was idiotic and thoroughly unhelpful, good for nothing but to give Oliver and Mrs. Purcell a hearty laugh. Who was April to take on that sort of work? She had left school at fifteen to go into service. She read books, but not difficult ones, and only the occasional newspaper, mostly to learn about the war, eager for any word about the Lonsdale Pals. Mrs. Wilson had warned her often enough to remember her place and not to follow Marjorie into trouble, but April somehow always forgot that advice until it was too late.

On the third day, April was startled to find Oliver waiting outside the shifting house when she and Marjorie emerged at half seven o'clock. Marjorie looked from her to him and back, bemused, and murmured, "I'll wait for you at the station, unless you'd like me to chaperone."

"Go on." April nudged her friend along as Oliver approached. "See you there."

Marjorie eyed him speculatively as they passed each other, and he gave her a courteous nod, which, considering everything, was rather kind of him. He halted in front of April and offered her an identical nod, which was disappointing. Didn't she merit something a trifle friendlier? "Miss Tipton."

"Mr. Corbyn." She gestured toward the gate. "My hostel is a long way off, and I'd rather not miss my tram."

"I'll walk with you partway," he said. She nodded and they set off together. "I gave your report to Mrs. Purcell. She's very grateful and she sends you her thanks."

Her heart dipped. "You weren't supposed to tell her—"

"What I mean is, she thanks the anonymous employee for providing her with such valuable information. She called the entire report 'illuminating.'"

"She said that?" April felt warmth filling her chest and a grin spreading over her face. "Even about the rough bits at the end?"

"Well—" He winced slightly. "Those bits she called humbling, but instructive. She appreciated the workers' honest opinions."

"Maybe too honest. I could have softened it up a bit."

"No, no. You did exactly what you should have done." They had reached the gate. Just ahead, the sentries were choosing men at random with a tap on the shoulder and taking them aside to be searched. The women were only peered at closely, on the assumption that their guilt would be written plainly on their faces if they were carrying anything they shouldn't. April and Oliver made it past the sentries unchallenged, and by the time they reached the park, April was still marveling over the revelation that Oliver believed she had acted properly for a change. Until then, he had probably thought her incapable of it.

They both spotted Marjorie at the same time, walking on the far side of the park with a few other girls from their hostel. If April hurried, she could catch up to them. "That's all I had to say, really," said Oliver, indicating her friends with a tilt of his head. "I guess this is where I leave you."

"I guess." April felt strangely disappointed. "Let me know if you—if Mrs. Purcell, I mean—need anything else."

She turned and hurried down the most direct gravel path to join her friends. "What was all that about?" Marjorie asked, glancing back toward Oliver, who was walking alone along the edge of the park closest to the arsenal.

"Nothing," April replied, but of course that wasn't true.

On their way to the arsenal two days later, when April and Marjorie saw Oliver up ahead in the park, Marjorie didn't wait for April to speak before heaving a sigh, shaking her head, and continuing on alone. April hesitated in front of Oliver's bench, and when he glanced up and nearly smiled, she sat down beside him. "I have a problem," she said.

"Yes, I know, I read your survey," he replied, offering a sympathetic grimace. "But Mrs. Purcell is on it. She's trying to do right by you girls."

"Good to know." April nodded, thoughtful. "But that's not the problem I meant."

He raised his eyebrows at her and nodded. "Go on."

"I behaved very badly to a fellow worker." She lowered her eyes to her lap, realized she was wringing her yellowed hands, and promptly stopped, grasping the bench on either side of her instead. "I've tried again and again to apologize, but he won't accept it. He's a very cold, dry, rigid sort of person."

He regarded her in utter disbelief. "Is that right." He withdrew his hands from his pockets and rested the prosthetic deliberately on his left leg. "Sounds like we have something in common."

For a moment she was confused, glancing between his shocked, angry expression and the prosthetic that he so rarely displayed. Then she thought about what she had said, and she felt all the blood drain from her face. "I didn't mean—I wasn't referring to—I wasn't making a sick joke about—" Throat constricting, she swallowed hard and shook her head. "I'm sorry. That's all I meant to say."

She rose, but before she could hurry away, he caught her by the wrist. "Don't run off. I'm sorry. I see now you didn't mean anything by it."

She whirled to face him, but didn't pull her wrist free. "I honestly didn't," she said, her voice low and fierce. "Why do you always have to assume the worst about me?"

He tugged her a bit closer. "Will you please sit down, Miss Tipton?"

She hesitated, then did as he asked. "I do admire you, you know," he said quietly after a long moment in silence. "You never once placed all the blame on your friend for giving me that feather, although you could have done. It was obvious that it was all her idea."

"I share responsibility for what happened. I should have tried harder to stop her."

"Do you think it would have mattered?"

"Maybe? I don't know." April shrugged and shook her head, sighing. "Once she's set a course, it's almost impossible to get her to change direction. She never sees the edge of the cliff straight ahead, even if you stand there pointing at it."

His mouth turned wryly. "I know people like that."

"Here at the arsenal, or—" She hesitated. "In the army?"

"Everywhere."

She shifted on the bench to face him. "Why did you enlist, anyway? If you don't mind me asking. My brother did it for the pay, and because all his friends were going."

"It was a bit of that for me," he replied, his gaze falling on his prosthetic. "King and Country, you know. An Englishman's honor, keeping our word to Belgium and France. My friends and I thought we had outgrown our village and we wanted some excitement and adventure, so we all joined up together. I admit I was nervous, but I didn't want to be left behind while all the lads I grew up with went off to claim battlefield glory. You remember how it was. Everyone thought it would be over by Christmas, and we wanted a taste of it before it was too late."

"I remember."

"I don't regret it, you know," he said, with unexpected intensity, but just as quickly, the fire faded. "Well, I regret this—" He raised his left hand an inch and let it fall back to his lap. "But not that I tried to do my duty and help my friends."

Just then, the warning bell above the main gate clanged.

"You'd better run," Oliver advised even as she bolted to her feet. She murmured a hasty goodbye and set off.

"Miss Tipton," he called after her.

She halted and turned around. "Would you please just call me April?"

"Oh, are we chums now?"

"I guess that's up to you. You're the one who was so angry."

He smiled. "All right, April, chum, when is your next football match?"

"Sunday," she replied, surprised. "Two o'clock at the Globe in Stratford."

"Sunday, two o'clock." He rose and tucked his hands into his pockets. "Maybe I'll come."

April shrugged as if it didn't matter one way or the other. "Maybe I'll look for you."

She turned and darted off before he saw from her hopeful smile and flushed cheeks just how much it did matter.

12

SEPTEMBER–OCTOBER 1916

HELEN

In the middle of August, the renowned medical journal *The Lancet* published the results of a rigorous five-month study of the effects of TNT on women munitions workers. Dr. Agnes Livingstone-Learmonth and Dr. Barbara Martin Cunningham, both munitions factory medical officers, had concluded beyond all doubt that TNT poisoned the women who directly handled it, as well as others who worked in the same building. They classified the women's symptoms as either "irritative," such as nasal congestion, sore throats, headaches, chest pains, abdominal pain, nausea, vomiting, constipation, diarrhea, and skin rashes; or "toxic," which included continuous bilious attacks, fainting, swollen feet and hands, fatigue, depression, and blurred vision. The most significant factors influencing how sick a worker became appeared to be the frequency and duration of exposure, as well as the individual's health in general.

Much to her annoyance and indignation, Helen did not learn of the study until nearly a fortnight after it was published. When she finally obtained a copy, she read it diligently, thankful to finally have some answers. She was not at all surprised to find striking similarities between the physicians' analysis and the simpler, vernacular report one of their own canary girls had put together.

In the conclusion of their report, the two physicians—both women, Helen was intrigued to see—had recommended measures factories could and ought to take to lessen the severity of TNT poisoning. They strongly urged improved ventilation to disperse the TNT dust, but they cautioned that respirators and masks, which provided excellent protection against airborne germs, could actually do more harm than good as a barrier against the yellow powder, since warmth and moisture worsened the irritation of the nose, throat, and sinuses. Veils were suggested as a more suitable alternative, and like the rest of a worker's uniform, they should be laundered between wearings. The doctors also encouraged serving plain nourishing food and bland drinks in factory canteens, coating the face with a protective powder, and practicing good "personal cleanliness," something factories could best encourage by constructing appropriate washing facilities on their grounds. They strongly recommended not to employ workers under eighteen or over forty in Danger Buildings, and to rotate workers in and out of other departments every twelve weeks. Ideally, TNT workers would be scheduled on three eight-hour shifts rather than two twelve-hour shifts, and every week, each worker would receive a thorough medical examination.

Some of the doctors' recommendations were already in place at Thornshire, but others, Helen knew, would be difficult to implement and might provoke resistance from the very workers they were meant to help. Thanks to her anonymous statistician, Helen knew that the Danger Building jobs appealed to many munitionettes because the risks merited higher pay. Would the canary girls' wages remain the same during their twelve-week rotation to less hazardous work, or would they be paid the going rate? Would the munitionettes currently assigned to less dangerous factories object to being transferred into the Danger Building while the regular workers rotated out? Would workers welcome a less grueling, eight-hour shift, or would they resent having their hours, and therefore their earnings, cut by a third? And with workers in such short supply, where would they find enough new hires for a proposed third shift?

Helen posed these questions and others to Arthur at the office whenever he agreed to grant her an audience. If days passed and he could not fit her into his busy schedule, she grew impatient with his repeated deferrals and queried him at home instead, over breakfast or dinner or even in bed. In her view, nothing was more important than the health of their workers, because on them, all else depended. She didn't want Arthur's ear because she was his wife and thereby entitled; she wanted it because the welfare supervisor ought to be able to speak to the arsenal manager whenever urgent matters required it, and not just when he had spare time.

Helen was surprised to learn, nearly three weeks after the fact, that the Ministry of Munitions and the Health of Munition Workers Committee had met with managers of factories undertaking TNT work, and as a result, a new section of the ministry had been created to develop and administer revised safety regulations. More astonishing yet, Arthur had not attended the meeting. "Why didn't you tell me?" she protested, striding into his office after ignoring Tom's halfhearted claims that Arthur was not available. "If you were too busy, I gladly would have gone in your place."

"My brother attended," he told her, his eyes on several documents spread out upon his desk. "He represented all of the Purcell Products factories."

"What did he learn?" She leaned forward to rest her hands flat on the desk, trying to catch Arthur's eye. "What are the new regulations? Surely I ought to be kept informed."

"Let me think." Arthur closed his eyes and rubbed his forehead with a pinching motion, his thumb on one temple, his fingers on the other. "Check with Oliver. He should have the report."

"*I* should have the report," Helen said crisply as she left the office to find the assistant secretary, who regarded her mildly when she snapped at him that she needed a turn with the report if anyone expected her to do her job. Oliver produced the report so promptly that she felt rather rotten for taking her anger out on him. "Please forgive my ill temper," she said. "I'm not angry with *you*." He accepted that too with his usual imperturbable decency.

She took the report to the nearest canteen to read over a cup of tea, and soon learned that most of the committee's regulations merely echoed the two doctors' recommendations. A few things stood out, however: Full-time doctors would be appointed to all large factories, while smaller factories would be assigned local physicians on a part-time basis. Women would be permitted to work a maximum of sixty hours per week, which in Helen's opinion still pushed the limits of human endurance. Lastly, munitions factories were urged to cancel all labor on Sundays except for repairs, maintenance, and other work of particular urgency.

That suggestion so astonished Helen that she had to set down the report and ponder it. Women already had Sundays off at Thornshire, but as for the men, roughly half had Sunday off, while the other half took Saturday. But it was inconceivable that all munitions production should cease altogether for an entire day, at every arsenal and factory in Britain. The demand for armaments was simply too great, the consequences of falling behind and failing to keep the armies well supplied too dire. Helen didn't suppose the Germans took Sundays off; how could the Allies afford to? She could not imagine Arthur endorsing such a policy, and his brother and father would never give the idea serious consideration.

But that was the arsenal manager's problem to sort, she told herself irritably as she recorded a few notes in her jotter, finished her tea, and returned to headquarters, where she left the report on Oliver's desk and retreated to her cupboard of an office to plan for implementing the new regulations. She was glad the Ministry of Munitions was taking the problem of TNT poisoning seriously, because in her opinion, it seemed to have a rather fatalistic attitude about other dangers—especially accidental explosions.

Some preventative measures were taken, of course. Workers were forbidden to bring matches or metal objects into the factories. Every building was required to have sufficient firefighting equipment at the ready. Each factory had to employ designated firefighting personnel or to organize volunteer firefighters from among their workers. All munitions factories must maintain a first aid facility with an adequate supply of medicines, bandages, and other items needed to treat serious burns and other injuries. But other than that, the prevailing attitude seemed to be that accidental explosions were tragic but inevitable in munitions work, and it was up to each and every worker to follow the rules, use caution, and report careless coworkers to the foreman for additional safety training. Helen didn't know what she would recommend in addition to all that, but she had a terrible feeling that it would all become tragically obvious in hindsight.

In her windowless office, she could not judge the passing of the hours by the fading daylight, but when the need to stretch her legs coincided with the distant clang of the shift bell, she checked her schedule, saw that she had no more appointments, and decided to head home. Taking her coat and bag in hand, she went upstairs to Arthur's office, rapped on the open door, and peered inside. "It's six o'clock," she said, giving him a start. For a moment she held her breath, shocked by how haggard he looked, the dark circles beneath his piercing green eyes, the general disarray of his wavy brown hair. "Shall we go home?"

"You go on—" he began, but a yawn interrupted him. "I have a report due at the Ministry of Munitions at noon tomorrow. Rather not leave it to the last minute."

Setting her bag and coat on the stand by the door, she went around the desk behind his chair and placed her hands on his shoulders. "Can't Tom or Oliver take care of it?" she asked, massaging his tense muscles, which were alarmingly full of hard knots of strain and worry.

He let out a soft groan, relaxing into her touch. "Tom and Oliver have mountains of their own work to attend to."

"Shall I stay and help you? I could take dictation. You could close those gorgeous eyes—"

"Bloodshot eyes, I think you mean."

"If we work together, we could finish the draft quickly and leave it for the typist to deal with first thing in the morning. The courier would still be able to deliver it to the ministry with time to spare."

"Thank you, darling, but you should go home, have a bite to eat, and get some sleep."

"How shall I sleep if I know you're here, toiling yourself into exhaustion?" She let her hands travel down his back, kneading the muscles, hiding her dismay at all the tension she felt in him. "In the time you've spent arguing, we could have finished two paragraphs, perhaps three."

"Irrefutable logic, as ever." He placed his right hand on her left and gazed wearily up at her. "You are so lovely, darling. If I weren't absolutely knackered—"

"And if I hadn't left the door open, and if Tom or Oliver might not wander in and interrupt us—" She bent down to kiss the top of his head, then smoothed his unruly hair. "It's all right, darling. I'm tired too."

"It's just as well, I suppose," he said, straightening in his chair, rearranging papers on his desk. "This would be a bloody terrible time to bring a child into the world."

Stung, she withdrew her hands to her sides. "What do you mean?"

"What do I mean?" He laughed bleakly and turned to peer up at her again. "Well, it's the end of the world, isn't it? Or at least the end of civilization. Chaos and bloodshed, violence and destruction, slaughter on a scale never before seen or even imagined. I wonder what animal will emerge to become master of the earth when all the men have killed one another?"

For a moment she could only look at him in heartsick silence. "Let's finish the report and get you home," she said gently, stroking his hair away from his brow, wishing she could wipe away the double furrow of worry carved there. Managing a smile, she gave his shoulders a quick pat, then retrieved her jotter and a pen from her bag and settled into a chair on the other side of his desk, determined to help him—against his will, if necessary.

Together they made quick work of the rest of his report. Afterward, Arthur would have found other tasks to complete, but Helen planted his hat firmly on his head, put him into his coat, threw on her own, and linked her arm through his to steer him out of the office, leaving the report on the typist's desk on the way out. The Silver Ghost had been mothballed, or whatever the equivalent was for a motorcar, to save petrol, so they caught the train and were home within the hour. Helen had telephoned ahead, so a hot, nourishing meal was waiting for them in the dining room. Helen chose light, amusing topics for conversation, reminiscing about their days back in Oxford. By the time the dishes were cleared away, Arthur seemed much like his old self again—an exhausted, world-weary version of himself, but still, and always, the man she loved.

That night she held him in her arms as he sank into sleep, stroking his hair, kissing his brow. Eventually she fell asleep too, but she woke several times to find him tossing and turning beside her, muttering unintelligibly. Each time she stroked his back and murmured soothingly until he grew calm and drifted off again.

Two days later, a Sunday, she proposed that he not go into the arsenal but spend the day with her instead. "After breakfast, we can take a restorative walk through Hyde Park," she said, smiling, reaching for his hand across the table. "This afternoon, we can attend the Canaries' match in Stratford."

"The what match?" he inquired absently, his eyes on the newspaper.

"The Thornshire Canaries. The arsenal football club." She squeezed his hand, and eventually he looked her way, offering an apologetic smile for his neglect. "Our side is playing the Workington Ladies from the National Shell Factory."

That got his attention. "Workington, in Cumbria? That's a rather long way to come."

"It's a rather important match. All the proceeds will benefit the Border Regiments' Prisoners of War Funds." Helen smiled, encouraged. "They're supposed to be quite good. I understand they don't have as many munitionettes' teams in their area since they don't have the concentration of factories we do here, so they have to travel far afield sometimes."

Arthur frowned thoughtfully. "I'm glad to know the Workington Ladies are coming here. I'd rather not have our girls travel so far and arrive exhausted for work Monday morning." His brow furrowed. "Where on earth do our players—the Canaries, you called them?"

"Yes. The Thornshire Canaries." She had told him all this when she organized their uniforms.

"Where do the Canaries play? Certainly not on that miniscule field on the arsenal grounds. I thought all the football pitches had been converted to military use."

"Not all of them. Stratford United has allowed our girls, and a few other local teams, to use the Globe until professional football resumes, which I presume means for the duration."

His eyebrows shot up. "The girls play on a professional pitch?"

"Yes, indeed."

"How marvelous." He smiled and shook his head. "To be honest, I didn't even know girls played football."

"I don't believe most of these girls *did* play, at least not regularly, until men's football was cancelled."

"Well, why shouldn't they play?" Arthur tossed off an amused shrug. "If we can have women munitions workers, why not women footballers?"

"Exactly so." Helen was very much pleased to see him in such good spirits. "You'll come with me, then? It should be an exciting match, and I know it would mean so much to our girls to see you in the stands and know they have your support."

Her heart sank as his smile turned regretful. "It sounds like great fun, darling, but I have important matters to attend to at the arsenal."

"On a Sunday? Are you certain? Can't it wait until tomorrow?"

It couldn't, he explained, because new machinery was being installed, and he and the shop foreman had to supervise the operation to make sure it was done properly. So Helen kissed him goodbye in the foyer when he left, and went on her morning walk alone, and rode to Stratford alone.

But when she arrived at the Globe and searched for a good seat in the stands, she realized that she would not have to watch the match alone, for there was Oliver Corbyn, a few rows back right in the middle, apparently also unaccompanied. She had seen him at two previous games that she recalled, but he had always been with friends, and she and he had only exchanged courteous nods from a distance.

His gaze was fixed intently on the Thornshire Canaries as both teams warmed up on their half of the field, but he rose politely when he saw Helen approaching. "May I join you," she asked, suddenly embarrassed, wondering if she should have left him alone, "or is this seat taken?"

"I was saving it for you," he said, gesturing, a corner of his mouth turning up.

"Liar," she scoffed, seating herself beside him. "So, what are our chances today, do you think?"

"The teams seem to be fairly evenly matched," he said, studying the Workington players as they ran through various exercises and drills. Unlike the Canaries, who wore black short trousers as men did, the Workington Ladies wore hip-length red jerseys and black skirts hemmed just below the knee.

"I would imagine those skirts would be an encumbrance in football," Helen remarked, looking from the team in red to the team in yellow. "Short trousers seem to allow a much greater freedom of movement."

"One would think so, but not all ladies feel comfortable in trousers. Or so I hear," Oliver replied, a bit sheepish. She hid a smile. "Some reporters can be a little unkind about so-called unfeminine apparel and the girls who wear it, but our Canaries are used to trousers, because of their munitions uniforms."

"The press covers these matches?"

"Sometimes. Local papers cover local teams. Big matches might be worth a mention in the more important papers. Most of the reporting is actually quite good. Descriptions of the play and players are exactly what you'd expect from a report about a men's game, and the women are respected as the accomplished athletes they are. Unfortunately, some reporters treat women's football as nothing more than an amusing lark, and write condescendingly about the players' hairstyles and charm instead of their strength and skill. A reader could almost mistake one for a story about a garden party rather than an athletic competition."

"No surprises there, I suppose." Helen scanned the pitch as the officials signaled that the start was near. "And is this considered a big match?"

Oliver shrugged as the whistle blew, his eyes on the center circle as the players in red kicked off. "It isn't the Dick, Kerr Ladies versus the Blyth Spartans, but the Workington Ladies have traveled far, which garners interest, and the Canaries are building a solid reputation in the Munitionettes' League."

"All the more reason for us to support our home team," Helen declared, facing forward again. She recognized some of the Thornshire players even at a distance—Lucy Dempsey, who had come to her about the players' uniforms and gear; Marjorie Tate, prowling in the goal as if she were looking for a fight; Peggy, tall and broad-shouldered, a slender version of her mother, Mabel, who stood on the sidelines; and several others who had spoken to her, briefly or at length, in passing on the arsenal grounds or in a quiet conference in her office. She watched the players moving the ball around the pitch, first yellow with the advantage, then red, and then yellow again. Lucy took a shot on the goal that went wide by mere inches, and Helen and Oliver joined in the groaning lament of the crowd.

"Have you ever played football?" Oliver asked as the ball went out of bounds and a Canary set up for a throw-in.

"Me? Heavens, no. Tennis is my game." Helen watched, holding her breath, as a swift, flaxen-haired Canary stole the ball from a red-shirted girl a head taller than herself, worked it down the field, then passed it off to Peggy, who sent the ball soaring toward the goal—and into the arms of the Workington keeper, who made the flying catch and landed with a roll. "How about you?"

"All the boys in our town played growing up. I played at school, too, even played a bit in the army when we had the chance, when we weren't in the trenches. It's been a while, though." He offered a dry chuckle. "The last time I played, I was in France, and I still had two good hands."

She studied him from the corner of her eye, but he sounded strictly matter-of-fact, without a trace of self-pity. "Oh, so you're a player of international renown, then?" she teased.

To her relief, he smiled. "Hardly. I think you're confusing me with Mrs. Dempsey's husband."

"Oh? Does he play football too?"

Oliver laughed. "A little, now and then." When she peered at him, uncomprehending, he explained that Lucy Dempsey's husband was *the* Daniel Dempsey, Olympic champion and longtime popular center forward of Tottenham Hotspur, now serving with the Footballers' Battalion somewhere along the Somme.

On an impulse, Helen asked Oliver if any of the other players had husbands or sweethearts or other loved ones in the military, and naturally they nearly all did; if Helen had asked the same question of any random group of strangers on any street in London, she would have received the same answer. April, the swift midfielder, had an elder brother in the army; Marjorie three; Peggy was engaged to a private in the West Ham Pals; and on the list went, and those were just the ones Oliver knew about. For some of the women, the soldiers they had kissed farewell on their front steps or a train station would never return from the war.

No wonder the munitionettes worked so tirelessly through the long shifts and longer weeks, Helen realized. Every shell they made hastened the end of the war, speeding the return of their absent loved ones. And for the grieving widows and sisters and mothers, every shell was a blow for vengeance, smoldering and bitter and full of anguish and spite.

She shivered at the thought, and pulled up the collar of her coat to pretend it was the wind that had chilled her and not her own grim thoughts.

The match had not been long underway when Helen's questions about certain plays and penalties revealed how embarrassingly little she knew about the sport. Oliver never failed to answer, patiently and clearly, but she knew that he would hardly tell his boss's wife to be quiet and let him watch in peace. "I hope I'm not being a nuisance," she said apologetically, after he explained why April and other players seemed to run wherever they pleased while others tended to remain in one region or another.

"Not at all," he replied. "I enjoy showing off how clever I am. I rarely get the opportunity."

She laughed aloud, for she knew that wasn't true; if she hadn't seen for herself how much Arthur and Tom depended upon him, Arthur's praise alone would have convinced her. She was glad Oliver was willing to help her learn the subtleties of the sport, but she did wonder why he was such a dedicated fan of the Thornshire Canaries. It had to be something more than pure love of the sport, for there were other matches in and about football-mad London, men's matches, charity and recruiting events with teams from regiments that were still in training. Loyalty to Thornshire Arsenal might have prompted him to attend a match or two, but not every Sunday afternoon; no one was *that* fond of their workplace. The most logical explanation was that he was fond of a particular player, but she could not tell who, for he applauded everyone who deserved it, even those on the other side, if a play was especially brilliant.

The first half ended with Workington ahead at 2–1, but after the interval, April came blazing out of nowhere and scored from twenty-five yards away at a ridiculously shallow angle. The defense must have expected her to pass, for they didn't react until the ball was soaring through the air, leaving the keeper on her own. "Yes!" Oliver shouted, bolting to his feet. "Good show, Canaries!"

"Well done," Helen called, applauding her gloved hands. "Brava! Encore!"

Oliver looked at her, astonished, then burst out laughing. "I don't recall ever hearing that particular cheer at a football match before," he said, taking his seat again.

"Let's see if it works," Helen said, folding her hands in her lap and studying the pitch with great anticipation.

Perhaps it did indeed work, albeit not immediately, for with nearly five minutes left, Lucy scored from within the box, and Marjorie successfully fended off a barrage of shots to hold on to a 3–2 victory. "Shall we congratulate the team?" Helen asked Oliver, rising, as the spectators filed from the stadium.

"You go on," said Oliver, tucking his hands into his coat pockets. "I'll wait, if you'd like me to escort you to the station."

She emphatically declined, having taken up far too much of his afternoon already. He bade her goodbye and left so quickly that she wondered whether she had been mistaken to think that he was fond of one of the players. Why squander a perfectly good opportunity to spend time with the person one most admired? It occurred to her that she might ask Arthur, as he did so often enough, and her joy in the team's victory dimmed.

Even so, she enjoyed recounting the match for Arthur that evening over dinner. Her efforts to entertain him were rewarded when genuine interest showed through his fatigue, and his face lit up with delight when she described April's brilliant, game-winning goal. "Perhaps you'll join me at next week's match," she said tentatively, and smiled when he told her he would certainly try.

But very late that night, she woke abruptly, heart pounding, after Arthur shouted in his sleep and sat upright, breathing heavily, turning this way and that as if he were searching desperately for something lost in the blackout darkness. "Arthur, darling, what is it?" she asked, reaching out to lay a comforting hand on his back.

He flinched at her touch, then took a deep, shuddering breath. "It was—" He hesitated, and she could feel the perspiration on his skin beneath his thin pajama shirt. "I don't remember now. It's all fading away." He lay down again and pulled the covers up, shivering. "It was just a nightmare."

"How dreadful," she murmured, lying on her side next to him, resting her head on his shoulder, placing her hand on his chest. His heart was racing. "Try to sleep, darling. It will all look better in the morning."

He kissed her forehead in reply. She listened as his pulse steadied and his breaths became deeper, more even, and only when she was certain he was all right did she allow herself to drift back to sleep.

The next morning, Helen woke first, and in the dim light of the autumn sunrise that seeped around the edges of the blackout curtains, Arthur slept on, motionless except for the slow rise and fall of his chest. Suddenly it came to her: She would let him sleep, for he needed and deserved it. She would telephone Tom and inform him that Arthur would not be in until the afternoon, and possibly not at all. Arthur might rebuke her afterward, but at least he would be well-rested.

Carefully she slipped out from beneath the covers, pulled on her dressing gown and slippers, switched off the alarm clock, and quietly crept across the floor. She eased open the door without a sound, stepped into the hallway, and was closing it behind her when she heard bedsprings creak and her husband murmuring sleepily, "Helen?"

She sighed softly. "Yes, Arthur?" she said, opening the door again.

He propped himself up on his elbows and blinked at her. "Why didn't you wake me?"

She had one gambit left. "Go back to sleep, darling. You don't have to be up for hours."

"Then why do I smell breakfast?" He reached for his wristwatch on the nightstand beside the bed, peered closely, and threw back the covers. "Dash it all, it's half six. What happened to the alarm clock?"

"We must have forgotten to set it."

Grumbling wearily, Arthur climbed out of bed and began getting ready for the day. Her scheme thwarted, Helen did the same, disappointed and anxious for him. She waited for him to mention the nightmare, but it seemed that he had entirely forgotten it.

They rode to the arsenal together and parted ways at the door to the administration building. She had two appointments that morning, one with a young woman whose father demanded that she transfer out of the Danger Building, the second with a scrappy middle-aged woman who was eager to transfer in, so it was an easy fix to have them trade places. Then she began her rounds, and since it was a Monday, she visited all of the canteens, inspecting them for safety and cleanliness, chatting with the girls, reminding them that her office door was always open if they wished to speak with her in confidence. In the canteen kitchen nearest the fuse cap shop, she was surprised to discover three smartly dressed young women, hair meticulously bobbed and coifed, faces adorned with powder and lipstick, shrieking with laughter as they peeled potatoes and chopped onions. It was one of the oddest scenes she had ever observed at the arsenal.

Puzzled, she took the head cook aside. "What's their story?" she asked in an undertone. "They're positively giddy."

The cook heaved a sigh and planted her fists on her broad hips. "You're looking at the nieces of the Duchess of Wellington," she replied, her voice low. "They're part of that, what's it called, the Week End Relief Scheme. I was expecting them yesterday, but here they are, so I put them to work."

"Of course." Helen should have guessed they weren't regular workers from the way they treated their menial tasks as a smashing joke. More than a year before, Margaret, Lady Moir, the wife of the MP for East Aberdeenshire, and Winifrede, Lady Cowan, who was married to a high-ranking official with the Ministry of Munitions, had organized a scheme in which ladies of the leisure class could contribute to the war effort by taking over for the regular workers on the weekends. The objective was to relieve the munitionettes of the strain of working six or seven days a week while keeping the factories in continuous operation. Superintendent Carmichael could always find an appropriate place for a qualified, willing worker, but thus far she had been reluctant to place Week End Relief ladies in any job requiring actual skill or responsibility. "How can I trust them around explosives or heavy machinery," she had asked Helen earnestly, "when they're only here on a lark?"

Perhaps that was not quite fair, Helen supposed. A great many middle- and upper-class women had signed on to do their bit as soon as the need for women workers became evident, without needing a special program for the privileged to prompt them to get their hands dirty. As for the Week End Relief workers, many surely took their work seriously and sincerely wanted to do their bit. Not so the three young ladies presently enjoying themselves overmuch in the canteen kitchen. "None of them has ever chopped or peeled veg before," the cook told Helen, eyeing the fashionable young women with bewilderment and distaste. "I guess the worst they could do is cut off too much tater with the peel, or stab themselves in the finger."

Helen grimaced. "Let's hope neither of those misfortunes come to pass. Sorry their care fell to you, and on a Monday, when you weren't braced for it. Good luck."

"Thank you, missus. I imagine I'll need it today."

Helen finished her rounds, hoping the relief workers did more good than harm, and that their presence did not worsen the existing class tensions at the arsenal. She had noticed a recent push by the Ministry of Munitions to draw more middle- and upper-class women into munitions work, but as regular hires, not merely weekend relief. Upper-class women were a largely untapped source of labor, certainly, but the ministry seemed to have other motives. Documents it had sent to Thornshire revealed an assumption that better-educated women would learn skilled work more quickly, they would naturally assume leadership roles, and they would readily give up their places after the war when the soldiers returned home and sought civilian jobs. Working-class women, on the other hand, the vast majority of the munitionettes at Thornshire, had worked to support themselves and their families before the war, and during it, and would almost certainly need to work for a living afterward. They had become accustomed to the higher wages and greater respect "men's work" had garnered them, and Helen suspected they would not relinquish their hard-won positions so easily. They had taken on difficult work when their country desperately needed them. More than two years into the war, with no end in sight, was the government already planning to push them aside when peace finally returned to Britain?

At six o'clock that evening, weary from a long day, Helen again went to Arthur's office, helped him finish up some paperwork, and firmly steered him out of the office and home. The next day he insisted that he absolutely must work late, full stop, so she stubbornly waited with him, retrieving some files from her own office, making herself comfortable on the davenport by the window, and reading silently while he completed various forms and studied reports. They didn't get home until midnight, but she had made her point without saying a word: She intended to work as late as he did, so if he wanted her to go home for a good meal and essential rest, he had to go too.

After a few days, she was pleased to see positive results from her scheme. Arthur left work and got to bed earlier, he seemed less hoarse and haggard, and although nightmares woke him two more nights that week, he still managed to get more sleep than before.

Then, around midnight on 26 September, something else jolted them awake, a searing, bright light that pierced the edges of the blackout curtains, followed by the deep rumble of explosives, ominously close. "Stay back," Arthur cautioned, but she followed him to the window, and as he drew back the curtain, she shielded her eyes from the glare. "Air raid. Likely no more than four miles off."

"What is that light?" Helen asked, just as it abruptly extinguished. "Some terrible new bomb?"

"Same bombs, but I believe the zeppelin is using Very lights too."

She flinched at another distant explosion. "Very lights?" she echoed, voice quavering.

"Brilliant white flares. The zeppelin crew drop them to illuminate the ground below to aid in navigation, or release them into the air to create a dazzling, baffling glare that our searchlights can't penetrate." He paused to listen. "They're moving off."

"Are you sure?"

"As sure as one could be." He put his arm around her. "You're shivering. Back to bed."

It was obvious he wanted to keep watch at the window, but she refused to go without him, so they both returned to bed, and somehow they both slept soundly the rest of the night. It wasn't until much later the next day that they learned how terribly devastating the air raid had been. Three zeppelins had approached London through Kent, following a five-mile line from Streatham Common past Brixton Hill to Kennington Park, where they dropped their last bomb less than three miles southeast of their home. Along the way, the Germans had bombed the high road as well as residential streets alongside it, leaving massive craters in the streets and destroying shops, a children's playground, countless trees, and many small brick homes and front gardens belonging to middle-class families. Twenty-two people were killed and at least that many injured. Helen found little comfort in learning that two of the three zeppelins had been brought down by British ground defenses as they retreated, one at Billericay, twenty-four miles east of London, and the other at thirty miles northeast of that, at Little Wigborough, on the coast near the Blackwater Estuary. All on board had been killed, their charred and broken bodies found amid the wreckage.

Less than a week later, on the night of 1 October, Helen and Arthur witnessed an even more dramatic, appalling spectacle in the skies above London. They had stayed quite late at the arsenal while Arthur conferred with his elder brother on the telephone, and were aboard a crowded tramcar about three miles east of home near Blackfriars Bridge when they heard frenzied shouts and cries of alarm from passengers seated on the other side. "Oh! Oh! She's hit!" a man called out, and nearly everyone aboard pushed their way to the windows or craned their necks to scan the sky to the north. High above, looking along the straight shot of New Bridge Street and Farringdon Road, Helen spotted a tangle of searchlights concentrated on a ruddy glow, which, as her eyes adjusted, she realized with horror was a blazing zeppelin. Suddenly the searchlights turned off, one by one, as the airship drifted as if on a parallel course with their tram, a massive ball of red and orange flames, a terrible fallen angel plummeting from the heavens to earth, its lurid glow lighting up the streets and buildings below, tinting even the dark waters of the Thames.

Helen watched the burning airship drift away, utterly riveted, scarcely breathing, her chest about to burst from emotion—terror, triumph, she could not say. When at last the doomed craft vanished from sight, two, perhaps three minutes later, a hoarse shout went up from the passengers, and indeed it seemed as if every voice in London joined in, a swelling roar of relief, exultation, and loathing. Helen wondered where the zeppelin had crashed, and her heart trembled with worry for the citizens below, followed a moment later by a small stir of pity for the crew, who had certainly perished.

"Did you see it?" Arthur asked, his eyes wild as he stared into the darkness on the other side of the window. "Do you know what it portends?"

She looked from her husband's pale, haggard face to the night sky and back. "Victory for the Allies?" she replied, uncertain. "Germany going down in flames?"

"No, no." The look he threw her was impatient, desperate, anguished. "That is the fate of Great Britain if I fail to do my duty."

"That's not what it means," said Helen, frightened for him. "That was an enemy ship on a course to destroy our city, only to be brought down by our defenses instead."

The tram had halted at their stop, and Arthur took her hand so they would not be separated in the crush of departing passengers. "I see something else," he said, his voice close to her ear, his earlier frenzy fading. "If we fail and Germany wins, London will burn."

"Germany will not win," Helen retorted, voice catching in her throat. "The whole war is not upon your shoulders alone. You do realize that, don't you?"

"I do," he replied wearily. "Of course I do."

She studied him for a moment, not sure whether he truly did. Then she took his arm and they made their way home in the darkness of the blackout, unable to see more than a few paces ahead of them, trusting their memory of what was familiar only by daylight.

13

OCTOBER–DECEMBER 1916

LUCY

On the fourth Sunday of October, the Thornshire Canaries left London on an early train and rode nearly three hundred miles north to Blyth in southeast Northumberland for a match against the Blyth Spartan Ladies Football Club. Privately, Lucy and Daisy agreed that it was an excessively long way to travel for a football match, but the Blyth Spartans were developing a reputation as one of the finest women's teams in the United Kingdom. When they had issued their challenge—or sent their invitation, depending upon how one chose to read it—the majority of the Canaries had been eager to accept. When else might they have the opportunity to challenge themselves against so skilled a team? How could they hold their heads up if their fans thought they had declined out of fear? Each club would receive half the gate to donate to a charity of their choice, and given the Blyth Spartans' popularity, the match would likely draw a sellout crowd and raise a substantial amount of money. That was what finally persuaded Lucy the match was worth the trip, although she would have gone along with the team's decision either way.

She didn't want the boys to spend most of their day on a train, so over their protests, she left them behind with their grandmother, who promised them a day full of fun to make up for it. In the end, Lucy was glad she had spared her sons the ordeal of traveling for hours only to watch their mother's team receive a thorough drubbing from a far superior team at Croft Park before a virtually all-Northumberland crowd thousands strong. Lucy took some consolation in her single goal, midway through the second half, which at least put the Canaries on the board, but the 6–1 loss was humbling after so many victories. After the match, as they congratulated the victors with handshakes all around, Lucy was glad to learn that the Blyth Spartans were perfectly friendly girls, gracious and encouraging in victory, which took a bit of the sting out of the lopsided score. As they made their way back to the train station, even Marjorie, who took every loss as a personal affront, agreed that it had been a revelation to watch such talented women in action.

"I didn't know girls could play like that," confessed Peggy as they boarded the train. "And to think, at least half of them were canary girls, just like us."

"If *they* can play like that, *we* can," Marjorie declared, proud and undaunted. Daisy threw Lucy a doubtful look, but most of the other girls brightened, a few applauded, and someone called out, "Hear, hear!"

That burst of enthusiasm drained them of their last reserves of energy, so they settled into their seats, some nursing minor injuries or sore muscles, others opening the boxed dinners they had packed that morning. The return journey was long enough for them to cover every possible topic of conversation, including the state of the war, recent air raids, soaring costs and shortages of food, and news from their own beloved soldiers far away. Peggy practically glowed as she described a recent letter from her sweetheart, who had sent her a beautiful scarf he had purchased while on leave in Paris.

"My brothers don't send me anything," Marjorie grumbled cheerfully. "All our family's parcels go in the opposite direction."

For a while, their chatter circled around the best things to include in a parcel for their dear Tommies. Fish paste and tins of biscuits were a popular suggestion, along with any canned fruits one could find.

Clean, dry socks were always welcome, as were cigarettes. Several Canaries remembered then that they had brought their knitting along, and soon at least half the players, including Lucy, were hard at work on a sock or washcloth. Their friends looked on, either admiring their handiwork or silently congratulating themselves for forgetting their own needles and yarn so they could just sit back and relax.

Eventually the conversation turned to the new regulations the Ministry of Munitions had imposed at the arsenal, the upheaval they had caused, and the girls' general dissatisfaction that no workers had been consulted during the process. The Canaries were about evenly divided between those who disliked the shorter shifts and those who were relieved to work fewer hours, but no one was happy about the reduced wages. "Isn't that just like a boss?" said Mabel, scowling. "Tell us we're essential to the war effort, lure us in with good wages and promises, then cut costs without so much as a by your leave."

"But they haven't cut our hourly wage, just our hours," said Lucy. "That means they haven't cut costs at all. It's actually more expensive to run three shifts with more workers. They're not trying to save money. They're trying to reduce our exposure to the yellow powder."

Mabel turned her head aside, covered her mouth with the handkerchief she always carried, and broke into hoarse, wracking coughs. "All I know is that I take home less each week," she said in a strangled voice when she could speak again.

"Just wait until we shift out of the Danger Buildings," said Peggy ominously. "We'll take home even less, for twelve whole weeks."

A disgruntled murmur went up from the group, and Lucy had to admit that the reduced wages would be a hardship. Food was not scarce, but even household staples were becoming more expensive, seemingly by the day. She could eat her fill at the arsenal canteen two times a day, six days a week, but not so the rest of her family. Jamie and Simon were growing boys and were always hungry, and although Lucy's mother was a frugal shopper and a skilled cook, she found it a challenge to keep them all comfortably fed. In Brookfield, Alice and Eleanor had taken over her garden, and throughout the summer and early fall they had frequently sent boxes of produce, but with winter approaching, fresh fruits and vegetables would soon become scarce. Lucy was not sure what the family would do then, especially on reduced earnings.

"We need to talk to Purcell," Mabel declared. "'On Her Their Lives Depend,' they all say, when they want more girls to sign up. Now's his chance to prove what that means to him."

"He needs to look out for our health while he's at it," said Daisy, holding out her hands, palms down, displaying her lurid yellow skin. "Shells must be made, no argument there, but none of us should die for it."

"The lads risk their lives in the trenches," Mabel reminded them, as she often had before. "We risk ours in the Danger Building."

"Yes, but we should minimize that risk if we can," said Lucy.

"Did you hear about the canary girl from Eley's Cartridge Factory in Edmonton?" asked one of the fullbacks. "Her mother says they took fourteen pints of poisoned blood from her at the military hospital before she died."

"I thought she died because her liver, heart, and kidneys were so badly damaged," said another. "There was an autopsy. It was in the papers just last week."

"You're thinking of that poor girl from Leicester who worked at Standard Engineering," said Louise, shaking her head, her expression grim. "She left behind four children under seven. The inquest ruled it death by misadventure, but I don't know if that means the company will be obliged to provide anything for her family."

"All the more reason we need our weekly pay as it were promised us," said Mabel, looking around the group. "So we can put more aside in case of, what-you-call-it, misadventure."

"We need to have it out with the boss," said Marjorie, raising her voice to be heard. "How much money has the Purcell family made off this war, I wonder? They can send a bit more of it our way, for the risks we take."

Hearty cheers and applause met her words.

"It's settled, then," said Mabel firmly. "First thing tomorrow morning, I'll march right into his office and tell him how it's going to be."

"But how *is* it going to be?" asked Lucy, before they could get too carried away. "Shouldn't you work that bit out first? What is it you're going to ask for?"

"We should go through the union," said Daisy. "That's what it's there for. That's how the men get things done."

Mabel took a moment to mull it over before agreeing to try the union first. Since the Danger Building's representative was Mr. Vernon, Mabel told Daisy that they would speak to him together, since it was Daisy's idea and she worked for him in the Finishing Shop.

The following afternoon, Lucy was at her station inserting a detonator into a shell when Daisy returned from the quick conference Mr. Vernon had grudgingly allowed them. When Lucy caught her eye, her friend frowned and shook her head. Later, at football training, Daisy and Mabel reported that Mr. Vernon flatly refused to help them, for women were not and never would be allowed to join the union. "The union's purpose is to protect workers who've been here all along," he had fairly snarled. "It's not for *you* lot, who come in and work cheap and steal our jobs."

"We tried it your way," Mabel said to Daisy, not unkindly. "Now we go to Purcell directly."

"I wouldn't just burst into his office unless you want to get sacked," said Lucy. "Make an appointment. Show him we understand how business is done."

"April can get that appointment sorted for you," Marjorie declared, grinning mischievously and nudging her friend, who looked caught out. "She and his assistant secretary are great chums."

"Is that why he comes to our games?" exclaimed Peggy. "April, you silly sausage, you never said a word!"

April flushed pink beneath her yellow pallor. "There's nothing to say. He didn't come today, did he?"

"Oh, dearie," crooned Marjorie, feigning sympathy. "Did you have a lovers' tiff?"

"No! We didn't have a tiff. And we're not lovers." April glared at her grinning teammates, folded her arms, and turned her head to the window. "If that's how it's going to be, get the appointment yourselves."

It took some cajoling from them all, and a contrite apology from Marjorie, before April agreed to see what she could do. Two days later, she arrived at the shifting house just ahead of the warning bell with news that a meeting had been scheduled for that evening after their shift—but not with Mr. Purcell. "He can't see us. It's simply impossible," said April, tossing her coat into her cubby and quickly unbuttoning her dress. "But we got the next best thing—a meeting with Mrs. Purcell and Superintendent Carmichael."

Some of the girls groaned, disappointed, but Mabel, coughing, waved them to silence. "We'll convince the wife, she'll convince the husband," she said, scanning the crowd of workers pulling on their uniforms. "Lucy, you're coming too."

Lucy froze in the middle of tucking her mottled hair beneath her cap. "Me? Why?"

"Because Mrs. Purcell likes you, going back to when you asked her to organize our football kit. And the superintendent must like you too, since she started you in the Finishing Shop. You know how to talk to their kind. They'll listen to you."

Lucy busied herself with her cap and clogs to conceal her sudden anxiety. Mabel was assertive and loud; Lucy was shy. Mabel cared most about wages, while Lucy's greatest concern was health and safety. Mabel was a second mum to all the Danger Building workers, especially the working-class girls, and Lucy was Lady Prim, the footballer's wife. It was either the worst pairing of canary girls the Danger Building could have produced, or the most ingenious. "All right," Lucy said, a faint tremor in her voice. "But I can't stay late. I have to get home to my boys."

The hours passed, given over to monotonous, painstaking work, wretched yellow powder dusting her veil and gloves, and more glares than usual from Mr. Vernon, who surely suspected that the canary girls hadn't abandoned their cause despite the union's refusal to help them. Over lunch and tea, Lucy and Mabel discussed what they should say at the meeting, but there was no time to prepare as thoroughly as Lucy would have liked. She hoped that Mabel's confidence would see them through.

After her shift, Lucy washed carefully and changed into her own clothing, smoothing out the wrinkles as best she could. Then she combed out her hair and twisted it up into a Psyche knot, holding it in place with a small wooden comb Jamie had ingeniously carved and polished for her after he overheard her telling her mother that metal was forbidden. Mabel too had clearly taken care with her own appearance, for she wore a smart dark blue suit and hat and kid gloves, and had pinched some color into her cheeks, which had sunken slightly in the months Lucy had known her.

The other girls wished them good luck as they set off together for the canteen nearest the administration building. As they crossed the threshold, Mabel threw Lucy a sidelong look, eyebrows raised, which told her that Mabel too noticed that this canteen was brighter and cleaner than their own, perhaps due to the absence of yellow powder in the air.

Mrs. Purcell and Superintendent Carmichael were at a table for four near the coal fireplace on the far side of the room. Mrs. Purcell spotted them, brightened, and waved them over. "She's friendly enough *now*," Mabel said under her breath, pressing a fist to her mouth to stifle a cough. The superintendent regarded them with cool disapproval as they approached, her mouth pinched in a thin line.

They had all met before, so they exchanged greetings rather than introductions as they seated themselves, Mrs. Purcell and Superintendent Carmichael on one side of the table, Lucy and Mabel on the other, a pot of tea and four cups between them. Mrs. Purcell began by pouring them each a cup and thanking Mabel and Lucy for coming, almost as if she had requested the meeting rather than the other way around. "Mr. Purcell sends his regrets," she said, her smile turning down a bit as she set down the teapot and offered cream and sugar. "He's so frightfully busy that he hardly has time to catch his breath."

"I know what that feels like," said Mabel levelly.

They wasted no time in idle chat but got straight to it. Mabel began with the issue of wages, and after some back-and-forth, Superintendent Carmichael acknowledged that reducing the workers' hours caused enough financial difficulties without also expecting the Danger Building girls to accept reduced wages. The canary girls had not sought the twelve-week respites, and they ought to be rewarded for willingly accepting Danger Building work when so many others had declined. Mrs. Purcell proposed that the current Danger Building workers would continue to receive their current pay even during their required rotations out. The munitionettes who took their places would be paid Danger Building wages as long as their rotation lasted, but they would return to their regular pay when they returned to their original jobs.

Mabel and Lucy exchanged a look, pleasantly surprised, for they had not expected such reasonable terms. Mabel promptly accepted on behalf of all the Danger Building girls.

Next Lucy thanked the two supervisors for the new safety precautions, which were already showing a reduction in the workers' symptoms. The girls complained less frequently of nose, throat, and eye irritations, and absences related to illness were slowly declining. But Lucy had come prepared with a file of articles her brother George had collected from his colleagues in medicine and government, recent studies and observations about precautions other arsenals had tried, with encouraging results. Mrs. Purcell seemed very interested, and she promised to study them thoroughly and to pass on her recommendations to Mr. Purcell. "We both want—we *all* want—our workers to be safe and healthy," she said emphatically, and Lucy believed her.

Lucy was pleased with how well the meeting had gone, and she was just about to thank the supervisors and bid them goodbye when Mabel folded her hands on top of the table and peered sharply first at Superintendent Carmichael and then at Mrs. Purcell. "There's one more matter, just as important as the rest," she said, not sparing a glance for Lucy, who had no idea what she intended to say next. "We want some assurances we won't all be sacked soon as the war's over."

Superintendent Carmichael removed her glasses, sighed, and rubbed her eyes. "Now you've taken things too far," she said wearily, returning her glasses to the bridge of her nose. "It is not in my power, nor in Mrs. Purcell's, to make you any such promises."

"We're doing our bit for the war effort same as any soldier," Mabel said. "Mucking about with explosives, we risk our health and our lives every day, same as soldiers do."

"Indeed, and as we have established, you are well compensated for that," said the superintendent. "Excellent wages, nourishing meals, subsidized housing, recreational activities—"

"And it could all disappear overnight." Turning to Mrs. Purcell, Mabel added, "We've done everything the government and this arsenal has asked of us. We've proved ourselves to be trustworthy workers, skilled workers. We want assurances that those of us who want to stay on after the war will be allowed to. We won't be sacked with the snap of a finger just because we're women."

Mrs. Purcell nodded, her expression sympathetic and concerned, but the superintendent glared at Mabel and Lucy both, incredulous. "You would put yourselves before our veterans?" she asked querulously. "What do you expect us to tell our brave soldiers returning from the front—that *they* shall be unemployed so that *girls* may hold on to the jobs they were given *for the duration*?"

"Tell the soldiers whatever you like," said Mabel bluntly, sitting back in her chair with a scowl. "What am *I* going to tell my girls?"

"Mrs. Burridge, I respect your position and I would like nothing more than to be able to put your mind at ease," said Mrs. Purcell, cupping her hands around her teacup as if to warm them. "Unfortunately, no one, not even Mr. Purcell, can guarantee anyone a job at Thornshire Arsenal after the war."

Mabel frowned, puzzled. "What, not even the union men? Not even the returning soldiers?"

"Not even them." Mrs. Purcell sighed and folded her hands on the table. "We must assume that the need for munitions will plummet after the war, which I trust will end in victory for the Allies."

Lucy and Mabel both nodded. So did Superintendent Carmichael, but with a wary sidelong glance for Mrs. Purcell.

"You may know that this facility closed after the Panic of 1873." Mrs. Purcell looked around the table for confirmation, and they all nodded in reply. "For about forty years, this entire compound was shuttered, quiet, gathering dust behind locked gates. So you see, unlike other factories that were converted to war production for the duration, Thornshire Arsenal essentially has no pre-war purpose to revert to."

"What about sewing machines?" Mabel asked hoarsely, pausing to clear her throat. "Purcell Products made those here once, before the Panic. Why couldn't they again?"

"Perhaps they could. If not sewing machines, then something else." Mrs. Purcell spread her hands and sighed. "But I can't promise that Thornshire Arsenal will remain open in any fashion after the war. I can only assure you that your jobs are secure as long as the war lasts and the Allies need munitions."

"I see," said Mabel flatly. "Thank you for respecting us enough not to give us false hope."

"I understand that you need to plan for your livelihoods after the war," said Mrs. Purcell. "Any munitionette who decides it is in her best interest to take a job at another arsenal, one that is more likely to remain open in peacetime, shall be free to resign. I will personally see to it that she receives a leaving certificate."

"Mrs. Purcell," exclaimed Superintendent Carmichael. "Perhaps you don't understand the purpose of leaving certificates!"

"On the contrary, I understand perfectly. Their purpose is to limit turnover by discouraging workers from resigning, since no other factory shall be permitted to hire them without the current employer's consent."

"That's how it was explained to me in my training course," said Lucy. She liked Mrs. Purcell, and she didn't appreciate the scolding tone the superintendent used with her.

Mrs. Purcell thanked her with a nod before turning her attention to Mabel. "I know that's not the answer you'd hoped for, Mrs. Burridge, but is it acceptable, given the other concessions we've made?"

Mabel thought for a moment, and then extended her hand. "My girls will be happy enough when I explain to them about the wages, and I don't suppose many of them will be leaving to find jobs elsewhere."

Mrs. Purcell shook Mabel's hand, and then Lucy's, and eventually the superintendent grudgingly shook their hands too.

As Mabel predicted, the other canary girls were very pleased with the arrangements she and Lucy had made on their behalf, and not one decided to take Mrs. Purcell up on the offer of a leaving certificate and seek her fortune elsewhere. "The war might drag on for years yet," Marjorie pointed out, something they had all considered but no one hoped for, not even if it meant keeping their jobs. "Why give up good wages now, when for all we know, Thornshire might stay open somehow? Maybe we'll make sewing machines again, or motorcars—something, anything, that doesn't require yellow powder."

Many of the girls brightened at the idea, smiling and nodding and chiming in assent. Lucy nodded too, hopeful that the girls who wanted to keep their jobs after the war might have that opportunity. As for herself, she wanted nothing more than to return to her old life, her *real* life, minding her sons and keeping a warm, happy, loving home for them and for Daniel. She wanted holidays in Brookfield and cheerful family dinners around the long wooden table in her childhood home. She wanted all the people she loved best in the world to be safe and not too terribly far away. She would miss her new friends and football, but perhaps she could find a way to remain on the team.

But with the war dragging on and on, all harrowing misery on the Western Front and lonely apprehension at home, Lucy knew that production would continue at Thornshire Arsenal for the foreseeable future, perhaps for years—years she would spend without her beloved husband by her side.

Her worry and longing for Daniel stayed with her always, a dull ache in the back of her mind that she could never entirely ignore, even when she focused on work or football or her children. She devoured his letters as if she were starving for words, and she scanned the newspapers for any mention of the 17th Middlesex, which she knew was engaged in the fighting on the Somme, but little more than that.

To her indignation, some reporters remained as unduly critical of footballers as they had at the height of public criticism before the Footballers' Battalion was formed. Without provocation, reporters would drop scathing remarks about how the players were "finally" testing themselves in the Greater Game, reminding their readers, with blithe inaccuracy, that footballers had been reluctant to enlist compared to other sportsmen. But in the nearly two years that had passed since December 1914, the footballers themselves had begun refuting the snide commentary in the same papers that mocked them. Private Tim Coleman, the same Nottingham Forest forward who had accepted the surrender of two disgruntled Germans, called out a reporter who said that footballers had waited until they were "nearly forced to join" before enlisting. "The man who wrote that must be 'up the pole,'" Coleman had written, "as we have been on active service for nearly ten months, and have been in some very hot places, and have also taken our part in the great push." They had all joined up to do their bit and some had given all, he noted, listing the names of numerous players who had been killed or wounded. "The professional footballers have done their bit. I was hoping that this little 'tiff' would have been over for next season, but it still wags on."

More recently, Major Frank Buckley, the center halfback for Bradford City, had returned to England to recover from serious wounds to his arm, shoulder, and chest. Though he wore a sling and some shrapnel remained lodged in his lung, he attended charity football matches and sometimes spoke to the press during the interval, never failing to praise the men of the 17th Middlesex. "Even knowing them as I had before the war," he declared, "I had never realized that they would have devoted themselves to their duties so unflinchingly, or proved themselves such efficient soldiers. I am proud to have been one of the officers in such a battalion."

Major Buckley's words warmed Lucy's heart and gave her hope that critics would finally give the Footballers' Battalion the respect and admiration they deserved, especially considering the dangers and hardships they confronted on the battlefield.

On Monday evening in the last week of November, Lucy returned home at seven o'clock, just in time for a late supper with her mother and her sons, thanks to the shorter shifts. Afterward, she helped the boys with their schoolwork, saw them off to bed, and settled down in her favorite chair with a cup of tea and the newspaper. But the news was not relaxing in the least. While they had slept the previous night, several German airships had attacked the northeast coast of England, dropping bombs on Yorkshire and Durham. Two of the German airships had been brought down in flames by aeroplanes of the Royal Flying Corps and by naval guns, one zeppelin plummeting into the sea just off the coast, the other struggling to escape its pursuers only to split into two flaming masses over the Channel and tumble into the rough waters below.

The next day, Lucy was loading one last shell onto the trolley before heading to the canteen for the midday meal when the bell over the main gate pealed, barely audible above the clattering machines. She and Daisy exchanged a puzzled glance, for according to the clock on the wall, it was not yet noon. As the bell rang on, distant enough that most of the girls apparently did not hear it, Mr. Vernon hurried past, scowling.

"What is it?" Lucy called to him. "Is it a fire?"

"No, no, not a fire," he said over his shoulder, waving her off impatiently. "Carry on, carry on."

So she did, for another ten minutes more until her break. In the canteen, a few other girls mentioned hearing the bell and wondering what it was about, but it was not until later, on their way to their training field, that they learned from one of the lorry drivers that an aeroplane flying at a very high altitude, obscured by light mist, had dropped six bombs on West London near the Victoria railway station and Buckingham Palace, and over Belgravia as far as Brompton Road. Four people had been injured, one woman quite seriously, but there had been no deaths and only minor damage to buildings and streets. Yet the raid struck new terror into their hearts—an attack by a swift aeroplane in broad daylight upon the heart of London! It was unimaginable, and yet it had happened, and now they must fear death from above by day as well as by night.

A little more than a week later, on 9 December, Mabel called the team aside as they were entering the canteen for dinner. Her expression was so grim that Lucy knew at once that something terrible had happened, and cold apprehension settled over her.

"Our match on Sunday with the Barnbow Lasses in Leeds has been cancelled," Mabel said, then drew in a shuddering breath and lowered her voice. "Their captain telephoned me, and really, she said more than she should've. We know we're not supposed to talk about such things. It won't be in the papers. But the truth will come out so you might as well hear it now, but don't carry it any further."

"What is it, Mum?" asked Peggy. "Tell us."

"Two nights ago there was a terrible accident at the Barnbow Munitions Factory."

A gasp went up from the group. "What happened?" asked Marjorie.

"The night shift had just started. A four-and-a-half-inch shell had been set in place on the machine that revolves it so as to screw the fuse in tightly. Suddenly the shell exploded, setting off all the shells around it, and on and on . . ." Mabel grimaced and shook her head.

Lucy was almost afraid to know, but she asked, "Was anyone hurt?"

"Thirty-two women and girls killed, and three men, and many more injured," said Mabel. "The Lasses' captain said the night shift workers carried off the victims, carted away the debris, cleaned the blood off the floor, and got back to work within hours. Can you imagine? The munitionettes volunteered to go back to work in the same room where their fellow workers had died only hours before."

"They died for their country," said Marjorie. "And the girls who went back in—how fearless they are! We should all be so brave."

Murmuring agreement, they all went in for a somber dinner, knowing that it could have been Thornshire instead of Barnbow, that before the war was over it still could be Thornshire *and* Barnbow. And yet they could not unburden their aching hearts to the other canary girls, who were laughing and chatting at the tables all around them. Nor would they be able to confide in their families later at home. For the sake of morale and military secrecy, they would have to carry their secret grief and worry in their hearts for the duration.

Until then, Lucy knew there would be no heartrending eulogies in the national press for the Barnbow Lasses, only thirty-five obituaries in the local paper with loving descriptions of the lost daughter or sister or parent, all with the same date of death and a vague statement about their passing—"killed by accident," perhaps.

Someday, Lucy hoped, when the world was at peace and the truth could be told, their names would be remembered and etched into stone. Then all Britons would honor the Barnbow Lasses as heroes who had given their lives for their country on the munitions front.

14

December 1916–February 1917

APRIL

In mid-December, half of the munitionettes in the Danger Building, including April, were transferred to other jobs in the arsenal, scattered among different factories where they were obliged to learn new tasks and make new friends. Only half of the canary girls were put on their scheduled rotation out at a time so that the experienced workers who remained could train those rotating in. April was assigned to a gauging shop, where she sat at a long table with other munitionettes measuring fuses and other shell parts with an assortment of calipers to make sure they met specifications. The fuses that passed inspection were collected and sent on to assembly, but those that failed were scrapped. After a few days on the job, April's overlooker took her aside and remarked that her sharp eye and deft touch made her a natural for the work. "If you want to stay on after your rotation, I'll have a word with Superintendent Carmichael," she offered. "The small drop in wages is well worth it to avoid that yellow powder, or so my other girls say."

April thanked her and said she would consider it. She did prefer the gauging work, and her persistent cough and sore throat seemed to improve with every day she spent away from the yellow powder, but she missed her old friends and the camaraderie of the Filling Shop. She also felt a bit guilty knowing that another girl would have to take her place; it seemed cowardly to dump the burden of risk on someone else. Yet she longed to inhale deeply without feeling as if she were breathing through cheesecloth, and she wished the yellow tint would fade from her skin and for her hair to return to the bright, glossy flaxen locks she had once taken for granted. She wished for an end to the shocked stares and pitying glances from strangers when she was minding her own business in a shop or café. She wished Oliver could have seen her when she was pretty, before she became a canary girl, not that he had ever recoiled from her yellow-and-ginger hues as some lads did.

She and Oliver had been seeing more of each other ever since her shifts had been cut back to eight hours, allowing her more time in the evenings to spend as she pleased. Instead of hurrying to catch the tram back to the hostel after her shift and dropping exhausted into bed so she could do it all again the next day, she would walk with Oliver to the station so they could chat along the way, or they might stop to get a bite to eat, and one night they went to the theatre. As they got to know each other better, April learned that Oliver was kind and clever, with a dry sense of humor that took a bit of getting used to, but was actually quite funny. Marjorie thought Oliver was dull, but of course she would; she craved excitement and glamour, and nothing drew her eye like a lad in uniform. Oliver was equally unimpressed with Marjorie, but he didn't criticize her or air old grievances, which April thought spoke well of his character.

As the festive season approached, a light, feathery blanket of snow covered London, white and pure for a brief, beautiful few hours until it turned ash gray on the roofs and pavements and streets. Due to blackout restrictions, no bright lights illuminated parks and store windows for the third and gloomiest Christmas of the war, and only in their absence did April realize how much the season's merriment depended on their warm, jolly glow. The arsenal would not close for Christmas or Boxing Day, so April could not travel home to Carlisle to spend the holidays with her family. She hadn't made the trip the previous

Christmas either, but this year her brother Henry was coming home on leave, and she was terribly disappointed to miss his visit. He had promised to see her in London before he returned to France, which she was looking forward to as a bright spot in what she feared would be a lonely, melancholy season.

But even before her happy reunion with her brother, the holidays turned out to be merrier than she had expected. The girls who shared her hostel decorated the front room with greenery and ribbons, and their landlady prepared an especially delicious dinner on Christmas Eve, after which the girls exchanged small gifts, played games, and sang all their old favorite carols. On Christmas Day, Oliver invited her to go for a walk after work, but along the way he surprised her with a lovely dinner at a Barking restaurant known for its splendid plum pudding. He also surprised her with a gift—a warm, soft cashmere scarf and mittens in her favorite shade of Wedgwood blue. She in turn gave him a necktie, silk, with fine gold and black stripes. "My favorite team's colors," he remarked as he admired it.

"Thornshire Canary colors," she clarified.

He smiled at her, amused. "I believe that's what I said."

That was the first time he kissed her—not there in the restaurant, which would have been mortifying—but later, outside the station as she waited for her tram. Since then they had spent nearly every evening after work together, a couple of stolen hours when they could almost forget about the war, and yellow powder, and the rising cost of food, and everything except each other.

Sometimes April marveled that they had ever become fond of each other, not only because of their contentious beginning, but because they had come from such different places in life. If April had stayed in domestic service, and if Oliver had kept his job in his uncle's shop, they never would have met. And if by some strange chance they had, most people would have considered them unsuited for each other. Maybe some nosey parkers still did. But for young people in London, the war had overturned many customs they had once accepted without question. April and Oliver both worked in munitions at the same arsenal, and somehow that made them more equal. Old-fashioned ideas didn't matter anymore, not to them.

On a Friday evening in the middle of January, they were chatting about the Canaries' upcoming football match when April paused at a street corner and looked up at him, a bit shyly. "May I ask you a personal question?" she asked. "You don't have to answer if you don't want to."

The corner of his mouth turned wryly, and he held up his prosthetic hand. "You want to know how I came by this."

April shook her head. "That wasn't what I was going to ask. What I've been wondering is whether it—if it still hurts." She hesitated. "Since you brought it up, I *am* curious—but I wouldn't have asked."

He was silent for a moment. "It doesn't hurt, not in the way you mean," he eventually said. "The injury from the amputation healed long ago. But sometimes I'll feel a sharp pain or a dull ache, and I know my hand isn't there anymore, but it still hurts, sometimes for hours before it goes away." He shook his head, thrust his hands into his pockets, and lifted his chin to indicate the street ahead, and that they should cross. "You probably think I'm mad. I used to think so."

April matched his pace as they crossed the street. "I don't think you're mad."

"As for how it happened—" He grimaced. "I told you I enlisted with my friends from home. We knew our mums would likely try to stop us, so we lit out for London and joined up with the Fifth London Rifle Brigade. We were mobilized right away and sent out to a camp in Bisley for training. By November fifth, we were on our way to Le Havre, but we barely had time to look around before we were sent into the trenches around Ypres."

"That must have been dreadful."

"It was hell on earth." He fixed his eyes on the pavement straight ahead as they walked along. "No matter what you think you know about the trenches, everything you've heard, and everything you can imagine, is just a shadow of how truly wretched it was."

His voice had taken on a brittle edge. April was about to tell him that he needn't say anything more if he didn't want to, but he inhaled deeply and continued. "There were our trenches, and the German trenches, and no-man's-land in between—a nightmare landscape of barbed wire, wooden barriers, bomb craters, and the corpses of men and horses."

In some places where the 5th London Rifles had been posted, no-man's-land narrowed to only a few hundred yards. There, the British frontline trenches were so close to the enemy that on calm nights Oliver could hear the Germans talking or coughing, or smell their sausages cooking. The Germans also liked to sing, often together, as loudly as they could, patriotic songs to steel their hearts and annoy their enemies. Oliver had been subjected to so many performances of "Deutschland über Alles" that he could almost sing along from memory, not that he ever would. To drown out the Germans, and to show off a bit, he and his pals would sing "God Save the King" at the top of their lungs, as well as all their favorite Tommy songs like "Tipperary" and "Pack Up Your Troubles," but only the jolly ones. They were melancholy and homesick enough without sad, slow songs in a minor key deepening the gloom.

On the first Christmas Eve of the war, Oliver was shivering in his trench as snow drifted down upon the mostly frozen mud, dreaming of a warm fireside and his mother's roast goose with sage dressing. Suddenly he heard music on the air. On the other side of no-man's-land, the Germans were singing "O Tannenbaum" with some of the most splendid harmonies Oliver had heard outside of a church choir. When the German voices fell silent, Oliver looked around at his mates. "What do you say, boys?" he asked. "Shall we treat Fritz to a good old English carol?"

His friends grinned and agreed, so Oliver hummed the pitch and led them in a rousing version of "We Wish You a Merry Christmas." When they finished, they had to laugh at the applause they heard from the German lines. "Well done, Tommy!" someone bellowed from afar in a thick Prussian accent.

They traded carols back and forth for nearly an hour before one of their sentries announced that strange lights were appearing all along the German parapets. "They're bloody Christmas trees, with candles burnin' on 'em," he added a few minutes later, incredulous.

Curious and skeptical, but not wanting to risk their necks for a quick look, Oliver and his mates took turns with the wooden periscopes and confirmed the sentry's wildly unbelievable claim. Then, all at once, a clear, warm tenor voice drifted over the ruined earth, a melody Oliver knew well but with words unfamiliar.

They all paused to listen. "Stille Nacht," one of his pals murmured. "Silent Night."

They fell silent, listening. Oliver felt something loosen in his chest, and then a pang of longing so intense he almost couldn't breathe.

"He's singing bloody well, you know," someone remarked. They all shushed him.

"He's climbing on their parapet," the sentry cried, incredulous. "What the devil is he on about?"

"He's comin' towards us," the lad at the periscope shouted. "Still singing!"

"Is he armed?" Oliver called back. This Fritz either put a lot of faith in the Christmas spirit of peace and goodwill, or he was entirely mad. Either way, Oliver had to give him credit for courage.

"I don't see a weapon. His hands are raised." A pause. "Now he's just standing there in the middle of no-man's-land, singing."

"More Germans are climbing up on their parapets," the sentry called. "They're coming over. Hands up, no rifles."

Sure enough, Oliver heard more German voices joining in with the first, some lower in pitch, some trembling, all singing words that lingered on the verge of his understanding.

"What should we do?" a younger fellow asked, alarmed. "Should we shoot?"

"Don't be a bloody fool," the sergeant bellowed, storming toward them from an adjacent trench. "We can't shoot unarmed men!"

"Do they want to surrender?" Oliver tapped his mate on the shoulder and gestured for a turn at the periscope. It took a moment of peering into the darkness for his eyes to adjust. Between the Germans' Christmas trees and scattered ground fires burning, no-man's-land was lit up just enough for him to glimpse eight to ten German soldiers gathered in a clearing of sorts between the twisted rows of barbed wire. Even as he watched, heart thudding, palms clammy in his gloves, another man joined them, then two more. A commotion some distance away in his own trench alerted him that something was going on farther down the line, but he heard no whistles or rifle fire, so he kept his gaze fixed on the Germans. Soon thereafter, shouts of warning and consternation in English told him that British soldiers had set down their weapons and were trudging out to meet the Germans.

Oliver watched, astonished, as men from both sides, including the 5th London Rifles, including some of his own childhood pals, climbed from the trenches and walked out to meet the enemy in no-man's-land. They shook hands, exchanged names, greeted one another like long-lost school chums—these men who had been trying to kill one another only a few hours before.

Oliver watched only a little while longer before he went out to join them.

"We traded chocolates and cigarettes, and shared photos of our families and sweethearts," Oliver told April. They had nearly reached the station, but their steps had slowed until they finally halted altogether on the side of the pavement, Oliver reminiscing, April hanging on every word. "Many of them spoke rather good English, and some had even lived and worked in Britain before the war. I talked to one chap who used to be the head waiter at the Great Central Hotel in Marylebone. We shook hands, and wished each other a merry Christmas, and soon we were chatting like old friends."

"Goodness," April murmured, her breath emerging faint and white on the cold night air.

"All around me, I saw Tommy and Fritz trading souvenirs and food, a Scotsman in a kilt lighting the cigarette of a German in a *Pickelhaube*, Germans and Englishmen laughing and joking as they swapped helmets and posed for photographs together." Oliver paused and shook his head, lost in the memory. "Then someone—a player from one of the battalion teams, probably—brought out a football, and most of us joined in to play."

"You had a football match in no-man's-land?"

"It was more of a kickaround at first. The ground was cratered and uneven, and neither wide nor long enough for a proper pitch. But soon we fell into a friendly match, Tommy against Fritz, of course—"

"Of course."

"The chaplain from the East Lancashire Regiment was the official. We used our helmets to mark the goals, but we didn't keep score. At least I didn't." Oliver managed a thin smile. "Afterward, we passed around a bottle of rum one of our lieutenants had kindly provided, and I struck up a conversation with a German artilleryman. He said he hated the war, and all he wanted to do was live through it and go home to his wife and daughter. In parting, he said something I'll never forget. 'Tonight we have peace. Tomorrow, you fight for your country, I fight for mine. Good luck.' I wished him luck too. Maybe another hour passed, and then, a few men at a time, we all started drifting reluctantly back to our trenches."

April shook her head, marveling. If she hadn't heard it from Oliver himself, she wouldn't have believed it.

"Afterward, I heard that there were truces all up and down the line," said Oliver. "Some of them lasted for days, until furious officers shut them down. Fraternization is never encouraged, but unofficial truces . . ." He shook his head. "That just isn't done, not at Christmas, not ever."

April imagined not. "How long did your truce last?"

"Until midmorning, Boxing Day." Oliver glanced at the station, then offered April a rueful smile. "You're going to miss your train."

"I'll wait for the next one. What happened on Boxing Day?"

"A German soldier was on their parapet, carrying a bucket—I don't know what for. One of our snipers took him down. Not even a minute passed before the Germans hit us with everything they had. It was the worst barrage I'd ever been through." He inhaled, two deep, sharp breaths. "We just kept our heads down in the trenches and waited for it to end, and I for one was cursing whichever one of our boys took that shot. I didn't expect the truce to end the war, but what I wouldn't have given for one more day of peace." He paused and shook his head. "A few hours later, I had to go up top for a bit. I almost made it back, almost, but my arm was over my head, my hand grasping the ladder as I descended—and that's when I was hit."

"Oh, Oliver, no."

"My friends wrapped up the wound so I didn't bleed to death, but I was unconscious by the time the stretcher bearers arrived and got me back to the field hospital. They tried to save the hand, but it had got infected and had to go."

She reached out to touch his shoulder. "I'm so sorry."

"Yes, bad luck, wasn't it? But not all bad. I lived. I was sent back to England to recover and then I was discharged. I went home, started working in my uncle's shop again, but I couldn't bear the pitying looks and false cheer of people coming in and out, gawping at my limb and shaking their heads and clearly thinking my life was ruined, and that I'd be a burden to my poor parents for the rest of my life."

"But that's nonsense," April protested. "There you were, *working at your job*, and they thought you'd be a burden?"

"They assumed my uncle gave me the job out of pity, and that I wasn't really doing much."

"But you had the same job *before* the war!"

"That's also true."

"It doesn't make any sense!"

"Now you understand why I had to leave. I was angry and bitter, and I wanted to prove that I could still contribute to the war effort even if I couldn't hold a rifle. So I took the job at Thornshire Arsenal, and now I help make the weapons that the Allies need to win the war." He frowned. "Of course, I can't ever forget that our victory means the crushing defeat of the German soldiers I befriended at Christmas, many of whom didn't want to be there either. Then it occurs to me that they all might have been killed by now, and I wonder what the whole point of it was. If we common soldiers had just refused to continue fighting after our truce, could the war have ended in December 1914?"

"Could it have, do you think?"

He paused. "Doubt it. If we hadn't started up again on our own, like that sniper did on our part of the line, our officer would have commanded it, and if we disobeyed orders, they would have shot us themselves."

April shuddered. Oliver saw it, and he took her hand. "Sorry you asked?"

"Never, but I wondered—"

"What?" he prompted, when she didn't continue.

"You said you had to leave the trench. Why? Why would your officers order your battalion over the top when the Germans were bombing you so furiously?"

Oliver hesitated. "It wasn't an order, exactly, and it wasn't all of us. Just me."

She shook her head, uncomprehending.

"One of my pals from home, one of my best pals, had got it in his head to take out one of their big guns. He needed a better angle to take the shot, so he climbed out and crawled forward, but we could still see the soles of his boots hanging over the edge of the parapet when he was hit. We shouted at him to come back, and when he didn't, I knew he was either unconscious or dead. So I climbed up to haul him back in, and the other lads caught him as I handed him down. I was descending back into the trench when I got shot. The shattered rung of the ladder came away in my fist."

April shuddered, imagining it all too vividly. "Was your pal still alive?"

"Thankfully, yes. He'd been struck in the shoulder, but he survived. He healed up and returned to the battalion a few months later. He's still there, last I heard."

"You saved his life," April said in wonder. "Oliver, you're a hero. They should've given you a medal for—"

Her voice failed her as suddenly a brilliant, florid light illuminated Oliver's face and the passersby and the station behind them, for a moment transforming night into day. Before the strange light faded, a terrible, booming roar struck them, an unearthly growl that made everything around them—the ground, the buildings, the bare-limbed trees, everything—tremble ominously.

Instinctively, Oliver pulled her closer, shielding her with his torso as he looked around for the source of the explosion. April gasped and stared at the sky, horrified, as an enormous, lurid fireball rose above them, too distant for them to feel its heat, a churning, chaotic mass of colors—violet, indigo, blue, green, yellow, orange, and scarlet—eddying and swirling like a violent sunset. Terrified, April pressed her face against Oliver's chest, blood pounding in her ears, but when she dared peer back up to the sky and saw the fireball still churning, she realized that it couldn't have been an air raid. There had been only one explosion, not a series of dropped bombs, and there was no sign of a zeppelin or aeroplane above, no frantic sweep of searchlights from the ground.

April and Oliver exchanged a bleak look, as they both realized with sinking dread what must have caused the explosion.

Oliver took her hand again, and together they headed toward the bank of the Thames. "It wasn't Thornshire," he said as they broke into a run. "It's too far away, and in the wrong direction."

April nodded agreement. If it had been Thornshire, the pressure wave would have knocked them off their feet, stunned, as shattered glass and rubble fell all around them.

When they reached the riverbank, they saw a red, fevered glow to the west beneath a dark cloud of churning smoke only two or three miles away, and the harrowing reflection in the Thames of a building, perhaps even a compound, engulfed in flames. They heard the distant wail of fire engines and ambulances, and April smelled traces of ash and acrid chemicals, or she imagined she did.

"Good Lord," Oliver murmured close to her ear. "That's north of the river. Silvertown."

April's heart thudded. The chemical munitions factory in the East End, a compound surrounded by densely populated working-class neighborhoods. "Brunner Mond?"

Accustomed to secrecy, he nodded rather than echo the name as other horrified spectators clustered nearby, some of them exclaiming loudly, others with hands pressed to their mouths and faces pale with shock. "It's directly opposite the Royal Arsenal at Woolwich," Oliver said, his breath stirring the fine, loose strands of hair that had come loose from her coiled braid. "Those flames—if they're hot enough, if they reach far enough, they could set off Woolwich too."

April felt her chest constricting. How many workers had been killed? How many injured? "Can we do anything to help?" she asked as tears gathered in her eyes. "What can we do?"

He convinced her that the best they could do for now was to stay out of the way and let the emergency crews do their jobs unimpeded. Even if bystanders wanted to be useful, they likely wouldn't be allowed to approach the site. The risk of another explosion was too great, and the blocks around both Brunner Mond and the Royal Arsenal were probably being evacuated at that very moment. April nodded bleakly, and she held tightly to Oliver's hand as he led her away from the river and back to the station. As they worked their way through the crowd, amid the mingled sobs and curses and exclamations, she overheard hot, furious accusations that the Germans surely were to blame, German spies, saboteurs. But as a canary girl, April knew it was far more likely to have been a cruel, terrible accident, just like the one that had claimed the lives of the munitionettes in Barnbow.

"We'll learn more in the days to come," Oliver told her before they parted at the station. "The censors will restrain the press as much as they can, but the government can't pretend this didn't happen, not as it did with Barnbow, not when nearly the whole of London saw it or felt it."

Oliver's words proved true. The next day, the Ministry of Munitions issued a statement to the press reporting that an explosion had occurred "at a munitions factory in the neighbourhood of London," an astonishing understatement, as thousands of eyewitnesses had already confirmed that only a huge, rubble-strewn crater remained where the Brunner Mond chemical works in Silvertown had once stood. "It is feared," the press statement continued, "that the explosion was attended by considerable loss of life and damage to property." The official casualty report stated that sixty-nine had been killed and four hundred injured, but everyone who had seen the devastation declared that this was surely an undercount. Would-be gawpers could not judge for themselves, for the police and the military had closed off the area, and no one was allowed past the guards without a written order from the Ministry of Munitions or the War Office.

In the days that followed, April learned more from word of mouth than from the censored newspapers. Hundreds of homes and flats had been utterly destroyed within a square mile around the crater, and about two thousand people had been rendered homeless. The massive fireball had been seen at a distance of twenty-five miles, and the explosion had been heard eighty miles away. And despite the widely held belief that German saboteurs were responsible, the Thornshire Canaries heard from the Woolwich football club that it was entirely accidental. A fire had broken out on the top floor of the factory. In the few minutes before the flames reached the explosives and chemicals, the chief chemist had managed to get his workers out, but he himself had perished.

Five tons of TNT had exploded, sending a jet of crimson flame across the river toward Woolwich, igniting a gasometer near the Royal Arsenal. At first, the managers feared that the Royal Arsenal too might be destroyed. Many of the workers panicked and fled, but eventually the blaze across the river had subsided and the danger had passed.

Oliver had been right again when he said that the government would not be able to ignore the accident on the grounds of military secrecy. It was impossible to do so when so many men and women were going about the East End with their heads bandaged and arms in slings, when whole streets were lined with the charred shells of houses, when entire families who hadn't had much to begin with had lost everything. Over time, the victims' pleas for government assistance grew into more widespread public outrage. About a month after the disaster, the Ministry of Munitions issued another statement that "without admitting any liability," it would "pay reasonable claims for damage to property and personal injuries caused by the explosion."

April figured that was the least the ministry could do. It hadn't caused the explosion on purpose, but it was responsible, and it ought to make it right.

For weeks following the accident, it seemed to April that everyone at Thornshire was on edge, following safety precautions to the letter and snapping at coworkers whose carelessness would have earned them only a frown and a rebuke before. Several munitionettes quit, including Louise, the Canaries' best fullback, leaving them with only enough players for a side, with no substitutes. Occasionally they found another girl willing to step in for a match or two, but no one would commit over the long term.

Mabel could no longer play, even though she wanted to. It was wrenching to watch her health visibly declining week by week. She had resisted her rotation out of the Danger Building even after striking the deal with management to keep their wages as they were, but she had grudgingly complied when Superintendent Carmichael decreed that it was rotate out of the Danger Building or out of the arsenal altogether. But although April and the other canary girls had found their symptoms ebbing within a week of leaving the Danger Building, Mabel had shown no such improvement. Privately her teammates speculated that it could be because she was older, or because she had started in munitions before any of them. Whatever the reason, they trusted that she would recover eventually, maybe just at a slower pace. Peggy had already resolved to use every argument she could think of to persuade her mother not to return to the Danger Building after her rotation ended, but she knew it wouldn't be easy.

Football and friendships—and that included Oliver's—sustained April when the news of the war and hardships closer to home steadily worsened. Food costs continued to climb, and ordinary staples like bread, sugar, and potatoes had become scarce. The arsenal canteen did not reduce their services, probably due to Ministry of Munitions regulations, and her hostel increased fees only slightly, so April knew she had it better than most, especially compared to her friends with children to feed.

Then one morning in early February, Oliver was so eager to speak with her that she found him pacing on the edge of the park closest to the train station rather than sitting on a bench near the center or strolling the gravel paths. "Did you hear about the Americans?" he greeted her as Marjorie rolled her eyes and continued on alone to the arsenal.

"No," said April, studying him, curious. His face was lit up with eagerness, and he was smiling broadly rather than only turning up a corner of his mouth. "What happened?"

"They've severed diplomatic relations with Germany."

"What?" Even after hundreds of American citizens had perished in the sinking of the RMS *Lusitania*, the United States had not gone so far. "Are you sure? From the very start the Americans have insisted they're going to stay neutral."

"Absolutely sure. Everything changed after the German chancellor announced that Germany is scrapping its restrictions on submarine warfare. Passenger liners and merchant ships are fair game to them again, regardless of nationality."

April felt a wave of dread. It was already difficult enough for ships carrying food and supplies to reach the British Isles through the German blockade. If German U-boats resumed their attacks on any ship that crossed their path, what would become of the British people? Would any food get through? How many more innocent civilians would die, at sea and in Britain? "What does this mean?" April asked. "Severing diplomatic relations—what does that actually mean for us, for Britain?"

"That isn't entirely clear, not yet," said Oliver, his enthusiasm dimming only a trifle. "Yesterday President Wilson spoke to the American Congress for more than two hours. He said that although the United States doesn't desire any hostile conflict with Germany, if the Germans destroy any American ships, they will find themselves in a state of war."

"So the Americans haven't joined the war on the side of the Allies yet."

"No," Oliver said as they turned and walked toward the arsenal together. "But I do think they've taken one step closer to it."

April nodded, hardly daring to hope. She had never been the sort to look to someone else to come to her rescue, but after two and a half hard years of war, it was profoundly cheering to think that the Americans might soon join them in the fight. The United States was rich in resources, and they had their own scores to settle against Germany. Having the Yanks in it at long last might make all the difference.

15

MARCH–MAY 1917

HELEN

In the weeks following the disaster at Brunner Mond, Helen divided her time between implementing the new safety protocols and reassuring anxious munitionettes. It was a fine line she walked, wanting to comfort and yet determined to be scrupulously honest.

The new safety measures had been designed to mitigate or prevent TNT poisoning, and Helen was relieved to see steady improvements in the health of most of her canary girls. But most was not all. Some workers reported that their symptoms had improved only slightly by the end of their twelve-week rotation out of the Danger Building, and inexplicably, a few felt even worse. Helen could not in good conscience send those canary girls back to their former posts. She endured numerous strained discussions on the subject with Superintendent Carmichael, who lectured her about the blows to discipline and morale that would ensue if they did not abide by the agreement they had struck with the canary girls' representatives. "We cannot appear to favor one worker over another," the superintendent warned.

"Declining to send a clearly unwell worker into a situation that is likely to make her sicker isn't playing favorites," Helen replied. "It's playing it safe." They had a long list of volunteers willing to transfer into the Danger Building to give the canary girls a respite, with the understanding that they would return to their original jobs in twelve weeks and would not be required to volunteer again. Fear of the yellow powder had diminished as fewer workers had become seriously ill. Now what worried the munitionettes most was the perpetual threat of an accidental explosion that could level the entire arsenal. But the new safety protocols did not address that problem. Aside from reminding the workers to follow the safety precautions and touring the factories to reassure herself that they were doing so, there was frustratingly little Helen could do to prevent a devastating accident.

As the war dragged on, the unsettling dread that all her efforts as welfare supervisor might ultimately prove worthless troubled Helen deeply—not enough to wake her in the middle of the night, shaking and breathless, haunted by nightmares as Arthur often was, but enough to make her brood over mistakes and resolve not to repeat them. She had already absorbed the humbling lesson that sometimes, despite her good intentions, she simply failed to understand the realities of working women's lives.

Earlier in the year, when the poisonous effects of TNT were first confirmed, if a worker felt unwell, Helen had sent her home to rest for a few days, until a doctor cleared her to return. Helen had considered this to be a sensible, generous, compassionate approach, best for the worker and best for the arsenal. To her astonishment, workers had simply stopped showing up for their scheduled weekly physicals. Overlookers had observed canary girls who were obviously feeling unwell toiling away until they nearly fainted at their stations, insisting that they felt perfectly fine even as they were being carried to the infirmary. Their behavior utterly bewildered Helen until Lucy Dempsey enlightened her. For many canary girls, Lucy had explained privately, several unpaid days off was a hardship, not a respite, and finding a doctor to clear them to return to work was difficult if not impossible. If they felt unwell, a brief lie-down in the shifting house would usually suffice, and failing that, taking the rest of the day off was often enough to restore them. But three days without wages felt like a punishment for weakness, so they skipped their weekly physicals rather than lie to their doctors about their symptoms, pushing through the pain and misery and hoping no one noticed.

Helen was grateful for Lucy's insight and changed their policies accordingly. Henceforth, workers would be permitted to return to work after a single day's rest, and either their own doctor or arsenal infirmary staff could clear them.

There were some canary girls whom neither Helen nor Superintendent Carmichael could justify returning to the Danger Building, and unfortunately, one of their most skilled and diligent workers was one of them. In February, although Mabel Burridge hadn't worked with the yellow powder in two months, her breathing remained so labored that Helen ordered her to take an entire week off. Upon her return, she seemed fine at first, but in early March, after a routine physical, the arsenal's head physician insisted she take a fortnight's leave. He extended that by another week when she again failed her exam. Eventually the arsenal infirmary cleared Mabel to return to a less dangerous job, but that did not satisfy her. Coughing and wheezing behind her veil, her cheeks gaunt and eyes shadowed, she showed up at Helen's office at least once a week to request a transfer back to the Filling Shop.

Helen could almost admire her dogged persistence if it wasn't so misplaced. "Every job at Thornshire Arsenal is essential war work," she reminded Mabel in late March when she stormed into Helen's office to have it out. "Inspecting detonators is as crucial to the process as filling shells, and you receive exactly the same pay. Why are you so determined to get back into the Danger Building when it's obviously taking a toll on your health?"

"My girls need me," Mabel said, sunken eyes glittering with determination. "No one looks out for them the way I do. They're good girls, hard workers, but sometimes they get distracted, chatting and teasing and such, and they need me to keep them on task."

"I admire your commitment, but that's what the overlookers are for."

"The overlookers can't be everywhere at once, and they don't know the job the way I do."

Helen paused. "No, I don't suppose they do."

"I'm not a boastful woman, Mrs. Purcell, so I'm just stating the facts when I say I've lost count of how many accidents I've prevented at the last moment because I know where and when to keep watch. Even with this cough, I'm more useful to you in the Filling Shop than anywhere else."

Helen didn't doubt it. Mrs. Burridge would surely make an excellent overseer—but that was no solution, since any job in the Danger Building would expose her to the yellow powder. Kindly but firmly, she told Mrs. Burridge that she could return to the Filling Shop when the arsenal's head physician cleared her for it, and not one day sooner.

Mrs. Burridge might have argued the point longer except that she was due back at her station, so she departed grumbling and glowering. Alone again, Helen sank back into her chair to recover her composure before moving on to the next crisis. She reminded herself that Mrs. Burridge's own daughter had thanked her for standing up to her mother. "She can be relentless, I know," Peggy had murmured, glancing over her shoulder for eavesdroppers. "I understand that she's finished in the Danger Building, and that's the way it has to be. She won't listen to me, but she has to listen to you."

Helen wasn't so sure. Although she could offer suggestions, the hiring, firing, and scheduling of workers was actually Superintendent Carmichael's responsibility. Helen suspected Mrs. Burridge had pled her case to the superintendent, and when that failed, she had turned to Helen, whom she perceived as less of a stickler for the rules. There was some truth to that, but Mrs. Burridge underestimated Helen's determination to protect the munitionettes from harm as much as she possibly could, in circumstances that required them to take extraordinary risks every day.

Helen trusted that her work on behalf of the munitionettes truly benefitted them and was appreciated, but rather than taking pride in her accomplishments, she felt deep humility, both for how little she had understood working-class women's lives before Thornshire, and for how much she had yet to learn. She could not advocate for them effectively if they did not trust her, and trust grew out of mutual respect and regard. Touring each building once a week and attending football matches was a good start, but she could do more. So Helen began taking her tea and dinner in different canteens, hoping the workers would find her cordial and approachable, but not intrusive. She attended the munitionettes' musicales and theatricals, and frequently she joined the Lunchtime Knitters Club to make warm socks and scarves for the soldiers. Sometimes she convinced Arthur to accompany her to a recital or a football training, since he would not have to leave the arsenal grounds to attend, but usually he could not spare the time.

Then in early April came astounding, glorious news: In an address to a joint session of Congress, President Woodrow Wilson declared that "neutrality is no longer feasible or desirable where the peace of the world is involved." He urged Congress to "formally accept the status of belligerent which has thus been thrust upon it." Three days later, after intense debate, both the Senate and the House passed a war resolution by large majorities.

The United States had declared war on Germany.

Helen fervently hoped this meant that the Americans would soon send desperately needed supplies to Britain, including food, munitions, and raw materials. Arthur expected all of that and troops as well. Helen would be grateful for anything that might relieve Arthur of the great burden of responsibility he carried, but the thought of tens of thousands of fresh troops joining the fight on the side of the Allies filled her with the first true, sustained hope she had felt since the outbreak of war.

But her joy was fleeting.

One evening in mid-April, Helen and Arthur had just sat down to supper after a particularly exhausting day when the housekeeper bustled in to tell Helen that her sister Penelope was on the telephone. Helen hurried to the sitting room to answer, trepidation stirring, for her mother and sisters never telephoned so late in the middle of the week. "Penny, darling," she said breathlessly into the mouthpiece, clutching the candlestick stem in one hand and holding the receiver to her ear with the other. "How are you? Is Mother well?"

"Mother is fine, and so are Daphne and I." Yet Penelope's lovely voice, low and rich, trembled with grief. "But as you guessed, I'm afraid I have dreadful news."

Helen braced herself. "Tell me."

"Did you hear about the *Lapland*?"

"The ship?" asked Helen, bewildered. "I only know what was in the papers this morning. She was carrying Canadian troops to Liverpool when she struck a mine off the coast, but she made it into port. I understand there were minimal injuries."

"And one fatality. Roland—Mr. Aylesworth—"

"What? Oh, Penny, no!" Helen sank into a chair, stunned. "How can this be? Mr. Aylesworth, on a troop transport—"

"He was returning from diplomatic missions to Washington and Ottawa. He was below deck when the ship struck the mine, and some debris—well, he was struck rather badly, you see, and they say it was quick and he never felt any pain, but they *would* say that, wouldn't they—"

"Oh, my dear Penny." Helen inhaled shakily, closing her eyes. "Poor Margaret. How is she bearing up?"

"Bravely, of course. Uncomplaining. You know Margaret. Her only thought is for the children. Hestia has been told that her papa won't be coming home, but she asks for him anyway. Poor Hugh, he's still a babe in arms, he'll have no memory of his papa, and Roland did dote on him so—"

"I'll come to you at once," said Helen, standing. "Shall I come to you?"

"No, no, that's not necessary. You have Arthur to look after, and your war work." Penelope took a deep, shuddering breath. "Margaret's mother and sisters-in-law are here, and I'm here. Between the five of us, Margaret and the children shall be well looked after."

"But who is looking after you, dearest Penny?"

"I'll be fine. I have to be, for Margaret. I—I do love her so, Helen."

"I know you do, Penny." Helen felt tears gathering. "Are you sure I shouldn't come?"

"Thank you, but not now. Roland's family arrives tomorrow. They want only a small funeral in the family chapel, on account of the war—so many other families are grieving, you understand—"

"I do," Helen assured her. "But if you decide you need me, say the word and I'll be on the next train."

Their hearts heavy, they bade each other loving farewells and disconnected.

A fortnight passed. Helen wrote to Margaret to express their condolences, and she telephoned Penelope every few days to offer whatever loving support she could. Helen had met Mr. Aylesworth only once, but his death stunned her more than she would have expected. Perhaps she had assumed that, as a diplomat rather than an officer, he would never be truly in peril. How foolish her assumption seemed now, and the realization made her fear for her own husband all the more.

On Friday, 20 April 1917, London joyfully celebrated "America Day" in honor of the new alliance between the United States and the Allies, marking the occasion with parades, speeches, and solemn religious observances at St. Paul's Cathedral. Arthur had received tickets to the service, but he could not spare the time away from the arsenal, so he had declined the seats for both of them. Helen tried to conceal her disappointment, as well as her frustration that Arthur had not consulted her before sending their regrets. How inspiring and comforting it would have been to share a moment of hope and optimism with their fellow Britons and their new American allies!

Arthur knew her too well, and he was not fooled by her feigned disinterest. "I should have accepted the tickets. You could have taken your mother or one of your sisters." He frowned regretfully, interlacing his fingers through hers. "I'll make it up to you after the war, darling. I promise."

She spotted her opportunity and seized it. "You don't have to wait until after the war," she said. "You can make it up to me tomorrow.

The Thornshire Canaries are playing at Stratford, and the weather is supposed to be glorious. Come with me to the match, and we shall call it even."

"Darling, I can't take the time—"

"Yes, you can, and you should," she interrupted emphatically. "This ceaseless toil is not good for you, and it can't be good for the arsenal to have the head man perpetually exhausted."

"The men in the trenches don't get a day off."

"As a matter of fact, they do!" Helen exclaimed. "It's called *leave*. And even when the troops aren't on leave, they are rotated away from the front occasionally so they may rest, while fresh troops replace them. As Thornshire's welfare supervisor, I strongly recommend that you take leave tomorrow to rest and spend time with your devoted wife, who has been working very hard and deserves a bit of fun."

His lengthy pause told her he was wavering. "I don't believe your authority as welfare supervisor extends to the boss."

"Well, it should." She leaned forward, smiling, and clasped his hand in both of hers. "What do you say, darling? The war shan't be lost if you spend one Sunday away from the arsenal. Think of the boost to morale when the players see that the boss himself has come to cheer them on."

Somehow, something she said convinced him, and he agreed to go. Delighted, she kissed him deeply and promised he would not regret it.

Helen arranged for a late breakfast the next morning, and, just to be sure, she turned off the alarm clock so Arthur would sleep in. After breakfast, she invited him to accompany her on her daily walk, half afraid that he would suggest something mad like going into the office for a few hours and meeting her at the Globe at the start of the match. It was a lovely, perfect May morning, and although Arthur did not entirely lose his air of haggard distraction as they strolled through Hyde Park, the fresh air and sunshine clearly did him good.

Even on a Sunday, the train to Stratford was crowded with soldiers in khaki and workers, mostly women, traveling between factories and home. Helen and Arthur arrived at the Upstart Crows' stadium with a half hour to spare, and she was pleased to see how impressed he was with the size of the crowd already seated, at least seven thousand. A great many Thornshire workers, men and women alike, called out cheerful greetings to the Purcells as they found seats in the center, two rows back. They settled in and turned their attention to the pitch, where the Thornshire Canaries and the Handley Page Girls were running drills and warming up. The Handley Page Girls, from the Handley Page aircraft manufacturer in Cricklewood, wore charcoal gray pinafores over light blue, collarless blouses, allowing more freedom of movement than most of the skirted uniforms Helen had seen. None of them bore the yellow skin of canary girls, whereas half the Thornshire team did.

Helen was studying the Girls, assessing their skills and strength, when Arthur said, "I assume you know most of our players."

"Oh, yes, all of them," she replied, turning her attention to their own team. "Some better than others, of course." Beginning with Marjorie in the goal, she gave him the name and a brief description of each player, but only when she came to the forwards did she understand why their drills had seemed a bit off. "They have only ten players on the pitch, and no one in reserve," she said, bewildered. Even Mabel was absent. She hadn't played in ages, but even when she was on sick leave, she had attended the matches and coached from the sidelines.

"Who's missing?" Arthur asked, but she wasn't quite sure. A few of the canary girls on the team had become too ill to play, and the roster of substitutes was short and inconsistent. Helen had never known the Canaries not to take the pitch with a full side, and she wasn't sure what would happen if their eleventh player failed to show. Would they play one woman down, a decided disadvantage? Would they be required to forfeit? That would greatly displease the thousands of fans who had bought tickets and expected an exciting match. If they demanded refunds, it would be quite a blow to the charity to whom the gate had been promised.

From a distance, the Canaries seemed increasingly uneasy with their numbers, and with fifteen minutes to go, they gathered in their defending third and appeared to be discussing their options. On several occasions, a few players glanced up into the stands directly at Helen and Arthur before returning to their debate. "They know you're here," Helen murmured. "They're proud, and they don't want to disappoint you. They'd rather lose badly with only ten than not dare to make the attempt."

"I don't think they're looking at me," Arthur said, just as Lucy, April, and Peggy broke away from the group and hurried toward the stands. Were they coming to welcome the boss, or to explain why there would be no match that day?

The three Canaries halted in front of Arthur and Helen. "Good afternoon, Mr. Purcell," said Lucy, but the other two only spared him polite nods before turning appraising looks upon Helen. "Mrs. Purcell."

Arthur returned their greetings heartily, but Helen spoke more warily, bemused by their intense scrutiny.

"You might have noticed we're missing a girl," said Peggy.

Before they could reply, April put her head to one side and said to Helen, "You've watched us all season, practically. How well do you know the game by now?"

"Fairly well, I think," she replied.

"Have you ever played before?"

"Me? Play football?" Helen shook her head. "No, never."

"She's an excellent tennis player, though," said Arthur. "And she takes long, brisk walks every day. She's very fit."

As the three Canaries nodded, studying her, Helen realized where the conversation was heading. "Just a moment," she said, holding up a hand. "Before we get carried away—"

"We're running out of time," Marjorie shouted from the sideline, her hands cupped around her mouth. "Is she in or out?"

"In, of course," declared Arthur, putting his arm around Helen's shoulders and giving her a hearty squeeze.

"Give us a minute, Marjorie," Peggy called back.

"As you can see, Mrs. Purcell," said Lucy, "we need another player, and we'd be exceedingly grateful if you would step in. It's too bad you haven't played before, but you're clever and we know you've picked up the rules over the past few months."

"Oliver Corbyn says you have an excellent grasp of strategy," said April.

Helen found herself unexpectedly flattered by Oliver's approval, even as she was startled to learn that they had been talking about her. "I know the rules, but nothing about technique," said Helen. "I'm afraid I might do more harm than good."

"Nonsense," protested Arthur. "You're a natural athlete."

"There isn't a lot to remember," said Peggy, with false bravado. "Just remember which goal is ours, and which is theirs, and never put the ball in the wrong goal."

"Never use your hands," April added. "If the ball comes your way, kick it down the field toward a girl in a yellow jersey."

"That's really all you need to know to get started," said Peggy, studying her hopefully. "What do you say?"

It was so absurd, Helen had to laugh. "I don't have a uniform."

They assured her that extra uniforms for substitutes were in the changing room, freshly washed and ready for her to try on, and shoes too. Still uncertain, Helen turned to Arthur. "But darling, the whole purpose of this outing was for us to spend time together."

Arthur raised her hand to his lips and kissed it. "Believe me, the only thing I would enjoy more than watching this match *with* you would be to cheer you on as you play."

What else could Helen do then but take a deep breath and agree? She would rather reveal before thousands of people that she was a dreadful football player than admit to a few acquaintances that she was a coward.

While Peggy and April ran off to share the good news with the rest of the team, Lucy quickly led Helen to the changing room, where she tried on black short trousers, yellow jerseys, and football shoes until she found the best fit for each. Shin-guards, long black socks, and a cap finished the kit, and before she could fully comprehend what a dreadful mistake she was in all likelihood making, she was out on the pitch, heart thudding, stomach lurching, warming up with a bit of running and practicing a few kicks with the side of her foot, not the toe, which felt very odd but she did it anyway.

Peggy, who had taken over as team captain for her mother, told her where to stand for the start and gave her an encouraging grin. "Remember, however bad it gets, it's only ninety minutes, and then you're through it."

That was true enough. Anyone could bear ninety minutes of utter humiliation. Perhaps such a thing built character. At least she and Arthur would have a good laugh about it afterward, and she would have a marvelous story to share with the Marylebone ladies the next time they met for lunch.

Then the whistle blew, striking dread in Helen's heart. The match began, and there was no time to reconsider because the ball was in motion and all the other girls were running and passing and she had to keep up, straining her ears for the instructions the other Canaries called out to her. She sprinted for the ball when it came her way, got rid of it as swiftly as she could, and tried to slow down any opponent who dared approach Marjorie's goal by just generally getting in the way and interfering with her opponent's plans. In that, at least, Helen reckoned an inexperienced player was better than none at all.

At one point, as she caught her breath during a brief respite after the ball went out of bounds, it occurred to her that the Girls had no idea she had never played before. If she employed some of the acting skills she had acquired in school theatricals, it might take them a while to discover that she was the weakest player on the team, and to exploit that fact. But then the game resumed and she had no time to do anything but listen to her teammates, pass to the nearest teammate if the ball happened unluckily to come her way, and avoid scoring for the opposition.

The first half was nearly over, the score 1–0 in favor of the Handley Page Girls, when Helen realized that she was one of the fastest players on the pitch. She had only a novice's understanding of strategy and no footwork skills whatsoever, but she was a swift runner with excellent stamina thanks to years of tennis and brisk walks. She discovered she could almost always outrun her opponent when pursuing a loose ball, and if she got to the ball first, she could pass it to a teammate who actually knew what she was doing. The other Canaries seemed as pleasantly surprised by this discovery as she was.

"I told you she would be good," Helen overheard April tell Marjorie as they headed to the changing room for the interval, to her immeasurable delight.

Before she had fully recovered from the first half, they were back on the pitch for the second. Her legs felt like lead, and she was panting and perspiring as much as she ever had in any tennis match. Then Lucy made a brilliant run from midfield and scored, and only minutes after that, Peggy tapped one in, low and swift, from the top of the box. With only ten minutes left in the game, Helen concentrated on sending the ball in the opposite direction whenever a girl in a dark pinafore brought it anywhere near her. She was exhausted and thirsty and incapable of any stratagem more complex than that.

The whistle blew. A tremendous roar went up from the crowd, and suddenly Helen found herself surrounded by exultant girls in yellow jerseys, embracing her and laughing and running off to embrace other teammates. They had done it, she realized. They had won, 2–1. Somehow she had managed to survive the ninety minutes without looking like a complete fool. A partial fool, certainly, but not a complete fool.

The Canaries met the Handley Page Girls in the center circle to exchange handshakes and congratulations, and as they headed off to the changing room, exhausted and happy, Helen heard someone shout her name. She looked up into the stands to find Arthur standing with Oliver Corbyn, both of them applauding wildly, Arthur beaming proudly and laughing from sheer amazement.

As they changed back into their street clothes, the other Canaries thanked Helen for coming to their aid at the last minute, but she already felt them pulling away from her, becoming more formal, addressing her politely as Mrs. Purcell. After folding her perspiration-soaked uniform neatly, she brought the pile with the shoes on top to Lucy. "I'd be happy to launder it and return it to you at the arsenal," she offered. "It's no trouble."

"Oh, that's not necessary," said Lucy, smiling, as she accepted the pile and returned it to the large duffle bag. "I'll wash yours with mine."

Helen managed a smile and thanked her. It was only then, as her heart sank, that she realized she had hoped Lucy would tell her to keep the kit for the next time they needed her to fill in—better yet, to keep it for good because they wanted her to become a permanent member of the team. Of course that was a wildly ridiculous hope, and she was very glad that she had not spoken it aloud.

Arthur was waiting for her on the sideline in front of the stands, and when she emerged from the changing room, he hurried to meet her. "I'm so proud of you," he declared, kissing her on one cheek and then the other. "You astonish me. After all this time, I still marvel at you."

Her heart, which had felt so heavy only moments before, soared, full of light. It had been months, years even, since she had last seen him so happy. "Every muscle I possess aches," she told him, emphasizing each word. "I thought I was fit. Now I know better."

"Nonsense. I just saw you run for ninety minutes. You're perfectly fit." He offered her his arm, and together they left the stadium en route to the train station. "Do you suppose you'll play with the team again?"

"Oh, I don't know," she said lightly. "I suppose I would, if they ever ask me again."

"I should hate to miss that."

She squeezed his arm. "Then you shall have to attend every game with me just in case, whenever the Canaries play in London."

But the following week the Thornshire Canaries had a full side plus two substitutes, so they did not need her, and the next week they played in Birmingham, and Helen did not attend.

Some of the joy of the game had faded by the end of the month, when a wonderful surprise of an entirely different sort again filled Helen's heart with hope and anticipation. More than a year before, the Speaker of the House had established the Speaker's Conference on Electoral Reform to investigate how to make Parliament more representative of the will of the country. In January 1917, the Speaker's Conference had provided the Prime Minister with a concise nine pages of recommendations, which included conferring "some measure of woman suffrage." Helen had heard promises made and broken so many times through the years that she could not allow her expectations to rise too much, but then, four months after the report, Home Secretary George Cave introduced the Representation of the People Bill into Parliament—and it included provisions for limited women's suffrage to be enacted after the war.

Helen rejoiced, as did all of her suffragette friends, for although the bill proposed giving the vote only to certain women within specific, limited circumstances, it still marked a significant step forward, even though they would have to wait until the war ended for the law to come into effect.

That long-awaited day seemed terribly far off as the war dragged on, resources ran low, and casualties mounted, in both France and Belgium and at home.

16

JUNE–AUGUST 1917

LUCY

As the summer passed, it became increasingly evident that the Germans intended to starve the British into surrender before the United States could replenish their stockpiles. Lucy knew the country would rather go hungry than submit to Germany's wicked plan, but she dreaded to imagine how bad it might get before it got better. Lucy's mother cultivated vegetables where Lucy usually grew flowers in the front garden of the Dempsey home, and Lucy's sisters-in-law sent boxes of produce from Brookfield, but the gravest danger, for the Dempseys and all their neighbors, was the scarcity of bread. The Americans sent tons of wheat from the vast grainlands of the central United States, but the Germans targeted the wheat ships with impunity, sending the precious cargo to the bottom of the sea. Rumor had it that Great Britain's wheat stocks had dwindled to a nine-week supply, and fearful whispers about the possibility of famine circulated through neighborhoods and workplaces.

The arsenal canteens still managed to provide the workers with adequate meals, but Lucy had always been far more concerned for her sons than for herself. Like all growing boys, Jamie and Simon were always ravenous, and although they rarely complained, she worried that they might suffer long-term harm from their inadequate diet.

Sometimes she thought they might be better off if her mother took them to her home in Brookfield for the duration, but the boys liked their school and would miss their friends, and with Daniel so far away, she could not bear to be parted from her sons too.

Lucy knew her mother thought her grandsons would be safer in Brookfield, although she expressed her opinion only in gentle suggestions that she take them for a visit. After the school term ended, Lucy could not really justify refusing, especially since the boys were eager to go. She consented to a fortnight away, knowing they would have a merry time in the countryside with their cousins, wishing she could accompany them.

They had been gone only a few days one sunny Saturday morning in early July, bright and warm, with a faint haze high in the sky. The Thornshire Canaries were training on the arsenal field when suddenly they heard the low drone of engines overhead. They all stopped where they were, shaded their eyes with their hands, and searched the skies. "What a lot of aeroplanes," Peggy exclaimed, just as Lucy glimpsed them coming into view, more than twenty aircraft in a fan formation, surprisingly low and moving slowly toward the north.

Something about it struck Lucy as dangerously amiss. "Are those German planes?" she asked, turning in place, her gaze fixed on the formation as it passed over the arsenal.

"Don't be ridiculous, Lady Prim," scoffed Marjorie, supremely confident. "They must be our airmen carrying out maneuvers. If they were Germans, we would hear our own guns firing upon them."

"How splendid they are," cried Daisy. "And what a comfort to know that London is protected by such mighty aerial defenses! I wonder if the Handley Page Girls built them?"

Marjorie, Daisy, and a few others were discussing that possibility when a low, dull boom sounded somewhere to the west, followed seconds later by the faint rattling of windows in the nearest buildings.

Suddenly Lucy spied Mrs. Purcell running toward them from the edge of the field, waving her arms for attention. "Take cover!" she shouted. "Take cover!"

Her words were drowned out by another distant explosion, and then the percussive booming of several ground guns fired in rapid succession. Lucy heard shrieks as the players scattered, fleeing the open field for the shelter of the nearest buildings, pressing themselves against the brick or stone and covering their heads with their arms. Terrified, Lucy crouched into a corner where a concrete staircase met a stone foundation, but even as she strained her ears for the drone of aeroplanes and the roar of distant bombs, she wondered fleetingly whether it wasn't useless to seek protection from strong, solid walls when a direct hit on the arsenal could set off a massive explosion fueled by their own shells and yellow powder, reducing everything around her to rubble. She heard women sobbing, men shouting, and two, no, three more distant explosions, more defensive ground fire. For five minutes, ten minutes, the roar of guns and bombs continued. Then, as if the violence had paused for breath, all fell silent.

Lucy remained in place a while longer before she cautiously rose and peered out from her corner. Other workers, pale and shaken or red-faced and angry, were emerging onto the streets and open ground, gazing warily up at the sky, seeking out friends, muttering curses against the Germans or asking one another what had happened, what they knew. Then the bell at the main gate pealed, urging them back to work. Lucy hesitated, wondering how the foremen could be sure the raid was over, but lingering in the yard wasn't likely any safer, so she dusted herself off and made her way back to the Danger Building.

The censors must have decided that it was pointless to attempt to conceal what so many Londoners had observed, for the evening papers were remarkably forthcoming. At about half nine o'clock that morning, twenty-two German aeroplanes in two separate groups had appeared in the skies off the Isle of Thanet and the east coast of Essex. After dropping several bombs in Thanet, the raiders had proceeded in a fan formation toward London on a course roughly parallel to the north bank of the Thames. They had approached London from the northeast, then changed course and crossed over the city from northwest to southeast, dropping seventy-six bombs along the way, with most of the destruction concentrated in the East End. The Central Telegraph Office received a direct hit; temporary structures on the roof were reduced to rubble and sawdust, but the interior of the building was only slightly damaged. Fortunately, no one there was killed, since the War Office had warned them of the coming attack in time for everyone to shelter in the basement. Another bomb had set fire to a house near St. Bartholomew's Hospital, but thankfully the hospital itself, and the great many wounded soldiers among its patients, was unharmed. The most horrifying news was that an infant school had been struck, killing many of the children, all of them only four to seven years of age. In all, fifty-seven people had been killed and 193 wounded.

"Hold on," said Marjorie indignantly as they shared information Sunday afternoon while they warmed up for their football match. "The War Office warned the Central Telegraph Office that the German planes were on the way? Why didn't they warn Thornshire Arsenal?"

"Why didn't they warn *everyone*?" April asked. "If the teachers had known, they would have taken those poor children into the school's basement, and maybe they would have survived."

The other players chimed in, their voices rising, angry and indignant.

"I don't know why we weren't warned," said Lucy, "but I know who might be able to find out." She glanced significantly to the stands, where Mrs. Purcell sat with her husband, holding his hand and smiling, eager for the match to begin. Afterward, when the Purcells came over to congratulate the Canaries on their 3–2 victory, Lucy managed to have a private word with Mrs. Purcell, who looked intrigued by her question and promised to look into it.

A few days later, Mrs. Purcell met Lucy outside the canteen, clad impeccably as always in a lovely suit of apricot silk and a straw hat with an upturned brim. "I have some information," Mrs. Purcell said, falling in step beside Lucy as she headed off to football training, "but you didn't hear this from me."

Lucy smothered a laugh. "Very well." Although Mrs. Purcell outranked her at the arsenal, Lucy was five years older, and sometimes Mrs. Purcell reminded her of her girlfriends back in Brookfield.

"The good news is that the government intends to strengthen London's defense with more aeroplanes and ground guns," said Mrs. Purcell. "The army's needs must be satisfied first, of course, but London is next in the queue. Soon they expect our defenses to be so formidable that the Germans will not dare to make any more of these daylight raids."

"I'd rather they didn't make any raids, day or night."

"You and I both. As for notifying the public when an air raid is imminent, Saturday's attack has apparently not changed anyone's mind at Whitehall. They will continue to issue private warnings to particular departments and buildings, but they have no plans at this time to issue warnings to the general public."

"But why not?" Lucy protested. "Think of the lives they would save!"

"The Home Secretary disagrees. Last week, his office issued four private warnings, but only one attack actually occurred. His concern is that if public warnings were issued, a similar percentage of those would also prove to be unfounded, and eventually the public would disregard all warnings altogether."

Lucy frowned. "Maybe the problem is that our defenders need to improve their average. Get better spotters, or better intelligence, or whatever it is they do."

"That's an excellent point, and I sincerely hope they're working on it. But that's not the only reason for their silence." Mrs. Purcell grimaced. "Someone somewhere has calculated that every false alarm would bring all of our factories to a grinding halt for at least four hours. This, of course, would significantly reduce munitions production. As a result, though lives might be saved in London, it would mean a far greater loss of life at the front than if the armies were fully supplied."

"Oh my goodness." Lucy's heart gave such a thud that she pressed a hand to her chest. The government wanted the unwitting munitionettes to stick to their posts, toiling blithely away as bombs rained down upon them. Were their lives really of so little consequence? "The government is putting the soldiers' lives before ours."

"I suppose they are, but shouldn't they?"

"Before *our* lives, perhaps, we munitionettes," Lucy conceded. "We accepted the possibility of death or injury or illness when we took these jobs. But what about the children, the elderly, those who are true civilians in every sense? Don't they deserve fair warning?"

"Perhaps in their case, it's best to evacuate London altogether."

They had reached the training field, and they paused at the edge where the grass met the gravel. "That's not possible for everyone," said Lucy, with a pang of guilt, for it *was* possible for her. Should she tell her mother to keep the boys in Surrey? Her own loneliness would be a small price to pay for their safety. "Most people have nowhere to go."

"Fair point." Mrs. Purcell's gaze had shifted to the Canaries, who were laughing and calling out to one another as they kicked the ball around. For a moment she looked wistful, but then she turned back to Lucy, smiled briefly, and said, "I'll leave you to it, then."

As she set off toward the administration building, Lucy called, "Why don't you join us out here sometime?"

Mrs. Purcell halted and glanced back at her. "I beg your pardon?"

"Why don't you kick the ball around with us sometime?" Lucy urged. "I'm sure you're terribly busy, but if you can ever spare the time, you'd be welcome."

"Would I?"

"Of course. We haven't forgotten how you answered the call on such short notice when we needed you for that match with Handley Page Girls. We would have asked you again, but we didn't want to impose."

Mrs. Purcell uttered a single laugh. "Oh, well, I don't mind a little imposition now and again."

"In that case, why don't you join us after you get yourself a proper kit?" Lucy gestured to Mrs. Purcell's lovely suit. "It's a very becoming ensemble, but you can't play football in a hobble skirt."

"One can't do much of anything in a hobble skirt, not even one with hidden pleats." Mrs. Purcell mimed kicking a football. Her range of motion was greater than Lucy would have expected, but still not enough for sport. "Thank you for the invitation. It would be lovely. I—I'll let you know."

Lucy gave her a parting nod and hurried off to join her teammates, bemused. Should they have invited Mrs. Purcell to join the team before this? She had seemed so reluctant when they had begged her to complete their side for the Handley Page Girls match. Lucy clearly recalled that she had agreed to play only at her husband's urging. Lucy had not faulted her for that; she would give anything to spend ninety minutes with Daniel, and she could well understand why Mrs. Purcell would rather be in the stands with her husband than on the pitch. Was it possible Lucy had misinterpreted the entire scene? Surprise could have accounted for Mrs. Purcell's hesitation, since their request had come at the very last minute, or nerves, since she had never played before. The more Lucy mulled it over, the more their failure to invite Mrs. Purcell to play again seemed neglectful and unkind, especially considering all that the welfare supervisor did on the munitionettes' behalf.

Perhaps the Canaries could still make it up to her.

Lucy decided to propose inviting Mrs. Purcell to join the team as an alternate, but at their next training, she arrived at the arsenal field to find the other Canaries gathered around April, and even from a distance she could tell they were concerned and upset. As she hurried toward them, Daisy saw her and jogged over to meet her halfway. "It's Marjorie," she said. "She got a telegram from her mum."

Lucy's heart plummeted. "A telegram?"

Daisy nodded bleakly. "Two of her brothers were killed in action. The third is missing."

Lucy pressed her hand to her mouth, heartsick. "Oh no. Poor Marjorie." She drew in a shaky breath. "Where is she?"

"April says she left with Superintendent Carmichael as soon as she read the telegram. That was about an hour ago." Daisy shook her head, helpless, as they hurried to join the others. "We assume she went back to the hostel, and from there, maybe home, to be with her family."

April overheard the last. "I wanted to go with her to the hostel, but the superintendent told me to remain at my post," she said, an edge to her voice. "Marjorie shouldn't be alone right now."

"Was she upset?" asked Lucy.

"No, that's just it. She was like stone. I spoke to her but she didn't say a word." Tears appeared in April's eyes, but she blinked them away. "It was Superintendent Carmichael who told me about her brothers—but not their names, not who were killed and who's only missing. They were all in the Cheshire Regiment together. The last time Marjorie mentioned them, she said they were in Belgium, near Ypres."

Lucy saw from her companions' faces that they all shared the same sympathetic dread, but no one knew what to do or how to help their bereaved friend. Someone rolled a football into the center of the circle and they kicked it around halfheartedly, not saying much, thinking of Marjorie and their own beloved soldiers far away, how they might be the next to receive a telegram bearing the worst news imaginable. It was almost a relief when the break ended and they could return to work, where they might forget their worries in the repetitive, dangerous monotony of the machines and the shells and the yellow powder.

In the shifting house at the end of the day, they all urged April to give Marjorie their love when she saw her back at the hostel. But the next morning, April reported that Marjorie had already left for home by the time April arrived, and she had taken nearly all her things with her.

"She's coming back, isn't she?" asked Peggy.

"I think so," said April, pensive. "After the funerals, maybe? I suppose there will be funerals . . . unless her brothers were buried in Belgium?" She looked around, hoping someone would have the answer, but they all shook their heads, uncertain.

A week passed, and then another. Lucy's mother and her sons returned to Clapham Common, but Marjorie did not return to the arsenal and no one heard from her. April said that she desperately wanted to write to her longtime friend, but she didn't know where to send a letter. She knew only that Marjorie's family lived in Warrington, about sixty miles west of Alderlea.

At last, thought Lucy, a way to help.

She sought out Mrs. Purcell in her office, explained their concerns, and asked if she could provide Marjorie's address. Mrs. Purcell thought for a moment, then leafed through some files, removed a blue form, and copied an address onto a small card. "I admire your wish to help your poor friend," she said, handing the card to Lucy, "but—"

"I know," said Lucy, managing a halfhearted smile. "I didn't get this from you."

She turned to go, and was halfway to the staircase when she remembered something else. Hurrying back, she addressed Mrs. Purcell from the doorway of her office. "As to what we were discussing the other day—would you be interested in joining the football club?"

Mrs. Purcell's eyebrows rose. "Because Miss Tate is absent? I'd be delighted, except I could never play keeper. I would be terrified, and you'd lose every match."

"Oh, no, I didn't mean you'd replace Marjorie. You can play halfback as you did before, if you felt comfortable there. Peggy has been filling in as keeper." Lucy hesitated before adding, "We do hope Marjorie will be back in the goal soon."

"As do I, but do you truly believe she'll return? Superintendent Carmichael filled out the paperwork for a leave of absence, but we haven't heard anything from Miss Tate since she left. Unfortunately, the superintendent can't hold the job for her indefinitely."

Lucy held up the card with the Tate family's address. "Give us a week, time for a letter there and a reply back."

"Let's make it two," said Mrs. Purcell.

April wrote to Marjorie that evening after work and posted her letter the next day. A fortnight passed, but Marjorie did not respond. "I'm worried that Marjorie is going to lose her job," April confided to Lucy on the train to Bristol, where they would face the Chittening Factory Ladies in a match at Avonmouth. "And I'm worried that we're going to lose Marjorie."

Lucy did too, but she suspected that Marjorie would lose something more important than her job and her football club if she did not return to Thornshire.

By the time the match ended in a 2–2 draw, Lucy had concluded that the only way to be sure that Marjorie was all right was to send a delegation to Warrington. She proposed the idea to the team as they rode back to London, and all agreed that she was right. Marjorie's silence had become so worrisome that they would not rest easy until they looked in on her. April should go, as Marjorie's closest friend among the Canaries. Lucy offered to accompany her.

The others exchanged looks of surprise and misgivings. "Are you sure you'd want to?" Peggy ventured. "Marjorie hasn't exactly been kind to you through the years."

"All the more reason I should go," said Lucy. "If I'm there, of all people, she'll have absolutely no doubt that the entire club is very concerned."

Everyone conceded that she made a fair point, so the next day, Lucy spoke to Mrs. Purcell and Superintendent Carmichael, who granted her and April two days off. The following day, Lucy packed a small bag, kissed her sons goodbye, thanked her mother, and set out for the train station. She met April at Euston, and together they boarded the train to Warrington. Some of their fellow passengers did a double take at the sight of their yellow skin; others recoiled or averted their eyes. Two middle-aged gentlemen offered them their seats in the crowded carriage, and an elderly woman seated beside April smiled kindly and thanked them for their service to the war effort. Lucy nodded politely in reply, uncomfortably conspicuous in her yellow skin and mottled hair. She had worn gloves, but she almost wished she had worn a veil as well.

As they traveled northwest, Lucy and April spoke quietly about what they might expect when they found Marjorie, and what they might say to her, how they might persuade her to return to Thornshire, if that did indeed seem to be for the best. Perhaps her grieving family needed her closer to home. Perhaps Marjorie had already found new war work in Warrington, and she was perfectly content—mourning her brothers, of course, but otherwise fine, simply too busy to write.

April, who knew her best, found this explanation very unlikely. "It's just not like her to go an entire month without a word," she insisted. "Either she's too heartbroken to write, or she's fallen ill from the yellow powder."

"If she's heartbroken, we'll try to comfort her," said Lucy. "If it's TNT poisoning, we can make sure she's getting proper care."

April nodded, looking only slightly less anxious than before.

They disembarked at Warrington Bank Quay, and after asking a ticket agent for directions, they made their way to the Tate family home on foot, Lucy carrying her bag, April a small pasteboard suitcase. They weren't sure if they would need to spend the night, but they had agreed they should come prepared.

They soon discovered that the address on Mrs. Purcell's card belonged to a three-story red-brick building on Sankey Street, part of a row that ran the length of the block. Marjorie's home was two flights up above a confectioner's, between a tailor's shop with a "For Let" sign in the window and another business that had been converted to an army recruitment office. They climbed the stairs, exchanged an encouraging look, and knocked on the door.

A lad of about fourteen opened it, his hair thick and wavy beneath a tweed cap, the same honey-gold color Lucy reckoned that Marjorie's had been before she became a canary girl.

"Hello, you must be Jack," said April warmly. "We're friends of Marjorie's. Is she in?"

"She's through here, in her room." Jack opened the door wider and beckoned them inside. "I know you're her friends, since you've got yellow skin same as her. So you've come all the way from London?"

"Yes, we have," said Lucy, thinking of Jamie, so close to this boy's age. "Have you ever been there?"

"No. I want to see it, and Marjorie said she'd let me visit, but—" He broke off and shrugged, as if he knew it would never happen now, and the reasons were too obvious to mention. As he turned to lead them into the front room, Lucy and April exchanged a look. Had Marjorie already decided not to return to Thornshire?

Just then, a tall woman with dark blonde hair fading to gray entered the room through another doorway. "Oh, hello," she said, wiping her hands on a dishtowel. She managed a smile, but her eyes were red and puffy. "Have you come to see Marjorie, or were you friends of Frank and Even?"

"We're friends of Marjorie," said Lucy, though with regard to herself, Marjorie might disagree. "Please accept our condolences for your losses. We're so sorry."

Mrs. Tate pressed her lips together and nodded, closing her eyes for a moment as if to hold back tears.

"I'm April Tipton." April then gestured to Lucy. "And this is Lucy Dempsey. We work with Marjorie at Thornshire Arsenal."

Marjorie's mother nodded. "Have you come to fetch her back, then?"

"We've come to offer our condolences, and to see if she's all right," said Lucy. "We weren't sure what her intentions are."

"She hasn't responded to my letters," said April. "We're all quite worried."

Marjorie's mother peered at her. "You're the lass who was in service with her at Alderlea."

April nodded. "Yes, that's right. Is she—do you know if Marjorie's planning to come back to the arsenal?"

"Maybe," said Jack glumly, "if you can get her out of her room first."

"What do you mean?" asked Lucy.

"She shut herself in there after the funeral and hasn't left since, except to do her business out back—"

"That's enough, Jack," said Mrs. Tate, cheeks flushed. "Oh, girls, I don't know what to do. She won't speak to us. She won't join us at the table. I leave plates outside her door and sometimes she picks at the food but more often than not she doesn't."

"Is she ill?" asked Lucy. "Sometimes, canary girls like us—"

"Oh, no, no, it isn't that. She's heartsick, I think, but that's all."

"She doesn't cry, though," said Jack, "least not so we can hear it, not like our mum."

"Jack," his mother admonished quietly, cheeks flushing deeper.

"She hasn't cried at all?" asked April, puzzled. "Not even at her brothers' funeral?"

"No, no, she's strong, our Marjorie." Mrs. Tate took a deep, tremulous breath and squared her shoulders. "But I do think it's well past time she joined the world again. We all have to carry on, don't we? We can't just shut ourselves away. My sons wouldn't want that."

"I'm sure they wouldn't," said Lucy gently.

"Will you girls be so kind as to coax her outside for a bit?" Mrs. Tate pleaded. "Some fresh air and sunshine would do her a world of good."

Lucy and April exchanged a quick glance. "Leave it to us," said April.

Mrs. Tate led them through the room to a narrow hallway, where she stopped at the first closed door and gave it three quick raps. "Marjorie, dearie?" she called. "Some kind lasses from Thornshire have come to visit. Won't you come out to see them?"

There was no reply, only heavy silence.

Mrs. Tate knocked again. "Dearie, please. They've come such a long way."

"Marjorie, come on out," said April, a trifle sharply. "I didn't sit on a train for hours to talk to a closed door."

Lucy turned to Mrs. Tate. "I think a cup of tea wouldn't go amiss. Would you put the kettle on, please?"

"Oh, of course," said Mrs. Tate, flustered to have forgotten such a simple courtesy. She hurried away, and Jack eyed the visitors speculatively before following after.

April pounded twice on the door with her fist. "Marjorie, the boss said we have to give your wages directly to you."

Lucy gasped and gestured sharply for silence.

"What?" April said in an undertone. "If it gets her to open the door—"

"And when she finds out you lied, she'll slam it in our faces." Resting her hand on the doorknob, Lucy called, "Marjorie, we're coming in." Bracing herself, she turned the knob and eased the door open.

Marjorie was sitting on the end of the bed, legs crossed, gazing out the window. She didn't even turn her head as Lucy led April into the room. "Marjorie?" Lucy greeted her tentatively.

"Lady Prim," Marjorie replied flatly, without turning her head away from the window. "Let me guess. You don't really have my wages."

"Sorry," said April, not sounding at all sorry. "If you want your wages, you'll have to come to Thornshire and collect them yourself."

"Are you coming back?" asked Lucy. "Mrs. Carmichael has been holding your job for you, but she can't for much longer."

"We need you on the pitch too," said April. "Peggy has been filling in for you in the goal, but—"

Marjorie turned to face them. "Peggy? Why Peggy?"

"She volunteered," said Lucy, shrugging. "She's tall and has a long reach. It's not like we have a lot of options."

"That's true enough," said April. "We have so few substitutes, Lucy asked Mrs. Purcell to join the club."

Marjorie stared at her friend in utter disbelief, then fixed Lucy with a glare. "You can't be serious." Her eyes were raw and red-rimmed, with dark shadows beneath from lack of sleep. "She's fast, fair enough, but she hardly knows the sport."

"She's learning," Lucy defended her. "And Peggy is doing rather well as keeper."

April pulled a face. "Well, that's a matter of opinion."

"At least Peggy is *there*," said Lucy. "She hasn't abandoned the team."

"What's that supposed to mean?" snapped Marjorie. "If you've got something to say, Lady Prim, just say it."

Lucy spread her hands and shrugged. "What is there to say? You lost your brothers, and that is a terrible, tragic thing. My heart goes out to you. What I don't understand . . ." She let her voice trail off.

"What? What don't you understand?"

"Well, it's not just that you've run off without a word for the team, when we're hoping to qualify for the Munitionettes' Cup. That's fine. After all, in the end, football is just sport."

Marjorie regarded her balefully. "Said the footballer's wife."

"I'm not saying football isn't important. It is. But if you abandon the team, we'll manage without you." Lucy steeled herself, choosing her words carefully, knowing that what she said next might make all the difference. "But the thing is, our soldiers *won't* manage without you. You're the girl behind the man behind the gun, remember? On us their lives depend."

Marjorie barked out a hollow laugh, and a tear trickled down her cheek. "Fat lot of good that did my brothers."

"Your brother who's missing—what if he's been captured?" said April. "The sooner the war ends, the sooner he can come home."

Marjorie frowned at her bleakly, and another tear spilled over, and then another. "If he's still alive," she said, her voice breaking. "If he's not lying dead in a trench in Belgium."

Swiftly, April sat beside her on the bed and embraced her friend, and Marjorie allowed it. "Then take revenge," she said, her voice low and angry. "Make them pay."

"Or you can stay here," said Lucy, "and turn the white feather."

Marjorie tore herself from April's embrace and bolted to her feet. "How dare you? You can't possibly understand how I feel."

"No, I probably can't," said Lucy. "But I'm sure that hiding in your room until the pain goes away isn't the answer. I don't think revenge is either," she added, with a glance for April. "The best we can do is to help our armies win the war as soon as possible so that all the killing can stop and the men we love can come home."

Marjorie's tears were flowing freely now, her breath coming in ragged gasps. "My brothers never will."

"Don't give up hope for your third brother," Lucy implored. "He may yet come home."

"Archie," said Marjorie. "It's Archie who's missing."

As she broke down in aching sobs, April put her arms around her, murmuring soothingly. Lucy stood back at a respectful distance, hands clasped behind her back. She had done her part by infuriating Marjorie so much that her stoic mask had shattered. Now she could begin to grieve, but that didn't mean she would tolerate comforting embraces from her unwilling rival.

After a time, Marjorie took a deep, steadying breath and wiped her eyes. "Mrs. Purcell convinced the superintendent to hold my job for me?"

Lucy nodded, and April said, "Yes, for now. Mrs. Carmichael's patience is wearing thin, though."

"Best not keep her waiting, then. And there's a football tournament, you said?"

"The Munitionettes' Cup," said Lucy. "It won't be held until spring, but we have to win our next few games to qualify for the first round. We're sure we can do it."

"Reasonably sure," April added.

Marjorie's eyebrows rose. "With Peggy in the goal, and Mrs. Purcell on the pitch?"

Lucy spread her hands, resigned. "I won't pretend it'll be easy, but what choice do we have?"

"Obviously you'll have a choice." Marjorie squared her shoulders. "You'll have me."

She waved them out of her room so she could wash and dress and pack her things. Concealing their delight and relief until they had closed the door behind them, Lucy and April made their way to the kitchen, where Mrs. Tate poured tea and offered them some bread and cheese with a bit of pork, apologizing for the meager rations. They ate sparingly, assuring her that they were so well-fed at the arsenal that they weren't very hungry.

Within an hour, the three munitionettes were on their way to Warrington Bank Quay, where they boarded the train to London. They passed the first hour chatting, sharing the news of the arsenal and especially of the Canaries, but after that, Marjorie fell soundly asleep, her head on April's shoulder. Undoubtedly she had slept poorly in the weeks since her brothers had been killed. Lucy hoped with all her heart that somehow Archie's life had been spared, and that his family would soon hear from him, or have word of him. But as April had said, only victory would end the war, stop the slaughter, and let the men come home.

They parted at Euston, Lucy heading toward Clapham Common, April and Marjorie nearly as far in the opposite direction. When Lucy reached her own front gate, her sons startled her by bursting out of the house and racing to meet her. "Mum!" Simon cried, flinging his arms around her waist. "The American soldiers are in London! Did you see them?"

"Why, no," Lucy exclaimed, looking to Jamie for confirmation. "American troops are here? Passing through on their way to France, I assume?"

Jamie nodded, beaming. "They're going to parade from Wellington Barracks past Trafalgar Square through Whitehall and Westminster, and after they pass the American Embassy, they're going to march before Buckingham Palace and the King!"

"May we go, Mum?" Simon begged, seizing her hands and jumping up and down. "Please, Mum? It's going to be so grand, with bands and soldiers and everything!"

"I'm sorry, darlings, but I can't take you," said Lucy, truly regretful, for it was sure to be a jubilant spectacle, and she could use a bit of joy and optimism herself.

"I shall take them," said her mother, who also emerged from the house, though at a more sedate pace than her grandsons. "I already promised them."

"Then it's all settled." Still holding Simon's hand, Lucy reached for Jamie's and smiled upon them both. "You must promise to remember every detail and tell me all about it afterward."

"We will," said Jamie, and Simon nodded vigorously.

And she in turn would share the wonderful news with Daniel. Help was on the way, and victory must soon follow.

17

SEPTEMBER-OCTOBER 1917

APRIL

For several weeks after Marjorie's return to Thornshire, whenever a shift was especially grueling or a foreman particularly demanding, she would fix April with a wry look and deadpan, "I can't believe I let you drag me back here for *this*." But April knew she didn't really mean it. On several occasions, April had overheard Marjorie declare to the other canary girls that she was glad to be back in the Danger Building so she could make more weapons to strike back at the enemy who had killed her beloved brothers. Yet aside from the occasional ironic remark, Marjorie's former lighthearted nonchalance was gone. She had become mirthless and reticent in the Danger Building, steely-eyed and fierce on the pitch. The Canaries won match after match with her in the goal, snatching balls out of the air with flying leaps that defied gravity and ended in hard, rolling landings that made April flinch and look away, appalled by her friend's rash indifference to injury. Marjorie seemed to have no regard for her own safety anymore, and in their work, that attitude could prove fatal. April worried for her.

She was equally worried for Mabel, who had become so infirm that she had been obliged to resign from the arsenal altogether. At first, she had still attended the Canaries' football matches when they played on their home ground, sitting on the sidelines, coughing and wheezing as she cheered them on, but by early September she could no longer manage even that. Peggy frequently updated Mabel's friends about her condition, noting her small improvements and good spirits, but Peggy's bleak eyes belied her hopeful words. Nearly every day, a canary girl would send Peggy home with an encouraging note or a small gift for her mother, which Peggy assured them were always gratefully received. Mabel never wrote back, though, which after Marjorie's silence, April took as a foreboding sign.

There were other curious signs that she had no idea how to interpret. One morning in the middle of September, Lucy arrived at the shifting house looking pensive and bemused. Later, at dinner, April was seated at a table with Marjorie and Peggy across the aisle from Lucy and Daisy when she heard Lucy say, "I witnessed the most curious scene this morning at Clapham Junction station."

Marjorie gave April and Peggy a significant look, and they paused their own conversation to eavesdrop.

"My train was delayed, so I was waiting, one eye on the clock, when I saw two special trains pull in on opposite sides of the same platform," said Lucy, her yellow hands resting on the table around her teacup. "The northward-bound train was packed with German prisoners."

"My goodness," said Daisy, eyes widening.

Lucy nodded. "They must have come directly from the battlefield. They were filthy, disheveled, and unshaven, looking more like a gang of hardened convicts than soldiers. I admit I pitied them."

Marjorie caught April's eye. "Pity for Germans," she mouthed silently, her expression scornful. In reply, April threw her an exasperated look and quietly inched her chair closer to Lucy's.

"The southbound train carried our Tommies," Lucy was saying. "They were neatly shaved, hair trimmed and combed, faces scrubbed and smiling, uniforms in perfect order."

"They must have been new recruits," said Daisy. "Off to the front for the first time."

"Yes, that was my thought too. Only a narrow platform divided the two groups physically, but they could not have been more far apart in condition and spirits. Our boys were eager and jovial, the Germans unkempt and despondent, though neither knew what fate awaited them at the end of their journey."

Daisy leaned forward and rested her arms on the table, hanging on every word. "What did they do when they saw one another? Did they shout and curse? Did they rush off the trains and brawl on the platform?"

"You'd imagine so, wouldn't you, but no. The Germans were under guard, so they *had* to stay in their railcars—"

"Of course."

"But when they saw our soldiers, they smiled and waved from the windows and shouted, '*Kameraden!*' Hearing that, the Tommies called back, 'Good old Jerries!' Then they climbed out of their carriages, crossed the platform, and threw packets of tobacco and chocolate through the windows to the Germans!"

"That's astonishing!" gasped Daisy, wide-eyed.

"That's fraternization," snapped Marjorie, giving Lucy and Daisy a start. "They should be ashamed of themselves."

"Why?" asked Lucy, genuinely puzzled. "There's no personal hatred between the enlisted men of either side. They're all caught up in a war none of them have done anything to cause, and they fight only because their commanders and their heads of state order them to. It's little wonder that this common unhappy fate would evoke a sense of comradeship among them."

April nodded, thinking of the Christmas truce Oliver had experienced, a bit of peace and goodwill in the midst of war. It had not lasted, nor had it ever been repeated, but for a moment, enemies had become friends. The generals had been apoplectic with fury, but what harm had the brief respite done? Everyone had resumed killing one another all too soon.

"It would be different if Tommy and Fritz were fraternizing on the battlefield," said Peggy, as if she had overheard April's thoughts. "But these Germans can't hurt anyone now. What's the harm in giving them a bit of cheer? Wouldn't you want the same done for our lads if they were prisoners?"

Marjorie hesitated for the barest of moments, but then her expression hardened. "The Germans are the enemy. Have you forgotten what horrible things those prisoners might have done to our lads before they were captured?" She gestured sharply to indicate the whole arsenal. "Have you forgotten why we're here?"

"I know why we're here," said Lucy evenly. "It was a strange scene. I have mixed feelings about it too. That's why I was unburdening myself to Daisy." She extended her hand, palm up, to indicate her friend on the other side of the table. April felt heat rise in her cheeks at the gentle reminder that Marjorie had interrupted a private conversation.

Marjorie abruptly rose and snatched up her tray. "If you had lost someone you loved in this bloody war, you wouldn't be so eager to make friends with the Hun."

Lucy blanched beneath her yellow pallor, but Marjorie was already striding away and didn't see the effect of her words. "That was uncalled for," said Peggy, and Daisy reached across the table to take Lucy's hand. "She knows your husband is in the thick of it with the Seventeenth Middlesex."

Lucy managed a tight smile. "She's upset. She wasn't thinking of that." Then she too rose and cleared away her dinner tray.

Later, as April, Marjorie, and two other girls from their hostel were passing through Gate Four, April heard someone call her name. When she glanced over her shoulder she saw Oliver jogging toward her. His smile warmed her heart, and she told her companions to go on ahead while she waited for him to catch up.

"I have to get back to the office, so I can only walk you to the park," he said, apologetic.

"That's fine," she said, concealing her disappointment as they turned and passed between the sentries at the gate. "Marjorie had a rough day. I should spend some time with her."

"Are you going out?" He glanced overhead, where the moon had risen, nearly full, in a clear, cloudless sky, without a breeze to stir the cool autumn air. If they were not at war, April would have sighed over the evening's romantic beauty, but conditions were ideal, from the Germans' perspective, for an air raid. London had suffered two attacks earlier that week under similar skies.

"No, we're heading straight back to the hostel," she assured him. She wished he would take her hand as they walked along together, but there were too many other Thornshire workers around, and as he put it, he disliked drawing attention to himself.

"Because people would look at us and think you can do better than a housemaid?" she had challenged him once.

"No," he had replied levelly. "We both know it's far more likely that people would think you can do better than a cripple."

April thought that ridiculous and had told him so. He shrugged, and they never mentioned it again, but April had not forgotten it.

They reached the edge of the park, and to April's surprise, Oliver took her hand and squeezed it. "Get home safe, all right?" he said, bending down to give her a swift kiss on the cheek.

"Careful," she said. "Someone might see you and think we're sweethearts."

He grinned, and her heart warmed to him all over again. "People can think whatever they like. I'll see you at your match tomorrow." With that, he nodded and hurried back to the arsenal.

She watched him for a moment, wondering what was keeping him late at the office. In July, Winston Churchill had been appointed Minister of Munitions, and little more than a week ago, he had announced that the leaving certificates workers so loathed would be abolished on 15 October. Maybe Mr. Purcell needed Oliver to come up with another, less tyrannical scheme to encourage workers to stay. Or maybe they had received word that the armies would be making another push soon, and they were preparing to ramp up production. Or maybe it was something horribly dull, like filing paperwork. As she turned to go, April hoped that whatever it was, Oliver would much rather be sitting at a café with her, or sitting beside her in a darkened theatre, anticipating a moment when they might find themselves alone and share a long, slow, breathtaking kiss.

She gave herself a little shake and hurried after Marjorie and the other girls. She caught up to them just as they were boarding the tram. "We thought you'd run off with your fellow," said Marjorie archly as April dropped into the seat beside her.

"He *is* handsome," said one of the other girls, Ethel, glancing back the way they had come as if hoping for another look at him. "Pity about the arm."

"Don't pity him," said April, smiling sweetly, but with a brittle edge to her voice. "He doesn't like it, and neither do I."

She pretended not to notice the raised eyebrows and significant looks the other girls exchanged. At least her words had the desired effect, for they stopped teasing her. Conversation turned to other matters—arsenal gossip, the next day's football match, and the vexing rumors that their landlady intended to raise the weekly rate for room and board at their hostel. Early in the war, they had paid 12 shillings for a shared room and two meals a day, but now they were paying 16s 6d, and another tenant from the night shift had overheard their landlady say that 17s was coming soon.

"We'll need to ask for a rise in our wages, if it comes to that," said April, gazing out the window to the darkened streets.

"We should have a word with our newest teammate," said Marjorie, covering a yawn. "Helen will tell her husband what's what."

"Don't ask for a pay rise only on behalf of the Canaries," protested Ethel. "Ask for all of us."

Marjorie spread her hands, feigning uncertainty. "If there's enough money to go around, sure, but—"

"Hush a moment," April broke in, holding up a hand, gaze fixed on the darkness beyond the window. "Do you hear that?"

"I hear *you*, and I hear the tram," said Ethel, but the tram quieted as it slowed on its approach to their station, and they all fell silent, listening. Somewhere on the street, someone was blowing a whistle—a policeman's whistle. Uneasiness settled over the carriage, and as soon as the tram halted, everyone rushed to disembark. The Thornshire girls were separated in the crush, but they quickly reunited on the pavement and hurried on toward their hostel. They had gone little more than a block when they heard the shrill whistle again, growing louder and louder behind them. Suddenly a policeman passed them in the street on a bicycle, blowing his whistle, wearing white placards on his chest and back bearing an inscription in bold red print: "Police Notice—Take Cover."

April had heard that the government had changed its policy against issuing air raid warnings to the general public, but this was the first time she had witnessed it in action. Heart thudding, she and her companions quickened their pace until they were nearly running, but only tentatively in the blackout, the route illuminated by nothing more than moonlight and daubs of luminescent paint on curbs and obstacles. April and Marjorie pulled ahead of the other two, but they waited at street corners for their friends to catch up. Omnibuses and motorcars increased their speed, hastening their occupants to safety; other pedestrians were darting in all directions, fleeing for shelter; shops and businesses were closing their doors and shuttering their windows. A tense, ominous hush descended upon the street, and as the hostel came into view, April sprinted for the front door, with Marjorie right behind her. The door opened as they approached, and they raced inside past their landlady, who stood with one hand on the doorknob, the other waving them on through to the kitchen. There the door to the coal cellar stood open; April flew down the stairs and squeezed in with the other tenants, pressing close to make room for her friends.

They waited, straining their ears for the all-clear signal, starting when the door flew open and more munitionettes tumbled in, followed last of all by their landlady. For a while silence prevailed, broken only by the occasional murmur of anxious voices. Then, at roughly half eight o'clock, the heavy pounding and roar of bombs and gunfire shook the building and the earth all around. April's heart thudded so intensely, blood pounding in her ears, that she could take no more than quick, shallow breaths. She knew that they were threatened not only by German bombs, but by the falling shrapnel of their own ground guns, which fired fierce, terrible barrages into the skies above London, hoping fortune would speed their shots to their unseen targets. The German aeroplanes flew too high to be seen from the ground, so taking proper aim was impossible. April guessed that the defense forces counted on luck to bring the enemy down, or they hoped to shake the Germans' nerves so badly that they turned around and sped back across the Channel.

"I never imagined I'd say this," Marjorie murmured close to April's ear, as the explosions diminished, an encouraging sign that the raiders were moving off, "but I almost miss the zeppelins."

April agreed. If an attack had to come, better a zeppelin than an aeroplane any day. A zeppelin could be spotted at a distance and tracked as it drifted over London, but swift, high-flying aeroplanes could be anywhere, right above you or miles away, and you would never know until the bombs fell.

An hour passed before the all-clear sounded. Only then did they climb wearily from the coal cellar, dust themselves off, stretch their cramped limbs, and sit down to the supper their landlady had prepared for them, long since grown cold. Soon thereafter, as she settled into bed, April doubted that she would be able to sleep out of fear that another attack might come at any moment, but eventually exhaustion overcame her.

Fortunately, the next day was Sunday, so she was able to sleep in and get a proper rest before she had to rise and attend to the usual tasks of her single day off, rushing to finish before setting out for her football match. Oliver attended, she was glad to see as she searched the stands for him, and so did Mrs. Purcell's husband, as well as dozens of Thornshire munitionettes and thousands of other spectators who missed cheering on their favorite men's teams and were satisfied to watch the ladies, for the duration. After the match, a 4–2 victory for the Canaries that moved them one step closer to qualifying for the Munitionettes' Cup, Oliver took April out to a café for a bite to eat. He escorted her back to the hostel afterward even though it was miles out of his way.

"Lie low tonight," he said after kissing her goodnight at the door, glancing warily up at the brilliant sunset lighting up the darkening sky.

"I intend to lie perfectly snug in my bed," she said, blushing a little at his grin. She had not meant it flirtatiously, but there it was. "The Germans better not trouble us two nights in a row. I need to be well-rested for work tomorrow."

But the Germans did come, only minutes after she had fallen asleep, sending her and Marjorie and the other tenants scrambling for their dressing gowns and slippers and racing downstairs to the coal cellar. This time the assault lasted a full ninety minutes, and the pounding of the ground guns seemed louder than ever before, but the munitionettes tried to doze, closing their eyes and resting their heads on one another's shoulders, jolted awake whenever an explosion struck too near and the building trembled around them. Eventually the all-clear sounded and they were able to return to bed, but April woke the next morning groggy and with a faint headache behind her right eye. All that day, she felt as if she were dragging herself through her shift, and instead of meeting the other Canaries on their training field after lunch and tea, she rested in the shifting house, lying on her back on a bench with her eyes closed. She fervently hoped that dense clouds, or better yet, thunderstorms, would roll in by late afternoon, filling the skies with dangerous winds and lightning, unsettling the German pilots so much that they kept their planes grounded.

But it was not to be. Another beautiful clear, moonlit sky greeted her when she exited the shifting house that evening, and as she rode the tram back to the hostel, she steeled herself for another terrifying night of flinching as bombs slammed into London and the ground guns flung deadly fire back. And so it happened, just as she and her friends were finishing dinner, so that several girls snatched up the last of their bread and tea and carefully brought them along, down into the cellar. The guns sounded fainter that night, and the walls around them trembled less than usual, so they reckoned that the attack must have focused on another part of the city. This eased their anxiety somewhat, though they sympathized with their fellow Londoners who were bearing the brunt of it. To keep their spirits up, they chatted and even joked in the intervals between distant explosions, and eventually they were able to return to bed.

Fortunately, the Germans left them alone for the rest of the night, so in the morning April woke feeling somewhat better rested than usual, though far from refreshed. As she and Marjorie hurried off to work, she overheard other passengers on the tram say that although the previous night's attack had seemed less intense in their area than previous raids, the scale of it had actually been larger than any the Germans had attempted before. Three separate groups of enemy aeroplanes had crossed the Essex coast and approached London roughly fifteen minutes apart, but most of the planes had been driven back by the British guns. The few that slipped past their defenses had dropped five bombs on Shoreditch, but most of the damage had been inflicted on the West End. Victoria railway station had apparently been the raiders' intended target, but it had escaped harm, unlike Grosvenor Road near Buckingham Palace, where numerous houses had been destroyed and many residents had suffered serious cuts from shattered glass. A fourth group of aeroplanes had flown separately over Kent, dropping five or six bombs on the outskirts of town, but all had fallen on marshland. The windows of a few cottages had been smashed, and one man had received cuts to his face from flying shards, but there had been no serious injuries and no deaths. No one aboard the tram seemed to know whether there had been any fatalities in London, but April had little doubt that some unfortunate residents had been killed, for so it had been for every other raid over the past week.

Disembarking the tram, she hurried to the Thornshire park, eager to find Oliver and put her mind at ease that he too had made it through the night unharmed. But he was not there. It was a perfectly sunny autumn day with only a slight cool mist in the air, exactly the sort of morning when workers were inclined to linger and grouse about problems they hoped Oliver could solve, but she didn't see him anywhere. Heart thudding, she pressed a hand to her stomach, took a steadying breath, and tried to reason her fear away. No one on the tram had mentioned any fatalities in London, and no bombs had struck Oliver's neighborhood as far as she had overheard, so he was almost certainly fine. Likely the same business that had kept him late at work a few evenings ago had obliged him to report to Mr. Purcell's office early today. Yes, that was almost certainly it.

She worried about him all the same, of course, and wished she could invent an excuse to dash over to the administration building just to see for herself that he was all right.

Hours later, she was in the washroom scrubbing her hands and face before tea when she heard the hum of anxious conversation coming through the doorway to the canteen. She quickly toweled off and hurried inside, where she found several girls passing around newspapers, others peering over their friends' shoulders to read, and still more discussing something obviously dreadful in tense or tearful voices. Finding Marjorie and Peggy seated on the same side of their usual table, she hurried over without bothering to collect her tea things first. "What's happened?" she asked, bracing herself. "What's the matter?"

Marjorie glanced up from the newspaper spread open on the table before them. "A munitions accident, somewhere," she said grimly, rotating the paper around to face April. "Sounds bad."

April sank into the chair across from her, eyes fixed on the headline, and as she took in the first words, her hand flew to her mouth to smother a gasp.

MUNITIONS EXPLOSION
MUCH DAMAGE TO A FACTORY

PRESS BUREAU, 1.5 P.M.

The Ministry of Munitions announces that a serious fire and an explosion have occurred at a munitions factory in the North of England.

Much damage has been caused to the factory, but up to the present no deaths have been reported, although injuries have been sustained by a number of workers.

"Oh no," April murmured, reading the brief statement again, quickly skimming the rest of the page for more details, thinking of all the munitionettes from the North of England they had met on the pitch over the past few years. "This can't be all the press knows. Where did it happen? And when? Maybe another newspaper will tell us."

"They all say the same thing," said Marjorie, tapping the words "Press Bureau" with her finger. "It's a formal statement from the Ministry of Munitions, not an interview. It's a wonder the ministry made any announcement at all."

"It must have been a truly terrible accident, like the one at Brunner Mond in January," said Peggy. "Only when a disaster's too big to conceal from the public do they make an announcement like this."

April read the article again and found a slender thread of hope to seize. "No deaths have been reported."

"So they say," Marjorie scoffed.

April glanced from Marjorie to Peggy and saw the same skepticism in both of their faces. "Those poor girls, whoever they are," she murmured, sitting back in her chair, heartsick.

"It'll come out," said Peggy. "It may take months, maybe not until the end of the war, but you can't keep something like this secret forever."

Indeed, rumors began wending their way to Thornshire within a matter of days, rumors that were eventually confirmed by reliable secondhand accounts.

The disaster had occurred at the White Lund Factory in north Lancashire, a 250-acre complex of around 150 closely packed one-story buildings that out of wartime necessity had been swiftly constructed of wood with felt roofs. The shells made there were filled with amatol, a mixture of ammonium nitrate and TNT, which was melted and poured into the shell casings. It was incredibly dangerous labor, and worker deaths happened monthly—men perishing from accidents, women from TNT poisoning.

On 1 October, at about half ten o'clock in the evening, a massive explosion occurred in one of the melt plants, setting off a series of explosions in adjacent buildings that continued through the night until six o'clock the next morning, then sporadically until three o'clock the following afternoon. Terrible fires spread from one structure to another, setting off explosions that began with an ominous, demonical hissing and a leaping blue flame, followed quickly by tremors and reverberations that shattered plate glass and windows for miles around. Loaded shells went off, shrieking into the air and exploding over nearby Lancaster and Morecambe, sending the terrified residents fleeing to the waterfront in their nightclothes. The factory fire brigade was quickly overwhelmed, and firefighters from all over the region were summoned to battle the ferocious blaze, but the last fires were not extinguished until early in the morning of 4 October.

Miraculously, only ten people were killed in the disaster, most of them while fighting the blaze—five firefighters, five munitions workers, all of them men. A great many more workers were injured, some of them seriously, by falling debris, flying shrapnel, burns, and the panic of the crowd. In the towns, local residents were hurt by falling masonry, shattered glass, accidents incurred as they fled for safety, and exposure to the cold. It was believed that the casualties would have been much worse except that most of the workers were in the canteen on their dinner break when the alarm was sounded after the first explosion. In the end, nearly the entire compound was destroyed, except for a few brick buildings along the perimeter.

Even before the ashes cooled, rumors and wild speculation abounded that German agents had sabotaged the factories, or that a zeppelin had bombed it. Eventually the workers and their foremen agreed that the most likely explanation was an unknown worker's carelessness with matches and cigarettes.

"That could've been us," April said to Marjorie one chilly, overcast morning as they made their way from the tram to the arsenal, crossing the park, where the brilliant autumn hues were already beginning to fade and the stiff wind whipped the ends of their scarves and sent dried leaves scuttling over the gravel paths.

"It might yet be us," Marjorie replied, her voice strangely emotionless.

But was it strange, really? After all, munitionettes were expected to brave the potential for disaster every day, to keep a stiff upper lip even as their skin turned yellow and their friends became too ill to work, to remain calm when an accident occurred, and, like the girls from the Barnbow Munitions Factory in Leeds, to help remove the debris, mop up the spilled blood, and get back to work. Unless a factory was utterly destroyed, as seemed to be the case with White Lund, they were required to time in for their next shift and carry on as if nothing had happened, even if their own dear friends had lost their lives only a few paces away from where they worked. The soldiers faced worse every day, and as Mabel used to declare, the lads risked their lives in the trenches, and the munitionettes risked theirs in the arsenal.

Whenever April thought of how close Oliver had come to never returning from the battlefield, she felt lightheaded from relief, so narrowly had she escaped terrible loss. *Her* loss, indeed, she mocked herself whenever the thought crossed her mind. She hadn't known him then. Oliver would have suffered the loss, he and his family, his friends and loved ones. She wasn't even sure what she meant to him, not really, and it was silly and sentimental to think of her own feelings when the times required courage and sacrifice. She knew this, and yet the thought of Oliver being absent from the world was too dreadful to contemplate.

He was not in the park again that morning, but there hadn't been any air raids upon London the night before, so she was merely disappointed, not worried, as she and Marjorie quickly made their way around the compound to Gate Four.

She tried to put Oliver out of her thoughts entirely as she worked alongside Marjorie in the Filling Shop, packing yellow powder into shell casings, refusing to think of what the TNT was doing to her lungs and throat and blood. She would be rotated to less dangerous work soon enough. They all had to do their bit until the war was won.

Later that afternoon, she was working away, keenly focused, when Marjorie spoke her name. Glancing up, she realized the other canary girls around their table had paused in their work to watch the corridor as Superintendent Carmichael, her jaw set in grim resolution, a telegram in her hand, strode past the broad open doorway to the Filling Shop—

And kept going. A collective sigh went up from the table, but only a few girls promptly resumed their duties. The others exchanged wary glances. Someone in the Danger Building—not anyone in their shop, thank goodness, not any of them—was moments away from receiving devastating news.

They all knew the superintendent would soon escort the unfortunate woman past their doorway, back through inspection to the washroom and the shifting house. No one who received one of those telegrams was allowed to finish her shift, even if she felt up to it. As far as April knew, no one had ever wanted to.

They waited, their eyes fixed on the open doorway. Some held their breath, some murmured prayers. Others counted their blessings that this time, they had been spared.

They waited.

Superintendent Carmichael appeared first, passing the open doorway without glancing into the Filling Shop. Close behind her came Daisy, her arm around the waist of another woman, keeping her upright as she stumbled forward, her eyes bleak and unseeing, her face pale with shock beneath the yellow pallor—

"Lucy," Marjorie murmured, a catch in her throat.

They all watched, helpless, silent with stunned misery, until the three women passed the doorway and disappeared from view.

18

October–December 1917

HELEN

Helen learned about Lucy Dempsey's heartbreaking news only after Superintendent Carmichael returned from delivering the dreadful telegram and escorting Lucy to the main gate. "I wish you would have told me first," Helen said, bounding up from her chair and leaving the bemused superintendent staring after her. If Helen caught up with her teammate, she could escort her to her tram station, or even all the way home, rather than leave her alone with her grief. But she was too late. When Helen reached the park in front of the main gate, Lucy was nowhere to be seen.

Frowning in consternation, Helen returned to her tiny office to find the superintendent gone but a memo placed neatly in the center of her desk, the significant details of Lucy's telegram transcribed from memory upon it. On 24 September, Corporal Daniel Dempsey had been wounded on the front lines. Immobilized by his injury, he had been taken prisoner before stretcher bearers could evacuate him to a field hospital.

Dire news indeed, but at least Lucy's husband was alive, or he had been at the time of his capture, which meant that hope remained for his safe return home.

Helen hurried to Arthur's office and paced in his antechamber until he finished a phone call with the Ministry of Munitions. After she explained Lucy's circumstances, he agreed that she could use the Purcell name and contacts to seek more information.

It took a few days, and the promise of a coveted bottle of wine to a surly clerk at the War Office, but eventually Helen learned all that had been recorded about Corporal Dempsey's status. On 24 September, the 17th Middlesex was on the front line north of Givenchy, enduring heavy shelling, which the enemy used as cover for a raid. Obscured by a heavy mist, a small party of Germans slipped into the British trenches, seized munitions, and exchanged fire before they were driven off, carrying off bombs, crates of explosives, and boxes of machine-gun ammunition. Thirteen men of the Footballers' Battalion had been wounded, four of them seriously, and two were missing, including Corporal Dempsey. Witnesses observed that he had taken shrapnel to the leg and, unable to withdraw, had been dragged back to the German lines, presumably for interrogation.

Heartsick for Lucy, Helen hoped that the report would still offer her some measure of comfort. Helen had heard from other wives and sisters of soldiers that bad news, distressing though it could be, was easier to endure than interminable uncertainty.

Superintendent Carmichael had urged Lucy to take a few days off, so it wasn't until the following week that Helen was able to share what she had learned. Lucy absorbed the news stoically, eyes shadowed and cheeks hollow, her yellow skin and patchily bleached dark hair evidence of her own sacrifices for the war effort—and now she might lose her husband too. Disregarding the arsenal's desperate need for workers, Helen offered Lucy more time off, but her fellow Canary shook her head. "I need to occupy myself with useful work," she said quietly. "It's the only thing that distracts my thoughts away from . . . all the dire possibilities."

Helen understood, but just to be sure, she asked, "Are you sure you wouldn't benefit from more time at home with your sons?"

Again Lucy shook her head. "I sent them to Surrey to stay with my mother for the duration." She glanced down at her lap, realized she was wringing her hands, and abruptly stopped, spreading her hands flat on her lap and then balling them into fists. "I've been weighing whether to get them out of the city for months, what with these dreadful air raids showing no signs of easing, but this finally convinced me. I may have lost Daniel, but I won't lose them too—" Her voice choked off.

"You don't know that you've lost your husband, not for certain," said Helen, her voice low and soothing. "He was alive and conscious when his companions last saw him."

"Yes, I know. Believe me, I've been holding on to that hope with all my might." Lucy inhaled shakily. "I suppose everything depends upon how serious his wounds are, and how the Germans treat him."

In parting, Helen promised to tell Lucy immediately if she managed to wring any more details out of the War Office, and Lucy thanked her and returned to work. When Helen saw her at football training later, she seemed as determined on the pitch as ever, her expression stoic, her kicks so forceful that Helen reckoned she was taking out her anger and grief on the ball. Afterward, as Helen headed back to the administration building and the other players returned to their various factories and shops, she glimpsed Lucy and Marjorie walking together. Just as they rounded a corner, she thought she saw Marjorie put her arm around Lucy's shoulders. Helen was so astonished by the unexpected gesture that she halted and stared, but the pair had slipped out of sight, and she couldn't be sure she hadn't imagined it. In her time at Thornshire, Helen had observed Marjorie treating Lucy with respect on the pitch but with kindness nowhere. Perhaps Marjorie's own sorrows had rendered her able, at long last, to empathize with someone else's suffering.

At the end of a long day, Helen tidied her desk, gathered her coat and purse, and went to coax Arthur out of his office so they could go home to a late dinner. She had done her bit for King and Country, and she had provided her munitionettes with the resources they sought and the comfort they needed. If only she could tend to Arthur half as well, she would feel like an adequate wife. But although she plied him with nourishing food and practically forced him to rest, he was perpetually exhausted and overstressed, toiling relentlessly at the arsenal, rising early and staying up late to attend to paperwork at home, waking in the middle of the night tormented by terrible dreams. She felt powerless to comfort him, watching anxiously as the war aged him before her eyes, pleading with him to take better care of himself, succeeding only rarely. She prayed for an end to the war, not only to bring the senseless slaughter to an end, but to spare her husband's life. She was terribly afraid that he might actually work himself to death if she did not fight like mad to prevent it.

But she had learned by now that confessing her worries to Arthur accomplished little. The only scheme that worked was to approach him with fond humor, gentle teasing, and loving requests. When the Thornshire Canaries had a match in Stratford or elsewhere around London, he would take a Sunday afternoon off to watch Helen play. In late summer and early autumn, she had occasionally been able to coax him into accompanying her on her morning walks through Hyde Park, but his interest had plummeted after an air raid on the last night of September, one of several that had struck harrowingly close to home. When they set out for their walk the following morning, they discovered that the Germans had dropped a bomb in the Serpentine, the curved lake that narrowed in the west to become the Long Water of Kensington Gardens. As they strode past, they had been stunned and sickened to see dozens if not hundreds of dead fish floating on the water, killed by the concussion. Arthur uttered a string of sharp, disparaging remarks about the Germans' utter disregard for life, but then he abruptly fell silent, and she had barely been able to pry another word from him for the rest of the walk.

On the few occasions they had visited the park after that, he had steered them away from the Serpentine, and for the past fortnight he had declined to go walking with her altogether. She wanted to blame his refusals on the increasingly cold and rainy autumn weather, and on his urgency to leave for the arsenal, but even when she offered to set out earlier and the weather wasn't all that bad, still he would not join her. In the absence of a more logical explanation, she concluded that the sight of the pale, bloated fish bobbing lifeless on the Serpentine, which they had both long admired for its serene beauty, had profoundly unsettled him, and every glance evoked a memory that made him shudder.

So, as October passed, since Helen could no longer enjoy her husband's company and conversation on her morning walks, she instead used the time to plan her day, and to plan, too, bits of encouraging news to share with him later that evening, anything that might ease the worry lines in his face, if only for a moment.

But with British casualty lists lengthening, the demand for munitions rising, and foodstuffs such as tea, sugar, butter, and bacon becoming scarcer by the day, it was often quite a stretch to grasp hold of anything that qualified as good news. Yet she could nearly always find something. The Thornshire Canaries were enjoying an excellent run and were one victory away from winning a spot in the Munitionettes' Cup tournament. The new TNT safety protocols seemed to be working; canary girls were still falling ill, but less seriously and in lower numbers than before. Soldiers and resources from the United States were steadily bolstering the Allies throughout France and Belgium, which must inevitably shift momentum in their favor. Surely they were closer to the end of the war than the beginning.

Yet there were days when all the news was sorrowful, and there was nothing Helen could do to remedy it.

One morning at the end of October, Helen arrived at the administration building to find Oliver waiting outside her office, his expression grim. "What is it?" she asked, heart sinking, as she led him into her office. "Just tell me, quickly. Get it over with."

"I'm afraid it's bad news," he said—stalling, despite her request. "One of our former employees died yesterday of TNT poisoning. The coroner confirmed it."

Helen closed her eyes and took a steadying breath. "Who?"

"Mabel Burridge."

"Oh no."

"I'm sure you're aware that she's been ill for months."

"Yes, but she always had such a strong will. I suppose I thought she would pull through somehow." Helen paused. "Are you sure? Are you absolutely sure it was Mabel?"

Oliver nodded. "Her daughter Peggy told me on her way in this morning."

"Peggy came to work?" said Helen, incredulous. "But she's entitled to bereavement leave."

"I told her. She knows. She doesn't want it."

Helen shook her head, words failing her. She did not know Mabel well, except as a respected adversary at the negotiating table, but she knew Peggy, and her heart ached for her teammate. "I'll talk to her later," she said. She thanked Oliver, and after he left, she shut the door, removed her coat and hat, and sank into her chair. Folding her arms on top of her desk, she closed her eyes and rested her head on them, tamping down her grief and hopelessness, searching for the proper words of comfort, something that didn't sound clichéd and empty from too much repetition throughout this wretched, terrible war.

When she spoke with Peggy later that afternoon, she was relieved to see that her teammate was bearing up well, all things considered, and that her friends were surrounding her with love and support. Helen told her privately that Purcell Products would pay for her mother's burial expenses and provide a small life insurance payment to her dependents, two longstanding employee benefits that had predated the war. Peggy gratefully accepted, but when Helen again reminded her about bereavement leave, Peggy refused to take any more than two days for the service and burial. She needed the distraction of work and the wages, she explained, turning her head and clearing her throat, then breaking into a fit of hoarse coughing that sent Helen running to fetch her a cup of water. A harrowing thought came to her as Peggy accepted the cup and sipped between coughs—Would the daughter soon follow the mother into an early grave?

Peggy took Friday and Saturday off, as they had arranged. No one expected to see her at the match on Sunday, but she arrived halfway through their warm-up, her eyes puffy and tired but dry. Her teammates embraced her and offered their condolences, and she managed a smile at their surprise that she had felt up to what was expected to be a grueling match. "I couldn't let the side down, could I?" she said, eyeing the team on the other half of the pitch. "They look like they know what's what. You need me if we're going to make it into the tournament."

"And well we know it," said Lucy, and the other Canaries murmured agreement.

April had sewn black armbands for them all to wear, and as she handed one to Peggy, she said, "It's been decided. We're dedicating this match to your mum."

Peggy's eyes glistened with fresh tears as she slipped the black cloth around her upper arm. "I expect you all to play your best, then," she said, and everyone smiled or applauded or laughed through their tears, remembering Mabel.

It was a hard-fought match, one that left Helen with bruises and scrapes and a fresh blister on the ball of her right foot, but the Thornshire Canaries seized victory, 4–3, with a brilliant goal from Peggy on an assist from Lucy with three minutes left in the match. An impressive defensive effort, if she did say so herself, held off their rivals until the whistle blew.

How fitting it was, Helen thought as her teammates went mad with elation, cheering and embracing and tossing their caps into the air, to honor their club's founder by securing a place in the tournament for the first-ever Munitionettes' Cup.

And how grateful she was that Arthur had been there to witness it, and even now was on his feet in the stands, applauding and beaming. If only that light in his eyes would remain after they left the Globe, and if only his delight in their victory would keep his worries at bay long enough for him to get a proper good night's sleep, without his heavy responsibilities jolting him awake at all hours. But as thankful as she was for these brief respites, they were never enough to fully restore him.

Then, in late November, astonishing news from the Western Front relieved the sinking dread that had settled over Britain, like a fresh wind sweeping away storm clouds. On the morning of 20 November, more than one thousand Allied guns had launched an assault against German defenses and artillery positions, and as tanks had pushed forward, the infantry had begun an advance along a six-mile frontage. By nightfall, the British Army had advanced nearly four miles, taking four thousand German prisoners and capturing or destroying more than one hundred enemy guns.

At long last, the Third Army had broken through the Hindenburg Line.

When the news reached Britain a few days later, London church bells pealed in celebration for the first time since the outbreak of war. It was the most wonderful news Helen could recall in ages, perhaps since the United States had joined the war. Although the work of the arsenal continued and she could not spare a moment to join the crowds rejoicing in the streets, she relished the jolly music of the bells, too long silent, and as fatigued as she was, she could not keep from smiling.

Of course the merriment could not last; she had not expected it to, but neither had she expected their soaring hopes to come crashing down again so soon and so suddenly. On 5 December, Russia, now fully under the control of Lenin and his Bolsheviks, signed a ceasefire agreement with the Central Powers. The Russian foreign ministry had invited its counterparts in Britain, France, and Italy to join in the negotiations for a fuller peace, but the Allies were said to have rejected the overture with furious, stony silence. Ten days later, the Russians and Central Powers agreed to a thirty-day armistice, which would be automatically renewed unless one of the parties notified the other of its intention to resume hostilities. When the armistice went into effect on 17 December, Russia essentially withdrew from the war.

That same week, a cloud of despondency darkened the skies over London. The British advance on the Western Front, which they had celebrated so joyfully scarcely three weeks earlier, had become a retreat as the German counterattack forced their troops back so far that nearly all the ground they had gained was lost. Thousands of Tommies had been killed or wounded, and thousands more taken prisoner. Helen hardly knew what to make of dismaying assertions in the press that the Allies were in a worse position now than they had been at any time since the war began. And yet the British people were urged to steel their nerve and strengthen their morale, for Britain would never go the way of their erstwhile ally Russia, and the only way to reach the peace that awaited them at the end of the war was to go through it.

Helen understood the point of such encouragement, for if the choice was to proceed or to surrender, they really had no choice at all. And yet it was difficult to see how one was meant to take heart when an important ally had bowed out of the conflict, casualty lists relentlessly lengthened, pantries were growing bare, especially in the homes of the poor, and food was increasingly hard to find. Each week, an average of fourteen ships bringing food and raw materials to the British Isles were sunk by enemy submarines. Helen had become accustomed to the sight of long queues winding outside the doors of provision shops, the anxiety plain on the customers' faces as they waited to be served, wondering whether anything would be left by the time their turn came. Demands for compulsory rationing were increasing from all corners of government and commerce, but the Food Controller, Lord Rhondda, opposed such measures, insisting that rationing would be difficult to administer and the heaviest burden would fall upon the shoulders of the poor. In the meantime, he assured the press that food and drink remained plentiful, and all that could be desired for the festive season would be in the shops. Even so, citizens were encouraged to be "sparing" in their observances so the abundance would continue into the New Year.

Helen was skeptical of any government statement that included the words "sparing" and "abundance" in such close proximity, but she decided to call Lord Rhondda's bluff, if a bluff it was. If food rationing would be imposed at the end of the year, as others in the government hinted it would, she would take steps now to ensure that there would be plentiful, nourishing meals in the Thornshire canteens in the future—confirming agreements with their suppliers, stocking up on non-perishable items, reviewing kitchen procedures to reduce waste. Then—because she knew her munitionettes deserved it, and because she doubted the much-lauded plenty extended to most working-class homes, and simply because she *could*—she organized holiday feasts to be served at dinner and tea in the canteens for Christmas Eve and Christmas Day. On Boxing Day, each worker would take home a generous box of food and drink to share with their families. She didn't tell Arthur until after all the arrangements were made, the bills were paid, and it was too late to cancel, but fortunately, he agreed that it was an excellent way to thank their workers for a job well done in arduous circumstances.

"I'm glad you think so," said Helen, reaching for his hand across the dinner table, mealtimes being one of the few occasions when she had his undivided attention. "In addition to the feast, I've arranged for the arsenal's most overworked employee to go on holiday from Christmas Eve through New Year's Day."

His eyebrows rose. "A paid holiday?"

"Not quite. Special circumstances." She leaned forward and clasped her other hand around the one she already held. "Darling, the employee is you."

He frowned and heaved a sigh. "Helen—"

"Arthur." She held up a hand to forestall argument. "You need time off. It's all been arranged. Tom and Oliver have cleared your calendar, and they'll manage everything in your absence. And before you propose to work so they may go on holiday instead, they decided between themselves that Oliver shall take off Christmas Eve and Tom Christmas Day. They are as resolved as I am that you must take a holiday."

"Darling, if you think Christmas in Warwickshire with my family will be restful—"

"Goodness, no, not that. Oxford, darling. Our house in Oxford. The staff is preparing for our arrival even now. My mother and Daphne will pass the holidays with us there. Penelope, Margaret, and the children shall spend Christmas with Margaret's family and join us later in the week. We'll have a lovely, long-overdue reunion." She hesitated.

"If you'd like, we could invite your father and siblings and their families to join us—"

"No, no," he broke in, shaking his head. "That won't be necessary. I'm sure they've already made their own plans by now." He clasped a hand to his brow, rubbed his eyes, and wearily added, "Let's keep this a Stahl family gathering, just to ensure peace of mind."

"Then we're agreed?" she asked, hardly daring to believe it. "You'll take a week off, and we'll spend the holidays in Oxford?"

"Agreed," he said, managing a wan smile. "I'm too exhausted to debate you, which I'm sure you'll interpret as conclusive evidence that I need a rest."

"I do indeed, but even so, I always win our debates in the end."

"Not always."

"Often enough."

He shrugged and nodded, conceding the point. Rising, still holding his hand, she drew closer and bent down to give him a fond, lingering kiss.

He would get all the rest he needed; she would see to it. They would celebrate the fourth Christmas of the war with as much joy and hope and gratitude as they could muster, and on New Year's Eve, they would bid farewell to an age of war and strife, and look forward to a year that might, and surely must, bring victory and peace.

19

December 1917–February 1918

LUCY

It was the fourth Christmas of the war, and the first Lucy had ever spent without a single member of her family near. Her mother had begged her to join the Evanses in Brookfield, but the exigencies of the war would not allow Thornshire Arsenal to shut down even for the day. It broke Lucy's heart to tell her loved ones, especially Jamie and Simon, that she would not be among them to exchange gifts, sing carols, enjoy a delicious if smaller-than-usual feast of their favorite holiday dishes, and to break open Christmas crackers and don the paper hats and laugh at the silly riddles enclosed.

"But Edwin will be home on leave," her mother protested when it was her turn on the telephone.

"Yes, I know he will. He promised to see me in London on his way to Surrey."

"But did you know that Dieter Brandt has been released from Lofthouse Camp? He's back at the bakery, and even with sugar rationing, his breads and rolls are as delicious as ever."

Lucy had to laugh. "Do you really think I wouldn't come just to see the family, but I'd come for sweets?"

"You shouldn't be alone on Christmas," her mother chided. "Not when I know you worry night and day for your Daniel. Not when you have a loving family an easy train journey away."

"I won't be alone," Lucy assured her, struggling to keep her voice from breaking. Yes, she ached for Daniel—and the thought of returning to their childhood village, where every street and meadow and woodland reminded her of him, of all they had shared, of all she might have lost, pained her beyond reckoning. "I'll be at work, but surrounded by friends."

"That's not a proper Christmas, filling shells at an arsenal."

"No, it isn't, but it's wartime, and we all have to make the best of things."

"I wish you'd resign from that dreadful place," her mother said, impassioned. "That horrible yellow powder has stolen your beauty and God forbid it should take your life too. The stories I hear about explosions—"

"Mother, please. Jamie and Simon—"

"Are outside playing in the snow with their cousins. Do you really think I would speak so freely if they could overhear?"

"No, of course you wouldn't." Lucy pressed her lips together and squeezed her eyes shut against tears. "Keep their spirits up, won't you? Don't let them worry too much about their father. Remind them that at least now he's out of the trenches, away from the fighting."

"I'll do my best, dear, but they're clever boys and they know what a prison camp is." Her mother sighed. "I can't bear the thought of you coming home from work to an empty house on Christmas. Why don't I bring the boys up to London on Boxing Day? We'll see you in the evenings, and we'll have all Sunday together, and I can take them back to Brookfield on Monday."

For a moment Lucy hesitated, overwhelmed by longing. But then she reminded herself that the danger was too great, that her loneliness was far less important than their safety. She had sent her children out of London to escape the air raids, but after Mabel died, she had found another reason for them to stay away—not from the city, but from herself, and she prayed she had not acted too late. What terrible contamination might she have inadvertently brought home to her mother and sons throughout the long months she had worked in the Danger Building? She always scrubbed herself thoroughly in the washhouse after removing her soiled uniform at the end of a shift, and she changed back into her own clean clothing afterward, but that vile powder got everywhere and she might have missed some. Or what if she had breathed infinitesimal grains into her lungs, only to exhale them at home, where her mother and sons might have taken them in? The thought that she might have unwittingly poisoned the people she loved best in all the world terrified her—which was why the boys could not come home, nor could she go to them, not as long as she worked at the arsenal. And she had to keep working at the arsenal or they would lose their home, and how could she bear the shame if Daniel returned—*when* Daniel returned—to find that she had failed him in this one essential duty? He must have a home to return to, *this* home, with the cozy study where he worked on blueprints at the architect's desk his father had made for him, where his Olympic gold medals were displayed in two glass cases. She would prepare his favorite meals in the kitchen and he would kick the football around with the boys in the garden, and everything would again be as it once was—happy, peaceful, perfect.

If not, what had all their sacrifices been for?

Although the Winter Solstice had passed, the days were still short and the nights cold and long, so Lucy left for the arsenal before sunrise and returned home after nightfall. Her only glimpse of thin winter sunshine came during her breaks, when she and the other Canaries ate swiftly and hurried out to their frozen practice field for a kickaround, and on Sundays, her only day off. On the rare occasions when she went marketing, she found reassuring pleasure in the unexpectedly ample, enticing displays in storefront windows, an abundance she had not seen since before the war. Nor had proprietors neglected the familiar festive touches of the season. Shops were adorned with holly; poulterers decorated their geese and turkeys with cheerful red ribbons, as did butchers with their cuts of beef and mutton. Lucy's mouth watered at the charming arrangements of apples, oranges, and nuts at the fruiterer's, and although she was not a drinker, she marveled at the profusion of wine, whiskey, and brandy at the grocer's. Everything was very dear, of course, but it seemed that everyone had money to spend, thanks to steady government wages for war work, and the increased allowances paid to soldiers' wives and other dependents. Her own allowance had increased, not quite enough for her to resign from the arsenal, but enough to tempt her, just as the marvelous shop displays tempted her to splurge. Yet she resisted, since she had only herself to feed at home, and because the Food Ministry urged prudence during the festive season. With rumors circulating that food rationing would soon be imposed, Lucy worried that the abundance in the shops was an illusion, holiday feasting before midwinter famine.

At least no one went hungry at Thornshire Arsenal. From Christmas Eve through Boxing Day, the canteens served splendid meals—Irish gammon, cabbage and potatoes, roast goose, plum pudding, among other tantalizing dishes—for dinner as well as tea. A jolly mood prevailed elsewhere too, as munitionettes sang carols while they worked, and friends exchanged small gifts in the shifting house at the end of the day. Lucy had knit soft, warm scarves for her fellow Thornshire Canaries in their team colors, yellow and black, and her friends declared themselves absolutely delighted with them.

At dinner on 26 December, Daisy entertained the Canaries with a secondhand report of a football match that had taken place on Christmas Day at Deepdale Stadium, home ground of the Preston North End. Daisy's cousins had joined more than ten thousand other spectators for a spectacular match between the Dick, Kerr Ladies of Preston and their crosstown rivals, the Coulthards Ladies of the Arundel Coulthard Foundry.

"Ten thousand spectators?" said Marjorie, skeptical.

"Ten thousand and then some," said Daisy. "They raised more than six hundred pounds for the local auxiliary war hospital."

A gasp of astonishment went up from the group. Even Marjorie looked impressed.

"Dick, Kerr commanded the pitch from the first whistle," Daisy continued as her teammates inched their chairs closer, enthralled by her descriptions of astonishing goals, brilliant passing, and wildly cheering fans. In the end, Dick, Kerr had triumphed, 4–0, but the lopsided score misrepresented how exciting the game had been thanks to the masterful skill and intriguing personalities of several of the Dick, Kerr players.

"Their center forward is a gorgeous blonde named Florrie Redford," said Daisy. "They're always putting her photo in the papers."

"I was a gorgeous blonde once," said Marjorie mournfully, tugging at a brittle, gingery curl.

"And will be again," April reassured her. "After the war. Give it time."

"My cousins say their inside left player, Jennie Harris, is an expert dribbler and very quick on her feet," said Daisy.

"She's no match for our Peggy, I'm sure," said Marjorie, "and our Helen can outrun anyone."

"I don't think you're hearing me," said Daisy, with a warning shake of her head. "My cousins said they've never seen a girls' team like Dick, Kerr before. After the most stubborn critics in England watch them play, they never again argue that girls can't or shouldn't play football."

"Doesn't every team in the Munitionettes' League prove that girls can play football?" asked Lucy indignantly, evoking emphatic nods from her friends.

"Agreed, but the Dick, Kerr Ladies go above and beyond," said Daisy. "Their outside left, Lily Parr, is only fifteen years old, and they say she has a kick like a Division One back."

"Fifteen years old?" echoed Marjorie, incredulous. "Is she even allowed to work in munitions? What do they make at Dick, Kerr, anyway? Tea biscuits?"

That earned her a laugh, but Daisy shook her head. "Munitions, same as us. They used to make locomotives and tramcars, but they converted at the start of the war."

"Do you suppose we'll face the Dick, Kerr Ladies when we play for the Munitionettes' Cup?" asked Lucy.

"Not if the Blyth Spartans knock them out of the running first," said Peggy. Everyone nodded soberly, remembering their own humbling 6–1 defeat all too well. "We'll surely face one of those teams eventually—or both of them, if we make it far enough."

"I don't know about the rest of you," said Marjorie, stretching her arms overhead as if warming up for the goal, "but I intend to make it to the finals."

They all sat for a moment, pondering Daisy's report, and then they quickly rose and cleared away their trays and dishes. Minutes later, they were out on the frozen field, heedless of the cold and the snowflakes drifting lazily on the wind all around them, running and dribbling and taking shots on their makeshift goal. The first round of the tournament wouldn't come until late February, but they must keep their skills sharp. The Dick, Kerr Ladies and the Blyth Spartans surely weren't neglecting their training.

Thus Christmas passed, the loneliest Lucy had ever known, and the New Year began, bleak and bitter cold. In the shops and markets, the abundance of the festive season proved as fleeting as she had feared. After years of asking the British people to practice voluntary rationing, the government had finally imposed official restrictions, but only on sugar. Every household was issued a ration card and was required to register at a local grocer, where they could purchase their allotted share and not an ounce more. Some of Lucy's friends praised the measure, because at last they would be assured of getting their portion regardless of where they found themselves in the queue, while others speculated that this was surely a sign of additional, harsher rationing yet to come. Lucy thought that more rationing might actually be desirable, in service to the common good, especially if it included price fixing. Without government intervention, shopkeepers could charge whatever they liked for scarce products, so the rich paid extra to obtain whatever they wanted while the poor did without. Just as some of Lucy's munitionette friends had never eaten so well until they were provided nourishing meals in the arsenal canteen, some less fortunate Londoners had been unable to afford sugar until rationing put everyone on an equal footing. Even the King and Queen had ration cards, which made it very awkward for anyone else to grouse about the scheme.

In the meantime, meat and poultry were increasingly difficult to find—which made her wonder what necessity she now took for granted would be the next to become unexpectedly rare and precious. Not tea, she fervently hoped. She could hardly drag herself out of bed some mornings without the promise of a reviving cup of tea at breakfast.

She wondered what Daniel was provided to drink in his German prison—tea, ersatz coffee, filthy water—but as soon as the thought came to her she shoved it away, before her breath caught in her throat and tears filled her eyes and she collapsed in grief. She could not bear to think of what he might be suffering. Sometimes it was all she could think about.

If only she had some news of him—another report from the War Office, a letter via the Red Cross, anything to assure her that Daniel lived, and might yet come home to her.

January dragged on, as cheerless as the holidays had been hopeful. Lucy felt weary to her bones, plagued by coughs and headaches, dread and grief. She longed for her children and her husband and springtime and rest. London was rife with rumors that the Germans, determined to terrorize the British and break their will, were preparing their most massive bombardment yet, fleets of dozens of their largest aeroplanes armed with their most destructive bombs. Lucy did not know where the rumors had started or how credible they might be, only that the threat of an impending raid seemed to be in everyone's thoughts. As the nights of the full moon approached, dread infused every conversation. Even in the daytime, traveling between home and work or training on the arsenal field, Lucy found herself adopting the common habit of glancing up at the sky or pausing in mid-conversation to strain her ears for the distant drone of German aircraft.

At the end of her shift on the last Monday of January, Lucy stayed a few minutes late at her station to finish up one last shell before racing to the shifting house to wash and change clothes. She quickly slipped into her coat and tugged on a warm knit cap and scarf and mittens, but she and one other girl from the Filling Shop were still the last to leave. Together they stepped out into a clear, frigid, moonlit night, their breath ghosting white on the air as they hurried to Gate Four, commiserating over their belated departures, which might oblige them to wait for later trains. Then, with exasperatingly poor timing, one of the sentries selected Lucy for a random questioning. If she were a man, it would have been a pat-down search, but a less invasive delay was still a delay.

"Bad luck, Lucy," her friend called from the other side of the gate. "Do you want me to wait?"

"No, go on ahead," she called back. "No sense in both of us missing our trains."

"Sorry, miss," the sentry said gruffly, looking her up and down as if she might have a shell poking out of her pocket.

"Not at all," she said briskly, rubbing her mittened hands together for warmth. "You're just doing your job."

A long moment later, he waved her on through the gate. She hurried through the arsenal park as quickly as she dared, mindful of the dark patches of ice on the gravel paths, but her friend had disappeared into the blackout. Fortunately, she did not have to wait long for the next tram, and it was only half full so she easily found a seat, but by the time she reached Barking station, her usual train had already departed. She had to wait another thirty minutes for the next one, pacing for warmth, ignoring the gnawing hunger in her stomach and the double takes of passersby who had apparently never seen a canary girl before, a rather odd thing in a neighborhood so close to an arsenal. She ignored too the raised eyebrows and furrowed brows of fellow would-be passengers whenever she covered her mouth and bent over in a hoarse, prolonged bout of coughing. Her next rotation out of the Danger Building was only days away, but it could not come too soon. She dared to hope that after the war, her symptoms would vanish and her skin and hair would regain their former loveliness, though she knew not everyone fully recovered.

Suddenly a memory came to her, something she had all but forgotten—her interview at Thornshire Arsenal, when Superintendent Carmichael had remarked that she would have hired Lucy to work in her own office, if only she'd had a vacant post. Would it be worthwhile to reintroduce herself to the superintendent and ask if anything had become available since then? She could afford a small pay cut, especially with the boys away in Surrey. At one time Lucy would have been ashamed to abandon Danger Building work to other, braver girls, but with the onset of winter, she had felt herself sinking, and she was not sure how much more exposure to the yellow powder she could endure. Even walking to the train station from the arsenal winded her, and football training had become a struggle. The winter pause in the Munitionettes' League had come just in time. She could only hope that after rotating out, she would recover her strength in time for the tournament.

She caught herself and laughed aloud, incredulous. She was worried about football, when her very life could be at stake.

It was nearly eight o'clock by the time she boarded her train, famished, exhausted, her toes numb with cold that had crept up from the pavement through the soles of her boots. A kind older gentleman gave her his seat, and she sank into it gratefully, murmuring thanks. Muffling a yawn, she settled in for a long ride, hoping that the rocking of the train wouldn't lull her to sleep and cause her to miss her stop at Victoria station, where she changed trains for Clapham.

She closed her eyes, just for a moment, but they flew open again when she imagined she heard the shrill blast of a policeman's whistle. But it was only the brakes squealing as the train pulled into an underground station. As they slowed and halted at the platform, Lucy took in the dismaying sight of hundreds of people packing the chamber—elderly couples, women with infants in their arms, children—huddled against the walls or sitting on the stairs, most of them wrapped in blankets over their coats. A few policemen and special constables milled about in the limited space between them, keeping order, offering assistance. A distant, dull, barely audible roar confirmed what she had already guessed: The air raid Londoners had dreaded for a week had finally come. And yet she observed no trace of fear on the faces of the sheltering people, just a weary, matter-of-fact acceptance, while many of the children slept peacefully in their parents' arms or on their laps. As her train started up again and the faces dissolved into a blur of motion, she was riveted by how dissimilar the scene was to the terror and fear she observed—and had herself felt—during air raids when she had been caught aboveground, or had sheltered in a simple basement or cellar. No doubt these people considered themselves as safe as one could be under assault in wartime, so far beneath the surface. She hoped they were not mistaken.

As she rode on, the awareness that the city was under attack kept her awake and alert. At every station, in the interval between halting and proceeding, she observed different crowds, but the same weary resignation, even boredom. Very rarely she glimpsed smiles and conversation and laughter, and she marveled at them. No one there or anywhere in Britain, not even the most prescient scholar of international politics and warfare, could have imagined four years ago that they would find themselves where they were now. Had they really become so accustomed to the unimaginable? But what choice did they have? It was keep a stiff upper lip or go mad, carry on or surrender.

At last her train reached Victoria station. She disembarked swiftly, but struggled to make her way through the crowds to the platform where her connection would depart. Finally, after some increasingly frantic weaving and dodging, she made it to the platform and boarded with seconds to spare.

Victoria to Clapham Junction was the shortest leg of her commute, and it was heartening to know that she was nearly home, but she could hear the bombs better now, and a frisson of dread passed through her. She realized, too late, that it might have been safer to wait out the raid at Victoria and continue on after the all-clear had sounded. Instead, when she disembarked at her stop, she found what seemed to be a secure corner between two solid walls and waited for the terrible pounding to cease, knowing that she had as much to fear from British shrapnel falling back to earth as from the German bombs. She was not alone; dozens of other travelers waited for the bombing to cease, apprehension and fatigue as plain on their faces as undoubtedly they were on her own.

At last, not long after eleven o'clock, the cannonade fell silent. Lucy was eager to get home and go to bed, but she lingered in the station with the other passengers, awaiting the all-clear, the official signal from military authorities that the raiders had departed British skies. But the minutes passed, and midnight approached, and still the all-clear had not been sounded. The passengers conferred quietly, impatiently, all of them still listening keenly for the signal that it was safe to step out onto the streets. Had the officials forgotten? Had the ground spotters lost sight of the aircraft and had no idea where they were? Or worse yet, was another wave of enemy aeroplanes already on the way, determined to unleash another barrage upon the unsuspecting citizens, who assumed the danger had passed?

Eventually the crowd began to disperse, as passengers decided the risk was worth taking and left the station. A few others settled down on benches or on the floor to doze while they awaited the all-clear. Lucy, lightheaded from hunger and illness and fatigue, tallied up the hours until she would have to hurry off to the arsenal again. If she did not get to bed soon, she would be in no condition to work the next day, and might be a danger to herself and others.

"Best not go out quite yet, miss," a police officer warned as she approached the exit. "I could find you a blanket and keep watch over you and the other ladies. Get a bit of sleep while you're waiting."

She thanked him, but she had made up her mind. It was not far to go, and she could run if she had to.

The darkened streets were deserted as she strode briskly toward home, her pace quickening as she thought of her own front door, the cold supper she had prepared that morning, her warm, comfortable bed. She had reached the corner of her block when she heard a distant explosion. Heart in her throat, she tucked her bag tightly under her arm and ran the rest of the way, sprinting up her front walk, fumbling with the key in the lock, racing inside, shutting the door firmly behind her, dropping her things, snatching up her dinner on the way as she fled through the kitchen into the cellar. All the while, the roar of bombs and the incessant thud of guns grew louder and louder.

Alone in the dim cellar, she choked down her dinner, then lay down and tried to make herself comfortable on the pallet of blankets she had left for this purpose. Eventually she dozed off, shivering and alone in the dark, even as the house trembled around her.

She woke at half one o'clock to the all-clear.

It was madness, she thought as she climbed wearily upstairs and, after a cursory wash and brushing of teeth, collapsed into her own bed. She ought to evacuate to Brookfield and wait out the rest of the war with her children and family. She had done her bit. The arsenal would carry on without her just fine, as would the Thornshire Canaries. As for her lost wages, Daniel would forgive her if she lost the house. He would be less able to bear the loss of his wife and his children's mother.

In the morning, she rose at her usual hour, still exhausted, drank a cup of hot tea and hastily ate a thick slice of toast, all without sparing time to sit. Then she raced off to the arsenal, dozing on the train, learning from overheard conversation that forty-seven people had been killed in the previous night's raid, and 169 others had been injured. Many of these were women and children who had taken refuge in a printer's works in Long Acre, near Bow Street.

"Tragic indeed," a gentleman remarked to his companion, "but if one considers the millions of homes and businesses in London, the odds of any one particular building being struck by a bomb are actually quite small. We can all take comfort from that fact."

Perhaps he could, but Lucy couldn't.

A week later, Lucy had just arrived at the training field when Helen greeted her with a halfhearted smile. "Congratulations, Lucy," she said sincerely, shifting the football she carried to her left arm so she could shake Lucy's hand. "I'm happy for you. A bit forlorn for myself, I confess, but happy for women like you. My turn will come."

"I'm sorry, but I have no idea what you're talking about," said Lucy, bewildered. "Why am I to be congratulated?"

"The vote, of course," said Helen. "The Fourth Reform Act passed. Women over thirty who are householders or who are married to a man who owns property are allowed to vote in Parliamentary elections now."

Lucy shook her head, astonished. "I hadn't heard. I follow news of the war and little else these days. I thought election reform wouldn't be enacted until peacetime." Then she paused. "Daniel doesn't own our home outright. We have a mortgage. Does that matter?"

"An excellent question." Helen dropped the ball and passed it to Lucy. "I honestly don't know. I suppose you'll find out when you try to register at the polls. As for me, I have to wait another two and a half years to vote for my MP. It's ludicrous. How will I be more capable then than I am now?"

"Didn't you know?" teased Lucy, passing the ball back to her. "Wisdom descends from above the moment a woman turns thirty."

"Really? Well, I look forward to it." Helen attempted a bit of fancy dribbling, gave up, and passed the ball to Lucy. "At least I can vote in local elections now, since I'm over twenty-one."

"My mother owns property," Lucy mused aloud, juggling the ball back and forth from one knee to the other before lofting it higher and heading it over to Helen. "My father left her the building with our family home and his medical practice when he passed. She'll be tickled to hear she can vote now."

"Property restrictions have been abolished altogether for men," Helen said, frowning thoughtfully. "Virtually all men over twenty-one can vote. If you ask me, that's so they can maintain a male majority in the electorate. After the war, we suffragettes will have a thing or two to say about that."

Lucy nodded, deferring to Helen's expertise on the subject. What would happen to Great Britain, she wondered, if one day women voters outnumbered men? Would there be more justice, equality, and peace? She hoped so. But would the nation never again go to war? Unlikely, she reckoned. Englishwomen would honor their promises to their allies as faithfully as Englishmen did, and they would be equally determined to defend their island from attack. As for starting an unprovoked war, it was unfathomable to her that women would go along with that sort of idiotic brutality, not with the grief and loss of this war haunting their memories.

A fortnight later—a blessedly cloudy night with only a sliver of a moon—Lucy returned home from the arsenal to find a letter from the War Office waiting for her in the post. She had a moment of stricken terror before she remembered that announcements of soldiers' deaths came by telegram, not by letter. Even so, her heart thudded as, still standing in her coat and hat in the foyer, she opened the envelope and withdrew two typed pages. She had barely finished the first lines before her legs became too weak to hold her, and she sank to the floor, breathless, eyes fixed upon the stunning pages.

Daniel had been located. He was alive.

A sob escaped her throat as she read on. Daniel had been wounded before his capture by the Germans—this, she already knew—and although his injuries were very serious, he was being attended to in a prison hospital and was expected to survive. The Red Cross had taken a particular interest in his case after reports appeared in German newspapers of the capture of a renowned footballer and Olympic champion. Such triumphant boasting in the press about a particular prisoner of war treaded close to violating the Geneva Convention of 1864, which established the right to the impartial treatment of combatants. There was no evidence, as yet, that the Germans intended to parade Corporal Dempsey in public as a trophy, but to forestall any such disgraceful and inhumane treatment, the British government was determined to secure his release as soon as possible.

An organization known as the Swiss Medical Mission had brokered an agreement at The Hague between Great Britain and Germany for the exchange of disabled prisoners of war. First, medical officers at internment camps on both sides identified eligible prisoners on the grounds that they were unfit ever again to become combatants. The selected prisoners, an equal number of Britons and Germans, would then be transported to an observation camp, where they would be examined by physicians with the Swiss Medical Mission. According to the nature of a prisoner's physical disability, he would either be sent home or to neutral Switzerland, for internment in surroundings more conducive to healing and rehabilitation until the war's end.

At the time of the letter's writing, Corporal Dempsey had already passed the selection and was en route to an observation camp. Whether he would be sent on to England or to Switzerland had not yet been determined, but his family must rest assured that he would not be returned to the German prison camp.

More details would be forthcoming.

Lucy clutched the letter to her chest, tears streaming down her yellowed cheeks. Daniel lived. He would soon be out of German hands. He might be coming home to her soon, but if not that, she could take great comfort in knowing that he would be safe in Switzerland for the duration, and their family would be reunited after the war.

It was only later, after she had telephoned her mother and Daniel's father with the wonderful news, after she had eaten supper and washed her face and felt much restored, that a nagging worry buried beneath the good news came to the forefront of her thoughts.

She read the letter again.

As certain phrases leapt off the page, her hands trembled until the page shook so much she could scarcely read it. But the implicit warning was clear, as were all the terrible questions it prompted.

Only prisoners unfit ever again to become combatants.

The nature of a prisoner's physical disability.

What did these ominous words mean for Daniel? What had happened to him?

20

FEBRUARY–MARCH 1918

APRIL

The Thornshire Canaries rejoiced in Lucy's happy news, but they declared that even if Daniel returned home within a fortnight, which they all hoped he would, Lucy couldn't resign from the arsenal until after the Munitionettes' Cup. "You can't expect us to carry on without our star center forward," Marjorie teased, smiling.

"I wouldn't dream of it," said Lucy, her face alight with happiness and hope as April had never seen her before.

April had no doubt that Marjorie was entirely sincere in her joy for Lucy, but later, when Marjorie and April were alone, Marjorie confessed that she wished the Swiss Medical Mission could do as much for her brother Archie, who was still missing and unaccounted for.

The most obvious reason for Archie's prolonged absence and silence was too heartbreaking for April to admit aloud. "Maybe the Swiss can't help Archie because the Germans never captured him," she said instead. It was the first hopeful explanation that came to mind. "Maybe he's working with the resistance in occupied France or Belgium. Or maybe he *is* a prisoner, but he's in such fine health that he doesn't qualify for the disabled prisoner exchange."

"Yes, I suppose that could be so," said Marjorie, managing a wan smile.

The Canaries' shared delight in Lucy's relief and happiness must have invigorated them, for in the first round of the tournament, they soundly defeated the Associated Equipment Company Ladies at the AEC home ground in Beckton, 4–1, the same margin by which the Ladies had beaten them in their first meeting, when the Canaries were barely fledglings.

A few days later, April came down to breakfast to find the hostel dining room buzzing with anxious conversation. "Did you hear about the munitionettes in Leeds?" Marjorie greeted her when April joined her and two friends, Ethel and Rose, at their table.

April's stomach gave a lurch. "Another accident?"

"No, not that." Frowning, Marjorie leaned forward, resting her arms on the table around her bowl of porridge. "A few days ago, more than one thousand girls were sacked from a munitions factory. No notice, no nothing, just fare thee well, ladies, goodbye."

"One thousand workers?" exclaimed April. "But why? Has the war ended and they forgot to tell us?"

"The war's not over, more's the pity," said Ethel, "but munitions orders are down all over, what with the Russians pulling out. I never realized how many shells and such we British girls made for them until they stopped wanting 'em."

"Then the munitionettes weren't sacked so that lads who've come home from the war could take over?" asked April.

Marjorie shook her head. "Not from the sound of it. The jobs are just . . . gone."

"Goodness." April sat back in her chair. "What's going to become of those girls? Will they get benefits, at least?"

"They're asking for benefits," said Rose, brooding over her teacup. "Whether they'll get them is an open question."

"Most of these girls worked in textiles before the war," said Marjorie. "A secretary from one of their trade unions told the press that most of them don't want to go back to the clothing trade or the mills, not after enjoying such good wages on munitions work."

"And the ones who are willing to go back can't find jobs," said Rose. "What does this mean for us, I wonder?"

"It means you'd better make sure your work is up to scratch," said Marjorie flatly. "The lazy girls and troublemakers will be the first to go. Smile politely at the foremen and the superintendents, and make sure the boss thinks well of you."

"Easy for you to say," said Ethel, eyeing Marjorie and April enviously. "You play football with the boss's wife. You'll be the last to go."

"I don't intend to go at all," Marjorie retorted, finishing up the last of her porridge. "I'll be at Thornshire until the last shell rolls off the line. Whatever happens to the arsenal after that, I'll hang on as long as I can."

Unsettled, April took her spoon in hand and made herself swallow a few bites of porridge, but her appetite had fled. She wanted the war to end and peace to come, and she certainly wouldn't miss that awful yellow powder, but what would become of her and her friends after the men returned home and the factories reverted to their former purposes? April desperately did not want to return to domestic service, not after knowing better hours, higher wages, greater freedom, and far more respect as a munitionette. As much as she missed the beauty of Derbyshire, the thought of returning contrite and humble to beg Mrs. Wilson for her old job back made her clench her teeth in consternation. Returning to Alderlea wasn't even an option for Marjorie, not after she had fled in the night without giving proper notice, unfairly disgraced by that awful soldier's lies.

Suddenly April realized that she and Marjorie might never work together again. And what about the Thornshire Canaries? How could they have an arsenal football club without an arsenal to sponsor them, or a Munitionettes' League without munitionettes?

The following Sunday, the Canaries played the Handley Page Girls at Cricklewood, claiming a 5–2 victory that advanced them to the third round of the tournament. Afterward, players from both teams mingled as they exited the grounds, and conversation soon turned to the layoffs in Leeds and the rumors that more were pending, not only in the North but throughout Britain. They all adamantly agreed on two points: None of them wanted the war to last a single day longer than necessary, but they had to consider what victory would mean for their own livelihoods. The Handley Page Girls were guardedly optimistic that aircraft manufacturing would continue into peacetime, and even grow, as fascination with flight increased. They were less confident that women would be allowed to do the work, because foremen and bosses would surely prefer to hire returning veterans. "But who among us would begrudge a brave Tommy a job, after all they've been through?" the Handley Page Girls' captain asked, and every one of them chimed in agreement, April as loudly as anyone. This was her own elder brother they were talking about, and lads like Oliver. Of course they should have priority, after all they had done for King and Country.

But the munitionettes had sacrificed also, some with their very lives. Was it too much to hope that in peacetime there would be work and decent wages enough for everyone who wanted them?

In the meantime, it seemed to April that it was much too early to sack munitions workers and plan for peacetime employment, for the war raged on as intensely as ever. Air raids continued, devastating homes of innocent civilians as well as public roads and buildings. Food rationing expanded in London and the Home Counties to include butcher's meat, bacon, butter, and margarine as well as sugar.

Then, after months of negotiations, in early March the Bolsheviks formally established a separate peace with the Central Powers. All of Russia's former commitments to the Allies were now officially defunct, and since the Germans no longer had to battle the Russians in the east, they could concentrate all their forces on the Western Front. Although the Russians were no longer purchasing British munitions, April reckoned that the British military would need every shell and bullet and bomb their canary girls could possibly produce, now and for many months to come.

As the weeks passed, April's suspicions seemed spot on. On 21 March, the Germans launched a massive new offensive that the press declared would be "the greatest and most critical battle of the War," a phrase that struck April as both ominous and hopeful. The Germans apparently had concluded that their only chance of victory was to crush the Allies before the United States was fully able to bring its impressive resources to bear on the battlefield. Now Germany seemed intent on one last, desperate effort to break the British line, driving a wedge between the British and French armies and capturing Amiens, the crucial junction for both Paris and the English Channel.

By all accounts, the fighting was fierce and terrible. The official reports from the War Office to the press were deliberately vague and abrupt, stating only that the enemy was advancing and the Allies retreating. By the end of March, official reports from Germany quoted in the British press claimed that the Germans had taken forty-five thousand prisoners and had captured nearly one thousand guns. But April and the other Thornshire munitionettes could estimate the massive size of the battle by a source not available to the general public—by the surge in production that swiftly followed the launch of the German offensive. Shifts were extended, new workers recruited. The foremen urged them to work ever faster, to surpass their quotas and increase their efficiency, but without sacrificing quality or jeopardizing safety. Often April felt too exhausted to train for football, even knowing that the Munitionettes' Cup was at stake. After two more back-to-back victories, she realized that munitionettes from other factories must have been working at the same relentless pace as at Thornshire, because on the pitch their opponents seemed as fatigued as they were.

"At least the surge keeps us evenly matched," April remarked tiredly to Marjorie as they dragged themselves back to the hostel one Sunday after they had barely eked out a 1–0 win against the Hackney Marshes Ladies.

"There's enough hard work to go around," Marjorie agreed, muffling a yawn with her hand. "Think of it this way. The Ministry of

Munitions desperately needs our shells. That means Thornshire Arsenal still needs us. They wouldn't dare sack any of us canary girls now."

Maybe not now, April thought, maybe not today, but maybe only as long as this so-called Kaiser's Battle lasted. At best, the current surge in production likely offered munitionettes only a temporary reprieve, one she would just as soon have over and done, if it meant victory for the Allies.

But even she was caught by surprise at how swiftly the reprieve for women workers ended. She had expected it to endure well into spring, judging by the relentless pace at Thornshire, but in the last few days of March, word came that the Kynoch munitions factory at Stanford-le-Hope in Essex had dismissed eight hundred munitionettes. By the first week of April, hundreds of these displaced workers had made their way to Thornshire, clutching their Labour Exchange papers as they queued up for interviews outside the square, oversized low shed that served as Superintendent Carmichael's headquarters.

"We're taking on as many as we can," Helen confided to the team after a particularly messy football training, spring rains and thawed soil having rendered the arsenal field a sodden, muddy morass. "With any luck, the others will find work at other factories, although they may have to venture farther from home."

As they returned to work, Marjorie lowered her voice for April's ears alone. "That's promising," she remarked. "If Purcell is hiring new workers, he can't be planning to sack the experienced workers like us."

"I suppose not," said April, "unless the plan is to squeeze all the work out of the whole lot of us while they need us, and then cut us loose, all at once, when they're through."

Marjorie heaved a sigh. "You probably have it spot on, of all the rotten luck."

April gave her a sharp sidelong look. "Not that you want the war to drag on another year or two."

"Of course not. Not even another month. You should know better than that."

April did know better, really. They were all under tremendous strain—long hours, arduous shifts, chronic illness, sleep interrupted by air raids, worries for their brave soldiers on the Western Front, and now, fear for their livelihoods. April wanted to take heart, as Marjorie had, from the new hires, but her hopes diminished when she overheard the Finishing Shop foreman, Mr. Vernon, boast to one of his mechanics that the National Federation of Discharged and Demobilised Sailors and Soldiers had sent a deputation to the Ministry of Munitions to protest the veterans' unfair treatment in the munitions factories. It was the worst sort of injustice, Mr. Vernon said, while his mechanic nodded vigorously, that throughout Britain, brave men who had fought and had been wounded were being sacked and their places filled by women and girls. "The Ministry of Munitions will put things right, mark my words," Mr. Vernon declared, but then he noticed April listening and clamped his mouth shut in a scowl, glowering.

She hurried on her way, too astonished to challenge the foreman, not that she would have dared even if she had found the words. Did this so-called unfair treatment of men even exist? Where exactly were able-bodied veterans being sacked and replaced by less experienced, lower-paid girls? Certainly not at Thornshire. From everything she had seen, and everything she had heard from munitionettes at other factories, precisely the opposite was true. Women were being sacked by the hundreds, and if men were not hired to replace them, that was only because the jobs had been made redundant. Everywhere, returning soldiers were demanding their old jobs back, submitting petitions to factory managers and politicians and other prominent figures in a position to help them—but no one, aside from the occasional solitary advocate like Helen, was championing the munitionettes. The labor unions, which barely tolerated women doing "men's work" and had fought aggressively against women's advancement throughout the war, had quite unsurprisingly taken up the soldiers' cause.

To April it seemed that a losing battle loomed on the horizon, and they had not yet finished fighting the war.

"It's not fair," April lamented to Oliver one Saturday evening after work as they walked hand in hand to a favorite café. "I know we weren't in the trenches, but we too have served."

"You're the girls behind the man behind the gun," Oliver remarked, squeezing her hand. When she shot him an accusing look, thinking he was teasing her, he quickly added, "I mean that sincerely. We've been lucky at Thornshire. We've had many illnesses from TNT poisoning but only three deaths, and no explosions. There's a factory in Chilwell, in Nottinghamshire, that has already had seventeen."

"Deaths?"

"Explosions. Three deaths, and it could have been much worse." Oliver shook his head. "If you ask me, any worker, woman or man, who is injured, sickened, or killed on the munitions front should be counted among war casualties."

"The munitions front," April echoed thoughtfully. "Yes, that's what it is, isn't it? We girls *have* served. We *are* serving. We answered our country's call and enlisted in war work. Yet after serving King and Country so faithfully throughout this crisis, we girls are almost certainly going to be discarded and forced back into our old jobs—jobs we left for good reason, most of us—unless we prefer idleness and poverty. I'll say it again—it just isn't fair."

Oliver halted, bringing her to a stop too. "What *is* fair, then?" he asked, regarding her curiously. "A soldier goes through hell, somehow survives, mostly intact, then comes home and wants his old job back, wants his old *life* back. Are you going to tell him to shove off?"

"Of course not," said April, feeling heat rising in her cheeks. "Especially not if it was his job first. It's only that—" She drew in a breath and exhaled sharply, angrily. "It'll be a massive relief to be away from TNT, I'll give you that, but I don't want to go back to domestic service. I don't ever want to be a housemaid again, but if I get sacked, what else is there for me?"

She folded her arms and strode off before he saw the tears of frustration gathering in her eyes.

"April." Oliver quickly caught up to her and put his hand on her shoulder. She halted, but she would not look up at him. "You don't have to go back into service. No one can force you."

She snapped out a laugh. "You only say that because you've never met my mother."

A corner of his mouth turned up. "If she's half as formidable as you—"

"Not half. Twice."

"Then I'll be sure to watch my step," he said solemnly, but she heard the amusement in his voice. He turned her toward him and stooped a bit so she could not avoid meeting his gaze, unless she wanted to stand there with her eyes closed, which would have been ridiculous. "Listen, April. I sympathize with the lads coming home from the war. Obviously." With a tilt of his head, he indicated his prosthetic, entirely concealed from prying eyes by his coat sleeve and pocket. "I was one of those lads."

"I understand why you'd take their part," she said, a bit grudgingly. "I just want you to see that there's this whole other side to it."

"I do see that," he said emphatically. "You want to know what else I think?"

The last of her anger faded away. She was too weary to sustain it, and she wasn't really angry at Oliver anyway. "What's that?"

"I think if you want to keep your job, you should fight for it—and I wouldn't want to be the man who finds himself opposed to you."

21

APRIL–MAY 1918

HELEN

As the *Kaiserschlacht* continued, production at Thornshire Arsenal continued apace, without so much as a pause for Eastertide. Minister of Munitions Winston Churchill had issued an appeal to munitions workers and factory owners to voluntarily waive the holiday in order to keep up with the increased demand for war materiel, and with very few exceptions, munitions factories throughout Britain carried on with their duties. Some arsenals, including all of those belonging to Purcell Products Company, had already planned to forgo the holiday even before Mr. Churchill made his appeal. Helen was immeasurably proud of her munitionettes for their dedication and patriotism, working through their fatigue and illness, confident that every shell and bullet they made hastened the end of the war, which they all desired so fervently.

Helen's only regret was that the Thornshire munitionettes' impressive dedication made it all the more difficult to convince Arthur to take any time off. Despite her best efforts to tend to him, with every week his health visibly declined. He did not even attempt to deny it anymore, but stated frankly that he was willing to sacrifice his health and his very life if need be rather than fail in his duty. Equally stubborn, Helen declared that she would do everything in her power to see that it did not come to that. "Then I am already saved," he told her, smiling through his exhaustion.

Helen was grateful to note that the burgeoning spring alleviated her husband's melancholy, if only by a scant degree, his mood rising with the increasing sunlight and warmth of the lengthening days. To her great joy and relief, after a long absence, he once again found comfort and release in her embrace, moments of intimacy that promised better days to come, of the restoration of their own peace and happiness when the war was over.

She found encouragement elsewhere too—in the Thornshire Canaries' steady progress through rounds of the tournament, and in the more temperate weather and pretty scenery that brightened her morning walks. That season, the gardens of London were remarkable for a different kind of beauty than in previous years, for the government had strongly urged local gardeners to plant vegetables rather than flowers. For Londoners, it was a time-honored custom to spend a good portion of the Easter holiday planting hundreds of flower bulbs in one's home garden, but now the order of the day was to plant tubers instead, or, better yet, to take a plot or an allotment in the nearest open space and sow an assortment of nutritious vegetables. King George himself had set the example by decreeing that the flower beds surrounding the Queen Victoria Memorial at Buckingham Palace should forgo the traditional blaze of scarlet geraniums in lieu of potatoes, cabbages, parsnips, and carrots, less dazzling but far more essential. So it was in the Royal Parks as well. Two hundred acres of Kew Gardens had been set aside for the cultivation of vegetables, and when Helen walked through Kensington Gardens, the only flower on verdant display was an abundance of wild daffodils, thriving on the enclosed grassy lawns, their sunny yellow blooms swaying in the breeze. The delightful fragrances of spring blossoms and freshly turned soil invigorated her, evoking fond memories of her mother's garden at Banbury Cottage and her own beloved gardens at her home in Oxford. She and Arthur had not visited their estate since Christmas and she missed it terribly, but it seemed unlikely that they would be able to spare time away from the arsenal until the surge in demand for munitions subsided.

That day seemed very far off indeed as the *Kaiserschlacht* continued pounding away at the Allied armies on the Western Front. The need for fresh troops as well as armaments came into stark focus in the middle of April when Royal Assent was given in the House of Lords to the Military Service Act, which deemed every man in the United Kingdom between the ages of eighteen and fifty-one to be "duly enlisted in His Majesty's Regular Forces for general service with the Colours or in the Reserve for the period of the War, and to have been forthwith transferred to the Reserve."

To Helen, it sounded alarmingly as if every young man and a great many who had already reached middle age had been instantly, unwittingly conscripted. "It sounds that way, darling," said Arthur, pausing between mouthfuls of a very late dinner, "because that is in effect what has happened."

"But don't you find it distressing?" said Helen, moving food about on her own plate, untasted. "What about our fine old English principle of a man's right to choose for himself? Are they going to drag portly, gouty gaffers from their families and vocations, thrust a rifle in their arms, and send them off to the front lines, where even young and fit men struggle?"

"I hardly believe that's what will happen," said Arthur wryly. "Your portly, gouty gaffers will likely be pressed into service only to defend England against an invasion, in which case I reckon everyone who *can* fight *shall* turn out to fight the Hun, conscripted or not."

Helen supposed he was right. "Then what does this mean for men your age, practically speaking?"

Arthur heaved a sigh. "For me? Not much. My exemption for essential war work remains valid. However, if I thought I could make a greater contribution elsewhere—"

"What do you mean, elsewhere?"

He shrugged, avoiding her gaze. "Tom could step into my role at the arsenal easily enough. Oliver could take over for him, and with you and Superintendent Carmichael continuing your work—"

"You can't mean to enlist," breathed Helen, shocked. "No. Absolutely not. If you take a leave of absence from Thornshire, it should be to recover your health, not to rush headlong into battle."

He grimaced. "I hadn't considered that. I might not pass the physical exam in my current state."

She reached across the table and seized his hand. "Darling, I don't understand. You're already doing your bit, and it's killing you. Why on earth would you want to enlist and hasten the process?"

He tried to smile. "When you put it that way, it sounds quite mad."

"Arthur," Helen said, her voice trembling. "Please talk to me. Why would you even consider this?"

"I never meant it seriously. It was just an idle thought brought on by the new law." Arthur brought her hand to his lips. "I'm sorry I upset you, darling. Think nothing more of it."

But of course she would. How could she not? She wasn't fooled by the forced lightness in his voice. There was something he wasn't telling her, she was certain of it, and it pained her that he wouldn't confide in her.

Two days later, on Saturday, 20 April, a solemn memorial service was to be held at St. Paul's Cathedral in honor of London's war workers killed or injured while performing their duties. The Bishop of London would preside, Minister of Munitions Winston Churchill would attend, and tickets were in such demand that the organizing chairman claimed the cathedral could be filled five or six times over. Twelve hundred members of the Women's Auxiliary Corps would march from Wellington Barracks to the cathedral accompanied by a band of Scots Guards and a detachment of pipers. During the service, the Grenadier Guards, clad in their scarlet uniforms, would sound the Last Post and Reveille "as a tribute to those who in munition factories and other branches of war work have laid down their lives for their country."

Because munitions workers comprised the greatest number of the honored dead, factory owners were invited to attend so they might pay their respects on behalf of the industry. Since Helen and Arthur were the only Purcells in London, Arthur's father instructed them to attend as official representatives of the family and the company. On the appointed day, Helen finished her work early, but Arthur toiled away until the last possible moment, which meant they left the arsenal late and had to run for the tram, and run again for their train in Barking. They arrived to find the streets around the cathedral packed with sightseers eagerly awaiting the parade, so Arthur clasped Helen's hand firmly as he guided her through the crowd. As they slowly made their way to St. Paul's, Helen realized that many of the people gathering there were not sightseers at all but mourners—parents, siblings, or friends of the deceased. Amid the din of eager conversation, she was certain she heard muffled sounds of weeping.

When at last they reached the cathedral, Arthur pulled her closer and she took his arm, the better to keep her balance in the crush. Ascending the stairs, they passed an older couple dressed in solemn, well-worn black and charcoal gray, the woman clutching a handkerchief and a small bouquet of pink and white flowers, her husband with his jaw set grimly. Suddenly the man reached out and touched Arthur's shoulder. "Beggin' your pardon, sir," he said, removing his hat, "but if you're goin' in, would you be so kind as to do me and me wife a small favor?"

"Certainly, if I can," said Arthur, and Helen nodded.

The older fellow looked to his wife, who stepped forward and tentatively held out the bouquet to Helen. "Would you please leave these flowers on the altar for our Susan?" she asked, her voice choked with tears. "It don't have to be on the altar if that's not proper. Anywhere will do."

"Of course," said Helen.

But when she reached out to accept the bouquet, Arthur touched her arm to forestall her. "I assume you lost a daughter in munitions work?" he asked, his voice low and respectful.

They nodded. "Our oldest girl," the man said, a tremor in his voice, as his wife pressed her lips together to hold back a sob. "'Twas but a fortnight ago."

"Please accept our condolences. I'm truly very sorry for your loss." Arthur glanced to Helen, a question in his eyes, and she nodded in reply. "I think it would be best if you left the flowers for Susan yourselves," he said, retrieving the tickets from his breast pocket and holding them out to the man. "These will get you through the door."

The man regarded him, puzzled. "Don't you want to pay your respects too?"

"We can do that just as well out here." Arthur extended the tickets yet closer.

"Please take them," said Helen. "They really should have been offered to you first."

The couple gratefully accepted, thanked them profusely, and hurried off toward the entrance to the cathedral, tickets in hand. When they disappeared into the crowd, Helen turned to Arthur and kissed his cheek. "That was very decent of you," she said. Her heart overflowed with pride and affection, yet with an undercurrent of sadness mingled in. In his kindness and generosity, he resembled so much the idealistic scholar she had fallen in love with years before. She had not realized until that moment how much she had missed him.

Arthur shrugged, his eyes fixed on the cathedral. "I hope it will comfort them to see their daughter honored. It's a sacrifice she—and they—never should have had to make."

Abruptly he turned, but Helen caught his arm and hurried along beside him. They made their way down the stairs and back to the street, where they paused and turned to gaze back at the magnificent cathedral as the crowd milled about them. When Arthur closed his eyes and bowed his head, Helen realized that he was paying his respects, just as he had told the older couple he would, and she quickly did the same, silently reciting a prayer she remembered from her father's funeral.

Her thoughts turned to Mabel Burridge, and to the other Thornshire munitionettes who had perished over the winter. Her heart plummeted as she tallied the number of canary girls who were currently on sick leave, and those who remained at their posts, their health clearly failing them despite the increasingly strong preventative measures she imposed. How many more would die or suffer irreparable harm before the end of the war? Thornshire Arsenal was only one small munitions works, and their casualties had been relatively few thus far. Only recently, the Ministry of Munitions had announced that more than one million women were working in munitions industries at present. If Thornshire's sick, injured, and killed were extrapolated to all the munitions works throughout the British Isles, Helen scarcely dared imagine the extent of the suffering. The number might be impossible to calculate. But that did not absolve the government of the responsibility to try to make a full accounting and offer compensation wherever it was needed.

But first they had to win the war.

An air raid shattered their sleep that night, catching Helen and Arthur entirely by surprise since the sky was partly cloudy and the moon would not be full for another week. They fled to the shelter beneath their building, and returned upstairs around one o'clock to climb wearily back into bed. The next morning Helen woke late to discover that Arthur had already breakfasted and had left for the arsenal, to make up the work hours he had missed the previous day. Although she had not expected him to accompany her all the way to Birmingham for the Canaries' match against the Coulthards Ladies, she had hoped he would at least see her off at the station. Instead she ate her breakfast while studying the papers, packed her kit, and set off for the station, alone.

She was glad to meet up with the other Thornshire Canaries when she boarded the train to Birmingham. Lightheaded from fatigue, she dozed in her seat surrounded by her equally weary teammates, and when they arrived, she fortified herself with a strong cup of tea. But it was not enough. During their warm-up, she felt as if her legs were weighed down by sandbags, and from the first whistle, she found herself unable to shake off her lethargy. Nor was she the only Canary thus afflicted. The Coulthards Ladies countered their every attack with a steal, and it was all Helen could do to clear the ball from the defending third to give Marjorie a respite between shots on goal. During the interval, down 2–1, the winded Canaries rested and regrouped, while Marjorie, Peggy, and Lucy strategized for the second half.

"The long train ride wore us out," said Daisy, between gulps of water. "Otherwise they wouldn't be running circles around us."

"The Ladies have traveled nearly as far," Helen reminded her. "Birmingham is halfway between Preston and London, which is why our captains agreed to play here."

"Then it must be because of the yellow powder that we're more winded," said Daisy. "You can tell from their skin that they aren't canary girls. That gives them an advantage."

"Perhaps in your case," said Helen wryly. "What's my excuse?"

Daisy thought for a moment. "Air raids. They don't get air raids in Preston, do they? Of course the Coulthards Ladies are better rested."

"That must be it," said Helen, groaning as she rose from the bench to loosen up her muscles, just as Marjorie, Peggy, and Lucy called the team together to explain their new scheme for the second half.

It proved to be a simple but effective strategy, a matter of concentrating their defense to shut down the Ladies' exceptional striker and second forward, and sending April up with Lucy and Peggy to break open their defense. In the first five minutes, April drove in a goal after an assist from Lucy to even up the score, but the Ladies dug in and would not allow another. Helen and Marjorie were equally tenacious, so the match ended in a 2–2 draw.

"What does this mean for our place in the tournament?" Helen asked later as they boarded the train for London.

"The Ladies already have one loss, so we'll advance," said Peggy, sinking into her seat with a sigh. "That being said, our first loss will be our last. We'll be out of the tournament."

"That's not our only concern," said Lucy, her voice just loud enough to be heard over the murmurs of worry and frustration. "We barely held our own today, but back in December, the Dick, Kerr Ladies beat the Coulthards Ladies four–nil."

The teammates exchanged uneasy glances. "So you mean to say we should expect Dick, Kerr to be twice as good as Arundel Coulthard?" queried Marjorie, frowning. "Two–all compared to four–nil?"

"I wouldn't state it as a maths problem," Lucy replied, "but we should expect the Dick, Kerr Ladies to be a tough team to beat."

"But we already knew that," said Helen, hoping to reassure her teammates, some of whom looked rather anxious. "Their reputation precedes them."

"We weren't at our best today—except for April, who was marvelous." Peggy paused to smile at her, and as a ripple of applause broke out, April blushed and waved off the praise, embarrassed. "All I mean is, if we had played our best game, we would have won handily."

"I always play my best game," declared Marjorie, and although Helen spied some furtive looks passing between the others, no one contradicted her. "And I say we can take the Dick, Kerr Ladies on any pitch, anytime, and I hope we have the chance to prove it!"

Louder applause broke out, evoking glances—some indulgent, others wearily annoyed—from other passengers, who probably wished they had chosen a different carriage. Not that there had been much choice. The railcars were all packed full, mostly with soldiers coming home on leave or returning from it, mixed among other soldiers in uniforms so crisp and unmarred that they were surely new recruits. Some of Helen's unmarried teammates, especially the younger girls, enjoyed chatting or even flirting with the Tommies, and some exchanged addresses and promises to write, but Helen only smiled, exchanged polite greetings, and thanked them for their brave service. They glanced at her wedding band and turned their friendly attention elsewhere, and that was fine with her. There were certain days, bad days, when she could not glimpse a young man in uniform without the unsettling certainty that she was witnessing a fellow human being in his last, precious hours of life. It was an occasion so moving and terrible that only profound words would suffice, but she had none, so gentle smiles and sincere thanks were all she could offer.

Over the course of the war, Britain's train stations had become public stages for such piteous, heartrending scenes of grief and parting as soldiers left for training encampments or for the front that Helen had learned to brace herself and avert her gaze. In those last moments before boarding, a soldier's family would gather around him—pale, weeping wives with babies in their arms; older children clutching their mother's skirts, wide-eyed and bewildered; mothers and grandmothers, gray-haired, shawls in disarray, faces clouded over with worry; silver-haired or balding men, stoic and silent except when they cleared their throats or issued final words of advice to their sons. The young men would stand with their packs and rifles, some with their faces stony and jaws set, others grinning with the anticipation of adventure. The latter never failed to astonish Helen. Hadn't all young men learned by now, from the returning wounded and the stories of those who would never return, how utterly hollow and false were the promises of battlefield glory that had enticed young men exactly like themselves back in 1914? If not, the evidence was all around them, for elsewhere in that same station was another platform where anxious crowds awaited the arrival of the ambulance trains, parents and wives desperate to catch a glimpse of their own wounded, broken, beloved boys as they were transferred from the railcars to ambulances and on to convalescent hospitals.

How any soldier came through the war unscathed mystified Helen. It seemed inevitable that all would perish in the end, the death of the body or morbidity of the soul. Nothing mattered anymore but to hasten the end of the war so they could all wake from their enduring nightmare of misery.

But the war dragged on and on.

Yet even so, Helen found moments of grace—the blossoming spring, the sweet cries of a neighbor's newborn, bliss in her beloved husband's arms. She cherished the friendships she had discovered at Thornshire and in Marylebone. Neighbors seemed to look out for one another more than they once had, and even strangers offered a kind word in passing where silence would have sufficed before. They were all in this together, their nods and polite greetings seemed to say, and they would all get through it together. England would endure. London would survive the current crisis as it had other calamities throughout the centuries. They only had to carry on and never surrender.

Nor were they alone in their great struggle. Russia had forsaken them, but they had other allies—and on Saturday, 11 May, all of London celebrated the nation that had most recently joined the fight. A regiment of the United States Army, nearly three thousand strong, were to be reviewed by King George at Buckingham Palace, after which they would march through London on their way to Wellington Barracks to prepare for transport to France, where more than a half a million of their fellow Americans were already deployed. Arthur had already gone to the arsenal when Helen ventured out in the late morning with her Marylebone friends, and as they walked along, they spied the Stars and Stripes flying from the flagstaff of every government department building, institution, and club. Old Glory, an alternative title for the flag Helen had heard some Americans employ, also hung from the windows of innumerable offices and private residences. Hundreds of thousands of spectators thronged the streets carrying smaller copies in their hands or wearing them in their buttonholes, while hawkers with trays of tin versions milled through the throng crying, "Stars and Stripes! Old Glory! One penny each!" Unable to resist, Helen bought one for herself and another to give to Arthur later.

She would have been content to remain among the enthusiastic crowd, but her next-door neighbor, Evelyn, escorted their group to a suite of offices her husband owned in Grosvenor Gardens, not far from the American Embassy. The balcony on the second floor offered a perfect view of the street below, where between half twelve and one o'clock, the three battalions passed, led by bands from the Brigade of Guards. One remarkable feature of the procession was a contingent of veterans of the American Civil War, who carried a banner inscribed "Not for Ourselves, but for Our Country." Most striking of all, however, were the soldiers themselves, tall and slim, strong and fit, youthful but sternly serious. Helen did not glimpse a single face relaxed in a smile. The troops' martial demeanor was most impressive as they strode along, displaying no reaction whatsoever to the earsplitting cheers and frantic waving of flags of the tens of thousands of onlookers on the congested pavements.

"Except for their uniforms and good health, they bear little resemblance to the American soldiers who marched through London last August," remarked Beatrice, echoing Helen's thoughts. "They were a jollier set by far."

"If I recall correctly, those lads were engineers sent over early to prepare for the rest of the army's arrival," said Evelyn. "I presume they've been busy constructing quarters, building roads, laying power and telephone lines and such. These troops are infantry soldiers. Naturally they would be fiercer."

"And now they are crossing over by the thousands every day," said Violet, with great satisfaction. "Trained fighting men, well-rested and ready."

"Trained, but inexperienced," Beatrice noted.

"They'll acquire experience soon enough," the baroness replied.

"They shall indeed," said Helen. God help them.

Little more than a week later, she remembered those proud troops as she sat in her darkened home, her face pressed against Arthur's chest, his arm around her, flinching at the dull roar of German bombs and the incessant thudding of their own guns, wishing with all her might that the Americans had brought with them some means to keep the British sky clear of German aircraft—some marvel of engineering or new weapon or ingenious scheme, she did not care what.

It was a Sunday, and until that dreadful night it had been a lovely spring day blessed with clear skies and gentle breezes. The Thornshire Canaries had no match that afternoon, so Helen had convinced Arthur to enjoy a relaxing day off with her. As it happened, he spent several hours poring over documents at a table on the back terrace while she read a novel in the chaise nearby, but at least they were together and he had not stolen off to the arsenal when her back was turned. Later that evening, a waxing moon shone brightly and stars glimmered in a cloudless sky, with only a slight breeze from the east stirring the tree boughs in the garden. "Ominously lovely," Arthur remarked as they drew the blackout curtains and went to bed.

They woke shortly after eleven o'clock to the harrowing noise of the air raid warning. Bolting out of bed, they threw on some clothes and raced downstairs to the cellar. The servants quickly joined them, taking their usual places on the piles of blankets the housekeeper had placed there for comfort, while Helen and Arthur sat side by side on the bottom stair. For an hour they endured the jarring roar of the ground guns, which at times shook the house so intensely that Helen felt as if she were being torn to pieces.

"Should we be any the worse off without the guns?" asked the cook, indicating the defenses with a wave of her hand. "What with their shrapnel raining down on us and the terrible noise? Might it not be better to sit here in silence, with only the German bombs to affright us?"

"Better? Hardly that," a footman scoffed. "If our ground guns weren't there to ward off the German pilots, we'd have more German planes in our sky and more German bombs dropped on our heads."

"More German planes?" a housemaid exclaimed. "Goodness, don't they crowd our skies enough as it is?"

"It's not fair," Helen blurted, her voice breaking. "We're civilians, not combatants, and how dare the enemy treat us as such? What a cruel and savage new form of warfare this is. No wonder we were entirely unprepared for it, and even now we remain utterly powerless to strike back to defend ourselves!"

"Calm yourself, darling," Arthur murmured, stroking her hair. "It'll be all right."

Suddenly conscious of the servants' stares, Helen muffled a sob, thoroughly ashamed of herself. "Forgive me," she said, taking a steadying breath. "What a ghastly display."

"No harm done, ma'am," said the housekeeper kindly. "You're only saying out loud what we're all feeling."

"Shall I fix us all a nice cup of tea?" asked the cook.

"Thank you, but no," said Helen quickly. "It's not safe."

"Oh, I don't know," said Arthur easily, lifting his arm from her shoulders and rising. "One might risk it. How should a few inches of wooden planks and beams make the cellar any safer than the kitchen, really? I'll fetch the tea."

"No, Arthur—" said Helen, reaching for him, but he was already climbing the stairs, and her fingertips only brushed the cuff of his trousers.

"I'll help you, sir," said the footman, hurrying after him.

The women exchanged bemused looks. "Very well," said Helen, forcing a smile. "Let the men see to it while we take our ease."

The others smiled tentatively in reply as the house trembled from the barrage.

Soon Arthur and the footman returned with the tea tray and a plate of biscuits. They ate and drank, conversing now and then about household matters or inquiring about one another's families, but mostly listening to the bombs and guns and waiting for the storm to pass.

Three hours later, the all-clear sounded, and after carrying the dishes to the kitchen, they all wearily went off to bed. Arthur climbed beneath the covers before Helen was finished undressing, and she thought he had immediately fallen asleep, but when she lay down beside him, she discovered that he was shaking. "Darling?" she murmured, alarmed. "What is it? What's wrong?"

"I'm sorry, darling. I don't want to keep you awake." He threw back the covers and would have risen except she clung to his arm. "I'll sleep in the guest room."

"Don't be ridiculous. You'll stay right here with me." Helen sat up and wrapped her arms around him. "Please, please, tell me what's wrong."

He was silent for a long moment, trembling, perspiration on his brow. "You're right, what you said down below. This sort of warfare is savage and cruel, and it treats civilians as if they were soldiers—and I am just as guilty as any German pilot raining bombs down upon London."

"How can you say such a thing? You have nothing in common with these dreadful raiders."

"How can you be so sure? How do we know that Thornshire Arsenal shells haven't killed innocent German civilians?"

"Well, I—I don't know for certain, but we aren't like that—"

"Aren't we? We don't know. We can't know." Throwing off the covers, he bolted out of bed and paced the room. "You don't know how torn I've been, required to produce the weapons our armies need, wracked by guilt that my creations inflict untold death and suffering upon my fellow man—soldiers, yes, but civilians too, and not just men but women and children and the elderly, and not only Germans but French and Belgians, prisoners in their own occupied territory!"

Helen clutched the bedcovers, her throat constricting with grief and worry. "It's wartime," she choked out. "It's all dreadful and I know it doesn't help to hear this but you're only doing your duty. What choice do you have?"

"I do have a choice," he retorted, his chest heaving with deep, ragged breaths. "As it is now, I build bombs. I take lives indiscriminately without ever risking my own."

Suddenly she understood. "That's why you want to enlist."

"Yes! If I must kill, the only honorable thing to do is to give them a fair shot at me in return." Abruptly he sat down on the edge of the bed and buried his head in his hands. She reached for him, but before she could touch him, he whirled upon her, face haggard, eyes wild with torment. "For years, I've resolved to do my duty to King and Country, as well as to my father. I've committed myself utterly to my work. I'm confronted daily by the demands of the Ministry of Munitions, the constant threat of German bombs or horrific workplace accidents, the alarming reports of rising worker illnesses and death—and the pervasive, haunting fear that if I fail, the war shall be lost, and all that I hold most dear lost with it."

"Oh, my dear Arthur." Tears streaming down her cheeks, Helen laid her hand on his cheek and kissed his brow.

He drew back, his expression bleak and haunted, a muscle working in his jaw. Then suddenly he embraced her, pulling her close, burying his face in her hair. "My dearest love," he said, his voice low and trembling with barely controlled anguish. "Forgive me."

There was nothing to forgive, but that was not what he needed to hear. "Of course. All is forgiven."

"I never wanted this."

"I know. None of us did."

"I can't lose you. I can't lose us."

"You won't," she assured him, with all the love and certainty she could give voice to him, her own dearest Arthur, her beloved. "You never will."

22

JUNE–AUGUST 1918

LUCY

In early June, Londoners welcomed reports that the United States 1st Division had finally captured Cantigny, a battle-scarred farming village on a high plateau in the Picardy region of northern France. General Pershing's first victory of the war had flattened a deadly German salient, but at the cost of more than one thousand U.S. casualties. It was a small battle, relatively speaking, but the victory was heartening even so, not only because it liberated more French territory, but because it showed the British, the French, and perhaps even the Americans themselves that the novice United States troops could fight and win. Lucy reckoned that if the Germans were not deeply worried, they were seriously underestimating the forces now arrayed against them.

By mid-June, the most intense fighting centered on the town of Château-Thierry on the Marne, where French explosives experts destroyed bridges and American troops provided covering fire to prevent the Germans from crossing the river. Yet although the German army's advance on Paris had been thwarted, their long guns continued to bombard the city with devastating shells. Lucy feared that if the Allies did not continue to hold off the German advance, Paris would fall, and in time, London might suffer the same fate.

Lucy was grateful that her sons, her mother, and all her Brookfield family remained out of harm's reach—at least, they had so far. She only wished she knew whether the same was true for Daniel. She trusted that he had not been returned to a German prison camp, as the letter from the War Office had promised. Whether he was still being held in a Swiss Medical Mission observation camp, and where that might be, or whether he had been transported to Switzerland or was on his way home to Britain, she had no idea. The War Office had said that more details would be forthcoming, but she had heard nothing from them in ages. What she longed for most was a letter from Daniel himself assuring her he was safe and sound, but such a letter never came.

She checked the post every evening as soon as she arrived home from the arsenal, but day after frustrating day, she heard nothing. The silence was especially difficult to bear on the last Tuesday of June, their thirteenth wedding anniversary. She had clung to a superstitious and foolish hope that surely a letter would come on that day of all days, but still there was no word from him, only aching loneliness.

"No news must be good news," Daisy reminded her the next day as they worked side by side in the Finishing Shop, after Lucy confessed her frustration and disappointment. "If anything had happened to your Daniel, you would have received a telegram."

Daisy was almost certainly right, so Lucy found what comfort she could in her friend's words.

Football proved to be a welcome distraction. Since midsummer, the days had been unusually warm, but Lucy and her teammates took to the pitch as vigorously as ever as the last few rounds of play for the Munitionettes' Cup began. On the first Sunday of July, the Thornshire Canaries met at the Globe early so Lucy and Peggy could explain a few new plays they had devised. They were determined to learn every step before their opponents, the Shell Girls from the National Shell Filling Factory in Chilwell, arrived from Nottinghamshire. A few of their most ardent fans, including Mr. Purcell and Mr. Corbyn, observed from the stands as Lucy and Peggy described the new stratagems, and more spectators trickled in as the Canaries walked through the plays in their street clothes. As soon as every player was confident with her role, they hastened into the changing room, certain their opponents would arrive at any moment and relieved to have preserved the element of surprise.

Lucy sensed a frisson of excitement and nervousness as the players donned their uniforms and gear. They had never faced the Shell Girls before, but although the Chilwell munitions works had been troubled by accidents and explosions as well as TNT poisoning cases through the years, the owner, Lord Chetwynd, was said to be very mindful of his workers' well-being. Out of his own vast wealth he had provided them with fine canteens; bands to play dance music at mealtimes; comfortable shifting houses and rest facilities; and many opportunities for sport and entertainment, including sumptuous baths, a swimming pool, well-appointed changing rooms, and spacious outdoor sports grounds. The Purcells kept the Canaries supplied with fine uniforms, balls, and other equipment, but there was nothing to be done about their small, uneven training field. Lucy could well imagine how fit and prepared the Shell Girls must be, with such superior resources at their disposal.

She expected to be thoroughly awestruck as the Shell Girls took the pitch, but when the Canaries emerged from the changing room for their warm-up kickaround and drills, their opponents were nowhere to be seen. Neither were the Shell Girls' supporters, judging by the overwhelming dominance of yellow-and-black pennants and scarves in the stands, with the few scattered appearances of red-and-white seeming entirely incidental.

"Maybe their train was delayed," said April as they ran laps of their half of the pitch.

They all agreed this was possible. As they took shots on the goal and Marjorie ran and leapt to block them, a lighthearted debate ensued regarding whether they should do the sporting thing and delay the start of the game to allow the Shell Girls time to properly warm up once they finally arrived. Marjorie was adamantly against it. Lucy thought they absolutely must, for the Canaries would appreciate the gesture if their places were reversed. Daisy, ever pragmatic, pointed out that the question would probably be decided by the officials. "Tournament rules might require a forfeit," she pointed out. "Does anyone know?"

No one did.

"There might be another match scheduled for the Globe immediately following ours, in which case we'd have to clear off," Peggy noted. "Then there's the officials. They might not be willing or able to stay late."

There was nothing to do but hope the Shell Girls arrived at the last minute.

As the Canaries moved on to passing drills, Lucy noticed that she and her teammates frequently glanced to the entrance hoping to spot the other team, but the minutes ticked away and still they did not appear. With five minutes to go before the opening whistle, the officials beckoned Peggy over for a conference. The Canaries continued to warm up, observing the sideline discussion from a distance but unable to hear a word. When Peggy jogged back, her teammates gathered around her to hear the verdict. "The officials say the pitch is ours for the next two hours," she reported. "We can accept the Shell Girls' forfeit, advance in the tournament, and call it a day. Or we can wait and start the match when they arrive, but the score will stick even if we don't get a full ninety minutes of play."

"Take the forfeit as a win," someone called out from the back. "Why look a gift horse in the mouth?"

"The rules may allow it, but that doesn't mean it's fair play," Helen protested, looking around for affirmation.

"I say we play them when they get here," declared Marjorie. "I don't care how fancy their sporting fields and changing rooms are. They're just working-class girls same as us, not Division One footballers. We can beat them, and I want the chance to prove it."

A few girls cheered and applauded. Craning her neck, Peggy caught Lucy's eye and raised her eyebrows in a question. Lucy made a few small gestures which she knew Peggy would recognize as the opening moves of their new plays. Peggy nodded, understanding her perfectly, and clapped her hands for attention. "Since we're warmed up and our opponents aren't here to watch, let's run through our new plays properly. If the Shell Girls show up before we finish, grand. We'll play. If they don't, we'll decide what to do next."

The players quickly took their places and ran through the plays until they could execute each one flawlessly. Their performance earned them applause from the few spectators who remained, for the crowd had thinned as time passed and the prospect of an actual match became increasingly unlikely. An hour after the match should have begun, Peggy called the team together in the center of the pitch. The sky had clouded over, the breeze was picking up, and the air smelled of rain. By a show of hands, the players agreed to accept the Shell Girls' forfeit and head home before the storm rolled in.

"I have a very uneasy feeling about this," Helen said as they trooped into the changing room. "A train delay of more than an hour on a Sunday afternoon? Doubtful."

"Maybe the Shell Girls thought we were playing at their field," said Marjorie, her voice muffled as she pulled her jersey over her head.

"If they had, they would have figured it out when neither we nor the officials showed up." Lucy turned to Helen. "Would any of your contacts know?"

"Possibly," said Helen, pensive. "I'll make inquiries."

A sinking dread fell over the room, taking all the pleasure out of their smashing new plays and the easy victory that had brought them one round closer to the Munitionettes' Cup. They left the Globe just as a light rain began to fall, and they parted company at the station as they boarded different trains and trams. Lucy made sure to check the platform display before embarking, but nothing suggested that there had been an accident on the route from Nottinghamshire, or any other delay of more than a few minutes.

At home, Lucy checked the post out of habit, remembering with chagrin that of course there was no post on a Sunday. Lonely and unsettled, she telephoned Brookfield and in turns spoke with her mother,

her sons, and her brother George, who lectured her so sternly about her ongoing TNT exposure that she almost wished she hadn't rung. Later, after the brief rain shower passed, she and Gloria met in the garden and knitted together until it was time to prepare dinner. "Looks like we're in for a calm night," Gloria said as they parted, gazing with satisfaction at the gray, overcast skies.

"Let's hope so," Lucy replied. London hadn't endured an air raid in more than a fortnight, and she wasn't sure how much longer their luck would hold out.

The next day when the Canaries gathered on their training field after dinner, Helen arrived a few minutes late, her expression stricken. "I've learned why the Shell Girls missed our match," she said.

"What happened?" asked Lucy, dreading the answer.

Helen looked around the circle as the others drew closer. "You didn't hear this from me," she said, "and please don't spread it around. There were hundreds of witnesses, so it's not likely to stay secret forever, but you know how the Ministry of Munitions likes to keep these things out of the press so word doesn't get back to the Germans—"

"Helen, what happened?" Marjorie broke in sharply. "An accident? Tell us."

Helen nodded and closed her eyes for a moment, pressing her hand to her heart as if to steady it. "On the evening of July first, just as the night shift was beginning, there was an explosion in the mixing house at the National Shell Filling Factory in Chilwell."

A gasp of horror went up from the group—hands pressed to mouths, eyes filling with helpless tears.

"As best they can determine now, the casualties include more than one hundred and thirty killed, a few seriously wounded, and a great many slightly wounded." Helen looked around the circle, her expression warning them to brace themselves. "I'm told—I'm told the carnage was beyond imagining."

Lucy took a steadying breath, lightheaded and sick to her stomach. "I saw something in the papers the other day," she said tremulously, "no more than a line or two, about a Midlands factory explosion. Sixty were feared dead. Is this—"

"The same accident," Helen finished for her, nodding. "Yes. One and the same."

A few of the girls sat down on the grass, staring bleakly; others put their arms around one another for support and comfort. "Do we know if any of the Shell Girls were hurt?" asked Marjorie.

"I wasn't given any names of casualties," said Helen, "and I never saw their team roster."

Lucy knew that even if none of the footballers had been injured or worse, the accident would have been enough to drive all thoughts of the match from their minds. "We should do something for them," she said. "Dedicate our next match to the injured and killed. Give the survivors all the proceeds from the gate. Something. Anything."

Her friends chimed in agreement, but Helen gave her head the slightest shake. "We cannot mention the Chilwell factory by name or the War Office would strenuously object. I trust that Lord Chetwynd will have relief measures well in hand. That said, I'm sure we could organize a benefit for a general fund for injured and ill munitions workers and their survivors."

"I'd like to help with that," said Lucy, and several others quickly volunteered too. From the long, stricken looks her teammates exchanged, Lucy knew they shared one chilling thought: As thorough as their own safety protocols were, and as scrupulously as everyone followed them, the devastating accident just as easily could have happened at Thornshire.

But April apparently disagreed. "There were seventeen explosions at the National Shell Filling Factory before this one," she said sharply. "Did Lord Chetwynd have those well in hand too?"

"Seventeen?" echoed Marjorie, incredulous. "Who told you that?"

"Someone who would know," April retorted, prompting a chorus of bewildered, heated questions from the others.

Everyone's nerves were fraught, Lucy realized, and little wonder.

"Let's not argue," she pleaded, raising her voice. "If there was negligence, that's for the Ministry of Munitions to sort out."

"Enough! Let's play football," Peggy shouted, tossing one ball and then another into their midst. "If ever I needed an official's whistle . . ."

She left the thought unfinished as the Canaries began to pass the ball around, some still looking very much upset for the munitions workers of Chilwell, others scowling, as if they'd like to have a sharp word with Lord Chetwynd. Whoever was to blame, if anyone was, Lucy hoped her mother would not hear of the terrible accident, or she would be on the telephone at once, pleading with her to leave Thornshire. If Lucy were perfectly honest with herself, she knew she ought to consider it. Her health continued to decline. Munitions factories throughout Britain were dismissing women workers by the hundreds. After nearly three years of munitions work, it was fair to say that she had done her bit. Why should she continue to risk her life, especially with the end of the war drawing ever nearer, and Daniel's homecoming surely not far behind?

As July passed, it truly did seem as if the tide of the war had turned. In the middle of the month, the Allies halted the dreadful *Kaiserschlacht*, after the German army attempted to encircle Reims only to be pushed back from the Marne, delivering Paris from the threat of invasion. On 20 July, London newspapers were exultant with reports that the German commander had ordered a retreat, and in the days that followed, the German armies were forced back to the positions from where they had launched the Kaiser's Spring Offensive months before. Then, at the end of July, word came that after weeks of brutal fighting, the Allied counteroffensive had finally driven the Germans out of the Marne region altogether and continued to pummel the German defensive line.

In Britain, hopes for victory soared, but in early August, the Central Powers dug in and renewed their offensives, stalling the Allied counterattack and inflicting massive casualties. Lucy struggled not to sink into despondency after allowing herself to hope that victory was within reach. After all, military authorities reported to the press that the front had been moved eastward roughly 280 miles, containing the Germans behind a line running along the Aisne and Vesle Rivers. There was good reason to believe the Allies would continue to advance until the Germans were driven out of France and back within their own borders. How long this might take was another question altogether.

One Thursday morning, Lucy was heading out the door in the pink light of early dawn, fortified with a strong cup of tea and running through football plays in her mind, when someone called her name. With a start, she glanced toward the voice and discovered Gloria waving urgently from the other side of the low fence that separated their two front gardens. "Lucy," she called out, holding out an envelope. "Thank goodness I caught you. This got mixed in with our post. I didn't see it until this morning."

Lucy felt her breath catch in her throat. Setting down her bag, she hurried over, took the letter, and glimpsed her own name and the return address of the War Office. Heart thudding, she exchanged a quick, stricken look with her friend and opened the flap. Withdrawing a single page, she read quickly, unable to breathe, her head swimming and heart pounding.

"Lucy," Gloria exclaimed, clutching the top of the fence as if she might leap over it. "Heavens, you've gone quite pale. What is it?"

"I—I need to sit down." Stunned, Lucy returned to her front steps and sank down upon them. Daniel, so close. She heard Gloria call her name, alarmed, but her gaze was fixed on the letter, and she was only dimly aware of her friend bolting away from the fence, through her own front gate, and down the pavement toward Lucy's. He was so close, so close he was almost home.

Suddenly Gloria was sitting beside her, supporting her with a strong arm around her waist. "Dearie, what is it? What's happened? Oh, I pray it isn't—"

"Daniel's alive and well," Lucy said, voice shaking. "He's in Britain. He's been in Britain a week already."

"Oh my goodness! That's wonderful news!" Gloria embraced her. "But—where is he? When is he coming home?"

"Derbyshire." Lucy checked the letter again to be sure, because it was too good to be true, and she was terrified that she had mistaken it. "In Derbyshire, at an estate called Alderlea." Alderlea. The name was familiar, but she couldn't place it; she had never visited Derbyshire except to attend a few of Daniel's football matches through the years.

"Why would they send him to a hospital so far away?"

"I—I don't know." Her thoughts were spinning like scattered leaves in a windstorm. She took a steadying breath and willed herself to calm. "Perhaps it's not a proper hospital, but a country house converted to a military hospital for the duration. I know of a few in Surrey." She gasped, seized the railing, and pulled herself shakily to her feet. "I must go to him at once."

Grasping the handle of her bag, she unlocked her front door and hurried back inside, Gloria following. "Are you in any condition to travel?" Gloria asked worriedly. "You seem distressed. Unwell."

Lucy almost laughed. She was a canary girl; she was always somewhat unwell. "I'm fine, just a bit . . . stunned." She thought for a moment. "I'll need a few things." She hurried off to her bedroom, dumped the contents of her bag onto the bed, and swiftly began filling it with clothing and other essentials.

Gloria watched her from the doorway, uncertain. "Do you even know where to go?"

"I'll check the maps at the station. I'll ask the ticket agent. How difficult could it be?" Lucy closed the bag and snapped the fastenings shut. She should pack something to eat. She darted past Gloria and off to the kitchen, where she quickly threw together a few sandwiches of cheese and bread, two staples that were not rationed, not yet, and so she kept her pantry well stocked. She wrapped the sandwiches carefully in paper and tucked them into her bag, snapping it shut again afterward.

"Is there anything I can do to help?" Gloria asked, trailing after her to the foyer.

"Thank you, no," said Lucy, pulling on her summer coat. "Wait. Yes. Would you please be a dear and telephone Thornshire Arsenal? Ask to speak to Mrs. Helen Purcell in administration. If you reach her, tell her I won't be in today, possibly not for the rest of the week, and tell her why." Lucy opened the door, and Gloria scurried outside ahead of her. "If you have to leave a message with someone else—" She paused to think as she fit the key in the lock. "Tell them I'll be absent, but say—say only that it's a family emergency." If they sacked her for missing work with no notice, Helen could likely get her her job back—if she wanted it back.

"I'll ring the arsenal right away," Gloria promised, giving her a quick hug. "Safe travels. Give my best to Daniel."

Tears of joy sprang into Lucy's eyes. "I shall," she said, blinking them away. She had not spoken to her beloved Daniel in years. To think that in a matter of hours, she would be able to speak to him, to hear his voice, to clasp his hand in hers, to kiss him, to hold him!

Lucy fairly ran to Clapham Junction, clutching her bag, thoughts in a whirl. The ticket agent, a cheerful auburn-haired girl of about nineteen looking very smart in her navy blue uniform and cap, helped her construct a route as far as the train could carry her—short jaunts from Clapham to Victoria, King's Cross, and St. Pancras, and then a long ride all the way to Sheffield. Lucy purchased the tickets and boarded, and caught all of her connections with time to spare. Her last train was passing through Cricklewood when she realized, too late, that she could have telephoned Alderlea before she set out. She could have asked the nurses to tell Daniel that she was on her way, or perhaps she could have spoken to Daniel himself. She could have rung her mother's home in Brookfield to share the happy news.

It was just as well she had not thought of it. Telephone calls would have delayed her at home, and she would be with Daniel all the sooner this way.

She was too full of anticipation to doze, and she had forgotten to pack anything to read, so instead she gazed out the window at the passing scenery, the green hills and meadows, the quaint villages and farmers' fields, summer gold and ripening like a promise. At noon she ate one of her sandwiches, amusing herself by imagining her sons' delight when they reunited with their father. She wondered how long Daniel would need to stay in hospital, how soon she might bring him home.

Her heart sank a little when she realized that as long as she continued working at Thornshire, Daniel might be better off in Brookfield, away from the German bombs, from the yellow powder, from his yellow-skinned, mottle-haired wife.

Suddenly a wave of embarrassment and worry swept over her. In her many letters, she had never mentioned the drastic changes to her appearance since she had become a canary girl. Her mother had said that Lucy's war work had stolen her beauty, a painful truth that Lucy had not been obliged to fully reckon with, what with her husband so far away, and his opinion of her beauty being the only one that mattered. What would Daniel think of her, seeing her like this after years apart? Would he be worried and heartbroken? Would he be repulsed? Tears gathered in her eyes, but she dabbed them with a handkerchief and angled her body toward the window so the other passengers wouldn't see. If any kindhearted soul tried to comfort her, she would lose her composure altogether.

It was midafternoon by the time she disembarked at Sheffield, calmer and resolved. She'd had time to think, and to remember how much Daniel adored her, how he always had, ever since they were children. They had planned to grow old together, and she had never doubted that he would still love her after she became wrinkled and stooped and gray. Why should she fear that he would stop loving her now, all because of a persistent cough and sallow skin and strange hair?

Inquiring at the station, she learned that an omnibus made the round trip to Alderlea three times a day, a service for medical staff and families visiting patients, managed by the War Department but funded entirely by Lord Rylance, the master of the estate. On the long, winding drive through the countryside, Lucy's seatmate, a matronly nurse, shared the heartrending story of the bereaved lord and lady who had devoted themselves to the care and rehabilitation of wounded soldiers after their eldest son and heir, Harrison Rylance, an officer with the Derbyshire Yeomanry, had lost his life at the Battle of Scimitar Hill in Gallipoli.

"We take on some of the most difficult cases," the nurse added proudly. "Amputees, shell shock, head wounds, facial mutilations. Oh, those are heartbreaking, but we do our best for them. Lady Rylance converted her finest sitting room into a workshop and hired the best lady sculptors and painters to make the masks these poor Tommies wear—paper-thin copper to replace their missing features, specially painted to match their skin tones. True works of art, easily as fine as anything from the Third London General Hospital, and they're famous for their excellent masks."

"Is that so?" Lucy replied faintly. "Then a soldier with a simple leg wound, even a very serious one, wouldn't be treated here?"

"Oh, no, not likely," said the nurse, shaking her head. "Not unless he's a Derbyshire lad, and he requested Alderlea in particular, to be close to his family." She paused to peer at Lucy quizzically. "So, what sort of nursing did you do at that arsenal, exactly? From the look of it, you must have worked rather too closely with the canary girls, if you'll forgive me for saying so."

"I'm not a nurse," Lucy said, taken aback. "I'm a munitions worker. I'm coming to Alderlea to visit my husband."

"Oh?" The nurse's smile faltered. "Oh. I misunderstood. Well, as you'll soon see, Alderlea is a lovely, tranquil spot, the very ideal for rest and rehabilitation." She managed a tight smile, glanced idly about for a moment, then retrieved a small book from her satchel and soon appeared engrossed in it.

Lucy felt sorry for the chagrined woman, who clearly never meant to speak so freely to a patient's wife, but her apprehension for Daniel surged, sweeping away every other emotion.

Before long they approached Alderlea, a breathtakingly beautiful three-story Elizabethan mansion of gray slate and buff sandstone nestled among broad, rolling lawns at the edge of a splendid grove of ancient alder trees, which no doubt had inspired the estate's name. Lucy's heart thudded in expectation mingled with worry as the omnibus drove around the back of the mansion and halted near the rear entrances. Clutching her bag, she disembarked as swiftly as she could, murmuring apologies to the passengers seated in the rows before her, whom she hurried past before they even had time to stand. Once outside, she inhaled deeply the cool, fresh air, comfortingly reminiscent of Brookfield, and followed the signs to patient inquiries. The orderly there offered her directions to Daniel's room, which, she realized as she turned down the hallway, was on the amputees' ward.

She halted, braced herself with a hand against the wall, and reminded herself to breathe. It could be worse. As the nurse had unwittingly warned her, it could have been far, far worse. He was her Daniel. She loved him. He was alive. That was all that mattered.

And she was so desperate to see him she could not wait a moment longer.

She hurried to the nurses' station, where a cherubic trainee checked a schedule and told her Daniel was outside in the western gardens. Lucy thanked her and fled down the stairs and outside.

She spotted him from a distance, knowing every curve and line of his form so well that even seated, with his back to her, she could not have mistaken him for anyone else. Setting her bag on a bench, she approached him at a sedate walk, scarcely able to breathe for the pounding of her heart. He must have heard the muffled crunch of her boots on the gravel path, for he glanced over his shoulder at the sound, and when he could not see who approached, he reached for the wheels on either side of his seat—wheels, he was in a wheelchair—and deftly maneuvered himself around to face her.

They both halted at the same time, each gazing in stunned disbelief at the other.

Her eyes lingered on his face—that wonderful, beloved face—for a long, utterly still moment, but eventually her gaze traveled downward, and her stomach dropped. His right leg was missing below the knee.

When she glanced up, she discovered that Daniel was gazing at her in wonder. "Lucy," he said, his voice as warm and dear as she remembered. "You're here. How—"

But she did not let him finish the thought. With a gasp, she ran to fling her arms around him, weeping for joy, for loss, for the years they would never get back, and for all those that she had once feared they would never have, but now dared hope they would.

23

AUGUST–SEPTEMBER 1918

APRIL

One Monday evening after work, April and Marjorie relaxed in the dining room of their hostel enjoying a late supper—although enjoying might have been overstating things, since rationing and food scarcity had steadily diminished the quality of their subsidized meals. After a long, exhausting shift, it was difficult to soothe one's gnawing hunger with a bowl of bland fish soup thickened with barley and seasoned with a sprinkling of dried herbs, ersatz coffee made from roasted chicory root and rye, and a wheat roll so hard that they were obliged to chip off pieces with their knives or dunk the roll in coffee to soften it before they could chew it. Yet they ate without complaint, for they were more fortunate than most. An unsavory dinner was easier to bear than hunger pangs, and they could rely on luncheon and tea at the arsenal canteen to fill their stomachs.

Conversation distracted them from the bland food. They still marveled at Lucy's unexpected three-day absence from Thornshire, followed by her sudden reappearance at the Globe just in time for their Sunday match. Most astonishing of all was Lucy's revelation that her husband had returned to England, and that she had seen him, at Alderlea, of all places. Daniel couldn't be discharged from hospital quite yet, but he should be able to come home by the end of September. Lucy's glow of joy had dimmed when she acknowledged that he would never play football again.

"It's sad, but he's luckier than a lot of wounded veterans," Marjorie pointed out, stirring her soup. "He can be an architect sitting down."

"Marjorie!" April exclaimed.

"I'm not being flippant. All I meant is that he can go back to his pre-war occupation, unlike some." She peered at April, curious. "What about your Oliver? What did he do before getting his hand shot off?"

April sighed, exasperated. "Honestly, Marjorie, how can you be so glib about a soldier's injuries? And don't call him 'my Oliver.' I don't know whether he is."

Marjorie gaped. "Are you mad? Of course he's 'your Oliver.' You see him nearly every day at the arsenal. He comes to all your matches. You kiss and hold hands—"

"Marjorie—"

"Don't deny it. I've seen you."

"I don't deny it, but—" April searched for the words to explain how she felt. "If he is 'my Oliver,' shouldn't he tell me so?"

Marjorie paused. "You mean he hasn't?"

"No."

"He's never told you he loves you?"

"No, not ever."

Marjorie thought for a moment. "That's a bit odd, fair enough, but maybe he's just shy."

"Or maybe he doesn't love me."

"I really don't think that's it." Marjorie shook her head, brow furrowing. "I think he's 'your Oliver,' if you want him to be. It's obvious to everyone but you."

April didn't want to discuss it anymore, so she took up Marjorie's other question. "Before the war, Oliver worked in his uncle's shop, in the same village where he grew up."

"Is that so?" Marjorie mulled it over. "Well, he could always go back to that, if Thornshire closes."

"Do you think it will?" April leaned forward and lowered her voice. "What have you heard?"

Marjorie leaned forward too, but before she could reply, their friends Nellie and Eliza, canary girls from Woolwich Arsenal, set their dinner trays on the table, pulled out the two remaining chairs, and sank heavily into them, their expressions disconsolate.

"Hello, girls," said April, looking from one to the other, wary. "How was your day?"

"Perfectly dreadful," said Nellie flatly.

"The worst since Brunner Mond exploded across the river and almost took us down with them," Eliza added.

"Now you have me worried," said Marjorie. "What happened?"

Nellie stirred her fish soup, scowling. "When we clocked out today, the lady superintendent told us—us two and about four dozen other girls—that there won't be no work for us next week."

"Oh no!" said April. "That's awful. I'm so sorry."

"Did they give you a character?" asked Marjorie. "Will they give you a week's pay in place of the week's work, to hold you over until you find something new?"

Nellie and Eliza shook their heads, miserable. "A character, yes," Eliza clarified, "but a week's pay, no, and we don't know where we'll find new jobs now."

"There's always domestic service," said Nellie glumly. "Oh, how I hated cleaning up other people's messes! I swore I'd never do that again."

April knew exactly how she felt. "From what I hear," she ventured nonetheless, "it's better than the mills."

Nellie shook her head. "It couldn't be."

"Trust me, it is," said Eliza vehemently. "I've done both and I know. The pay is better in the mills, but that's the only advantage. All the women of my family have worked in the mills for generations. They're even worse than the Danger Buildings. All the toil and lung ailments, plus it was so loud, we couldn't even sing."

"You sing at Woolwich?" asked April, exchanging puzzled glances with Marjorie. "Do you mean in theatricals?"

"No, while we work," said Eliza. "Don't you? It helps pass the time."

"We don't, at least not on the day shift," said April. Singing seemed unwise for the Filling Shop, what with all that yellow powder in the air despite the vastly improved ventilation system Helen had mandated months before.

"Well, we won't be able to sing together anymore," said Nellie, gulping her ersatz coffee and setting the cup down with a bang. "And we even had our own song, one we made up ourselves. What do you say, Eliza? Should we sing it here, one last time?"

Eliza smiled wistfully and nodded. She rose and beckoned to the other Woolwich girls, who, April now realized, stood out for their expressions, a tearful mix of anger, resignation, and unhappiness. They met in the center of the room and whispered among themselves for a moment. Then someone hummed a note for pitch, they turned to face their audience, and their voices rose in unison in a lively, merry song.

Where are the girls of the arsenal?
Working night and day;
Wearing the roses off their cheeks
For precious little pay.
Some people call us "canaries"
Working for the lads across the sea.
If not for munition lasses,
Where would Britain be?
I ask you
Where would Britain be?
So tell me
Where would Britain be?

The dining room rang with cheers and applause and calls for an encore. The Woolwich girls obliged, but the second time through, their audience shouted out the line "I ask you," and even louder "So tell me." As applause rose again, and the Woolwich girls took their bows,

Marjorie leaned close to April and said, "We have to teach this to our canary girls."

April agreed, laughing, but her mirth quickly faded. If the Woolwich Arsenal was sacking munitionettes, how much longer could Thornshire hold out?

The applause faded, the singers returned to their seats, and conversation resumed, a low hum punctuated with occasional bursts of laughter. April wanted to assure Nellie and Eliza that everything would work out in the end, but all she could offer them was sympathy, which wouldn't pay their bills.

After they finished dinner, Nellie asked Marjorie and April not to divulge the bad news to their landlady, out of concern that she might evict them at the end of the month, if they had not found new war work. "She won't toss you out," said Marjorie confidently. "This may be a munitionettes' hostel, but she wants that rent money, no matter what work you do."

Marjorie left unsaid what they all knew was certain: If Nellie and Eliza couldn't find new jobs, any jobs, the landlady would evict them for unpaid rent instead.

It was only later, as they were getting ready for bed, that April remembered the conversation their Woolwich friends' arrival had interrupted. "Marjorie," she ventured, climbing beneath her covers while Marjorie lingered at the mirror, brushing her gingery curls. "Do you really think Thornshire might close?"

Her friend was silent for a moment, attending to a particularly springy lock. Then she set down her brush with a sigh. "This is what I think," she said, sitting down on the edge of her bed and resting her elbows on her knees. "The war seems to be winding down—and thank God for that—and the demand for shells has been falling for months. I don't know that Thornshire will close altogether, but they won't always make munitions, and that's the work we know."

"We can learn other tasks just as we learned to fill shells. Obviously no one wants to work with TNT one day longer than necessary, but we need work—safe jobs that pay a living wage. Don't they owe us canary girls that much, after all we've done?" April lay back and pulled the comforter up to her chin. "Do you think Helen will tip us off if layoffs are coming?"

"She might," said Marjorie thoughtfully, turning off the lamp. "But first she'll say, 'You didn't hear this from me.'"

April managed a smile.

"But hiring and firing is really the superintendent's responsibility," Marjorie went on. "Helen might not find out we're all about to be sacked until right before Mrs. Carmichael delivers the blow."

"But Mr. Purcell would know before anyone at Thornshire. Wouldn't he tell Helen?"

"Maybe not, if he thought Helen might tell *us*." Springs creaked as Marjorie shifted in bed. "Do you think your Oliver would warn you what's coming?"

April yawned. "If he *were* 'my Oliver,' maybe he would." But she was a lot less certain than Marjorie that he was. They had been seeing each other for months, but Oliver never said a word about their future—his future, yes, his hopes for after the war, but nothing very specific, and nothing that included her in the picture.

Was she just a wartime fling after all?

Early the next morning, April and Marjorie were leaving the dining room after breakfast just as Nellie and Eliza were entering. When Marjorie quietly asked if they had a plan for finding new jobs, they said they intended to call at the nearest Labour Exchange as soon as it opened on Monday, their first day of unemployment, in hopes of getting into the queue before the rush. "If that don't turn up anything," said Nellie, pulling a face, "our next stop might be the BWEA office."

Marjorie grinned. "Please tell me you're not serious."

When Nellie burst out laughing, Eliza and Marjorie joined in, and a bit belatedly, April did too. Only later, as she and Marjorie were hurrying to catch their tram, did she abashedly ask, "What's this BWEA office? An employment agency?"

"No, silly." Marjorie leapt onto the tram as it was pulling away and extended a hand to assist April aboard after her. "It's the British Women's Emigration Association. Surely you've seen their leaflets and posters. They've been around for years, but lately they've been busier than usual, and little wonder."

Her words prompted a vague memory. "Wasn't a lady handing out their leaflets at the arsenal park last week?"

"Last week? She or one of her chums has been there nearly every day since the factories started sacking munitionettes. Not that you'd notice, since you're always in such a rush to see your Oliver."

"He's not my—" The tram hit a bump and April bit her tongue. "What does the BWEA want, anyway?"

"They want to help us, of course." Marjorie rolled her eyes. "They resettle surplus women abroad, mostly in Canada, Australia, and New Zealand, I think."

"Surplus women?" April echoed, clinging tightly to the bar opposite Marjorie, since the tram was packed as usual and they couldn't find seats.

"Their words, not mine. You know what they mean—the vast crowds of unmarried women fated to become a burden to their families and society." Marjorie paused to think. "I suppose there'll be more 'surplus women' around than ever before, since Britain has lost practically a whole generation of young men to this bloody war."

April suppressed a shudder. An entire generation lost—what a terrible thought. Yet it wasn't entirely true. It couldn't be. Some young men would eventually return from the war, and she hoped her brother Henry would be among them. He had been away so long, and had only come home once on leave. In all that time he had reported only a few minor scrapes and bruises, although in his last letter he had mentioned several of his chums falling desperately ill from influenza. April fervently hoped he wouldn't catch it. If his luck would hold out a few months longer, maybe he would survive the war and return to Carlisle none the worse for the experience.

"So these so-called surplus women," April said. "They go abroad to—to do what, exactly? Find husbands?"

"Yes, that, but also to find work." Marjorie shrugged. "The BWEA posters say there are loads of jobs throughout the commonwealth."

April fell silent, thinking. It sounded like a marvelous adventure, to travel across the ocean to a new country, full of opportunities, sparkling with possibilities. But a worry or two nagged at her. Didn't these countries have their own women? What sort of jobs were on offer? If it was all domestic service, she could do that here in England, and still visit her family a few times a year. "Have you thought about emigrating?" she asked Marjorie tentatively, not wanting to seem too interested herself.

"Me? Certainly not. I've been working too hard to defend Britain from the Hun to pack up and leave now. Besides," Marjorie added airily, "I'll never be a surplus woman. Once this war is over and all my Tommy pen pals come home, I'll likely marry one of them, when I'm good and ready."

April grinned, imagining a gang of eager Tommies all showing up at the hostel at the same time with bouquets and proposals, beaming adoringly at Marjorie and scowling at their rivals. "How will you ever choose among them? Most handsome, best prospects—"

"Best kisser," Marjorie added emphatically.

They both burst out laughing, and after that, all they had to do was glance at one another and the mirth would bubble up again, until they had to avoid eye contact and bite their lips together rather than annoy the other passengers. Only when they disembarked did they allow themselves to double over with laughter, wiping away tears. April knew the joke wasn't even that funny, but it felt so good to laugh.

As they approached the arsenal park, April looked around until she spotted a woman holding leaflets, standing on the corner nearest the tram stop and farthest from the arsenal's main gate. April hung back a bit, waiting until Marjorie was busy chatting with other canary girls arriving for their shifts. Then she darted over to take a leaflet from the BWEA woman, a gray-haired matron with spectacles who reminded her of her favorite schoolteacher. Folding the leaflet twice, April tucked it into her bag to read later, in private, and hurried to catch up with her friends.

In the shifting house, the Filling Shop, and the canteen, the munitionettes chatted about the latest news of the war and the worrisome rumors of layoffs, but on the training field, the only subject of conversation was football: the Munitionettes' Cup standings, the Canaries' upcoming match, and how to improve as much as they could in a very few days with better drills and clever strategy. Their most recent victory had qualified them for the semifinals, and they had been eager—and more than a little nervous—to learn who their opponents would be. Two days later, Peggy greeted them on the training field fairly bursting with the news that the Blyth Spartans had defeated the Dick, Kerr Ladies 4–2, knocking the celebrated team out of the tournament. The Canaries' next match would pit them against the Spartans, a side that had once defeated them 6–1.

"We're a different team now," Peggy reminded them, panting, as they ran sprints up and down their small training field. "They won't thrash us so horribly this time."

"They won't thrash us at all," Marjorie declared, bolting ahead of her captain to cross the finish line. "We'll do the thrashing this time."

April hoped she was right, but a nervous flutter filled her stomach every time she thought of those swift, strong footballers in their blue-and-white vertically striped, collared jerseys and matched striped caps. They were such kind, admirable girls, gracious in victory and perfectly friendly after a match, but they were fierce and merciless on the pitch. The Thornshire Canaries would have to play better than they ever had just to hold their own. It wouldn't help that the semifinal match would be played at St. James' Park in Newcastle-upon-Tyne, prestigious grounds nearly three hundred miles north of London, but only thirty miles south of Blyth, giving the Spartans an advantage even though St. James' Park wasn't their home pitch. Although some of the Canaries' most faithful supporters and fans had promised to attend, the crowd would no doubt skew as strongly blue-and-white as if the match were at Croft Park.

In the first few days of the week, their trainings went splendidly. Marjorie dazzled in the goal. Peggy's and Lucy's passes were crisp and precise, their attacks powerful and true. Helen ran as if she had a gale at her back, speeding her along the pitch, and she had a steely-eyed look as she swept in to steal the ball that made even April nervous, and April considered her a friend. As for herself, she ran and passed with a swiftness and agility that were unmatched by any of her teammates, if she did say so herself. She only wished her brother could be there to see her play in a semifinal match of a real tournament. Henry had taught her to play football years before, and she knew he would be exceedingly proud of her. As for Oliver, he attended every training that week and seemed impressed with her progress, even if his highest compliment was that she had "done well out there." Oliver was not given to effusive praise—but if his words were few in number, at least she knew he meant every one of them.

On the Thursday before the match, things took a dismal turn. Training was an absolute disaster. Marjorie hurt her wrist, which sent a frisson of panic through the team, even as she assured them, wincing, that she would be fine by Sunday. Every one of Peggy's shots soared over the top of the goal, while Lucy couldn't get anything past Marjorie, despite her wrist injury. Helen—bold, witty, fearless Helen, who had marched with the suffragettes and been jailed for it—had come down with an acute case of nerves. Queasy and pale, she seemed distracted and ran sluggishly—except when she fled to a trash bin to heave her dinner into it. "I'll be fine by Sunday," she said weakly, echoing Marjorie. "I'm fine now." Regardless, Peggy ordered her to sit down and put her head between her knees, and Helen seemed all too grateful to obey.

"Get your heads straight, girls," Peggy chided them kindly as they trooped back to work, discouraged. "Don't let the Spartans intimidate you before we even take the pitch."

Sound advice indeed, April thought, but easier said than done.

Later, back at the hostel, April was sitting cross-legged on her bed mending a torn seam on her football jersey when Marjorie returned from collecting her post. "I've a letter from my mum," she said, her voice distant, her eyes wide and stunned. She shut the door behind her but remained standing, one hand clutching a white envelope, the other pressed flat against her stomach. "It's my brother Archie. They've found him."

April's heart cinched. "And?"

"He's alive." A smile flickered in the corners of Marjorie's mouth, but her eyes were dazed, unfocused. "He's alive, and he's been in France all this time."

She looked so pale that April set her sewing aside, leapt off the bed, and hurried to put an arm around her friend and steer her to her bed, settling her down gingerly upon it. April ran to fetch her a glass of water, which Marjorie sipped, pausing now and then to take a deep, shaky breath.

"What happened?" April asked gently, stroking her friend's back. "You said Archie's in France. I thought he went missing in Belgium."

In reply, Marjorie handed her the letter.

The story it told was so astonishing that April had to read it twice over. Archie had been injured in the same attack that had killed his two brothers, taking minor shrapnel wounds and a severe blow to the head. A Belgian patrol found him days later, wandering alone close to enemy lines. He was unarmed, carrying no gear, wearing only the tatters of a blood-soaked French infantry soldier's uniform and the coat of a French officer. The Belgians took him to a field hospital, where his visible wounds were tended. Medics determined that although he could hear and seemed to understand simple questions in French and English, he could not speak, or write, and he could not remember his name or any detail of his life before he had been taken in by the patrol.

Remnants of insignia on the French officer's coat indicated a particular battalion, so a nurse arranged for a photographer to take Archie's picture and send it to the corps headquarters. Months passed before a response came, and it was disappointing: They were unable to identify the injured soldier, or match up the tattered coat with any missing officer, or even to explain why a man from that battalion would have been where the Belgian patrol had found him. They promised to keep the photograph in case any relevant cases were reported.

By that time, Archie had recovered enough to be evacuated safely. Assuming Archie was French, the chief medical officer had sent him to a convalescent hospital on the outskirts of Tours. In a fifteenth-century chateau hundreds of miles from the front, Archie regained his strength but not his memory. The nurses noticed that he seemed to understand very little of what was said to him, but appeared lost in a fog of melancholy.

The months passed, and still the soldier's voice and memory did not return. As part of his convalescence, he was encouraged to try various handicrafts and trades, and after he displayed considerable aptitude for gardening, he was offered a position on the groundskeeper's staff. He seemed grateful for the work, for something to do. His physical injuries had healed, but he obviously was in no condition to return to the front, and the staff had no idea where else to send him.

Then, more than a year after he had been found wandering in Belgium, a children's choir from a local convent's orphanage performed for the patients in the chateau's chapel. Archie attended the concert, and he listened impassively as the choir sang several French children's tunes. But when they began a traditional English round, "Come Follow Me," Archie suddenly joined in the song.

After that, the stunned, astonished staff began addressing him in English. In the weeks that followed, his voice and memories came back to him piecemeal. His photograph was given to British military authorities, but there were many missing and unaccounted-for Tommies, and identifying this particular unknown soldier was neither quick nor easy. It was Archie himself who solved the mystery when he remembered his own name, as suddenly and strangely as he had remembered the children's song. One day, another Englishman newly hired on the groundskeeper's staff introduced himself as Archie, and as they shook hands, Archie remarked, "Well, how about that—I'm an Archie too."

And then it all came back to him—his name, his family, his home, his childhood—but not the attack that had caused his injury and his brothers' deaths. Indeed, he was unaware that they had died, although he had likely witnessed their deaths. The doctors were obliged to break the news to him, but only after they were confident the shock would not cause a relapse. How he had ended up attired in a French uniform and coat was a mystifying question no one could answer.

"I see your mother expects Archie home by the end of September," said April, returning the letter to her friend. "He'll be back in Warrington within a fortnight. Oh, Marjorie, this is such wonderful news!"

Marjorie turned to her and smiled, tears of joy shining in her eyes. "Yes, can you believe it? I almost can't. It's simply too marvelous!"

They embraced, laughing aloud and bouncing on the edge of the bed, overcome by happiness and relief. For more than a year they had hardly dared believe Archie lived, and now he would be coming home.

The next day, when Marjorie shared her good news with her friends at Thornshire Arsenal, their shared happiness and relief seemed to energize the Canaries, for the blunders of the previous day were forgotten, their usual skill and confidence restored. By their final training on Saturday—a lighter session emphasizing technique over exertion, the better to rest their legs—April reckoned that they were as ready to face the Spartans as it was possible to be.

Match day began early for April and Marjorie, with a quick wash and a hasty breakfast before they hurried off with their bags to catch the tram to the train station. Before long they reached the platform at King's Cross where they met up with the other Canaries, and together they boarded a carriage and settled in for the long ride to Newcastle-upon-Tyne. Oliver was taking the same train, or so he had promised, but April had asked him to ride in a different car so she could sit with her teammates and concentrate on the match ahead.

"Your feelings aren't hurt, are they?" she had asked tentatively the previous evening as they had crossed through the arsenal park after her shift.

"Not at all," he had replied, interlacing his fingers through hers. "If that's what you need to do to play your best, I'd be a poor friend indeed to object."

Stung, April had smiled tightly and thanked him. So that's what she was after all—a friend. Heart heavy, she thought she ought to slip her hand from his grasp, but she didn't want to make a fuss, so she carried on as if nothing were amiss.

It was just as well Oliver was riding in a different carriage, April thought ruefully as the train sped northward. It would have been impossible to conceal her disappointment over so many miles.

Hours later, the Canaries arrived in Newcastle-upon-Tyne to find St. James' Park awash in lovely autumn sunshine, a steady but not too strong breeze stirring the pennants at the top of the stands. The Canaries had discussed their assignments and strategies on the train, so there was little conversation in the changing room as they put on their uniforms and organized their gear. April's heart thudded heavily in her chest as she laced up her shoes. Battling nerves yet again, Helen raced off to the toilet to empty her stomach. Marjorie was cheerful, her eyes alight with anticipation as she pulled her long socks over her shinguards and tucked her curls beneath her cap. April decided to pretend she felt the same way, and with any luck, the feeling might actually take hold by the time the whistle blew.

But when the Canaries emerged onto the pitch for their warm-up, April's stomach dipped at the sight of the strong, swift women in blue-and-white stripes, running drills and sinking shots into the net with seemingly effortless grace. "Keep your heads," Peggy reminded them as they took their places on their half, and soon, April felt the familiar pre-match routine settling her, until excitement won out and she could finally revel in the moment. How marvelous it was that she, a canary girl, a former housemaid from Carlisle, a girl nobody had ever expected to amount to much, was playing football, a sport she loved, before a crowd of tens of thousands of cheering spectators! If anyone had told her four years before that she would be running up and down Newcastle United's pitch with a massive grin on her face, awaiting the start of a girls' football tournament that would raise thousands of pounds for charity, she never would have believed it. But there she was, taking the center midfielder's position, crouched and poised to run, shifting her weight lightly from one foot to the other awaiting the start.

The whistle shrilled. The Blyth Spartans' brilliant striker, Bella Reay, the eighteen-year-old daughter of a local coal miner, kicked off.

The forty-five minutes that followed sped past in a blur of swift motion, sharp passes, daring runs, and alarming or thrilling shots on the goal depending upon the side, all accompanied by thunderous cheers from the crowd. After the first scoreless twenty minutes, in a brief respite after the ball soared out of bounds and Daisy ran to retrieve it, it occurred to April that the Spartans had improved since their two teams had last faced one another, but so had the Canaries. If anything, the Canaries' dramatic improvement seemed to have caught the Spartans off guard, but they were too skilled and too clever not to overcome their surprise. With ten minutes left in the half, the Spartans' halfback stole the ball from Lucy and sent it up to Reay, who spun and wove her way through the defense to pound a goal into the upper right corner, just beyond Marjorie's outstretched hands. About five minutes later, their right midfielder made a blistering run along the right side, keeping the ball close until she passed up to her forward. Helen flew between them to intercept the ball and sent it off to Daisy, but after a tussle, a defender stole it and fed it to Bella Reay just past midfield. Marjorie prepared to challenge the attack, balancing on the balls of her feet, knees bent, hands spread, but at the last possible moment Reay passed the ball off to her left forward, who was waiting stealthily in the box. Marjorie was still turning around as the Spartans' forward drove the ball into the net behind her.

The Canaries managed to prevent the Spartans from scoring again in the first half, but when the whistle blew, they headed off to the changing room discouraged, winded, and relieved that it hadn't been any worse.

"We have another forty-five minutes to make our mark," Peggy reminded them as they massaged tired muscles and gulped water. "We're only down by two. We're not out of this yet."

"That's right," Marjorie declared, clapping her hands. "We can do this, girls. Remember our training. Stay on your man. Stick to the basics. Talk to one another. They can't beat us unless we let them, and I say we don't let them!"

"That's right," someone shouted.

"Think about everyone who ever told you girls can't play football," Peggy said, looking around at the circle of tired, hopeful faces. "Imagine them sitting out there right now, all smug and superior, watching us play—and prove them wrong!"

"Hear, hear," cried April, and as everyone cheered and applauded, Daisy stuck two fingers in her mouth and let out a piercing whistle. Everyone laughed, and suddenly all was well again, and they were ready to face whatever the second half might bring. Win or lose, they were footballers, and no scowling, whinging, naysaying critic could ever take that away from them.

Smiling as if they were several goals ahead, they checked their shoelaces, adjusted their shin-guards, exchanged proud nods and hand clasps, and followed Peggy back out onto the pitch. They took their places and awaited the official's whistle.

And so the second half began.

Marjorie's and Peggy's inspiring words might have carried them on to victory, but the Blyth Spartans wanted to prove that girls could play football as much as the Canaries did, and in the fullness of ninety minutes, they made the stronger argument. The Spartans attacked fiercely from the kickoff, and nearly made a goal on the opening play, a cross from the right forward to Bella Reay that would have hit the net except for a brilliant flying save from Marjorie. The Canaries rallied and went on the offense, with Lucy in particular making many daring attempts to break into the Spartans' defending third, only to have each attempt thwarted by a halfback line as hard as steel. Ten minutes into the half, the Spartans scored on a penalty, bringing the score to a demoralizing 3–0.

April could feel the game slipping away from them, but she dug in and fought for every loose ball, every steal. Once she sent the ball up to Peggy, who dribbled a few yards before firing a shot that sailed past the Spartan keeper's fingertips only to hit the upright and careen away. The Spartans won the battle for the rebound at the top of the box. The ball found the swift feet of Bella Reay, who outran Helen in the race toward the goal. Daisy tried to intervene, but Reay made a deft spin and darted around her to score yet again.

Four–nil.

"Rally, Canaries," Marjorie called from the goal, clapping her gloved hands, as they returned to their positions for the kickoff. "Let's get on the board."

With only five minutes left in the game, that was about the best they could hope for. April glanced up into the stands and caught Oliver's eye. He nodded and gave her an encouraging smile, applauding, and he shouted something that she couldn't quite make out, but it heartened her even so. They had to score, just one goal to avoid a shutout.

But the Spartans were determined not to let that happen. Despite their best attempts, the Canaries found themselves almost entirely confined to their own half, rarely able to break out. Whenever Peggy or Lucy managed to cross midfield, the Spartans' halfback line not only broke up their attacks but carried the play ahead, allowing their forwards to maintain constant pressure on the Canaries' defense. But Helen and Daisy refused to let them penetrate too deeply, foiling most of the Spartans' attempts on goal. Marjorie took care of the rest.

As the seconds slipped away, the Canaries exchanged grim, determined glances. They all wanted to get one goal on the board before it was over, and they knew Lucy was their best chance. Again and again they fed her the ball, or passed it to Peggy so Lucy could slip past their midfielders and await a long drive that she could turn and pound into the net, but the Spartans figured out their scheme and kept Lucy covered despite every attempt to elude her defenders.

A Spartan attack went wide and out of bounds. Grateful for the breather, April bent over, hands on her knees, and scanned the pitch, searching in vain for some key to unlock the Spartans' defense. Straightening, she saw Marjorie exchanging a few words with Helen as she set up for the goal kick. A frisson of excitement ran through her to see a familiar gleam in Marjorie's eye, a certain set to her jaw that no one but her closest friend would have detected. Something was up, and while Marjorie's schemes off the pitch sometimes went spectacularly awry, as far as football was concerned, April knew to trust her judgment.

Marjorie signaled for the goal kick, and approached the ball so fiercely that everyone jogged back a few paces—but instead of sending the ball deep, Marjorie drove it to the left just before midfield, where Helen was waiting. After a moment of stunned shock, April sprang into action, flanking Helen to clear the way of defenders as she ran swiftly down the side, keeping the ball close.

Suddenly a Spartan midfielder sped in out of nowhere directly in Helen's path. April realized with sinking dismay that she would never get there in time.

But Helen kept dribbling toward the goal, veering neither to the left nor the right, as the midfielder barreled toward her. Seconds before they would have collided, Helen kicked the ball between the Spartan's legs, dodged to the left, and picked up the ball on the other side, behind her confounded opponent. Helen drew back her leg as if she intended to fire a shot straight into the center of the net—but as the Spartan defenders instinctively reacted to block the shot they expected, Helen passed the ball to Lucy, who was darting past her on the right. With the defense momentarily in disarray, Lucy drove the ball into the high right corner.

Four–one.

As their fans leapt to their feet, cheering and applauding and frantically waving their yellow-and-black scarves and pennants, the Canaries went wild with joy, shouting and laughing and embracing as they ran back to their half of the field, everyone wanting to hug Lucy and Helen most of all.

"What was that move, Helen?" Peggy shouted, laughing, as they lined up to resume play, but then she answered her own question. "A perfect nutmeg!"

Even Bella Reay seemed to be enjoying the scene as she lined up the ball at midfield and kicked off to her left forward. Seconds later, time ran out, the whistle blew, and the match went to the Spartans. As for the Canaries, they were out of the competition for the Munitionettes' Cup, but they had played hard for a full ninety minutes, never losing heart, and they had put up a goal on the scoreboard.

They congratulated the victors, waved to their supporters, and trooped off to the changing room, not quite as miserable as April would have expected them to be in defeat. Peggy gathered them together for a moment before sending them off to wash up and change clothes. "My mum would have been proud of each and every one of you," she declared, looking around the circle with shining eyes.

It was all any of them needed to hear.

The excitement of the game was fading by the time April left the changing room, lugging her soiled kit in her bag, wondering when and if she would ever wear it again. She spotted Oliver waiting on the sideline. "I'll catch up," she told her friends, and headed toward him.

He met her halfway, and when she wordlessly held out her bag, he took it, his gaze never leaving her face. "You were marvelous out there," he said, shaking his head in wonder.

"We lost," she reminded him glumly, as disappointment began to sink in.

"It was a moral victory," he said as they turned and walked side by side to the gate, trailing after her teammates. "You made them work for every goal, you kept the margin of victory much closer this time."

"I suppose."

"You weren't shut out."

"That's true."

"They'll probably go on to win the entire tournament, so you lost only to the very best."

A smile tugged at the corners of her mouth. "I guess if you have to lose, that's the way to do it."

They walked along in silence as they left St. James' Park and turned down the pavement toward the train station.

"That rule you have about riding in the same carriage," said Oliver.

"That's only on the way to a match, not on the way back afterward, correct?"

"Correct," said April, managing a weary laugh. "You can sit beside me all the way to London if you like."

"I'd like nothing more."

That brought a warm glow to her heart that spread to her cheeks, but she replied, "If you say so."

"I do say so." He gave her a curious sidelong look, followed by a little nudge with his elbow. "You ought to be proud, making it to the semifinals in your first tournament. Better luck next year, eh?"

"I don't know that there will be a next year for us," said April, a sob catching in her throat. "Will there be another Munitionettes' Cup for the Thornshire Canaries if Thornshire Arsenal closes? Will there be a Munitionettes' League at all if there are no more munitionettes?"

Oliver fell silent, studying her with kind sympathy. "I suppose we'll have to wait and see."

"I suppose." Suddenly April halted. "Would you tell me if the arsenal was closing, if I was going to be sacked?"

"I would if I knew for certain," said Oliver. "Right now, all I can say is that it doesn't look promising, but I think you knew that already."

April heaved a sigh. "Of course I did."

"Won't you be relieved to stop working with TNT? That's what you've said—"

"Yes, but that's not the point. I don't want to go back to domestic service." April lifted her hands and let them fall to her sides, frustrated. "But what choice will I have? I'll need to find work, or else I might have to emigrate to New Zealand or Australia as a surplus woman."

Oliver peered at her, bewildered. "What are you talking about?"

"Well, there are these organizations—you've seen the leaflets at the arsenal park, surely—that help surplus women emigrate to other countries where they can find jobs, and—"

"I know *that*. What I don't understand is why you call yourself a surplus woman, of all the wretched phrases, when you already have a fellow right here in Britain who wants to marry you?"

For a moment, April couldn't breathe. "Which fellow do you mean?" she asked, feigning puzzlement, though the tremor in her voice gave her away.

He smiled, his gaze full of love and amusement. "I think you know."

He set down her bag, wrapped her in his embrace, and kissed her until she fairly soared with happiness.

24

OCTOBER–NOVEMBER 1918

HELEN

In early October, the Blyth Spartans faced the Bolckow Vaughan Ladies of Middlesbrough's Bolckow, Vaughan & Co. ironworks in the Munitionettes' Cup Final. The Spartans again had something of a home pitch advantage, since they played frequently at St. James' Park and Middlesbrough lay more than fifty miles to the south. More than fifteen thousand spectators filled the stands for what was expected to be a thrilling match, since the Bolckow Vaughan captain, Winnie McKenna, was considered Bella Reay's equal as a goal-scorer.

But as Helen had learned in her brief career as a footballer, an exciting match did not always mean an abundance of goals.

"A nil–nil stalemate," Marjorie exclaimed, incredulous. "With Bella Reay and Winnie McKenna on the same pitch?"

The Canaries were in the arsenal canteen, some seated, others standing and peering over their teammates' shoulders at a newspaper spread open on the table before them, their attention fixed on a single paragraph tucked amid the sport reports.

"Both fullback lines must have been exceptionally strong," Helen remarked. "It sounds as if the game was still quite exciting."

"If it had been us, we would have put a goal or two on the board," said Marjorie.

"Maybe we'll get another chance next season," said Daisy, hopeful, but the furtive glances the others exchanged suggested that they were less certain there would be another season, for them or any munitionettes' team. Helen braced herself for pointed questions about Thornshire—and muffled a sigh of relief when none came her way.

"So what happens now?" asked Lucy. "Will they share the title?"

No one knew, and they had to wait another week for an announcement that the Munitionettes' Cup Final would be replayed the following Saturday at Ayresome Park in Middlesbrough, not far from the Bolckow Vaughan ironworks. Daisy proposed that the Canaries attend the match together, since the Danger Building was no longer running weekend shifts and they all had the day off. The outing sounded like great fun, but Helen demurred, not because she wanted to avoid her teammates' questions about the arsenal's future, although she did, but because she had a physician's appointment she simply couldn't put off any longer.

Since the Bolckow Vaughan Ladies would have a clear home pitch advantage in the rematch, Helen reckoned they were favored to win. Fortunately, she was not a gambler, as she learned the following Monday when her friends enthusiastically described the thrilling action for those who had not attended. Finishing one another's sentences in their excitement, they explained how the Spartans' defense had entirely contained Winnie McKenna, whereas Bella Reay had amazed the twenty thousand spectators with dazzling runs and incredible shooting. When the final whistle blew, the Blyth Spartans had won the match, 5–0, to claim the first Munitionettes' Cup.

"I hope it's not the first and only," said April, and the others nodded or chimed in agreement. A few girls sent furtive glances Helen's way, as if they hoped she might enlighten them, but she had nothing to share.

A fortnight after the final, Helen and the Canaries gathered with other Thornshire Arsenal munitionettes, the Burridge family, and many of their friends at a small church in Poplar to celebrate Peggy's wedding. Earlier that autumn, her fiancé had been given a medical discharge from the army after a terrible bout with pneumonia, but now he was as right as rain, and the long-delayed ceremony could finally take place.

"They thought he'd had the influenza, but it turned out it was only pneumonia," Peggy had explained one day during their tea break as she had handed out the invitations, pretty ivory cards inscribed in a very fine hand. "Can you imagine anyone would ever say *only* pneumonia?"

"In this case, yes," April had replied, admiring her invitation. "I've heard the Spanish Flu is absolutely awful."

"But it's not Spanish at all," Marjorie had said. "The American soldiers brought it over with them from the U.S."

"That's a fine thing to say about our allies," Daisy had teased, nudging her.

"I think you mean, that's a fine thing for our allies to do," Marjorie had retorted, nudging her back.

Everyone had laughed. No one was terribly concerned about this new influenza, wherever it had come from. After four long, difficult years of war, a common illness seemed of little consequence.

On the morning of Peggy's wedding, Helen, Lucy, April, and Marjorie attended the bride in the ladies' lounge, assisting her into her gown, dressing her hair, arranging her veil, chatting and laughing and blinking away tears of joy. After so much loss and sorrow, so many unhappy partings, it was truly wonderful to celebrate a new beginning.

If the bridesmaids had any misgivings—and not all of them did—it was that Peggy had decided to honor their status as proud, loyal munitionettes by having them wear spotlessly clean, well-pressed Danger Building uniforms, complete with trousers, blouses, and jackets, with fresh flowers adorning their caps. Helen had dressed as a munitionette only once before, when she had entered the Danger Building on her reconnaissance mission, and she did not find the ensemble becoming in the least, especially for a formal affair.

"At least we're allowed to wear our own shoes instead of those wooden clogs," mused April as the bridesmaids studied themselves in a full-length mirror, their expressions ranging from amusement to muted dismay.

Helen glanced over her shoulder, and when she saw Peggy happily engaged in conversation with her aunt, she replied, "Nor did she forbid us to wear metal. We should indeed be thankful for small favors."

"It's Peggy's day," said Lucy. "This makes her happy, and I think we look quite charming."

"It could have been worse," said Marjorie. "Peggy might've asked us to wear our football kit."

Soon thereafter, they took their bouquets in hand and lined up for the procession. "I hope you don't come down with a bad case of nerves like you did in the semifinal," Marjorie teased Helen in a whisper as they waited for the organist to play their cue.

"As it happens, it wasn't nerves," Helen whispered back, but before she was obliged to say anything more, the music reached a crescendo and the ceremony began.

It was a truly lovely, joyful day, and much welcome for it. Yet for every sign that the war was drawing to a close—soldiers returning from the front, dwindling munitions orders—there were other, crushing indications that the Germans meant to drag the fighting out until the last man.

On 10 October, the Irish mail boat *Leinster*, on her way from Kingstown to Holyhead, was torpedoed by a German U-boat twelve miles out to sea. The exact number of passengers on board was uncertain, making the death toll difficult to determine, but it was believed that roughly six hundred passengers and crew had perished, including twenty-one postal workers, while only two hundred souls were saved. All of Britain was outraged by the attack, which many said equaled in atrocity the sinking of the *Lusitania*. Arthur James Balfour, Secretary of State for Foreign Affairs, denounced the attack as "an act of pure barbarism." Of the perpetrators, he declared, "Brutes they were when they began the War, and brutes they remain."

The words chilled Helen as she read them, for the Earl of Balfour did not seem to distinguish between the German sailors who had committed the terrible act, the German military as a whole, and all Germans everywhere. It was another episode in a disturbing trend. Only a few months before, a petition with 1,250,000 signatures calling for the British government to intern every enemy alien without distinction of any kind, and to take drastic steps to eradicate all German influence in British government circles and society, had been delivered with great pomp and ceremony to the Prime Minister's residence at 10 Downing Street. Anti-German sentiment had worried Helen throughout the war, but with outrage flaring in the aftermath of the sinking of the *Leinster*, the safety of her mother and sisters seemed more precarious than it had in years.

Helen often suspected she was more concerned for their safety than they were for their own. Mother was busier than ever, tending to Banbury Cottage and her many volunteer organizations. Daphne had begun her studies at Oxford, amid rumors that female students' qualifications might be formally recognized in the near future. Penelope, ever loyal and loving, remained Margaret's steadfast companion, no longer a paid servant but rather a member of the family. Although her exact role and title were not precisely defined, she had become Margaret's partner in managing the estate and raising the children, who called her Aunt Penny and absolutely adored her. As far as Helen knew, neither Margaret's family nor her late husband's disapproved of the unconventional domestic arrangement, which had clearly brought Margaret and the children much comfort during their bereavement. With so many men lost to the war, households comprised entirely of women and children were not at all unusual, and as a wealthy widow, Margaret had no need to marry again. In fact, she had firmly stated her intention never to do so. Given the circumstances, Penelope confided to Helen, she and Margaret dared to hope that society would not disparage two surplus women who had decided to live peacefully and quietly together, troubling no one.

Helen loathed that wretched phrase—"surplus women," indeed—but she adored her sister and Margaret, and she wished them much joy and contentment together.

Only two days after the attack on the *Leinster*, word broke that the German chancellor had accepted U.S. President Woodrow Wilson's Fourteen Points as a foundation for negotiating peace. "This is significant," said Arthur emphatically as he and Helen pored over the papers at breakfast. "For the Germans even to acknowledge that they'd consider a negotiated truce is a remarkable admission that they're no longer confident of victory."

But all too soon, Helen and Arthur's hopes for an imminent end to the war were dashed when the press reported Supreme Allied Commander Ferdinand Foch's declaration that the Allies would accept nothing short of unconditional surrender. The Germans seemed as if they would never capitulate, regardless of the overtures they made, and Helen was exhausted from having her hopes for peace raised and then dashed over and over again.

And yet, in apparent contradiction, the munitions industry seemed to be wrapping things up. Helen did not know what to make of it. First, following his father's orders, Arthur had reduced work hours at Thornshire Arsenal across the board. Next he had cancelled most weekend shifts, including Danger Building work. Then, in the third week of October, he summoned Helen, Superintendent Carmichael, Tom, and Oliver to a confidential meeting, arranged at practically the last minute without any explanation of its purpose.

Helen was the first to join her husband in the conference room. "This is all very mysterious," she remarked, taking the seat at his left hand. "You didn't breathe a word this morning."

"I endeavor—with mixed success—to keep arsenal business out of our breakfast conversation."

"Only when it's bad news," she pointed out. He nodded in acknowledgment but said nothing more.

Helen regarded him curiously, eyebrows raised, inviting him to elaborate. When he volunteered nothing, she muffled a sigh and settled back to wait for the others.

When everyone had taken their places at the table, Arthur did not leave them long in suspense. Orders had come down from above that Thornshire Arsenal was to reduce its workforce by a quarter in November, and a quarter more the following month, with additional reductions to be determined later according to the course of the war.

Helen had been expecting an announcement of this sort, and her heart sank as she thought of the munitionettes, many of whom had become her friends, who were about to be sacked and did not know it. Her only consolation was that Arthur had instructed Tom, and not herself, to consult with the foremen and determine which workers should be let go. Superintendent Carmichael would deliver the bad news. Helen's assignment was to comfort the dismissed munitionettes, explain how they would receive their separation benefits, direct them to various employment resources, and reassure those who remained.

"How am I to reassure them if they ask me whether more layoffs are coming?" asked Helen. "Everyone at the arsenal will wonder. You cannot expect me to lie, knowing that another twenty-five percent will be sacked in December."

Arthur shook his head, brow furrowed. "Of course I wouldn't ask you to lie. They deserve honesty and fair dealing for their years of loyal service to us, and to King and Country."

They deserved all that and more, Helen thought, but she didn't blame Arthur. None of this was his fault.

"I volunteer to help select the munitionettes who ought to be released from their duties first," said Superintendent Carmichael, nodding to Arthur and Tom. "I know the girls' histories thoroughly, and I would be happy to lend my expertise to this regrettable but necessary task."

Perhaps a bit too happy, thought Helen. Surely the superintendent knew that after the last munitionette left, she would have made herself redundant.

In the days that followed, the unlucky workers were selected, and the bad news was delivered to the day shift on the last Friday of October. Helen had deliberately not read the list, knowing she would be tempted to use her influence with Arthur to spare her friends at the expense of workers she knew less well, and that would be both unfair and unprofessional. She lingered in her office after the shift ended, just in case anyone wanted to air her grievances or plead for her job back. If Mabel were there, Helen thought wistfully, she would demand a sit-down at the negotiating table, and she would advocate fiercely for her canary girls. But not even Mabel could have saved their jobs this time. As the need for munitions went, so went the need for munitionettes.

But why must Thornshire make munitions or nothing at all? Surely there would be manufacturing needs in peacetime. They had an ample, skilled workforce. They could retool the various shops. All they needed was a product and a scheme—and a sound argument to put before Arthur's father.

A knock on the open door jolted her from her reverie. "Helen?" asked Lucy from the doorway. "May I have a moment?"

Helen's heart plummeted. Lucy had been let go. "Of course," she said, rising and gesturing to the chair in front of her desk. They sat down together. "I'm so sorry. Even though we all sensed that this was coming—at least, I believe we all did—it must be quite a blow."

"Oh yes," said Lucy, shaking her head sadly. "Those poor girls."

"Oh, you're still with us, then?"

"Yes, for now." Lucy studied her, curious. "Didn't you know?"

Helen shook her head. "It wasn't my task—and thank goodness for that, because I don't think I could have endured it."

A hint of a smile appeared on Lucy's face. "I don't think you're cut out to be a boss, then."

"Certainly not. I'll stick to halfback." Helen interlaced her fingers and rested them on the desk. "What can I do for you, Lucy?"

"Well, about the layoffs . . ." Lucy hesitated. "I'm afraid April was let go."

"Oh no! But she's such a good worker."

"They all are," said Lucy. "Everyone deserves her place here. But some need the wages more than others." She met Helen's gaze squarely. "I intend to resign, and I want April to be given my place."

Helen studied her, bewildered. "But what about you? What will you do for work?"

"Daniel's coming home soon, and I want to be with him, and with our boys. Our family has been separated too long. Besides, Daniel is eager to get back to work, and he may need my help to do so, especially as he . . . adjusts to his new circumstances." Lucy gave a small shrug. "He's always provided very well for us as an architect, and I'm sure he will again, but April—"

"April has no one else to provide for her."

"No, and indeed she has to provide for others, for her mother and siblings."

Helen admired Lucy's selflessness, but she tried to think of objections others might raise. "April has never worked in the Finishing Shop. She doesn't know your job."

"Not yet, but she's very clever and hardworking, as you know, and I'm certain she can learn. I've already asked Mr. Vernon, and he's agreeable."

Helen smiled. "He didn't grumble and scold and try to talk you out of leaving?"

"He did, actually," said Lucy, laughing, "but after I convinced him that my mind was made up, he said that he'd tolerate April as my replacement, since I endorse her."

"Since Mr. Vernon consents, it should be a simple matter to transfer April to the Finishing Shop. And if Mr. Vernon couldn't persuade you to stay, I know any attempt of mine would be futile." Helen rose and reached across the desk to shake her friend's hand. "Best of luck to you, Lucy. We'll miss you around here."

"Thank you, Helen." Lucy clasped both of her hands around Helen's for a moment, smiling warmly, but then she gave a little start. "I suppose I ought to run after April and give her the good news."

"Yes, please do," Helen urged. She didn't want April to be unhappy a moment longer than necessary.

After Lucy hurried off, Helen began the paperwork to discharge

Lucy and transfer April to the Finishing Shop. She felt a twinge of misgiving, knowing that another round of layoffs was coming and that April may have been granted only a temporary reprieve. Still, even if that's all it was, she would earn wages and acquire skills that could help her find work in peacetime, after the arsenal closed.

If Thornshire Arsenal closed. Where was it written that it must retreat to dormancy rather than become something new?

On 28 October, word reached London that Austria had surrendered. It was all so exhilarating that Helen could hear munitions workers shouting and cheering on the arsenal grounds from her windowless office. Soon thereafter, not even the disappointing news that Germany still refused to surrender could demoralize them. Without their staunchest ally by their side, how could the Germans reasonably hope to continue fighting much longer?

Less than a fortnight later, on Sunday, 10 November, Helen and Arthur went down to breakfast together, poured their tea, and opened the papers only to discover the most stunning, glorious news imaginable.

Prince Maximilian of Baden had announced that Kaiser Wilhelm II, having lost the support of his military leaders, had abdicated and had fled into exile to the Netherlands. Two prominent political leaders had each declared themselves in charge of the provisional German government, and while their two factions argued in the Reichstag, their followers marched and fought one another in the streets.

"We're surely coming to the end now," Arthur said, reaching across the table for her hand, his face both haggard and optimistic. Helen hoped with all her heart that it was true. In October, the Allies had signed an armistice with the Ottoman Empire to end the fighting in the Middle East. Only days later, the Austro-Hungarian Empire had signed an armistice with Italy. All that remained was for Germany to bow to the increasingly inevitable.

At last, later that evening, the joyous, long-awaited news came:

The Allies and Germany had agreed to an armistice to be signed early the following morning officially ending the hostilities on land, at sea, and in the air. At the eleventh hour of the eleventh day of the eleventh month, the war would end at last.

The child Helen carried would be born into a world at peace.

She couldn't wait to tell Arthur.

Epilogue

December 26, 1920

LUCY

Boxing Day 1920 proved to be a splendid day for a football match—chilly and brisk, with blue skies and sunshine playing hide-and-seek behind scattered clouds. Jamie and Simon were so excited on the train from London to Liverpool that Lucy and Daniel could hardly get them to sit still. "We wish you were playing today, Mum," Simon said at least twice along the way.

"My football days are over," Lucy would reply, smiling fondly, but the truth was, she wished she were playing too. In the two years since she had left munitions work, her skin had lost its yellow hue and her hair had grown back as dark and silky as before the war, but intermittent stomach ailments and headaches still troubled her, and bouts of coughing wracked her in cold, damp weather, or in smoky rooms, or if she overexerted herself. A kickaround in the garden with the boys was about all she could manage, and as the boys grew, even that was becoming too much. Both boys had clearly inherited their father's athletic skill, especially Jamie, who often said that he wanted to be a footballer and an architect like his father.

Neither of their sons ever said they wished to be a soldier—which Lucy and Daniel privately agreed was a tremendous relief.

Lucy would never forget that moment in the garden at Alderlea in the last months of the war, when she and Daniel had looked upon each other after more than three years apart—three long, hard, painful years that had transformed them both utterly. After the overwhelming joy of their first embrace, Daniel had cleared his throat, and his arms had loosened around her, though he had not entirely let go. "I should wonder," he said thickly, "if you would still want me, broken as I am."

Incredulous, tears trickling down her cheeks, Lucy pulled away just far enough to meet his gaze. "I could ask you the same question," she managed to say, choking on tears and laughter. "How could you still want me, with my horrid yellow skin, mottled hair, and wretched cough?"

"Do you really need to ask?" said Daniel, bewildered. "How could I not want you, having learned what my life is without you?"

They embraced again, and kissed, and promised that each was as beloved as ever to the other. And after that, they never wondered again. All that mattered was that they loved each other and they were together again. That was everything. Daniel would never play football again, and he felt the loss keenly, but within a few months of the Armistice, his architecture career was nearly restored to the heights he had achieved before the war. They were the lucky ones, and well they knew it.

They disembarked at Liverpool. Daniel managed the stairs and the gap rather deftly, aided by his prosthetic leg, his cane, and Simon's sturdy shoulder to lean on. They took a cab to Goodison Park, where Daniel had once led Tottenham Hotspur to several wins against Everton, but where today they and several of the former Thornshire Canaries were reuniting as spectators for a special Boxing Day charity football match between the Dick, Kerr Ladies and the St. Helens Ladies. Tickets had sold out well in advance. Lucy had heard that more than fifty thousand people would be in attendance, and more than ten thousand had been turned away. If the rumors proved true, the match would shatter all records for attendance at a women's football match—but Lucy needed to see only a few particular football fans for her day to be complete.

She glimpsed one of them standing in the queue near the front gate. "Peggy!" she cried out, hurrying ahead of her family to embrace her dear friend, but carefully, because Peggy had a toddler balanced on her hip. The lurid yellow hue had left her skin too, and instead of a long braid, she wore her reddish-brown hair in a stylish, wavy bob. They laughed and teared up a bit and exchanged a few words, with promises to catch up inside the stadium in just a few minutes.

Lucy hurried back to Daniel and the boys, who had taken their places in the queue, which had lengthened behind them and now included April and Oliver. The two former teammates caught sight of one another at the same time, cried out for joy, and broke out of the queue to embrace. Lucy hadn't seen April since she and Oliver had married the previous spring in a lovely ceremony in Carlisle, in the same church where April's parents had wed and she and all her siblings had been baptized. Her brother Henry had walked her down the aisle, looking dashing and proud in his army dress uniform. A year before the wedding, the couple had opened a shop in the town selling books, stationery, toys, and other gifts, and in her most recent letter, April had described their plans to take over the space next door to add a teashop. She had said nothing of her other work-in-progress; Lucy concealed her surprise and delight at the unmistakable signs that April was expecting. She would wait for her friend to tell her. Perhaps April planned to announce the happy news to the whole team at once, after they were reunited inside.

When the Dempseys found their seats, Lucy was delighted to see many former coworkers awaiting them, some with their husbands and a young child or two. After a flurry of joyful meetings and fond teasing, Lucy took her place with her own family and was delighted to find herself seated beside Helen. Her very active eighteen-month-old son wriggled and laughed and fussed on her lap until Arthur, seated on Helen's other side, swept him up in his arms and walked him around a bit to settle him down. Helen was the Thornshire Canary Lucy saw most often, as Helen and Arthur divided their time between their lovely Marylebone home and their estate in Oxfordshire. Lucy and Helen would meet for lunches or strolls through Hyde Park whenever Helen was in London, and occasionally the Dempseys visited the Purcells at their country house. There was a broad, sweeping lawn where the boys could kick the football around while Daniel and Arthur cheered and coached from the sidelines, and a tennis court where Helen had taught Lucy to play.

Often Helen would pass along news of their former teammate and Lucy's fellow Finishing Shop chum Daisy, who had assumed the role of welfare supervisor at Thornshire Automotive Works as it underwent the transition from munitions work to domestic manufacturing. The retooled Purcell Products factory was a smaller concern than Thornshire Arsenal had been, but it still employed more than two thousand workers, at least a quarter of whom were women, all overseen by Superintendent Carmichael, who apparently intended to remain a permanent fixture at Thornshire whatever incarnation it took.

"Do you ever miss your old job?" Lucy had asked Helen once, soon after her son's first birthday.

"Goodness no," Helen had replied, laughing, shifting her son from one hip to the other. "I have enough to do minding this wee lad, and keeping up with my work for the Women's Party. Someone has to make sure we focus on feminist issues and not anti-German nationalism. We had enough of that during the war."

"Have you considered running for Parliament, now that women are permitted?"

Helen laughed lightly. "My word, wouldn't that be a lark?"

Lucy had noted, with great interest and curiosity, that her friend hadn't said yes, but she hadn't declined, either.

A murmur of excitement passed through the crowd as the players emerged from their separate changing rooms and jogged out onto the pitch. The voices rose to a cheer as thunderous applause rained down upon both teams, but Lucy's gaze was fixed upon the Dick, Kerr

Ladies in their narrowly striped black-and-white jerseys, matching caps, and short blue trousers—all but one, who wore a solid blue jersey and gloves—

"There she is," Helen cried out, seizing Lucy's arm. "There's Marjorie!"

The erstwhile Thornshire Canaries rose, applauded wildly, and shouted their former keeper's name. From the goal, where she deftly fended off blistering shots from her new teammates, Marjorie spared her friends one broad, happy grin, and then immediately returned her attention to the pitch, all steely-eyed, canny determination.

A few rows ahead, Peggy turned around and called to her friends, "Do you suppose any of her Tommy pen pals have turned out to watch Marjorie play today?"

Her words met with a wave of laughter.

"She's narrowed it down to three," replied Daisy, climbing into the stands, having arrived just in time with her fiancé, a junior foreman at Thornshire Automotive. "She told me she's having far too much fun to choose between the last of them so soon, and she'll marry when she's good and ready."

"That sounds like Marjorie," said April, shaking her head in fond amusement. Beside her, Oliver nodded, his expression so comically exasperated that April burst out laughing.

Suddenly Lucy felt Simon tug on her coat sleeve. "Mum," he cried, fairly bouncing in his seat as the warm-up ended and the players took their positions. "It's about to begin!"

"I know, darling," she said, exchanging a fond smile with Daniel over their youngest son's head. "Isn't it all wonderful?"

And then the whistle blew.

AUTHOR'S NOTE

In the match at Goodison Park on Boxing Day 1920, the Dick, Kerr Ladies defeated the St. Helens Ladies 4–0, with right back Alice Kell completing a hat trick in the second half for an exciting finish. The match raised £3,115 for charity, the equivalent of £148,347.20 (US$195,151.48) in 2022. About 53,000 spectators attended the match, with an estimated 15,000 more fans turned away at the gate for lack of room. The attendance set a record for women's football that was not broken until March 2019, when an audience of 60,739 watched Atlético Madrid host Barcelona at the Wanda Metropolitano.

Even after the soldiers returned from the war and men's football resumed, women's football continued to grow in popularity, with attendance at women's matches averaging 12,000 eager fans. Unfortunately, this alarmed the Football Association, the governing body for football in the UK, responsible for overseeing all aspects of professional and amateur play in its jurisdiction. The FA had tolerated women's football during the war, since men's professional football had essentially gone on hiatus and women's matches usually raised funds for worthy causes. After the war, however, the FA became increasingly concerned that women's football could draw interest away from men's leagues, reducing attendance, and thereby revenues, for Football League matches.

Nearly a year after the record-breaking Boxing Day match, on 5 December 1921, the FA banned women's football from all of its affiliated grounds and forbade its members from serving as referees or linesmen at women's matches. To justify their decision, the FA released a statement declaring that football was "quite unsuitable for females and ought not to be encouraged," citing opinions from doctors who agreed that football posed serious physical risks to women.

Women footballers responded with outrage. "The controlling body of the FA are a hundred years behind the times and their action is purely sex prejudice," declared Jessie "Jean" Boultwood, captain of the Plymouth Ladies. But the women footballers' protests were to no avail. The FA's edict stood, essentially outlawing women's professional football in England. Women could still play at the recreational level on small, non-FA pitches or rugby grounds before reduced crowds, but most women's teams disbanded instead. A few, like the Dick, Kerr Ladies, traveled abroad to play. In 1922, the team embarked on a tour of North America, playing nine men's teams and drawing audiences of up to 10,000 in the United States, where women's football had not yet caught on. The tour did not take the Dick, Kerr Ladies to Canada, as the English FA had instructed Canada's Football Association not to allow the team to play on any of its pitches.

Unfortunately, other nations' football associations began to follow the FA's example by instituting their own bans: Norway in 1931, France in 1932, Brazil in 1941, and West Germany in 1955. "This aggressive sport is essentially alien to the nature of woman," the Deutscher Fußball-Bund declared, according to the *Guardian*. "In the fight for the ball, the feminine grace vanishes, body and soul will inevitably suffer harm . . . The display of the woman's body offends decency and modesty."

But the final whistle had not sounded for women's football quite yet. After England's national men's team won the World Cup in 1966, a movement began to revive women's football in the UK, and in 1969 the Women's Football Association was founded. Yet the FA still refused to lift their decades-old ban on women's professional football. It was only under pressure from UEFA, the Union of European Football Associations, that the FA finally relented, and in 1971 it lifted its restrictions on women playing on FA-affiliated grounds. At long last, after fifty years of exclusion, women footballers could play professionally in the UK.

The munitionettes faced similar challenges as the nation demobilized after the war. Approximately one million women had worked in British munitions factories during World War I, hailing from all regions of the country and drawn from all classes, although the vast majority were working-class women who had held other jobs before taking on war work. For many munitionettes, this was their first time living away from home, earning wages much higher than they had ever earned before, granting them greater social freedom and independence than they had ever known or perhaps had even imagined. Munitionettes became a powerful, visible symbol of the modern woman—confident, strong, patriotic, making her own decisions, spending her own money, and challenging gender roles and assumptions every day.

Munitionettes were essential workers in a time of national crisis, and they firmly believed that they were not merely earning a living, but were directly engaged in the war effort. Many were understandably upset when they were expected to quit their jobs and return to their former occupations or homemaking when the soldiers returned and the factories closed down or transitioned to peacetime manufacturing. On 19 November 1918, only a week after the Armistice, six thousand women munitions workers, mostly from Woolwich Arsenal, marched on Parliament to express to the Prime Minister and the Ministry of Munitions their demand for "immediate guarantees for the future." On 3 December, nearly six hundred newly unemployed munitionettes marched on 10 Downing Street and demanded an audience with Prime Minister Lloyd George, seeking "the immediate withdrawal of their discharges." A small delegation was granted a meeting with officials from the Ministry of Munitions, but the women left unsatisfied, and munitionettes continued to demonstrate in the streets outside the building for hours afterward.

It is impossible to calculate how many hundreds or thousands of munitions workers, women and men alike, were killed, seriously injured, or poisoned as a result of their war work. During 1915 alone, there were nearly 160,000 reported industrial accidents, most of which involved workers between thirteen and eighteen years of age. According to the BBC, 400 cases of toxic jaundice due to TNT poisoning were recorded during the war, and one-quarter of those were fatal. Without question, while the lurid yellow hue eventually faded from the canary girls' skin after the women left TNT work, many continued to suffer serious health problems for the rest of their lives. They were "the Girls Behind the Man Behind the Gun," but one could reasonably ask who had stood behind *them*, and *with* them, in their time of need. Organizations such as the BBC and the Imperial War Museum have done excellent, meaningful work in collecting and preserving the stories of many munitionettes and canary girls so their contributions may be better understood, their significant contributions recognized, and they themselves duly honored.

Canary Girls is a work of fiction inspired by history, as are all of my historical novels. The Blyth Spartans did defeat the Bolckow Vaughan Ladies in 1918 to win the Munitionettes' Cup, but I have altered the timing of the matches and the pairing of sides to better suit my story, and to include fictional teams. Thornshire Arsenal is a fictional munitions works based upon real factories that existed in Great Britain at the time. Other locations in the novel, such as Alderlea and Brookfield, are also fictional, as are most of the characters, with the exception of certain historical figures (including the royal family and various politicians and military officers) and notable football players such as Horace Bailey, Jennie Harris, William Jonas, Lily Parr, Bella Reay, Florrie Redford, Walter Tull, Norman Arthur Wood, and Vivian Woodward. Horace Bailey was a member of England's 1908 and 1912 Olympic teams, but he played primarily for Leicester Fosse and Birmingham City rather than Tottenham Hotspur. He and all historical figures who appear in this novel are used fictitiously and with all due respect intended.

ACKNOWLEDGMENTS

I wrote *Canary Girls* at my home in Dane County, Wisconsin, which I respectfully acknowledge as the ancestral homeland of the Ho-Chunk Nation.

I offer my deepest gratitude to Maria Massie, Rachel Kahan, Emily Fisher, Ariana Sinclair, Kaitlin Harri, Laura Cherkas, Francie Crawford, Roland Ottewell, Kyle O'Brien, and Elsie Lyons for their essential contributions to *Canary Girls*. Geraldine Neidenbach, Marty Chiaverini, Michael Chiaverini, and Heather Neidenbach were my first readers, and their comments and questions were invaluable and very much appreciated. My brother, Nic Neidenbach, was an excellent technical consultant, while my sons, Nick and Michael Chiaverini, generously shared their knowledge of Philoctetes and football, respectively. Many thanks to my former teammates, the fantastic footballers of Just for Kicks and Ignition, for the inspiration and fond memories. I'm grateful to you all.

Thanks also to the staff at the University of Wisconsin–Madison Memorial Library, where I found most of the books that proved essential to my research. Those I found most helpful include:

Braybon, Gail, and Penny Summerfield. *Out of the Cage: Women's Experiences in Two World Wars*. London and New York: Pandora Press, 1987.

Burnett, John, ed. *Useful Toil: Autobiographies of Working People from the 1820s to the 1920s*. London and New York: Routledge, 1994.

Foxwell, Agnes K. *Munition Lasses: Six Months as Principal Overlooker in Danger Buildings*. London and New York: Hodder and Stoughton, 1917.

Hamilton, Peggy, Lady. *Three Years or the Duration: The Memoirs of a Munition Worker, 1914–1918*. London: Owen, 1978.

MacDonagh, Michael. *In London During the Great War: The Diary of a Journalist*. London: Eyre and Spottiswoode, 1935.

Newman, Vivien. *We Also Served: The Forgotten Women of the First World War*. Barnsley, South Yorkshire: Penn & Sword History, 2014.

Riddoch, Andrew, and John Kemp. *When the Whistle Blows: The Story of the Footballers' Battalion in the Great War*. Centennial Edition. Scotts Valley, CA: CreateSpace, 2015.

Roberts, Elizabeth. *A Woman's Place: An Oral History of Working-Class Women, 1890–1940*. Oxford and New York: Basil Blackwell, 1984.

Williams, Jean. *A Game for Rough Girls?: History of Women's Football in Britain*. London: Routledge, 2003.

Woollacott, Angela. *On Her Their Lives Depend: Munitions Workers in the Great War*. Berkeley and Los Angeles: University of California Press, 1994.

Yates, L. K. *The Woman's Part: A Record of Munitions Work*. New York: George R. Doran Company, ca. 1918.

I consulted several excellent online resources while researching and writing *Canary Girls*, including the BBC (bbc.com), the British Newspaper Archive (britishnewspaperarchive.co.uk), the Imperial War Museums (iwm.org.uk), the *Guardian* (theguardian.com), Newspapers.com (newspapers.com), the Library of Congress (loc.gov), and Ancestry (ancestry.com). Although I had been aware of the Christmas Truce for many years prior to writing *Canary Girls*, I was inspired to include the historical event in this novel by the Four Seasons Theatre productions of *All Is Calm: The Christmas Truce of 1914* in 2019 and 2021.

Most of all, I thank my husband, Marty, and my sons, Nick and Michael, for their enduring love, steadfast support, and constant encouragement. *Canary Girls* is my third novel written during the pandemic. I could not have finished this book without you, and the endless supply of hugs, encouragement, laughter, and hope you gave me when I needed them most. I'll forever be grateful for your courage, optimism, resilience, and humor in difficult times. All my love, always.

ABOUT THE AUTHOR

JENNIFER CHIAVERINI is the *New York Times* bestselling author of many acclaimed historical novels and the beloved Elm Creek Quilts series. A graduate of the University of Notre Dame and the University of Chicago, she lives with her husband—and, occasionally, their two college-student sons—in Madison, Wisconsin.